LECTURES ON THE CURRY-HOWARD ISOMORPHISM

STUDIES IN LOGIC

AND

THE FOUNDATIONS OF MATHEMATICS

VOLUME 149

Honorary Editor:

P. SUPPES

Editors:

S. ABRAMSKY, *London*
S. ARTEMOV, *Moscow*
D.M. GABBAY, *London*
A. KECHRIS, *Pasadena*
A. PILLAY, *Urbana*
R.A. SHORE, *Ithaca*

ELSEVIER

AMSTERDAM • BOSTON • HEIDELBERG • LONDON • NEW YORK • OXFORD
PARIS • SAN DIEGO • SAN FRANCISCO • SINGAPORE • SYDNEY • TOKYO

LECTURES ON THE CURRY-HOWARD ISOMORPHISM

Morten Heine SØRENSEN
University of Copenhagen
Denmark

Paweł URZYCZYN
Warsaw University
Poland

ELSEVIER

AMSTERDAM • BOSTON • HEIDELBERG • LONDON • NEW YORK • OXFORD
PARIS • SAN DIEGO • SAN FRANCISCO • SINGAPORE • SYDNEY • TOKYO

Elsevier
Radarweg 29, PO Box 211, 1000 AE Amsterdam, The Netherlands
The Boulevard, Langford Lane, Kidlington, Oxford OX5 1GB, UK

First edition 2006

Copyright © 2006 Elsevier B.V. All rights reserved

No part of this publication may be reproduced, stored in a retrieval system
or transmitted in any form or by any means electronic, mechanical, photocopying,
recording or otherwise without the prior written permission of the publisher.

Permissions may be sought directly from Elsevier's Science & Technology Rights
Department in Oxford, UK: phone (+44) (0) 1865 843830; fax (+44) (0) 1865 853333;
email: permissions@elsevier.com. Alternatively you can submit your request online by
visiting the Elsevier web site at http://elsevier.com/locate/permissions, and selecting
Obtaining permission to use Elsevier material.

Notice
No responsibility is assumed by the publisher for any injury and/or damage to persons
or property as a matter of products liability, negligence or otherwise, or from any use
or operation of any methods, products, instructions or ideas contained in the material
herein. Because of rapid advances in the medical sciences, in particular, independent
verification of diagnoses and drug dosages should be made.

Library of Congress Cataloging-in-Publication Data
A catalog record for this book is available from the Library of Congress

British Library Cataloguing in Publication Data
A catalogue record for this book is available from the British Library

ISBN-13:   978-0-444-52077-7
ISBN-10:   0-444-52077-5
ISSN: 0049-237X

For information on all Elsevier publications
visit our website at books.elsevier.com

Printed and bound in The Netherlands

06 07 08 09 10   10 9 8 7 6 5 4 3 2 1

Preface

The Curry-Howard isomorphism, also widely known as the "propositions-as-types" paradigm, states an amazing correspondence between systems of formal logic and computational calculi.[1] It begins with the observation that an implication $A \to B$ corresponds to a type of functions from A to B, because inferring B from $A \to B$ and A can be seen as *applying* the first assumption to the second one—just like a function from A to B applied to an element of A yields an element of B. Similarly, one can argue that proving the implication $A \to B$ by actually inferring B from A resembles constructing a function that maps any element of a domain A into an element of B.

In fact, it is an old idea—due to Brouwer, Kolmogorov, and Heyting—that a constructive proof of an implication from A to B is a procedure that transforms proofs of A into proofs of B. The Curry-Howard isomorphism formalizes this idea so that, for instance, proofs in the propositional intuitionistic logic are represented by simply typed λ-terms. Provable theorems are nothing else than non-empty types.

This analogy, first discovered by Haskell Brooks Curry in the 1930's, has turned out to hold for other logical systems as well. Virtually all proof-related concepts can be interpreted in terms of computations, and virtually all syntactic features of various lambda-calculi and similar systems can be formulated in the language of proof theory. For instance, quantification in predicate logic corresponds to dependent product, second-order logic is connected to polymorphism, and proofs by contradiction in classical logic are close relatives of control operators, e.g. exceptions. Moreover, various logical formalisms (Hilbert style, natural deduction, sequent calculi) are mimicked by corresponding models of computation (combinatory logic, λ-calculus, and explicit substitution calculi).

Since the 1969 work of William Howard it has been understood that the proposition-as-types correspondence is not merely a curiosity, but a fundamental principle. Proof normalization and cut-elimination are another model of computation, equivalent to β-reduction. Proof theory and the theory of computation turn out to be two sides of the same field.

---

[1] Other slogans are *formulas-as-types, proofs-as-programs*, etc.

Recent years show that this field covers not only *theory*, but also the *practice* of proofs and computation. Advanced type theory is now used as a tool in the development of software systems supporting program verification and computer-assisted reasoning. The growing need for efficient formal methods presents new challenges for the foundational research in computer science.

For these reasons the propositions-as-types paradigm experiences an increasing interest both from logicians and computer scientists, and it is gradually becoming a part of the curriculum knowledge in both areas. Despite this, no satistfactorily comprehensive book introduction to the subject has been available yet. This volume is an attempt to fill this gap, at least in the most basic areas.

Our aim is to give an exposition of the logical aspects of (typed) calculi (programs viewed as proofs) and the computational aspects of logical systems (proofs viewed as programs). We treat in a single book many issues that are spread over the literature, introducing the reader to several different systems of typed λ-calculi, and to the corresponding logics. We are not trying to cover the actual software systems (e.g. Coq) based on the formulas-as-types paradigm; we are concerned with the underlying mathematical principles.

Although a great deal of our attention is paid to constructive logics, we would like to stress that we are entirely happy with non-constructive meta-proofs and we do not restrict our results to those acceptable to constructive mathematics.

An idea due to Paul Lorenzen, closely related to the Curry-Howard isomorphism, is that provability can be expressed as a dialogue game. Via the Curry-Howard isomorphism, terms can be seen as strategies for playing such games. While for us proofs and programs are more or less the same objects, we would hesitate to claim the same thing about dialogue games or strategies for such games. Nevertheless, there is an emerging picture that the Curry-Howard isomorphism is perhaps about three worlds, not two, or that all three worlds are one. We also go some way to convey this paradigm.

An important subject, the linear logic of J.-Y. Girard, is not treated here, although it is very close to the subject. In fact, linear logic (and related areas—geometry of interaction, ludics) opens an entire new world of connections between proofs and computations, adding new breathtaking perspectives to the Curry-Howard isomorphism. It deserves a broader and deeper presentation that would have to extend the book beyond reasonable size. After some hesitation, we decided with regret not to treat it at all, rather than to remain with a brief and highly incomplete overview.

As we have said above, our principal aim is to explore the relationship between logics and typed lambda-calculi. But in addition to studying aspects of these systems relevant for the Curry-Howard isomorphism, we also attempt to consider them from other angles when these provide interesting insight or useful background. For instance, we begin with a chapter on the type-free

λ-calculus to provide the necessary context for typed calculi.

The book is intended as an advanced textbook, perhaps close to the border between a textbook and a monograph, but still on the side of a textbook. Thus we do not attempt to describe all current relevant research, or provide a complete bibliography. Sometimes we choose to cite newer or more comprehensible references rather than original works. We do not include all proofs, though most are provided. In general we attempt to provide proofs which are short, elegant, and contain interesting ideas. Though most proofs are based on solutions known from the literature, we try, whenever possible, to add a new twist to provide something interesting for the reader. Some new results are included too.

The book may serve as a textbook for a course on either typed λ-calculus or intuitionistic logic. Preferably on both subjects taught together. We think that a good way to explain typed λ-calculus is via logic and a good way to explain logic is via λ-calculus.

The expected audience consists mainly of graduate and PhD students, but also of researchers in the areas of computer science and mathematical logic. Parts of the text can also be of interest for scientists working in philosophy or linguistics.

We expect that the background of the readers of this book may be quite non-uniform: a typical reader will either know about intuitionistic logic or about λ-calculus (and functional programming), but will not necessarily be familiar with both. We also anticipate readers who are not familiar with any of the two subjects.

Therefore we provide an introduction to both subjects starting at a relatively elementary level. We assume basic knowledge about discrete mathematics, computability, set theory and classical logic within the bounds of standard university curricula. Appendix A provides some preliminaries that summarize this knowledge. Thus the book is largely self-contained, although a greater appreciation of some parts can probably be obtained by readers familiar with mathematical logic, recursion theory and complexity.

We hope that the bibliographical information is reasonably up to date. Each chapter ends with a Notes section that provides historical information and also suggests further reading. We try to include appropriate references, but sometimes we may simply not know the origin of a "folklore" idea, and we are sorry for any inaccuracies.

Each chapter is provided with a number of exercises. We recommend that the reader try as many of these as possible. Hints or solutions are given for some of the more difficult ones, hopefully making the book well-suited for self-study too. Some exercises are used as an excuse for including something interesting which is not in the mainstream of the book. The reader is also expected by default to do all omitted parts of proofs by herself.

Earlier versions of this text (beginning with the DIKU lecture notes [450]) have been used for several one-semester graduate/Ph.D. courses at the Department of Computer Science at the University of Copenhagen (DIKU), and at the Faculty of Mathematics, Informatics and Mechanics of Warsaw University. Summer school courses have also been held based on it.

Acknowledgments

Just like many other milestone ideas, the Curry-Howard isomorphism was discovered to various extents by different people at different times. And many others contributed to its development and understanding. The *Brouwer – Heyting – Kolmogorov – Schönfinkel – Curry – Meredith – Kleene – Feys – Gödel – Läuchli – Kreisel – Tait – Lawvere – Howard – de Bruijn – Scott – Martin-Löf – Girard – Reynolds – Stenlund – Constable – Coquand – Huet – · · · – isomorphism* might be a more appropriate name, still not including all the contributors.[2] But Curry and Howard were undoubtedly the two persons who gave the idea the crucial momentum. And, after all, the name has been in use for many years now. Therefore we decided to stay with "Curry-Howard isomorphism", being aware that, as often happens, it does not properly acknowledge all due credits.

Many people have contributed to this work, one way or another. We would like to thank Sergei Artemov and Dov Gabbay for encouragement. Our wives and Andy Deelen deserve special thanks for patience.

Urzyczyn would like to acknowledge support from the University of Copenhagen and the Villum Kann Rasmussen Fonden (VELUX Visiting Professor Programme 2003–2005) for his long-term visits in Copenhagen. Special thanks are due to Fritz Henglein for arranging all this.

Sørensen wishes to acknowledge support from the University of Copenhagen and flexibility from his former employers at IT Practice A/S.

We are indebted to the late Albert Dragalin, and to Hugo Herbelin, Ulrich Kohlenbach, Erik Krabbe, Daniel Leivant, Favio Miranda, Henning Niss, Henning Makholm, Femke van Raamsdonk, Christian Urban, for their comments, explanations or suggestions which helped us to prepare our 1998 notes and to make corrections and improvements in the present version.

Various chapters of this book at the final stages of its development were read and commented by a number of our colleagues to whom we wish to express our sincere gratitude: Marc Bezem, Peter Dybjer, Roy Dyckhoff, Andrzej Filinski, Herman Geuvers, J. Roger Hindley, Neil Jones, Assaf J. Kfoury, Grigori Mints, Wojciech Moczydłowski, Aleksy Schubert, Helmut Schwichtenberg, Valentin Shehtman, Dmitrij Skvortsov, Daria Walukiewicz-Chrząszcz, Konrad Zdanowski.

During preparation of the book we received help from numerous people, who patiently responded to our questions, provided references, commented on various fragments, etc. Although we are certainly unable to col-

---

[2] See [92] for the proper name of the Skolem-Löwenheim theorem.

lect the full list (and we will certainly make some unforgivable omissions, for which we apologize in advance), we want to mention a few names in addition to those above: Thorsten Altenkirch, Henk Barendregt, Wiktor Bartol, Marcin Benke, Johan van Benthem, Stefano Berardi, Stanisław Betley, Viviana Bono, Wojciech Buszkowski, David Chemouil, Jacek Chrząszcz, Patryk Czarnik, Dirk van Dalen, René David, Marcin Dziubiński, Maribel Fernandez, Nissim Francez, Michał Gajda, Jean-Yves Girard, Fritz Henglein, Sachio Hirokawa, Martin Hofmann, Zbigniew Jurkiewicz, Leszek Kołodziejczyk, Agnieszka Kozubek, Lars Kristiansen, Robert Lane, Sławomir Lasota, Ralph Matthes, François Métayer, Mirosława Miłkowska, Ray Mines, Ieke Moerdijk, Roman Murawski, Linh Anh Nguyen, Damian Niwiński, Filip Noworyta, C.-H. Luke Ong, Christine Paulin, Jeff Pelletier, Jacek Pomykała, Tomasz Połacik, Adrian Rice, Fred Richman, Jakub Sakowicz, Eric Schechter, Joanna Schubert, Jakob Grue Simonsen, Dimiter Skordev, Andrzej Skowron, Sergei Slavnov, Sergei Soloviev, Katharina Spies, Richard Statman, Michał Stronkowski, Jerzy Tiuryn, René Vestergaard, Hongwei Xi, Marek Zaionc, Piotr Zakrzewski, Marek Zawadowski, Artur Zawłocki, Ting Zhang. Many others deserve thanks too, in particular the students who attended our courses in Copenhagen and Warsaw, and whose feedback in class was so useful. The LaTeX packages **xypic** by Kristoffer Rose and **prooftree** by Paul Taylor helped us to typeset diagrams and proofs.

Needless to say, we are solely responsible and truly sorry for all the remaining bugs, from little mistakes to serious errors, that will undoubtedly be discovered in the book.

M.H.S. & P.U., March 15, 2006

Contents

1 Type-free λ-calculus   **1**
   1.1   A gentle introduction   1
   1.2   Pre-terms and λ-terms   3
   1.3   Reduction 10
   1.4   The Church-Rosser theorem 12
   1.5   Leftmost reductions are normalizing 14
   1.6   Perpetual reductions and the conservation theorem 18
   1.7   Expressibility and undecidability 19
   1.8   Notes 22
   1.9   Exercises 23

2 Intuitionistic logic   **27**
   2.1   The BHK interpretation 28
   2.2   Natural deduction 32
   2.3   Algebraic semantics of classical logic 34
   2.4   Heyting algebras 37
   2.5   Kripke semantics 43
   2.6   The implicational fragment 47
   2.7   Notes 48
   2.8   Exercises 50

3 Simply typed λ-calculus   **55**
   3.1   Simply typed λ-calculus à la Curry 56
   3.2   Type reconstruction algorithm 60
   3.3   Simply typed λ-calculus à la Church 63
   3.4   Church versus Curry typing 65
   3.5   Normalization 67
   3.6   Church-Rosser property 70
   3.7   Expressibility 72
   3.8   Notes 73
   3.9   Exercises 74

4 The Curry-Howard isomorphism — 77
- 4.1 Proofs and terms — 77
- 4.2 Type inhabitation — 79
- 4.3 Not an exact isomorphism — 81
- 4.4 Proof normalization — 83
- 4.5 Sums and products — 86
- 4.6 Prover-skeptic dialogues — 89
- 4.7 Prover-skeptic dialogues with absurdity — 94
- 4.8 Notes — 96
- 4.9 Exercises — 100

5 Proofs as combinators — 103
- 5.1 Hilbert style proofs — 103
- 5.2 Combinatory logic — 108
- 5.3 Typed combinators — 110
- 5.4 Combinators versus lambda terms — 113
- 5.5 Extensionality — 116
- 5.6 Relevance and linearity — 118
- 5.7 Notes — 122
- 5.8 Exercises — 123

6 Classical logic and control operators — 127
- 6.1 Classical propositional logic — 127
- 6.2 The $\lambda\mu$-calculus — 132
- 6.3 Subject reduction, confluence, strong normalization — 137
- 6.4 Logical embedding and CPS translation — 140
- 6.5 Classical prover-skeptic dialogues — 144
- 6.6 The pure implicational fragment — 150
- 6.7 Conjunction and disjunction — 153
- 6.8 Notes — 156
- 6.9 Exercises — 157

7 Sequent calculus — 161
- 7.1 Gentzen's sequent calculus LK — 162
- 7.2 Fragments of LK versus natural deduction — 165
- 7.3 Gentzen's Hauptsatz — 169
- 7.4 Cut elimination versus normalization — 174
- 7.5 Lorenzen dialogues — 181
- 7.6 Notes — 189
- 7.7 Exercises — 191

8 First-order logic — 195
- 8.1 Syntax of first-order logic — 195
- 8.2 Informal semantics — 197
- 8.3 Proof systems — 200
- 8.4 Classical semantics — 205
- 8.5 Algebraic semantics of intuitionistic logic — 207
- 8.6 Kripke semantics — 212
- 8.7 Lambda-calculus — 215
- 8.8 Undecidability — 221
- 8.9 Notes — 224
- 8.10 Exercises — 225

9 First-order arithmetic — 229
- 9.1 The language of arithmetic — 229
- 9.2 Peano Arithmetic — 232
- 9.3 Gödel's theorems — 234
- 9.4 Representable and provably recursive functions — 237
- 9.5 Heyting Arithmetic — 240
- 9.6 Kleene's realizability interpretation — 243
- 9.7 Notes — 245
- 9.8 Exercises — 247

10 Gödel's system T — 251
- 10.1 From Heyting Arithmetic to system **T** — 251
- 10.2 Syntax — 253
- 10.3 Strong normalization — 256
- 10.4 Modified realizability — 260
- 10.5 Notes — 265
- 10.6 Exercises — 266

11 Second-order logic and polymorphism — 269
- 11.1 Propositional second-order logic — 270
- 11.2 Polymorphic lambda-calculus (system **F**) — 276
- 11.3 Expressive power — 280
- 11.4 Curry-style polymorphism — 284
- 11.5 Strong normalization — 287
- 11.6 The inhabitation problem — 290
- 11.7 Higher-order polymorphism — 296
- 11.8 Notes — 299
- 11.9 Exercises — 300

12 Second-order arithmetic — 303
- 12.1 Second-order syntax ... 303
- 12.2 Classical second-order logic ... 305
- 12.3 Intuitionistic second-order logic ... 308
- 12.4 Second-order Peano Arithmetic ... 311
- 12.5 Second-order Heyting Arithmetic ... 314
- 12.6 Simplified syntax ... 315
- 12.7 Lambda-calculus ... 318
- 12.8 Notes ... 323
- 12.9 Exercises ... 323

13 Dependent types — 325
- 13.1 The language of λP ... 326
- 13.2 Type assignment ... 328
- 13.3 Strong normalization ... 333
- 13.4 Dependent types à la Curry ... 335
- 13.5 Correspondence with first-order logic ... 337
- 13.6 Notes ... 339
- 13.7 Exercises ... 340

14 Pure type systems and the λ-cube — 343
- 14.1 System λP revisited ... 343
- 14.2 Pure type systems ... 344
- 14.3 The Calculus of Constructions ... 346
- 14.4 Strong normalization ... 348
- 14.5 Beyond the cube ... 351
- 14.6 Girard's paradox ... 352
- 14.7 Notes ... 357
- 14.8 Exercises ... 358

A Mathematical background — 361
- A.1 Set theory ... 361
- A.2 Algebra and unification ... 365
- A.3 Partial recursive functions ... 366
- A.4 Decision problems ... 368
- A.5 Hard and complete ... 370

B Solutions and hints to selected exercises — 373

Bibliography — 403

Index — 431

Chapter 1

Type-free λ-calculus

The λ-*calculus* is a model of computation. It was introduced a few years before another such model, *Turing machines*. With the latter, computation is expressed by reading from and writing to a tape, and performing actions depending on the contents of the tape. Turing machines resemble programs in imperative programming languages, like Java or C.

In contrast, in λ-calculus one is concerned with functions, and these may both take other functions as arguments, and return functions as results. In programming terms, λ-calculus is an extremely simple higher-order, functional programming language.

In this chapter we only cover the *type-free* or *untyped* λ-calculus. Later we introduce several variants where λ-terms are categorized into various *types*.

1.1. A gentle introduction

Computation in λ-calculus is expressed with λ-*terms*. These are similar to the nameless function notation $n \mapsto n^2$ used in mathematics. However, a mathematician employs the latter notation to denote functions as mathematical objects (defined as sets of pairs). In contrast, λ-terms are formal expressions (strings) which, intuitively, express functions and applications of functions in a pure form. Thus, a λ-term is one of the following:

- a *variable*;
- an *abstraction* $\lambda x M$, where x is a variable and M is a λ-term;
- an *application* MN (of M to N), where M and N are λ-terms.

In an abstraction $\lambda x M$, the variable x represents the function argument (or *formal parameter*), and it may occur in the *body* M of the function, but it does not have to. In an application MN there is no restriction on the shape of the *operator* M or the *argument* N; both can be arbitrary λ-terms.

For instance, the λ-term $\mathbf{I} = \lambda x\,x$ intuitively denotes a function that maps any argument to itself, i.e. the identity function. As another example, $\mathbf{K} = \lambda x \lambda y\,x$ represents a function mapping any argument x to the constant function that always returns x. Finally, $\mathbf{IK}$ expresses application of the function $\mathbf{I}$ to the argument $\mathbf{K}$.

In mathematics we usually write the application of a function, say f, to an argument, say 4, with the argument in parentheses: $f(4)$. In the λ-calculus we would rather write this as $f4$. The use of parentheses cannot be entirely eliminated though. For instance, the notation $\lambda x x y$ would be ambiguous, and we should instead write either $(\lambda x x)y$ if we mean an application of $\mathbf{I}$ to y, or $\lambda x(xy)$ to denote an abstraction on x with the body xy. In the latter case, it is customary to use *dot notation*, i.e. to write $\lambda x.xy$ instead. Similarly we may need parentheses to disambiguate applications; for instance, $\mathbf{I}(\mathbf{KK})$ expresses application of $\mathbf{I}$ to $\mathbf{KK}$.

If λxM denotes a function, and N denotes an argument, the "value" of the application $(\lambda xM)N$ can be calculated by substituting N for x in M. The result of such a substitution is denoted by $M[x := N]$, and we formalize the calculation by the β-reduction rule: $(\lambda xM)N \to_\beta M[x := N]$. For instance,

$$(\mathbf{IK})z = ((\lambda x x)\mathbf{K})z \to_\beta x[x := \mathbf{K}]z = \mathbf{K}z = (\lambda y \lambda x y)z \to_\beta \lambda x z.$$

This process of calculating the value of an expression is similar to common practice in mathematics; if $f(n) = n^2$, then $f(4) = 4^2$, and we get from the application $f(4)$ to the result 4^2 by substituting 4 for n in the body of the definition of f. A programming language analogue is the *call-by-name* parameter passing mechanism, where the formal parameter of a procedure is replaced throughout by the actual parameter expression.

The variable x in a λ-abstraction λxM is *bound* (or *local*) within M in much the same way a formal parameter of a procedure is considered local within that procedure. In contrast, a variable y without a corresponding abstraction is called *free* (or *global*) and is similar to a global variable in most programming languages. Thus, x is bound and y is free in $\lambda x.xy$.

Some confusion may arise when we use the same name for bound and free variables. For example, in $x(\lambda x.xy)$, there are obviously two different x's: the free (global) x, and the bound (local) x, which is "shadowing" the free one in the body. If we instead consider the λ-term $x(\lambda z.zy)$, there is no confusion. As another example of confusion, $(\lambda x.xy)[y := x]$ should replace y in $(\lambda x.xy)$ by a free variable x, but $\lambda x.xx$ is not the desired result. In the latter term we have lost the distinction between the formal parameter x and the free variable x (the free variable has been *captured* by the lambda). If we use a bound variable z, the confusion disappears: $(\lambda z.zy)[y := x] = \lambda z.zx$.

A local variable of a procedure can always be renamed without affecting the meaning of the program. Similarly, in λ-calculus we do not care about the names of bound variables; the λ-terms $\lambda x\,x$ and $\lambda y\,y$ both denote the identity

1.2. Pre-terms and λ-terms

function. Because of this, it is usually assumed that *terms that differ only in their bound variables are identified*. This gives the freedom to choose bound variables so as to avoid any confusion, e.g. variable capture.

1.2. Pre-terms and λ-terms

We now define the notion of a *pre-term* and introduce λ-terms as equivalence classes of pre-terms. The section is rather dull, but necessary to make our formalism precise. However, to understand most of the book, the informal understanding of λ-terms of the preceding section suffices.

1.2.1. DEFINITION. Let Υ denote a countably infinite set of symbols, henceforth called *variables* (also *object variables* or *λ-variables* when other kinds of variables may cause ambiguity). We define the notion of a *pre-term* by induction as follows:

- Every variable is a pre-term.
- If M, N are pre-terms, then (MN) is a pre-term.
- If x is a variable and M is a pre-term, then $(\lambda x M)$ is a pre-term.

The set of all pre-terms is denoted by Λ^-.

REMARK. The definition can be summarized by the following *grammar*:

$$M ::= x \mid (MM) \mid (\lambda xM).$$

In the remainder of the book, we will often use this short style of definition.

Pre-terms, as defined above, are fully parenthesized. As the pre-term $(\lambda f((\lambda u(f(uu)))(\lambda v(f(vv)))))$ demonstrates, the heavy use of parentheses is rather cumbersome. We therefore introduce some notational conventions, which are used informally whenever no ambiguity or confusion can arise.

1.2.2. CONVENTION.

(i) The outermost parentheses in a term are omitted.

(ii) Application associates to the left: $((PQ)R)$ is abbreviated (PQR).

(iii) Abstraction associates to the right: $(\lambda x(\lambda y P))$ is abbreviated $(\lambda x \lambda y P)$.

(iv) A sequence of abstractions $(\lambda x_1(\lambda x_2 \ldots (\lambda x_n P) \ldots))$ can be written as $(\lambda x_1 x_2 \ldots x_n . P)$, in which case the outermost parentheses in P (if any) are usually omitted.[1]

---
[1] The dot represents a left parenthesis whose scope extends as far to the right as possible.

EXAMPLE.

- $(\lambda v(vv))$ may be abbreviated $\lambda v(vv)$ by (i).
- $(((\lambda xx)(\lambda yy))(\lambda zz))$ may be abbreviated $(\lambda xx)(\lambda yy)(\lambda zz)$ by (i), (ii).
- $(\lambda x(\lambda y(xy)))$ is written $\lambda x \lambda y(xy)$ by (i), (iii) or as $\lambda xy.xy$ by (i), (iv).
- $(\lambda f((\lambda u(f(uu)))(\lambda v(f(vv)))))$ is written $\lambda f.(\lambda u.f(uu))(\lambda v.f(vv))$.

1.2.3. DEFINITION. Define the set $\mathrm{FV}(M)$ of *free variables of* M as follows.

$$\begin{aligned}
\mathrm{FV}(x) &= \{x\}; \\
\mathrm{FV}(\lambda x P) &= \mathrm{FV}(P) - \{x\}; \\
\mathrm{FV}(PQ) &= \mathrm{FV}(P) \cup \mathrm{FV}(Q).
\end{aligned}$$

EXAMPLE. Let x, y, z be distinct variables; then $\mathrm{FV}((\lambda xx)(\lambda y.xyz)) = \{x, z\}$ There are two *occurrences* of x: one under λx and one under λy. An occurrence of x in M is called *bound* if it is in a part of M with shape λxL, and *free* otherwise. Then $x \in \mathrm{FV}(M)$ iff there is a free occurrence of x in M.

We now define *substitution* of pre-terms. It will only be defined when no variable is captured as a result of the substitution.

1.2.4. DEFINITION. The *substitution of* N *for* x *in* M, written $M[x := N]$, is defined iff no free occurrence of x in M is in a part of M with form λyL, where $y \in \mathrm{FV}(N)$. In such cases $M[x := N]$ is given by:[2]

$$\begin{aligned}
x[x := N] &= N; \\
y[x := N] &= y, \text{ if } x \neq y; \\
(PQ)[x := N] &= P[x := N]\, Q[x := N]; \\
(\lambda x P)[x := N] &= \lambda x P; \\
(\lambda y P)[x := N] &= \lambda y P[x := N], \text{ if } x \neq y.
\end{aligned}$$

REMARK. In the last clause, $y \notin \mathrm{FV}(N)$ or $x \notin \mathrm{FV}(P)$.

1.2.5. LEMMA.

(i) *If* $x \notin \mathrm{FV}(M)$ *then* $M[x := N]$ *is defined, and* $M[x := N] = M$.

(ii) *If* $M[x := N]$ *is defined then* $y \in \mathrm{FV}(M[x := N])$ *iff either* $y \in \mathrm{FV}(M)$ *and* $x \neq y$ *or both* $y \in \mathrm{FV}(N)$ *and* $x \in \mathrm{FV}(M)$.

(iii) *The substitution* $M[x := x]$ *is defined and* $M[x := x] = M$.

(iv) *If* $M[x := y]$ *is defined, then* $M[x := y]$ *is of the same length as* M.

---

[2]In our meta-notation, substitution binds stronger than anything else, so in the third clause the rightmost substitution applies to Q, not to $P[x := N]\,Q$.

1.2. PRE-TERMS AND λ-TERMS

PROOF. Induction on M. As an example we show (i) in some detail. It is clear that $M[x := N]$ is defined. To show that $M[x := N] = M$ consider the following cases. If M is a variable y, then we must have $y \neq x$, and $y[x := N] = y$. If $M = PQ$ then $x \notin \mathrm{FV}(P)$ and $x \notin \mathrm{FV}(Q)$, so by the induction hypothesis $P[x := N] = P$ and $Q[x := N] = Q$. Then also $(PQ)[x := N] = P[x := N]\,Q[x := N] = PQ$. Finally, if M is an abstraction, we may have either $M = \lambda x P$ or $M = \lambda y P$, where $x \neq y$. In the former case, $(\lambda x P)[x := N] = \lambda x P$. In the latter case, we have $x \notin \mathrm{FV}(P)$, so by the induction hypothesis $(\lambda y P)[x := N] = \lambda y P[x := N] = \lambda y P$. □

1.2.6. LEMMA. *Assume that $M[x := N]$ is defined, and both $N[y := L]$ and $M[x := N][y := L]$ are defined, where $x \neq y$. If $x \notin \mathrm{FV}(L)$ or $y \notin \mathrm{FV}(M)$ then $M[y := L]$ is defined, $M[y := L][x := N[y := L]]$ is defined, and*

$$M[x := N][y := L] = M[y := L][x := N[y := L]]. \qquad (*)$$

PROOF. Induction on M. The main case is when $M = \lambda z Q$, for $z \notin \{x, y\}$. By the assumptions

(i) $x \notin \mathrm{FV}(L)$ or $y \notin \mathrm{FV}(Q)$;
(ii) $z \notin \mathrm{FV}(N)$ or $x \notin \mathrm{FV}(Q)$;
(iii) $z \notin \mathrm{FV}(L)$ or $y \notin \mathrm{FV}(Q[x := N])$.

For the "defined" part, it remains, by the induction hypothesis, to show:

- $z \notin \mathrm{FV}(L)$ or $y \notin \mathrm{FV}(Q)$;
- $z \notin \mathrm{FV}(N[y := L])$ or $x \notin \mathrm{FV}(Q[y := L])$.

For the first property, if $z \in \mathrm{FV}(L)$, then $y \notin \mathrm{FV}(Q[x := N])$, so $y \notin \mathrm{FV}(Q)$. For the second property, assume $x \in \mathrm{FV}(Q[y := L])$. From (i) we have $x \in \mathrm{FV}(Q)$, thus $z \notin \mathrm{FV}(N)$ by (ii). Therefore $z \in \mathrm{FV}(N[y := L])$ could only happen when $y \in \mathrm{FV}(N)$ and $z \in \mathrm{FV}(L)$. Together with $x \in \mathrm{FV}(Q)$ this contradicts (iii). Now $(*)$ follows from the induction hypothesis. □

A special case of the lemma is $M[x := y][y := L] = M[x := L]$, if the substitutions are defined and $y \notin \mathrm{FV}(M)$.

1.2.7. LEMMA. *If $M[x:=y]$ is defined and $y \notin \mathrm{FV}(M)$ then $M[x:=y][y:=x]$ is defined, and $M[x:=y][y:=x] = M$.*

PROOF. By induction with respect to M one shows that the substitution is defined. The equation follows from Lemmas 1.2.5(iii) and 1.2.6. □

The next definition formalizes the idea of identifying expressions that "differ only in their bound variables."

1.2.8. DEFINITION. The relation $=_\alpha$ (α-conversion) is the least (i.e. smallest) transitive and reflexive relation on Λ^- satisfying the following.

- If $y \notin \mathrm{FV}(M)$ and $M[x := y]$ is defined then $\lambda x M =_\alpha \lambda y . M[x := y]$.
- If $M =_\alpha N$ then $\lambda x M =_\alpha \lambda x N$, for all variables x.
- If $M =_\alpha N$ then $MZ =_\alpha NZ$.
- If $M =_\alpha N$ then $ZM =_\alpha ZN$.

EXAMPLE. Let x, y be different variables. Then $\lambda x y . x y =_\alpha \lambda y x . y x$, but $\lambda x . x y \neq_\alpha \lambda y . y x$.

By Lemma 1.2.7 the relation $=_\alpha$ is symmetric, so we easily obtain:

1.2.9. LEMMA. *The relation of α-conversion is an equivalence relation.*

Strictly speaking, the omitted proof of Lemma 1.2.9 should go by induction with respect to the definition of $=_\alpha$. We prove the next lemma in more detail, to demonstrate this approach.

1.2.10. LEMMA. *If $M =_\alpha N$ then $\mathrm{FV}(M) = \mathrm{FV}(N)$.*

PROOF. Induction on the definition of $M =_\alpha N$. If $M =_\alpha N$ follows from transitivity, i.e. $M =_\alpha L$ and $L =_\alpha N$, for some L, then by the induction hypothesis $\mathrm{FV}(M) = \mathrm{FV}(L)$ and $\mathrm{FV}(L) = \mathrm{FV}(N)$. If $M = N$ (i.e. $M =_\alpha N$ by reflexivity) then $\mathrm{FV}(N) = \mathrm{FV}(M)$. If $M = \lambda x P$ and $N = \lambda y . P[x := y]$, where $y \notin \mathrm{FV}(P)$ and $P[x := y]$ is defined, then by Lemma 1.2.5 we have $\mathrm{FV}(M) = \mathrm{FV}(P) - \{x\} = \mathrm{FV}(P[x := y]) - \{y\} = \mathrm{FV}(N)$. If $M = \lambda x P$ and $N = \lambda x Q$, where $P =_\alpha Q$, then by the induction hypothesis $\mathrm{FV}(P) = \mathrm{FV}(Q)$, so $\mathrm{FV}(M) = \mathrm{FV}(N)$. If $M = PZ$ and $N = QZ$, or $M = ZP$ and $N = ZQ$, where $P =_\alpha Q$, then we use the induction hypothesis. □

1.2.11. LEMMA. *If $M =_\alpha M'$ and $N =_\alpha N'$ then $M[x := N] =_\alpha M'[x := N']$, provided both sides are defined.*

PROOF. By induction on the definition of $M =_\alpha M'$. If $M = M'$ then proceed by induction on M. The only other interesting case is when we have $M = \lambda z P$ and $M' = \lambda y . P[z := y]$, where $y \notin \mathrm{FV}(P)$, and $P[z := y]$ is defined. If $x = z$, then $x \notin \mathrm{FV}(M) = \mathrm{FV}(M')$ by Lemma 1.2.10. Hence $M[x := N] = M =_\alpha M' = M'[x := N']$ by Lemma 1.2.5. The case $x = y$ is similar. So assume $x \notin \{y, z\}$. Since $M[x := N]$ is defined, $x \notin \mathrm{FV}(P)$ or $z \notin \mathrm{FV}(N)$. In the former case $M[x := N] = \lambda z P$ and $x \notin \mathrm{FV}(P[z := y])$, so $M'[x := N'] = \lambda y . P[z := y] =_\alpha \lambda z . P$.

It remains to consider the case when $x \in \mathrm{FV}(P)$ and $z \notin \mathrm{FV}(N)$. Since $M'[x := N'] = (\lambda y . P[z := y])[x := N']$ is defined, we have $y \notin \mathrm{FV}(N')$,

1.2. Pre-terms and λ-terms

and thus also $y \notin \text{FV}(P[x := N'])$. By Lemma 1.2.6 it then follows that $M'[x:=N'] = \lambda y.P[z:=y][x:=N'] = \lambda y.P[x:=N'][z:=y] =_\alpha \lambda z.P[x:=N']$. By the induction hypothesis $\lambda z.P[x:=N'] =_\alpha \lambda z.P[x:=N] = M[x:=N]$. □

1.2.12. LEMMA.

(i) For all M, N and x, there exists an M' such that $M =_\alpha M'$ and the substitution $M'[x := N]$ is defined.

(ii) For all pre-terms M, N, P, and all variables x, y, there exist M', N' such that $M' =_\alpha M$, $N' =_\alpha N$, and the substitutions $M'[x := N']$ and $M'[x := N'][y := P]$ are defined.

PROOF. Induction on M. The only interesting case in part (i) is $M = \lambda y P$ and $x \neq y$. Let z be a variable different from x, not free in N, and not occurring in P at all. Then $P[y := z]$ is defined. By the induction hypothesis applied to $P[y := z]$, there is a P' with $P' =_\alpha P[y := z]$ and such that the substitution $P'[x := N]$ is defined. Take $M' = \lambda z P'$. Then $M' =_\alpha M$ and $M'[x:=N] = \lambda z.P'[x:=N]$ is defined. The proof of part (ii) is similar. □

1.2.13. LEMMA.

(i) If $MN =_\alpha R$ then $R = M'N'$, where $M =_\alpha M'$ and $N =_\alpha N'$.

(ii) If $\lambda x P =_\alpha R$, then $R = \lambda y Q$, for some Q, and there is a term P' with $P =_\alpha P'$ such that $P'[x := y] =_\alpha Q$.

PROOF. Part (i) is by induction with respect to the definition of $MN =_\alpha R$. Part (ii) follows in a similar style. The main case in the proof is transitivity. Assume $\lambda x P =_\alpha R$ follows from $\lambda x P =_\alpha M$ and $M =_\alpha R$. By the induction hypothesis, we have $M = \lambda z N$ and $R = \lambda y Q$, and there are $P' =_\alpha P$ and $N' =_\alpha N$ such that $P'[x := z] =_\alpha N$ and $N'[z := y] =_\alpha Q$. By Lemma 1.2.12(ii) there is $P'' =_\alpha P$ with $P''[x := z]$ and $P''[x := z][z := y]$ defined. Then by Lemma 1.2.11, we have $P''[x := z] =_\alpha N =_\alpha N'$ and thus also $P''[x := z][z := y] =_\alpha Q$. And $P''[x := z][z := y] = P''[x := y]$ by Lemmas 1.2.5(iii) and 1.2.6. (Note that $z \notin \text{FV}(P'')$ or $x = z$.) □

We are ready to define the true objects of interest: λ-terms.

1.2.14. DEFINITION. Define the set Λ of λ-terms as the quotient set of $=_\alpha$:

$$\Lambda = \{[M]_\alpha \mid M \in \Lambda^-\},$$

where[3] $[M]_\alpha = \{N \in \Lambda^- \mid M =_\alpha N\}$.

---
[3]For simplicity we write $[M]_\alpha$ instead of $[M]_{=_\alpha}$.

EXAMPLE. Thus, for every variable x, the string λxx is a pre-term, while $\mathbf{I} = [\lambda xx]_\alpha = \{\lambda xx, \lambda yy, \ldots\}$, where $x, y, \ldots$ are all the variables, is a λ-term.

WARNING. The notion of a pre-term and the explicit distinction between pre-terms and λ-terms are not standard in the literature. Rather, it is customary to call our pre-terms λ-terms and remark that "α-equivalent terms are identified" (cf. the preceding section).

We can now "lift" the notions of free variables and substitution.

1.2.15. DEFINITION. The free variables $\mathrm{FV}(M)$ of a λ-term M are defined as follows. Let $M = [M']_\alpha$. Then

$$\mathrm{FV}(M) = \mathrm{FV}(M'). \qquad (*)$$

If $\mathrm{FV}(M) = \varnothing$ then we say that M is *closed* or that it is a *combinator*.

Lemma 1.2.10 ensures that any choice of M' yields the same result.

1.2.16. DEFINITION. For λ-terms M and N, we define $M[x := N]$ as follows. Let $M = [M']_\alpha$ and $N = [N']_\alpha$, where $M'[x := N']$ is defined. Then

$$M[x := N] = [M'[x := N']]_\alpha.$$

Here Lemma 1.2.11 ensures that any choice of M' and N' yields the same result, and Lemma 1.2.12 guarantees that suitable M' and N' can be chosen.

The term formation operations can themselves be lifted.

1.2.17. NOTATION. Let P and Q be λ-terms, and let x be a variable. Then PQ, $\lambda x P$, and x denote the following unique λ-terms:

$$\begin{aligned} PQ &= [P'Q']_\alpha, \text{ where } [P']_\alpha = P \text{ and } [Q']_\alpha = Q; \\ \lambda x P &= [\lambda x P']_\alpha, \text{ where } [P']_\alpha = P; \\ x &= [x]_\alpha. \end{aligned}$$

Using this notation, we can show that the equations defining free variables and substitution for pre-terms also hold for λ-terms. We omit the easy proofs.

1.2.18. LEMMA. *The following equations are valid:*

$$\begin{aligned} \mathrm{FV}(x) &= \{x\}; \\ \mathrm{FV}(\lambda x P) &= \mathrm{FV}(P) - \{x\}; \\ \mathrm{FV}(PQ) &= \mathrm{FV}(P) \cup \mathrm{FV}(Q). \end{aligned}$$

1.2.19. LEMMA. *The following equations on λ-terms are valid:*

$$\begin{aligned} x[x := N] &= N; \\ y[x := N] &= y, \text{ if } x \neq y; \\ (PQ)[x := N] &= P[x := N]\, Q[x := N]; \\ (\lambda y P)[x := N] &= \lambda y. P[x := N], \end{aligned}$$

where $x \neq y$ and $y \notin \mathrm{FV}(N)$ in the last clause.

1.2. PRE-TERMS AND λ-TERMS

The next lemma "lifts" a few properties of pre-terms to the level of terms.

1.2.20. LEMMA. *Let P, Q, R, L be λ-terms. Then*

(i) $\lambda x\, P = \lambda y.P[x := y]$, *if* $y \notin \mathrm{FV}(P)$.

(ii) $P[x := Q][y := R] = P[y := R][x := Q[y := R]]$, *if* $y \neq x \notin \mathrm{FV}(R)$.

(iii) $P[x := y][y := Q] = P[x := Q]$, *if* $y \notin \mathrm{FV}(P)$.

(iv) *If* $PQ = RL$ *then* $P = R$ *and* $Q = L$.

(v) *If* $\lambda y\, P = \lambda z\, Q$, *then* $P[y := z] = Q$ *and* $Q[z := y] = P$.

PROOF. An easy consequence of Lemmas 1.2.6 and 1.2.12–1.2.13. □

From now on, expressions involving abstractions, applications, and variables are always understood as λ-terms, as defined in Notation 1.2.17. In particular, with the present machinery at hand, we can formulate definitions by induction on the structure of λ-terms, rather than first introducing the relevant notions for pre-terms and then lifting. The following definition is the first example of this; its correctness is established in Exercise 1.3.

1.2.21. DEFINITION. For $M, \vec{N} \in \Lambda$ and distinct variables $\vec{x} \in \Upsilon$, the *simultaneous substitution of $\vec{N}$ for $\vec{x}$ in M* is the term $M[\vec{x} := \vec{N}]$, such that

$$x_i[\vec{x} := \vec{N}] = N_i;$$
$$y\,[\vec{x} := \vec{N}] = y, \ \text{if } y \neq x_i, \text{ for all } i;$$
$$(PQ)[\vec{x} := \vec{N}] = P[\vec{x} := \vec{N}]Q[\vec{x} := \vec{N}];$$
$$(\lambda y\, P)[\vec{x} := \vec{N}] = \lambda y.P[\vec{x} := \vec{N}],$$

where, in the last clause, $y \neq x_i$ and $y \notin \mathrm{FV}(N_i)$, for all i.

OTHER SYNTAX. In this book we define many different languages (logics and λ-calculi) with various binding operators (e.g. quantifiers). Expressions (terms, formulas, types etc.) that differ only in their bound variables are always identified as we just did it in the untyped λ-calculus. However, in order not to exhaust the reader, we generally present the syntax in a slightly informal manner, thus avoiding the explicit introduction of "pre-expressions."

In all such cases, however, we actually have to deal with equivalence classes of some α-conversion relation, and a precise definition of the syntax must take this into account. We believe that the reader is able in each case to reconstruct all missing details of such a definition.

1.3. Reduction

1.3.1. DEFINITION. A relation $\succ$ on Λ is *compatible* iff it satisfies the following conditions for all $M, N, Z \in \Lambda$.

  (i) If $M \succ N$ then $\lambda x M \succ \lambda x N$, for all variables x.
 (ii) If $M \succ N$ then $MZ \succ NZ$.
(iii) If $M \succ N$ then $ZM \succ ZM$.

1.3.2. DEFINITION. The least compatible relation $\to_\beta$ on Λ satisfying

$$(\lambda x.P)Q \to_\beta P[x := Q],$$

is called β-*reduction*. A term of form $(\lambda x P)Q$ is called a β-*redex*, and the term $P[x := Q]$ is said to arise by *contracting* the redex. A λ-term M is in β-*normal form* (notation $M \in \mathrm{NF}_\beta$) iff there is no N such that $M \to_\beta N$, i.e. M does not contain a β-redex.

1.3.3. DEFINITION.

  (i) The relation $\twoheadrightarrow_\beta$ (*multi-step β-reduction*) is the transitive and reflexive closure of $\to_\beta$. The transitive closure of $\to_\beta$ is denoted by $\twoheadrightarrow_\beta^+$.
 (ii) The relation $=_\beta$ (called β-*equality* or β-*conversion*) is the least equivalence relation containing $\to_\beta$.
(iii) A β-*reduction sequence* is a finite or infinite sequence
$$M_0 \to_\beta M_1 \to_\beta M_2 \to_\beta \cdots$$

1.3.4. REMARK. The above notation applies in general: for any relation $\to_\circ$, the symbol $\twoheadrightarrow_\circ^+$ (respectively $\twoheadrightarrow_\circ$) denotes the transitive (respectively transitive and reflexive) closure of $\to_\circ$, and the symbol $=_\circ$ is used for the corresponding equivalence. We often omit β (or in general $\circ$) from such notation when no confusion arises, with one exception: the symbol $=$ without any qualification always denotes ordinary equality. That is, we write $A = B$ when A and B denote the same object.

WARNING. In the literature, and contrary to the use in this book, the symbol $=$ is often used for β-equality.

1.3.5. EXAMPLE.

  (i) $(\lambda x.xx)(\lambda zz) \to_\beta (xx)[x := \lambda zz] = (\lambda zz)(\lambda yy)$.
 (ii) $(\lambda zz)(\lambda yy) \to_\beta z[z := \lambda yy] = \lambda yy$.
(iii) $(\lambda x.xx)(\lambda zz) \twoheadrightarrow_\beta \lambda yy$.
 (iv) $(\lambda xx)yz =_\beta y((\lambda xx)z)$.

1.3. REDUCTION

1.3.6. EXAMPLE (Some common λ-terms).

(i) Let $\mathbf{I} = \lambda x.x$, $\mathbf{K} = \lambda xy.x$, and $\mathbf{S} = \lambda xyz.xz(yz)$. Then $\mathbf{SKK} \twoheadrightarrow_\beta \mathbf{I}$.

(ii) Let $\omega = \lambda x.xx$ and $\Omega = \omega\omega$. Then $\Omega \to_\beta \Omega \to_\beta \Omega \to_\beta \cdots$.

(iii) Let $\mathbf{Y} = \lambda f.(\lambda x.f(xx))(\lambda x.f(xx))$. Then $\mathbf{Y} \to_\beta \mathbf{Y}' \to_\beta \mathbf{Y}'' \to_\beta \cdots$, where $\mathbf{Y}, \mathbf{Y}', \mathbf{Y}'', \ldots$ are all different.

1.3.7. REMARK. A term is in normal form iff it is an abstraction $\lambda x M$, where M is in normal form, or it is $x M_1 \ldots M_n$, where $n \geq 0$ and all M_i are in normal form ("if" is obvious and "only if" is by induction on the length of M). Even more compact: a normal form is $\lambda y_1 \ldots y_m. x M_1 \ldots M_n$, where $m, n \geq 0$ and all M_i are normal forms.

The following little properties are constantly used.

1.3.8. LEMMA.

(i) If $N \to_\beta N'$ then $M[x := N] \twoheadrightarrow_\beta M[x := N']$.

(ii) If $M \to_\beta M'$ then $M[x := N] \to_\beta M'[x := N]$.

PROOF. By induction on M and $M \to_\beta M'$ using Lemma 1.2.20. □

In addition to β-reduction, other notions of reduction are considered in the λ-calculus. In particular, we have η-reduction.

1.3.9. DEFINITION. Let $\to_\eta$ denote the least compatible relation satisfying

$$\lambda x.Mx \to_\eta M, \quad \text{if } x \notin \mathrm{FV}(M).$$

The symbol $\to_{\beta\eta}$ denotes the union of $\to_\beta$ and $\to_\eta$.

REMARK. In general, when we have compatible relations $\to_R$ and $\to_Q$, the union is written $\to_{RQ}$. Similar notation is used for more than two relations.

The motivation for this notion of reduction (and the associated notion of equality) is, informally speaking, that two functions should be considered equal if they yield equal results whenever applied to equal arguments. Indeed:

1.3.10. PROPOSITION. *Let $=_{ext}$ be the least equivalence relation such that:*

- *If $M =_\beta N$, then $M =_{ext} N$;*

- *If $Mx =_{ext} Nx$ and $x \notin FV(M) \cup FV(N)$, then $M =_{ext} N$;*

- *If $P =_{ext} Q$, then $PZ =_{ext} QZ$ and $ZP =_{ext} ZQ$.*

Then $M =_{ext} N$ iff $M =_{\beta\eta} N$.

PROOF. The implication from left to right is by induction on the definition of $M =_{ext} N$, and from right to left is by induction with respect to the definition of $M =_{\beta\eta} N$. Note that the definition of $=_{ext}$ does not include the rule "If $P =_{ext} Q$ then $\lambda x P =_{ext} \lambda x Q$." This is no mistake: this property (known as *rule ξ*) follows from the others. □

We do not take $\to_{\beta\eta}$ as our standard notion of reduction. We want to be able to distinguish between two algorithms, even if their input-output behaviour is the same. Nevertheless, many properties of $\beta\eta$-reduction are similar to those of β-reduction. For instance, we have the following.

1.3.11. LEMMA. *If there is an infinite $\beta\eta$-reduction sequence starting with a term M then there is an infinite β-reduction sequence from M.*

PROOF. First observe that in an infinite $\beta\eta$-reduction sequence there must be infinitely many β-reduction steps (cf. Exercise 1.6). These β-reduction steps can be "permuted forward", yielding an infinite β-reduction. Indeed, by induction with respect to $M \to_\eta N$, one can show that $M \to_\eta N \to_\beta L$ implies $M \to_\beta P \twoheadrightarrow_{\beta\eta} L$, for some P. □

OTHER SYNTAX. In the numerous lambda-calculi occurring in the later chapters, many notions introduced here will be used in an analogous way. For instance, the notion of a compatible relation (Definition 1.3.1) generalizes naturally to other syntax. A compatible relation "respects" the syntactic constructions. Imagine that, in addition to application and abstraction, we have in our syntax an operation of "acclamation," written as $M!!$, i.e. whenever M is a term, $M!!$ is also a term. Then we should add the clause

- If $M \succ N$ then $M!! \succ N!!$.

to our definition of compatibility. Various additional reduction relations will also be considered later, and we usually define these by stating one or more reduction axioms, similar to the β-rule of Definition 1.3.2 and the η-rule of Definition 1.3.9. In such cases, we usually assume that the reduction relation is the least compatible relation satisfying the given reduction axiom.

1.4. The Church-Rosser theorem

Since a λ-term M may contain several β-redexes, there may be several N with $M \to_\beta N$. For instance, $\mathbf{K(II)} \to_\beta \lambda x.\mathbf{II}$ and $\mathbf{K(II)} \to_\beta \mathbf{KI}$. However, as shown below, there must be some term to which all such N reduce in one or more steps. In fact, even if we make several reduction steps, we can still converge to a common term (possibly using several steps):

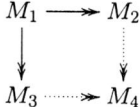

1.4. THE CHURCH-ROSSER THEOREM

This property is known as *confluence* or the *Church-Rosser property*. If the above diagram was correct in a stronger version with $\to$ in place of $\twoheadrightarrow$, then we could prove the theorem by a *diagram chase*:

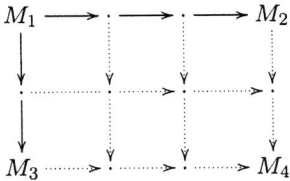

Unfortunately, our prerequisite fails. For instance, in the diagram

$$M_1 = \omega(\mathbf{II}) \longrightarrow \omega\mathbf{I} = M_2$$
$$\downarrow \qquad\qquad\qquad\qquad \downarrow$$
$$M_3 = (\mathbf{II})(\mathbf{II}) \dashrightarrow \mathbf{I}(\mathbf{II}) \dashrightarrow \mathbf{II} = M_4$$

two reductions are needed to get from M_3 to M_4. The problem is that the redex contracted in the reduction from M_1 to M_2 is duplicated in the reduction to M_3. We can solve the problem by working with *parallel reduction,* i.e. an extension of $\to_\beta$ allowing such duplications to be contracted in one step.

1.4.1. DEFINITION. Let $\Rightarrow_\beta$ be the least relation on Λ such that:

- $x \Rightarrow_\beta x$ for all variables x.
- If $P \Rightarrow_\beta Q$ then $\lambda x.P \Rightarrow_\beta \lambda x.Q$.
- If $P_1 \Rightarrow_\beta Q_1$ and $P_2 \Rightarrow_\beta Q_2$ then $P_1 P_2 \Rightarrow_\beta Q_1 Q_2$.
- If $P_1 \Rightarrow_\beta Q_1$ and $P_2 \Rightarrow_\beta Q_2$ then $(\lambda x.P_1)P_2 \Rightarrow_\beta Q_1[x := Q_2]$.

1.4.2. LEMMA.

(i) *If* $M \to_\beta N$ *then* $M \Rightarrow_\beta N$.
(ii) *If* $M \Rightarrow_\beta N$ *then* $M \twoheadrightarrow_\beta N$.
(iii) *If* $M \Rightarrow_\beta M'$ *and* $N \Rightarrow_\beta N'$ *then* $M[x:=N] \Rightarrow_\beta M'[x:=N']$.

PROOF. (i) is by induction on the definition of $M \to_\beta N$ (note that $P \Rightarrow_\beta P$ for all P), and (ii), (iii) are by induction on the definition of $M \Rightarrow_\beta M'$. □

1.4.3. DEFINITION. Let M^* (*the complete development of* M) be defined by:

$$\begin{aligned} x^* &= x; \\ (\lambda x M)^* &= \lambda x M^*; \\ (MN)^* &= M^* N^*, \text{ if } M \text{ is not an abstraction}; \\ ((\lambda x M)N)^* &= M^*[x := N^*]. \end{aligned}$$

Note that $M \Rightarrow_\beta N$ if N arises by reducing *some* of the redexes present in M, and that M^* arises by reducing *all* redexes present in M.

1.4.4. LEMMA. *If $M \Rightarrow_\beta N$ then $N \Rightarrow_\beta M^*$. In particular, if $M_1 \Rightarrow_\beta M_2$ and $M_1 \Rightarrow_\beta M_3$ then $M_2 \Rightarrow_\beta M_1^*$ and $M_3 \Rightarrow_\beta M_1^*$.*

PROOF. By induction on the definition of $M \Rightarrow_\beta N$, using Lemma 1.4.2. □

1.4.5. THEOREM (Church and Rosser). *If $M_1 \twoheadrightarrow_\beta M_2$ and $M_1 \twoheadrightarrow_\beta M_3$, then there is $M_4 \in \Lambda$ with $M_2 \twoheadrightarrow_\beta M_4$ and $M_3 \twoheadrightarrow_\beta M_4$.*

PROOF. If $M_1 \to_\beta \cdots \to_\beta M_2$ and $M_1 \to_\beta \cdots \to_\beta M_3$, the same holds with $\Rightarrow_\beta$ in place of $\to_\beta$. By Lemma 1.4.4 and a diagram chase, $M_2 \Rightarrow_\beta \cdots \Rightarrow_\beta M_4$ and $M_3 \Rightarrow_\beta \cdots \Rightarrow_\beta M_4$ for some M_4. Then $M_2 \twoheadrightarrow_\beta M_4$ and $M_3 \twoheadrightarrow_\beta M_4$. □

1.4.6. COROLLARY.

(i) *If $M =_\beta N$, then $M \twoheadrightarrow_\beta L$ and $N \twoheadrightarrow_\beta L$ for some L.*

(ii) *If $M \twoheadrightarrow_\beta N_1$ and $M \twoheadrightarrow_\beta N_2$ for β-normal forms N_1, N_2, then $N_1 \equiv N_2$.*

(iii) *If there are β-normal forms L_1 and L_2 such that $M \twoheadrightarrow_\beta L_1$, $N \twoheadrightarrow_\beta L_2$, and $L_1 \not\equiv L_2$, then $M \neq_\beta N$.*

PROOF. Left to the reader. □

REMARK. One can consider λ-calculus as an equational theory, i.e. a formal theory with formulas of the form $M =_\beta N$. Part (i) establishes *consistency* of this theory, in the following sense: there exists a formula that cannot be proved, e.g. $\lambda xx =_\beta \lambda xy.x$ (cf. Exercise 2.5).

Part (ii) in the corollary is similar to the fact that when we calculate the value of an arithmetical expression, e.g. $(4+2) \cdot (3+7) \cdot 11$, the end result is independent of the order in which we do the calculations.

1.5. Leftmost reductions are normalizing

1.5.1. DEFINITION. A term M is *normalizing* (notation $M \in \mathrm{WN}_\beta$) iff there is a reduction sequence from M ending in a normal form N. We then say that M *has* the normal form N. A term M is *strongly normalizing* ($M \in \mathrm{SN}_\beta$ or just $M \in \mathrm{SN}$) if all reduction sequences starting with M are finite. We write $M \in \infty_\beta$ if $M \notin \mathrm{SN}_\beta$. Similar notation applies to other notions of reduction.

Any strongly normalizing term is also normalizing, but the converse is not true, as **KI**Ω shows. But Theorem 1.5.8 states that a normal form, if it exists, can always be found by repeatedly reducing the leftmost redex, i.e. the redex whose λ is the furthest to the left. The following notation and definition are convenient for proving Theorem 1.5.8.

1.5. Leftmost reductions are normalizing

VECTOR NOTATION. Let $n \geq 0$. If $\vec{P} = P_1, \ldots, P_n$, then we write $M\vec{P}$ for $MP_1 \ldots P_n$. In particular, if $n = 0$, i.e. $\vec{P}$ is empty, then $M\vec{P}$ is just M. Similarly, if $\vec{z} = z_1, \ldots, z_n$, then we write $\lambda \vec{z}.M$ for $\lambda z_1 \ldots z_n.M$. Again, $\lambda \vec{z}.M$ is just M, if $n = 0$, i.e. $\vec{z}$ is empty.

REMARK. Any term has exactly one of the following two forms: $\lambda \vec{z}.x\vec{R}$ or $\lambda \vec{z}.(\lambda x P)Q\vec{R}$, in which case $(\lambda x P)Q$ is called *head* redex (in the former case, there is no head redex). Any redex that is not a head redex is called *internal*. A head redex is always the leftmost redex, but the leftmost redex in a term is not necessarily a head redex—it may be internal.

1.5.2. DEFINITION. For a term M not in normal form, we write

- $M \xrightarrow{l}_\beta N$ if N arises from M by contracting the leftmost redex.
- $M \xrightarrow{h}_\beta N$ if N arises from M by contracting a head redex.
- $M \xrightarrow{i}_\beta N$ if N arises from M by contracting an internal redex.

1.5.3. LEMMA.

(i) If $M \xrightarrow{h}_\beta N$ then $\lambda x M \xrightarrow{h}_\beta \lambda x N$.

(ii) If $M \xrightarrow{h}_\beta N$ and M is not an abstraction, then $ML \xrightarrow{h}_\beta NL$.

(iii) If $M \xrightarrow{h}_\beta N$ then $M[x := L] \xrightarrow{h}_\beta N[x := L]$.

PROOF. Easy. □

The following technical notions are central in the proof of Theorem 1.5.8.

1.5.4. DEFINITION. We write $\vec{P} \Rightarrow_\beta \vec{Q}$ if $\vec{P} = P_1, \ldots, P_n$, $\vec{Q} = Q_1, \ldots, Q_n$, $n \geq 0$, and $P_j \Rightarrow_\beta Q_j$ for all $1 \leq j \leq n$. Parallel internal reduction $\xRightarrow{i}_\beta$ is the least relation on Λ satisfying the following rules.

- If $\vec{P} \Rightarrow_\beta \vec{Q}$ then $\lambda \vec{x}.y\vec{P} \xRightarrow{i}_\beta \lambda \vec{x}.y\vec{Q}$.
- If $\vec{P} \Rightarrow_\beta \vec{Q}$, $S \Rightarrow_\beta T$ and $R \Rightarrow_\beta U$, then $\lambda \vec{x}.(\lambda y S)R\vec{P} \xRightarrow{i}_\beta \lambda \vec{x}.(\lambda y T)U\vec{Q}$.

REMARK. If $M \xrightarrow{i}_\beta N$, then $M \xRightarrow{i}_\beta N$. Conversely, if $M \xRightarrow{i}_\beta N$, then $M \xrightarrow{i}_\beta\!\!\!\twoheadrightarrow N$. Also, if $M \xRightarrow{i}_\beta N$, then $M \Rightarrow_\beta N$.

1.5.5. DEFINITION. We write $M \Rrightarrow_\beta N$ if there are $M_0, M_1, \ldots, M_n$ with

$$M = M_0 \xrightarrow{h}_\beta M_1 \xrightarrow{h}_\beta \cdots \xrightarrow{h}_\beta M_n \xRightarrow{i}_\beta N$$

and $M_i \Rightarrow_\beta N$ for all $i \in \{0, \ldots, n\}$, where $n \geq 0$.

1.5.6. LEMMA.

(i) If $M \Rrightarrow_\beta M'$ then $\lambda x M \Rrightarrow_\beta \lambda x M'$.

(ii) If $M \Rrightarrow_\beta M'$ and $N \Rrightarrow_\beta N'$, then $MN \Rrightarrow_\beta M'N'$.

(iii) If $M \Rrightarrow_\beta M'$ and $N \Rrightarrow_\beta N'$, then $M[x := N] \Rrightarrow_\beta M'[x := N']$.

PROOF. Part (i) is easy. For (ii), let

$$M = M_0 \xrightarrow{h}_\beta M_1 \xrightarrow{h}_\beta \cdots \xrightarrow{h}_\beta M_n \xRightarrow{i}_\beta M'$$

where $M_i \Rrightarrow_\beta M'$ for all $i \in \{0, \ldots, n\}$. Assume at least one M_i is an abstraction, and let k be the smallest number such that M_k is an abstraction. Then $M_k N \xRightarrow{i}_\beta M'N'$. By Lemma 1.5.3:

$$MN = M_0 N \xrightarrow{h}_\beta M_1 N \xrightarrow{h}_\beta \cdots \xrightarrow{h}_\beta M_k N \xRightarrow{i}_\beta M'N', \qquad (*)$$

where $M_i N \Rrightarrow_\beta M'N'$. If there is no abstraction among M_i, $0 \le i \le n$, then $(*)$ still holds with $k = n$.

For (iii) first assume $M \xRightarrow{i}_\beta M'$. We have either $M = \lambda \vec{z}.(\lambda y.P)Q\vec{R}$ or $M = \lambda \vec{z}.y\vec{R}$. In the former case, $M' = \lambda \vec{z}.(\lambda y.P')Q'\vec{R}'$, where $P \Rightarrow_\beta P'$, $Q \Rightarrow_\beta Q'$, and $\vec{R} \Rightarrow_\beta \vec{R}'$. By Lemma 1.4.2, $M[x := N] \xRightarrow{i}_\beta M'[x := N']$. In the latter case, $M' = \lambda \vec{z}.y\vec{R}'$. We consider two possibilities. If $x \ne y$, we proceed as in the case just considered. If $x = y$, let $\vec{S}$ and $\vec{S}'$ arise from $\vec{R}$ (respectively $\vec{R}'$) by substituting N (respectively N') for x. By (i), (ii) and Lemma 1.4.2 we then have $M[x := N] = \lambda \vec{z}.N\vec{S} \Rrightarrow_\beta \lambda \vec{z}.N'\vec{S}' = M'[x := N']$.

Now consider the general case. By the above and Lemma 1.5.3, we have

$$M[x := N] = M_0[x := N] \xrightarrow{h}_\beta \cdots \xrightarrow{h}_\beta M_n[x := N] \Rrightarrow_\beta M'[x := N'].$$

Also, $M_i[x := N] \Rrightarrow_\beta M'[x := N']$ holds by Lemma 1.4.2. □

1.5.7. LEMMA.

(i) If $M \Rrightarrow_\beta N$ then $M \xrightarrow{h}_\beta L \xRightarrow{i}_\beta N$ for some L.

(ii) If $M \xRightarrow{i}_\beta N \xrightarrow{h}_\beta L$, then $M \xrightarrow{h\;+}_\beta O \xRightarrow{i}_\beta L$ for some O.

PROOF. For (i), show that $M \Rightarrow_\beta N$ implies $M \Rrightarrow_\beta N$. For (ii), note that $M = \lambda \vec{z}.(\lambda x.P)Q\vec{R}$, $N = \lambda \vec{z}.(\lambda x.P')Q'\vec{R}'$, where $P \Rightarrow_\beta P'$, $Q \Rightarrow_\beta Q'$, and $\vec{R} \Rightarrow_\beta \vec{R}'$. Then $L = \lambda \vec{z}.P'[x := Q']\vec{R}'$. Define $O = \lambda \vec{z}.P[x := Q]\vec{R}$, and we then have $M \xrightarrow{h}_\beta O \Rrightarrow_\beta L$. Now use (i). □

1.5.8. THEOREM. If M has normal form N, then $M \xrightarrow{l}\!\!\!\to_\beta N$.

1.5. LEFTMOST REDUCTIONS ARE NORMALIZING   17

PROOF. Induction on the length of N. We have $M \Rightarrow_\beta \cdots \Rightarrow_\beta N$. By Lemma 1.5.7, the reduction from M to N consists of head reductions and parallel internal reductions, and the head reductions can be brought to the beginning. Thus $M \xrightarrow{h}_\beta L \xrightarrow{i}_\beta N$, where $L = \lambda \vec{z}.y\vec{P}$ and $N = \lambda \vec{z}.y\vec{P'}$, and where P_i has normal form P_i'. By the induction hypothesis, leftmost reduction of each P_i yields P_i'. Then $M \xrightarrow{l}_\beta N$. □

1.5.9. DEFINITION. A *reduction strategy* F is a map from λ-terms to λ-terms such that $F(M) = M$ when M is in normal form, and $M \to_\beta F(M)$ otherwise. Such an F is *normalizing* if for all $M \in \mathrm{WN}_\beta$, there is an i such that $F^i(M)$ is in normal form.

1.5.10. COROLLARY. *Define $F_l(M) = M$ for each normal form M, and $F_l(M) = N$, where $M \xrightarrow{l}_\beta N$, otherwise. Then F_l is normalizing.*

1.5.11. DEFINITION. A reduction sequence is called

- *quasi-leftmost* if it contains infinitely many leftmost reductions;
- *quasi-head* if it contains infinitely many head reductions.

1.5.12. COROLLARY. *Let M be normalizing. We then have the following.*

(i) *M has no infinite head-reduction sequence.*

(ii) *M has no quasi-head reduction sequence.*

(iii) *M has no quasi-leftmost reduction sequence.*

PROOF. Part (i) follows directly from the theorem. For (ii), suppose M has a quasi-head reduction sequence. By Lemma 1.5.7(ii), we can postpone the internal reductions in the quasi-head reduction indefinitely to get an infinite head reduction, contradicting Theorem 1.5.8.

Part (iii) is by induction on the length of the normal form of M. Suppose M has a quasi-leftmost reduction sequence. By (ii) we may assume that, from some point on, the quasi-leftmost reduction contains no head reductions. One of the reductions after this point must be leftmost, so the sequence is

$$M \twoheadrightarrow_\beta L_1 \xrightarrow{i}_\beta L_2 \xrightarrow{i}_\beta L_3 \xrightarrow{i}_\beta \cdots \qquad (*)$$

where $L_1 = \lambda \vec{z}.y\vec{P}$. Infinitely many of the leftmost steps in $(*)$ must occur within the same P_i and these steps are leftmost relative to P_i, contradicting the induction hypothesis. □

1.6. Perpetual reductions and the conservation theorem

Theorem 1.5.8 provides a way to obtain the normal form of a term, when it exists. There is also a way to *avoid* the normal form, i.e. to find an infinite reduction, when one exists.

1.6.1. DEFINITION. Define $F_\infty : \Lambda \to \Lambda$ as follows. If $M \in \mathrm{NF}_\beta$ then $F_\infty(M) = M$; otherwise

$$\begin{aligned}
F_\infty(\lambda\vec{z}.x\vec{P}Q\vec{R}) &= \lambda\vec{z}.x\vec{P}F_\infty(Q)\vec{R}, & \text{if } \vec{P} \in \mathrm{NF}_\beta \text{ and } Q \notin \mathrm{NF}_\beta; \\
F_\infty(\lambda\vec{z}.(\lambda xP)Q\vec{R}) &= \lambda\vec{z}.P[x := Q]\vec{R}, & \text{if } x \in \mathrm{FV}(P) \text{ or } Q \in \mathrm{NF}_\beta; \\
F_\infty(\lambda\vec{z}.(\lambda xP)Q\vec{R}) &= \lambda\vec{z}.(\lambda xP)F_\infty(Q)\vec{R}, & \text{if } x \notin \mathrm{FV}(P) \text{ and } Q \notin \mathrm{NF}_\beta.
\end{aligned}$$

It is easy to see that $M \to F_\infty(M)$ when $M \notin \mathrm{NF}_\beta$.

1.6.2. LEMMA. *Assume $Q \in \mathrm{SN}_\beta$ or $x \in \mathrm{FV}(P)$. If $P[x := Q]\vec{R} \in \mathrm{SN}_\beta$, then $(\lambda x.P)Q\vec{R} \in \mathrm{SN}_\beta$.*

PROOF. Let $P[x := Q]\vec{R} \in \mathrm{SN}_\beta$. Then $P, \vec{R} \in \mathrm{SN}_\beta$. If $x \notin \mathrm{FV}(P)$, then $Q \in \mathrm{SN}_\beta$ by assumption. If $x \in \mathrm{FV}(P)$, then Q is part of $P[x := Q]\vec{R}$, so $Q \in \mathrm{SN}_\beta$. If $(\lambda x.P)Q\vec{R} \in \infty_\beta$, then any infinite reduction must have form

$$(\lambda x.P)Q\vec{R} \twoheadrightarrow_\beta (\lambda x.P')Q'\vec{R'} \to_\beta P'[x := Q']\vec{R'} \to_\beta \cdots$$

Then $P[x := Q]\vec{R} \twoheadrightarrow_\beta P'[x := Q']\vec{R'} \to_\beta \cdots$, a contradiction. □

1.6.3. THEOREM. *If $M \in \infty_\beta$ then $F_\infty(M) \in \infty_\beta$.*

PROOF. By induction on M. If $M = \lambda\vec{z}.x\vec{P}$, apply the induction hypothesis as necessary. We consider the remaining cases in more detail.
CASE 1: $M = \lambda\vec{z}.(\lambda x.P)Q\vec{R}$ where $x \in \mathrm{FV}(P)$ or $Q \in \mathrm{NF}_\beta$. Then we have $F_\infty(M) = \lambda\vec{z}.P[x := Q]\vec{R}$, and thus $F_\infty(M) \in \infty_\beta$, by Lemma 1.6.2.
CASE 2: $M = \lambda\vec{z}.(\lambda x.P)Q\vec{R}$ where $x \notin \mathrm{FV}(P)$ and $Q \in \infty_\beta$. Then we have $F_\infty(M) = \lambda\vec{z}.(\lambda x.P)F_\infty(Q)\vec{R}$. By the induction hypothesis $F_\infty(Q) \in \infty_\beta$, so $F_\infty(M) \in \infty_\beta$.
CASE 3: $M = \lambda\vec{z}.(\lambda x.P)Q\vec{R}$ where $x \notin \mathrm{FV}(P)$ and $Q \in \mathrm{SN}_\beta - \mathrm{NF}_\beta$. Then we have $F_\infty(M) = \lambda\vec{z}.(\lambda x.P)F_\infty(Q)\vec{R} \twoheadrightarrow_\beta P\vec{R}$. From Lemma 1.6.2, we obtain $P\vec{R} \in \infty_\beta$, but then also $F_\infty(M) \in \infty_\beta$. □

1.6.4. DEFINITION. A reduction strategy F is *perpetual* iff for all $M \in \infty_\beta$,

$$M \to_\beta F(M) \to_\beta F(F(M)) \to_\beta \cdots$$

is an infinite reduction sequence from M.

1.7. Expressibility and undecidability

1.6.5. COROLLARY. F_∞ *is perpetual.*

PROOF. Immediate from the preceding theorem. □

1.6.6. DEFINITION. The set of λI-*terms* is defined as follows.
- Every variable is a λI-term.
- An application MN is a λI-term iff both M and N are λI-terms.
- An abstraction $\lambda x M$ is a λI-term iff M is a λI-term and $x \in \text{FV}(M)$.

The following is known as the *conservation theorem* (for λI-terms).

1.6.7. COROLLARY.
(i) For all λI-terms M, if $M \in \text{WN}_\beta$ then $M \in \text{SN}_\beta$.
(ii) For all λI-terms M, if $M \in \infty_\beta$ and $M \to_\beta N$ then $N \in \infty_\beta$.

PROOF. For part (i), assume $M \in \text{WN}_\beta$. Then $M \stackrel{l}{\twoheadrightarrow}_\beta N$ for some normal form N. Now note that for all λI-terms L not in normal form, $L \stackrel{l}{\to}_\beta F_\infty(L)$. Thus $N = F_\infty^k(M)$ for some k, so $M \in \text{SN}_\beta$, by Corollary 1.6.5.

For part (ii), suppose $M \to_\beta N$. If $M \in \infty_\beta$, then $M \notin \text{WN}_\beta$, by (i). Hence $N \notin \text{WN}_\beta$, in particular $N \in \infty_\beta$. □

1.7. Expressibility and undecidability

The untyped λ-calculus is so simple that it may be surprising how powerful it is. In this section we show that λ-calculus in fact can be seen as an alternative formulation of recursion theory.

We can use λ-terms to represent various constructions, e.g. truth values:

$$\text{true} = \lambda xy.x;$$
$$\text{false} = \lambda xy.y;$$
$$\text{if } P \text{ then } Q \text{ else } R = PQR.$$

It is easy to see that

$$\text{if true then P else Q} \twoheadrightarrow_\beta P;$$
$$\text{if false then P else Q} \twoheadrightarrow_\beta Q.$$

Another useful construction is the ordered pair

$$\langle M, N \rangle = \lambda x.xMN;$$
$$\pi_1 = \lambda p.p(\lambda xy.x);$$
$$\pi_2 = \lambda p.p(\lambda xy.y).$$

As expected we have

$$\pi_i \langle M_1, M_2 \rangle \twoheadrightarrow_\beta M_i.$$

1.7.1. DEFINITION. We represent the natural numbers in the λ-calculus as *Church numerals*:
$$\mathbf{c}_n = \lambda fx.f^n(x),$$
where $f^n(x)$ abbreviates $f(f(\cdots(x)\cdots))$ with n occurrences of f. Sometimes we write **n** for $\mathbf{c}_n$, so that for instance
$$\begin{aligned} \mathbf{0} &= \lambda fx.x; \\ \mathbf{1} &= \lambda fx.fx; \\ \mathbf{2} &= \lambda fx.f(fx). \end{aligned}$$

1.7.2. DEFINITION. A partial function $f : \mathbb{N}^k \to \mathbb{N}$ is λ-*definable* iff there is an $F \in \Lambda$ such that:

- If $f(n_1, \ldots, n_k) = m$ then $F\mathbf{c}_{n_1} \ldots \mathbf{c}_{n_k} =_\beta \mathbf{c}_m$.

- If $f(n_1, \ldots, n_k)$ is undefined then $F\mathbf{c}_{n_1} \ldots \mathbf{c}_{n_k}$ has no normal form.

We say that the term F *defines* the function f.

REMARK. If F defines f, then in fact $F\mathbf{c}_{n_1} \ldots \mathbf{c}_{n_k} \twoheadrightarrow_\beta \mathbf{c}_{f(n_1,\ldots,n_k)}$.

1.7.3. EXAMPLE. The following terms define a few commonly used functions.

- Successor: $\text{succ} = \lambda nfx.f(nfx)$.

- Addition: $\text{add} = \lambda mnfx.mf(nfx)$.

- Multiplication: $\text{mult} = \lambda mnfx.m(nf)x$.

- Exponentiation: $\text{exp} = \lambda mnfx.mnfx$.

- The constant zero function: $\text{zero} = \lambda m.\mathbf{0}$.

- The i-th projection of k-arguments: $\Pi_i^k = \lambda m_1 \ldots m_k.m_i$.

We show that all partial recursive functions are λ-definable.

1.7.4. PROPOSITION. *The primitive recursive functions are λ-definable.*

PROOF. It follows from Example 1.7.3 that the initial functions are definable. It should also be obvious that the class of λ-definable total functions is closed under composition. It remains to show that λ-definability is closed under primitive recursion. Assume that f is given by

$$\begin{aligned} f(0, n_1, \ldots, n_m) &= g(n_1, \ldots, n_m); \\ f(n+1, n_1, \ldots, n_m) &= h(f(n, n_1, \ldots, n_m), n, n_1, \ldots, n_m), \end{aligned}$$

1.7. EXPRESSIBILITY AND UNDECIDABILITY

where g and h are λ-definable by G and H. Define auxiliary terms

$$\begin{aligned} \text{Init} &= \langle 0, Gx_1 \ldots x_m\rangle. \\ \text{Step} &= \lambda p.\langle \text{succ}(\pi_1 p), H(\pi_2 p)(\pi_1 p)x_1 \ldots x_m\rangle; \end{aligned}$$

The function f is then λ-definable by

$$F = \lambda x x_1 \ldots x_m.\pi_2(x \, \text{Step Init}).$$

This expresses the following algorithm: Generate a sequence of pairs

$$(0, a_0), (1, a_1), \ldots, (n, a_n),$$

where $a_0 = g(n_1, \ldots, n_m)$ and $a_{i+1} = h(a_i, i, n_1, \ldots, n_m)$, so at the end of the sequence, we have $a_n = f(n, n_1, \ldots, n_m)$. □

1.7.5. THEOREM. *All partial recursive functions are λ-definable.*

PROOF. Let f be a partial recursive function. By Theorem A.3.8

$$f(n_1, \ldots, n_m) = \ell(\mu y[g(y, n_1, \ldots, n_m) = 0]),$$

where g and ℓ are primitive recursive. We show that f is λ-definable. For this, we first define a test for zero:

$$\text{zero?} = \lambda x.x(\lambda y.\text{false})\text{true}.$$

By Proposition 1.7.4, the functions g and ℓ are definable by some terms G and L, respectively. Let

$$W = \lambda y. \text{ if zero?}(Gyx_1\ldots x_m) \text{ then } \lambda w.Ly \text{ else } \lambda w.w(\text{succ } y)w.$$

Note that $x_1, \ldots, x_m$ are free in W. The following term defines f:

$$F = \lambda x_1.\ldots \lambda x_m.W\mathbf{c}_0 W.$$

Indeed, take any $n_1, \ldots, n_m$ and let $\vec{c} = \mathbf{c}_{n_1} \ldots \mathbf{c}_{n_m}$. Then

$$F\vec{c} \twoheadrightarrow_\beta W'\mathbf{c}_0 W',$$

where $W' = W[\vec{x} := \vec{c}]$. Suppose that $g(n, n_1, \ldots, n_m) = 0$, and n is the least number with this property. Then

$$W'\mathbf{c}_0 W' \twoheadrightarrow_\beta W'\mathbf{c}_1 W' \twoheadrightarrow_\beta \cdots \twoheadrightarrow_\beta W'\mathbf{c}_n W' \twoheadrightarrow_\beta L\mathbf{c}_n \twoheadrightarrow_\beta \mathbf{c}_{\ell(n)}.$$

It remains to see what happens when the minimum is not defined. Then we have the following infinite quasi-leftmost reduction sequence

$$W'\mathbf{c}_0 W' \twoheadrightarrow_\beta W'\mathbf{c}_1 W' \twoheadrightarrow_\beta W'\mathbf{c}_2 W' \twoheadrightarrow_\beta \cdots$$

so Corollary 1.5.12 implies that $F\vec{c}$ has no normal form. □

1.7.6. REMARK. A close inspection of the proof of Theorem 1.7.5 reveals that it shows more than stated in the theorem: For every partial recursive function $f : \mathbb{N}^m \to \mathbb{N}$, there is a defining term F such that every application $F\mathbf{c}_{n_1} \ldots \mathbf{c}_{n_m}$ is *uniformly normalizing*, i.e. either strongly normalizing or without normal form. The details of this claim can be found in Exercise 1.21.

1.7.7. COROLLARY. *The following problems are undecidable:*

(i) *Given $M \in \Lambda$, does M have a normal form?*

(ii) *Given $M \in \Lambda$, is M strongly normalizing?*

PROOF. For (i), suppose we have an algorithm to decide whether any term has a normal form. Take any recursively enumerable set $A \subseteq \mathbb{N}$ that is not recursive, and let f be a partial recursive function with domain A. Clearly, f is λ-definable by some term F. Now, for a given $n \in \mathbb{N}$, we can effectively decide whether $n \in A$ by checking whether the term $F\mathbf{n}$ has a normal form. Part (ii) now follows from Remark 1.7.6. □

1.8. Notes

The λ-calculus and the related systems of *combinatory logic* were introduced around 1930 by Alonzo Church [69, 70] and Haskell B. Curry [98, 99, 101], respectively. From the beginning, the calculi were parts of systems intended to be a foundation for logic. Unfortunately, Church's students Kleene and Rosser [271] discovered in 1935 that the original systems were inconsistent, and Curry [103] simplified the result, which became known as *Curry's paradox*. Consequently, the subsystem dealing with λ-terms, reduction, and conversion, i.e. what we call λ-calculus, was studied independently.

The notions of λ-binding and α-convertible terms are intuitively very clear, but we have seen in Section 1.2 that various technical difficulties must be overcome in order to handle them properly. This issue becomes especially vital when one faces the problem of a practical implementation. A classical solution [59] is to use a nameless representation of variables (so called *de Bruijn indices*). For more on this and related subjects, see e.g. [392, 395, 401].

The Church-Rosser theorem, which can be seen as a consistency result for the λ-calculus, was proved in 1936 by Church and Rosser [76]. Many proofs appeared later. Barendregt [31] cites Tait and Martin-Löf for the technique using parallel reductions; our proof is from Takahashi [470]. Proofs of the Church-Rosser theorem and an extension of Theorem 1.5.8 for $\beta\eta$-reductions can also be found there.

Church and Rosser [76] also proved the conservation theorem for λI-terms (which is sometimes called the *second Church-Rosser theorem*). Again, many different proofs have appeared. Our proof uses the effective perpetual strategy from [31], and the fact, also noted in [412], that the perpetual strategy always contracts the leftmost redex, when applied to a λI-term. More about perpetual strategies and their use in proving conservation theorems can be found in [406] and [361].

The λ-calculus turned out to be useful for formalizing the intuitive notion of effective computability. Kleene [267] showed that every partial recursive function was λ-definable and vice versa. This led Church to conjecture that λ-definability is an appropriate formalization of the intuitive notion of effective computability [72], which became known as *Church's thesis*.

The problems of deciding membership of WN_β and SN_β can be seen as variants of the halting problem. Church [71, 72] inferred from the former his celebrated theorem stating that first-order arithmetic is undecidable, as well as the undecidability of the *Entscheidungsproblem* (the "decision problem" for first-order logic), results that were "in the air" in this period [164].

Curry and his co-workers continued the work on *illative* combinatory logic [107, 108], i.e. logical systems including formulas as well as combinators and types. The calculus of combinators was then studied as an independent subject, and a wealth of results was obtained. For instance, the theorem about leftmost reductions is from [107]. Like many other classical results in λ-calculus it has been proved in many different ways ever since; our proof is taken from [470].

With the invention of computing machinery came also programming languages. Already in the 1960's λ-calculus was recognized as a useful tool in the design, implementation, and theory of programming languages [295]. In particular, type-free λ-calculus constitutes a model of higher-order untyped *functional* programming languages, e.g. Scheme [3] and Lisp [200], while typed calculi correspond to functional languages like Haskell [478] and ML [387].

The classic texts on type-free λ-calculus are [241] and [31]. First-hand historical information may be obtained from Curry and Feys' book [107], which contains a wealth of historical information, and from Rosser and Kleene's eyewitness statements [270, 420]. Other interesting papers are [28, 238].

1.9. Exercises

1.1. For a pre-term M, the directed labeled graph $G(M)$ is defined by induction.

- If $M = x$ then $G(M)$ has a single root node labeled x and no edges.

- If $M = PQ$ then $G(M)$ is obtained from the union of $G(P)$ and $G(Q)$ by adding a new initial (root) node labeled @. This node has two outcoming edges: to the root nodes of $G(P)$ and $G(Q)$.

- If $M = \lambda x P$ then $G(M)$ is obtained from $G(P)$ by

  - Adding a new root node labeled λx, and an edge from there to the root node of $G(P)$;
  - Adding edges to the new root from all final nodes labeled x.

For a given graph G, let $erase(G)$ be a graph obtained from G by deleting all labels from the variable nodes that are not final and renaming every label λx, for some variable x, to λ. Prove that the conditions $erase(G(M)) = erase(G(N))$ and $M =_\alpha N$ are equivalent for all $M, N \in \Lambda^-$.

1.2. Modify Definition 1.2.4 so that the operation $M[x := N]$ is defined for all M, N and x. Then prove that $M[x := N] =_\alpha M'[x := N']$ holds for all $M =_\alpha M'$ and $N =_\alpha N'$ (cf. Lemma 1.2.11).

1.3. Let $\vec{x} \in \Upsilon$ and $\vec{N} \in \Lambda$ be fixed. Show that Definition 1.2.21 determines a total function from Λ to Λ. *Hint:* Rewrite the definition as a relation $r \subseteq \Lambda \times \Lambda$ and show that for every $M \in \Lambda$ there is exactly one $L \in \Lambda$ such that $r(M, L)$. It may be beneficial to show uniqueness and existence separately.

1.4. Prove that if $(\lambda x P)Q = (\lambda y M)N$ then $P[x := Q] = M[y := N]$. In other words, the contraction of a given redex yields a unique result.

1.5. Show that M is in normal form if and only if M is either a variable or an abstraction $\lambda x M'$, where M' is normal, or $M = M'[x := yN]$, for some normal forms M' and N, and some x occurring free exactly once in M'.

1.6. Show that every term is strongly normalizing with respect to eta-reductions.

1.7. Which of the following are true?

(i) $\lambda x.Mx =_\beta M$, for any abstraction M with $x \notin \mathrm{FV}(M)$.

(ii) $\lambda x \Omega =_\beta \Omega$.

(iii) $(\lambda x.xx)(\lambda xy.y(xx)) =_\beta (\lambda x.x\mathbf{I}x)(\lambda zxy.y(xzx))$.

1.8. Prove the *weak Church-Rosser theorem:* For all $M_1, M_2, M_3 \in \Lambda$, if $M_1 \to_\beta M_2$ and $M_1 \to_\beta M_3$, then there is $M_4 \in \Lambda$ with $M_2 \twoheadrightarrow_\beta M_4$ and $M_3 \twoheadrightarrow_\beta M_4$. Do not use the Church-Rosser theorem.

1.9. Which of the following are true?

(i) $\mathbf{II}(\mathbf{II}) \Rightarrow_\beta \mathbf{II}$.

(ii) $\mathbf{II}(\mathbf{II}) \Rightarrow_\beta \mathbf{I}$.

(iii) $\mathbf{IIII} \Rightarrow_\beta \mathbf{III}$.

(iv) $\mathbf{IIII} \Rightarrow_\beta \mathbf{I}$.

1.10. Find terms M, N such that $M \Rightarrow_\beta N$ and $M \xrightarrow{i}_\beta N$, but not $M \xRightarrow{i}_\beta N$.

1.11. Find terms M, N such that $M \xrightarrow{i}_\beta N$ and $M \xrightarrow{h}_\beta N$ both hold.

1.12. Let $M \to_\beta F_\infty(M) \to_\beta F_\infty(F_\infty(M)) \to_\beta \cdots \to_\beta F_\infty^n(M)$, where $F_\infty^n(M) \in \mathrm{NF}$. Show that there is no reduction from M with more than n reduction steps. *Hint:* Generalize Lemma 1.6.2 to show that

$$l_\beta((\lambda x.P)Q\vec{R}) = l_\beta(P[x := Q]\vec{R}) + \epsilon(P) \cdot l_\beta(Q) + 1,$$

where $l_\beta(M)$ denotes the length of the longest reduction sequence from M and $\epsilon(P)$ is 1 if $x \notin \mathrm{FV}(P)$ and 0 else. Show that $l_\beta(M) = 1 + l_\beta(F_\infty(M))$, if $M \notin \mathrm{NF}$.

1.13. Show that there is no total computable $l : \Lambda \to \mathbb{N}$ such that, for all $M \in \mathrm{SN}_\beta$,

$$l(M) \geq l_\beta(M),$$

where $l_\beta(M)$ is as in the Exercise 1.12. *Hint:* That would imply decidability of SN.

1.14. Consider the *fixed point combinator:*

$$\mathbf{Y} = \lambda f.(\lambda x.f(xx))(\lambda x.f(xx)).$$

Show that $F(\mathbf{Y}F) =_\beta \mathbf{Y}F$ holds for all F. (Thus in the untyped lambda-calculus every fixpoint equation $X = FX$ has a solution.)

1.15. Let M be any other fixed point combinator, i.e. assume that $F(MF) =_\beta MF$ holds for all F. Show that M has no normal form.

1.16. Define the predecessor function in the untyped lambda-calculus.

1.9. EXERCISES

1.17. (B. Maurey, J.-L. Krivine.) Let $C = \lambda xy.(xF(\lambda z\,1))(yF(\lambda z\,0))$, where $F = \lambda fg.gf$. Show that C defines the function

$$c(m,n) = \begin{cases} 1, & \text{if } m \leq n; \\ 0, & \text{otherwise.} \end{cases}$$

How many steps are needed to reduce $C\mathbf{c}_m\mathbf{c}_n$ to normal form? Will the same hold if we define c using Proposition 1.7.4?

1.18. (From [74].) Find $\lambda\mathbf{I}$-terms P_1, P_2 (projections) such that $P_1\langle\mathbf{c}_m, \mathbf{c}_n\rangle =_\beta \mathbf{c}_m$ and $P_2\langle\mathbf{c}_m, \mathbf{c}_n\rangle =_\beta \mathbf{c}_n$, for all m, n.

1.19. Show that the following functions are λ-definable:

- For each n, the function $f(i, m_1, \ldots, m_n) = m_i$.
- Integer division, i.e. a function f such that $f(mn, m) = n$, for all $m, n \neq 0$.
- Integer square root, i.e. a function f such that $f(n^2) = n$, for all n.

1.20. Assume that $M \twoheadrightarrow_\beta z$. Show that if $M[z := \lambda x N] \twoheadrightarrow_\beta \lambda u Q$, where N is normal, then $\lambda u Q = \lambda x N$. Will this remain true if we replace $\twoheadrightarrow_\beta$ by $\twoheadrightarrow_{\beta\eta}$?

1.21. For $n \in \mathbb{N}$ put $d_n = \text{succ}^n(\mathbf{c}_0)$. Assume the following:[4]

Every primitive recursive function is definable by a term F such that all applications $Fd_{n_1}\ldots d_{n_m}(\lambda v.\,\text{false})\text{true}$ are strongly normalizing.

Prove the claim in Remark 1.7.6. *Hint:* Use Exercise 1.20.

1.22. Prove that β-equality is undecidable. *Hint:* See Corollary 1.7.7(i).

---

[4] The proof of this is beyond our reach at the moment. We return to it in Exercise 11.28

Chapter 2

Intuitionistic logic

The word "logic" is used with a variety of meanings, from common sense reasoning to sophisticated formal systems. In most cases, logic is used to classify statements as "true" or "false". That is, what we mean by logic is usually one of the many possible variants of two-valued *classical logic*.

Indeed, classical logic can by all means be seen as a standard wherever there is a need for precise reasoning, especially in mathematics and computer science. The principles of classical logic are extremely useful as a tool to describe and classify the reasoning patterns occurring both in everyday life and in mathematics.

It is however important to understand the following. First of all, no system of rules can capture all of the rich and complex world of human thoughts, and thus every logic can merely be used as a limited-purpose tool rather than as an ultimate oracle, responding to all possible questions.

In addition, the principles of classical logic, although easily acceptable by our intuition, are not the only possible reasoning principles. It can be argued that from certain points of view (likely to be shared by the reader with a computing background) it is actually better to use other principles. Let us have a closer look at this.

Classical logic is based on the fundamental notion of *truth*. The truth of a statement is an "absolute" property of this statement, in that it is independent of any reasoning, understanding, or action. A well-formed and unambiguous declarative statement is either true or false, whether or not we (or anybody else) know it, prove it, or verify it in any possible way. Here "false" means the same as "not true", and this is expressed by the *excluded middle* principle (also known as *tertium non datur*) stating that $p \vee \neg p$ holds no matter what the meaning of p is.

Needless to say, the information contained in the claim $p \vee \neg p$ is quite limited. Take the following sentence as an example:

There are seven 7's in a row somewhere in the decimal representation of the number π.

It may very well happen that nobody will ever be able to determine the truth or falsity of the sentence above. Yet we are forced to accept that either the claim or its negation must necessarily hold. Another well-known example is:

There exist irrational numbers x and y, such that x^y is rational.

The proof of this fact is very simple: if $\sqrt{2}^{\sqrt{2}}$ is a rational number then we can take $x = y = \sqrt{2}$; otherwise take $x = \sqrt{2}^{\sqrt{2}}$ and $y = \sqrt{2}$.

The problem with this proof is that we do not know which of the two possibilities is the right one. But here is a different argument: For $x = \sqrt{2}$ and $y = 2\log_2 3$, we have $x^y = 3 \in \mathbb{Q}$. We say the latter proof is *constructive* while the former is not.

Such examples demonstrate some of the drawbacks of classical logic. Indeed, in many applications we want to find an actual solution to a problem, and not merely to know that some solution exists. We thus want to sort out the proof methods that provide the actual solutions from those that do not. Therefore, from a very pragmatic point of view, it makes sense to consider a constructive approach to logic.

The logic that meets our expectations and accepts only "constructive" reasoning is traditionally known under the (slightly confusing) name of *intuitionistic logic*. To explain this name, one has to recall the philosophical foundations of intuitionistic logic. These may be expressed concisely and very simplified by the following principle: There is no absolute truth, there is only the knowledge and intuitive construction of the idealized mathematician, the *creative subject*. A logical judgement is only considered "true" if the creative subject can verify its correctness. Accepting this point of view inevitably leads to the rejection of the excluded middle as a uniform principle. As we learned from the noble Houyhnhnms [464]:

(...) reason taught us to affirm or deny only where we are certain; and beyond our knowledge we cannot do either.

2.1. The BHK interpretation

The language of *intuitionistic propositional logic*, also called *intuitionistic propositional calculus* (abbreviated IPC), is the same as the language of classical propositional logic.

2.1.1. DEFINITION. We assume an infinite set PV of *propositional variables* and we define the set Φ of *formulas* as the least set such that:

- Each propositional variable and the constant $\bot$ are in Φ;

- If $\varphi, \psi \in \Phi$ then $(\varphi \to \psi), (\varphi \vee \psi), (\varphi \wedge \psi) \in \Phi$.

2.1. THE BHK INTERPRETATION

Variables and $\bot$ are called *atomic* formulas. A *subformula* of a formula φ is a (not necessarily proper) part of φ, which itself is a formula.

That is, our basic connectives are: implication $\to$, disjunction $\vee$, conjunction $\wedge$, and the constant $\bot$ (absurdity). Negation $\neg$ and equivalence $\leftrightarrow$ are abbreviations, as well as the constant $\top$ (truth):

- $\neg\varphi$ abbreviates $\varphi \to \bot$;
- $\varphi \leftrightarrow \psi$ abbreviates $(\varphi \to \psi) \wedge (\psi \to \varphi)$;
- $\top$ abbreviates $\bot \to \bot$.

2.1.2. CONVENTION.

1. We often use the convention that implication is right associative, i.e. we write e.g. $\varphi \to \psi \to \vartheta$ instead of $\varphi \to (\psi \to \vartheta)$.

2. We assume that negation has the highest priority, and implication the lowest, with no preference between $\vee$ and $\wedge$. That is, $\neg p \wedge q \to r$ means $((\neg p) \wedge q) \to r$.

3. And of course we skip the outermost parentheses.

In order to understand intuitionistic logic, one should forget the classical notion of "truth". Now our judgements about a logical statement are no longer based on any truth-value assigned to that statement, but on our ability to justify it via an explicit proof or "construction". As a consequence of this we should not attempt to define propositional connectives by means of truth-tables (as it is normally done for classical logic). Instead, we should explain the meaning of compound formulas in terms of their constructions.

Such an explanation is often given by means of the so-called *Brouwer-Heyting-Kolmogorov interpretation*, in short *BHK interpretation*. One can formulate the BHK interpretation as the following set of rules, the algorithmic flavor of which will later lead us to the Curry-Howard isomorphism.

- *A construction of $\varphi_1 \wedge \varphi_2$ consists of a construction of φ_1 and a construction of φ_2;*
- *A construction of $\varphi_1 \vee \varphi_2$ consists of an indicator $i \in \{1, 2\}$ and a construction of φ_i;*
- *A construction of $\varphi_1 \to \varphi_2$ is a method (function) transforming every construction of φ_1 into a construction of φ_2;*
- *There is no construction of $\bot$.*[1]

---

[1] According to the Houyhnhnms (a construction of) a lie is "the thing which is not".

We do not specify what a construction of a propositional variable is. This is because the meaning of a propositional variable becomes only known when the variable is replaced by a concrete statement. Then we can ask about the construction of that statement. In contrast, the constant $\bot$ represents a statement with no possible construction at all.

Negation $\neg\varphi$ is understood as the implication $\varphi \to \bot$. That is, we may assert $\neg\varphi$ when the assumption of φ leads to absurdity. In other words:

- *A construction of $\neg\varphi$ is a method that turns every construction of φ into a nonexistent object.*

The equivalence of $\neg\varphi$ and $\varphi \to \bot$ holds also in classical logic. But note that the intuitionistic $\neg\varphi$ is stronger than just "there is no construction of φ".

The reader should be aware that the BHK interpretation is by no means intended to make a precise and complete description of constructive semantics. The very notion of "construction" is informal and can be understood in a variety of ways.

2.1.3. EXAMPLE. Consider the following formulas:

(i) $\bot \to p$;

(ii) $p \to q \to p$;

(iii) $(p \to q \to r) \to (p \to q) \to p \to r$;

(iv) $p \to \neg\neg p$;

(v) $\neg\neg\neg p \to \neg p$;

(vi) $(p \to q) \to (\neg q \to \neg p)$;

(vii) $\neg(p \lor q) \leftrightarrow (\neg p \land \neg q)$;

(viii) $((p \land q) \to r) \leftrightarrow (p \to (q \to r))$;

(ix) $\neg\neg(p \lor \neg p)$;

(x) $(p \lor \neg p) \to \neg\neg p \to p$.

All these formulas can be given a BHK interpretation. For instance, a construction for formula (i) is based on the safe assumption that a construction of $\bot$ is impossible. (In contrast, we do not have a construction of $q \to p$, because we cannot generally rule out the existence of a construction of q.) A construction for formula (iv), that is, for $p \to ((p \to \bot) \to \bot)$, is as follows:

> *Given a construction of p, here is a construction of $(p \to \bot) \to \bot$: Take a construction of $p \to \bot$. It is a method to translate constructions of p into constructions of $\bot$. As we have a construction of p, we can use this method to obtain a construction of $\bot$.*

2.1. THE BHK INTERPRETATION

The reader is invited to discover the BHK interpretation of the other formulas (Exercise 2.2).

Of course all the formulas of Example 2.1.3 are classical tautologies. But not every classical tautology can be given a construction.

2.1.4. EXAMPLE. Each of the formulas below is a classical tautology. Yet none appears to have a construction, despite the fact that some of them are similar or "dual" to certain formulas of the previous example.

(i) $((p \to q) \to p) \to p$;

(ii) $p \vee \neg p$;

(iii) $\neg\neg p \to p$;

(iv) $(\neg q \to \neg p) \to (p \to q)$;

(v) $\neg(p \wedge q) \leftrightarrow (\neg p \vee \neg q)$;

(vi) $(p \to q) \to (\neg p \to q) \to q$;

(vii) $((p \leftrightarrow q) \leftrightarrow r) \leftrightarrow (p \leftrightarrow (q \leftrightarrow r))$;

(viii) $(p \to q) \leftrightarrow (\neg p \vee q)$;

(ix) $(p \vee q \to p) \vee (p \vee q \to q)$;

(x) $(\neg\neg p \to p) \to p \vee \neg p$.

For instance, formula (iii) seems to express the same principle as Example 2.1.3(iv). Similarly, formula (iv) and Example 2.1.3(vi) are often treated as two equivalent patterns of a proof by contradiction. Formula (v) and Example 2.1.3(vii) are both known as *De Morgan's laws*, and express the classical duality between conjunction and disjunction.

Such examples show that the symmetry of classical logic disappears when we turn to constructive semantics, and this is obviously due to the fact that negation is no longer an *involution*, i.e. φ and $\neg\neg\varphi$ cannot be identified with each other anymore. Note however that formula (iii) expresses a property weaker than (ii), because we do not have a construction of (x).

Not very surprisingly, proofs by cases (vi) are not constructive. Indeed, this is a simple consequence of the unavailability of the excluded middle—we just cannot a priori split the argument into cases. But it may be a surprise to observe that not only negation or falsity may cause difficulties. Formula (i), known as *Peirce's law*, is purely implicational, but still we are unable to find a construction. Another example of this kind is formula (vii) expressing the classical associativity of equivalence. One can verify it using a binary truth-table, but from the constructive point of view this associativity property seems to be purely accidental.

Formula (viii) can be seen as a definition of classical implication in terms of negation and disjunction. Constructively, this definition does not work. We can say more: None among $\to, \bot, \vee, \wedge$ is definable from the others (see Exercise 2.26).

2.2. Natural deduction

In order to formalize the intuitionistic propositional logic, we define a proof system, called *natural deduction*, and denoted by $\mathrm{NJ}(\to, \bot, \wedge, \vee)$, or simply NJ. The rules of natural deduction express in a precise way the ideas of the informal semantics of Section 2.1.

2.2.1. DEFINITION.

(i) A *judgement* in natural deduction is a pair, written $\Gamma \vdash \varphi$ (and read "Γ proves φ") consisting of a finite set of formulas Γ and a formula φ.

(ii) We use various simplifications when we deal with judgements. For instance, we write $\varphi_1, \varphi_2 \vdash \psi$ instead of $\{\varphi_1, \varphi_2\} \vdash \psi$, or Γ, Δ instead of $\Gamma \cup \Delta$, or Γ, φ instead of $\Gamma \cup \{\varphi\}$. In particular, the notation $\vdash \varphi$ stands for $\varnothing \vdash \varphi$.

(iii) A formal *proof* or *derivation* of $\Gamma \vdash \varphi$ is a finite tree of judgements satisfying the following conditions:

- The root label is $\Gamma \vdash \varphi$;
- All the leaves are *axioms*, i.e. judgements of the form $\Gamma, \varphi \vdash \varphi$;
- The label of each mother node is obtained from the labels of the daughters using one of the rules in Figure 2.1.

If such a proof exists, we say that the judgement $\Gamma \vdash \varphi$ is *provable* or *derivable*, and we write $\Gamma \vdash_N \varphi$. For infinite Γ, we write $\Gamma \vdash_N \varphi$ to mean that $\Gamma_0 \vdash_N \varphi$, for some finite subset Γ_0 of Γ.

(iv) It is customary to omit the index $_N$ in $\vdash_N$. Note that in this way the notation $\Gamma \vdash \varphi$ becomes overloaded. It expresses the provability of a judgement and also denotes the judgement itself. However, the intended meaning is usually clear from the context.

(v) If $\vdash \varphi$ then we say that φ is a *theorem*.[2]

The proof system consists of an axiom scheme and rules. For each logical connective (except $\bot$) we have one or two *introduction* rules and one or two *elimination* rules. An introduction rule for a connective $\circ$ tells us how a conclusion of the form $\varphi \circ \psi$ can be derived. An elimination rule describes the

---
[2] In general, a *theorem* is a formula provable in a logical system.

2.2. NATURAL DEDUCTION

$$\Gamma, \varphi \vdash \varphi \ (\text{Ax})$$

$$\frac{\Gamma, \varphi \vdash \psi}{\Gamma \vdash \varphi \to \psi} \ (\to\text{I}) \qquad \frac{\Gamma \vdash \varphi \to \psi \quad \Gamma \vdash \varphi}{\Gamma \vdash \psi} \ (\to\text{E})$$

$$\frac{\Gamma \vdash \varphi \quad \Gamma \vdash \psi}{\Gamma \vdash \varphi \wedge \psi} \ (\wedge\text{I}) \qquad \frac{\Gamma \vdash \varphi \wedge \psi}{\Gamma \vdash \varphi} (\wedge\text{E}) \frac{\Gamma \vdash \varphi \wedge \psi}{\Gamma \vdash \psi}$$

$$\frac{\Gamma \vdash \varphi}{\Gamma \vdash \varphi \vee \psi} \ (\vee\text{I}) \ \frac{\Gamma \vdash \psi}{\Gamma \vdash \varphi \vee \psi} \qquad \frac{\Gamma, \varphi \vdash \vartheta \quad \Gamma, \psi \vdash \vartheta \quad \Gamma \vdash \varphi \vee \psi}{\Gamma \vdash \vartheta} \ (\vee\text{E})$$

$$\frac{\Gamma \vdash \bot}{\Gamma \vdash \varphi} \ (\bot\text{E})$$

FIGURE 2.1: INTUITIONISTIC NATURAL DEDUCTION NJ

way in which $\varphi \circ \psi$ can be used to derive other formulas. Observe that the natural deduction rules can be seen as a formalization of the BHK interpretation, where "construction" should be read as "proof". Indeed, consider for instance the implication. The premise $\Gamma, \varphi \vdash \psi$ of rule ($\to$I) can be understood as the ability to infer ψ from Γ if a proof of φ is provided. This is enough to derive the implication. The elimination rule ($\to$E) corresponds to the same idea: if we have a proof of $\varphi \to \psi$ then we can turn a proof of φ into a proof of ψ. In a sense, rule ($\to$E) can be seen as a converse of rule ($\to$I), and a similar observation (called the *inversion principle*, see Prawitz [403]) can be made about the other connectives. Rule ($\bot$E), called *ex falso sequitur quodlibet* (or simply *ex falso*) is an exception, because there is no matching introduction rule.

2.2.2. NOTATION. It is sometimes useful to consider fragments of propositional logic where some connectives do not occur. For instance, in Section 2.6 we discuss the implicational fragment IPC($\to$) of the intuitionistic propositional logic. The subsystem of NJ consisting of the axiom scheme and rules for implication is then denoted by NJ($\to$). This convention applies to other fragments, e.g. IPC($\to, \wedge, \vee$) is the positive fragment, called also *minimal logic*, and NJ($\to, \wedge, \vee$) stands for the appropriate part of NJ.

2.2.3. EXAMPLE. We give example proofs of our favourite formulas. Below, formulas φ, ψ and ϑ can be arbitrary:

(i)
$$\frac{\varphi \vdash \varphi}{\vdash \varphi \to \varphi} (\to I)$$

(ii)
$$\frac{\dfrac{\varphi, \psi \vdash \varphi}{\varphi \vdash \psi \to \varphi} (\to I)}{\vdash \varphi \to (\psi \to \varphi)} (\to I)$$

(iii) Here, Γ abbreviates $\{\varphi \to (\psi \to \vartheta), \varphi \to \psi, \varphi\}$.

$$\frac{\dfrac{\dfrac{\dfrac{\Gamma \vdash \varphi \to (\psi \to \vartheta) \quad \Gamma \vdash \varphi}{\Gamma \vdash \psi \to \vartheta} (\to E) \quad \dfrac{\Gamma \vdash \varphi \to \psi \quad \Gamma \vdash \varphi}{\Gamma \vdash \psi} (\to E)}{\Gamma \vdash \vartheta} (\to E)}{\dfrac{\varphi \to (\psi \to \vartheta), \varphi \to \psi \vdash \varphi \to \vartheta}{\varphi \to (\psi \to \vartheta) \vdash (\varphi \to \psi) \to (\varphi \to \vartheta)} (\to I)}}{\vdash (\varphi \to (\psi \to \vartheta)) \to (\varphi \to \psi) \to (\varphi \to \vartheta)} (\to I)$$

2.2.4. LEMMA. *Intuitionistic propositional logic is closed under weakening and substitution, that is, $\Gamma \vdash \varphi$ implies $\Gamma, \psi \vdash \varphi$ and $\Gamma[p := \psi] \vdash \varphi[p := \psi]$, where $[p := \psi]$ denotes a substitution of ψ for all occurrences of the propositional variable p.*

PROOF. Induction with respect to the size of proofs (Exercise 2.3). □

Results like the above are sometimes expressed by saying that the following are *derived* (or *admissible*) rules of propositional intuitionistic logic:

$$\frac{\Gamma \vdash \varphi}{\Gamma, \psi \vdash \varphi} \qquad \frac{\Gamma \vdash \varphi}{\Gamma[p := \psi] \vdash \varphi[p := \psi]}$$

2.3. Algebraic semantics of classical logic

In the next section we shall introduce an algebraic semantics for intuitionistic logic. To help the reader understand it better, let us begin with classical logic. Usually, semantics of classical propositional formulas is defined so that each connective is seen as an operation acting on the set $\mathbb{B} = \{0, 1\}$ of truth values. That is, we actually deal with an algebraic system

$$\langle \mathbb{B}, \vee, \wedge, \to, -, 0, 1 \rangle$$

2.3. ALGEBRAIC SEMANTICS OF CLASSICAL LOGIC

where $0 \vee 1 = 1$, $0 \wedge 1 = 0$, $0 \to 1 = 1$ etc. This system is easily ordered so that $0 < 1$, and we have the following property:

$$a \leq b \quad \text{iff} \quad a \to b = 1.$$

There is an obvious similarity between the algebra of truth values and the algebra of sets. The logical operations $\vee$ and $\wedge$ behave very much like the set-theoretical $\cup$ and $\cap$. The equation $A \cup B = \{x \mid (x \in A) \vee (x \in B)\}$ states one of many analogies. In a similar way negation corresponds to the complement $-A$ (with respect to a fixed domain) and implication to $-A \cup B$.

The notion of a *Boolean algebra* we are now going to introduce is a generalization of both the algebra of truth values and the algebra of sets. We begin though with a weaker and thus more general notion of a *lattice*.

2.3.1. DEFINITION. A *lattice* is a partial order $\langle A, \leq \rangle$ such that every two-element subset $\{a,b\}$ of A has both a least upper bound and a greatest lower bound in A. We use the notation $a \sqcup b$ for $\sup_A\{a,b\}$ and $a \sqcap b$ for $\inf_A\{a,b\}$. By analogy with set-theoretic operations, we refer to $\sqcup$ as *union* (or *join*) and to $\sqcap$ as *intersection* (or *meet*). A top (resp. bottom) element in a lattice (if it exists) is usually denoted by 1 (resp. 0).

2.3.2. LEMMA. *In a lattice, the following conditions are equivalent:*

(i) $a \leq b$;

(ii) $a \sqcap b = a$;

(iii) $a \sqcup b = b$.

PROOF. Immediate. □

2.3.3. EXAMPLE. Every linear order, in particular the set $\mathbb{B}$ of truth values, is a lattice. Every family of sets closed under set union and intersection is also a lattice. But the closure with respect to $\cup$ and $\cap$ is not a necessary condition for a family of sets to be a lattice. A good example is the family of all convex subsets of the Euclidean plane. (A set A is *convex* iff for all $a, b \in A$ the straight line segment joining a and b is contained in A.)

2.3.4. LEMMA. *The following equations are valid in every lattice:*

(i) $a \sqcup a = a$ *and* $a \sqcap a = a$;

(ii) $a \sqcup b = b \sqcup a$ *and* $a \sqcap b = b \sqcap a$;

(iii) $(a \sqcup b) \sqcup c = a \sqcup (b \sqcup c)$ *and* $(a \sqcap b) \sqcap c = a \sqcap (b \sqcap c)$;

(iv) $(a \sqcup b) \sqcap a = a$ *and* $(a \sqcap b) \sqcup a = a$.

PROOF. A routine application of the definitions. □

In fact, the properties listed in Lemma 2.3.4 give an axiomatization of lattices as algebraic systems of the form $\langle A, \sqcup, \sqcap \rangle$, cf. Exercise 2.6. We are mostly interested in lattices satisfying some additional properties.

2.3.5. DEFINITION.

(i) A lattice A is *distributive* iff the following equations hold in A:

(1) $(a \sqcup b) \sqcap c = (a \sqcap c) \sqcup (b \sqcap c)$;

(2) $(a \sqcap b) \sqcup c = (a \sqcup c) \sqcap (b \sqcup c)$.

(ii) Assume that a lattice A has a top element 1 and a bottom element 0. We say that b is a *complement* of a iff $a \sqcup b = 1$ and $a \sqcap b = 0$.

2.3.6. LEMMA. *Let b be a complement of a in a distributive lattice. Then b is the greatest element of A satisfying $a \sqcap b = 0$. In particular, a has at most one complement.*

PROOF. Suppose that $a \sqcap c = 0$. Then $c \leq b$, because

$$c = 1 \sqcap c = (a \sqcup b) \sqcap c = (a \sqcap c) \sqcup (b \sqcap c) = 0 \sqcup (b \sqcap c) = b \sqcap c.$$ □

2.3.7. DEFINITION. A *Boolean algebra* is a distributive lattice B with top and bottom elements, such that every element a of B has a complement (denoted by $-a$).

Boolean algebras are often presented as algebraic structures of the form $\mathcal{B} = \langle B, \sqcup, \sqcap, -, 0, 1 \rangle$. In this case the underlying partial order can be reconstructed with the help of the equivalence: $a \leq b$ iff $a \sqcap b = a$ (cf. Lemma 2.3.2).

2.3.8. EXAMPLE. Let X be any set. A *field of sets* (*over X*) is a non-empty family $\mathcal{R}$ of subsets of X, closed under set-theoretic union, intersection and complement (to X). Every field of sets is a Boolean algebra. Examples of fields of sets are:

(i) $\mathcal{P}(X)$, the power set of X;

(ii) $\{\varnothing, X\}$;

(iii) $\{A \subseteq X : A \text{ finite or } X - A \text{ finite}\}$.

Observe that the algebra $\mathbb{B}$ of truth values (forget about $\rightarrow$ for a moment) is isomorphic to (ii).

The following result is known as *Stone's representation theorem*.

2.4. HEYTING ALGEBRAS

2.3.9. THEOREM (M.H. Stone, 1934). *Every Boolean algebra is isomorphic to a field of sets.*

We omit the proof of Stone's theorem, as it would distract us too much from the mainstream of our considerations. The reader may spend some of her spare time doing Exercise 2.15, but we now turn to the Boolean algebra semantics for classical propositional logic.

2.3.10. DEFINITION. A *valuation* in a Boolean algebra $\mathcal{B} = \langle B, \sqcup, \sqcap, -, 0, 1 \rangle$ is any map v from the set PV of propositional variables to B. The *value* (in $\mathcal{B}$) of a formula φ with respect to a valuation v is defined by induction.

$$\begin{aligned}
[\![p]\!]_v &= v(p), \text{ for } p \in \text{PV}; \\
[\![\bot]\!]_v &= 0; \\
[\![\varphi \vee \psi]\!]_v &= [\![\varphi]\!]_v \sqcup [\![\psi]\!]_v; \\
[\![\varphi \wedge \psi]\!]_v &= [\![\varphi]\!]_v \sqcap [\![\psi]\!]_v; \\
[\![\varphi \to \psi]\!]_v &= -[\![\varphi]\!]_v \sqcup [\![\psi]\!]_v.
\end{aligned}$$

One writes $\mathcal{B}, v \models \varphi$ when $[\![\varphi]\!]_v = 1$, and $\mathcal{B} \models \varphi$ when $\mathcal{B}, v \models \varphi$, for all v.

It should be clear that the Boolean algebra semantics is a generalization of the ordinary two-valued semantics. Indeed, a formula is a classical tautology if and only if $\mathbb{B} \models \varphi$. In fact, this generalization is not essential.

2.3.11. THEOREM. *A propositional formula is a classical tautology if and only if $\mathcal{B} \models \varphi$, for all Boolean algebras $\mathcal{B}$.*

PROOF. The implication from right to left is immediate. To prove the other implication suppose that $\mathcal{B} \not\models \varphi$ for some $\mathcal{B}$. By Stone's representation theorem[3] we can assume that $\mathcal{B}$ is a field of sets over some X.

Since $\mathcal{B} \not\models \varphi$, there exists a valuation v in $\mathcal{B}$ with $[\![\varphi]\!]_v \neq X$. Thus, there is $x \in X$ such that $x \notin [\![\varphi]\!]_v$. Define a binary valuation w (a valuation in $\mathbb{B}$) so that $w(p) = 1$ iff $x \in [\![p]\!]_v$. Prove by induction that for all formulas ψ:

$$[\![\psi]\!]_w = 1 \quad \text{iff} \quad x \in [\![\psi]\!]_v.$$

Then $[\![\varphi]\!]_w \neq 1$. □

2.4. Heyting algebras

We will now develop a semantics for intuitionistic propositional logic. For this we analyze the algebraic properties of formulas with respect to provability. We begin by observing that provable implication behaves almost like an ordering relation on formulas, i.e. it is reflexive and transitive. More exactly, for every Γ we have:

---
[3] A hint for a direct proof is given in Exercise 2.14.

- $\Gamma \vdash \varphi \to \varphi$;
- If $\Gamma \vdash \varphi \to \psi$ and $\Gamma \vdash \psi \to \vartheta$ then $\Gamma \vdash \varphi \to \vartheta$.

This relation is however not anti-symmetric. But we can turn it into a partial order if we identify equivalent formulas. To make this precise, let Γ be a fixed set of propositional formulas (in particular Γ may be empty). We define a relation $\sim_\Gamma$ as follows:

$$\varphi \sim_\Gamma \psi \quad \text{iff} \quad \Gamma \vdash \varphi \to \psi \text{ and } \Gamma \vdash \psi \to \varphi.$$

It is not difficult to see that $\sim_\Gamma$ is an equivalence relation on the set Φ of all formulas, and that (we omit the subscript Γ in $\sim_\Gamma$):

$$[\bot]_\sim = \{\varphi \mid \Gamma \vdash \neg\varphi\} \quad \text{and} \quad [\top]_\sim = \{\varphi \mid \Gamma \vdash \varphi\}.$$

Let $\mathcal{L}_\Gamma = \Phi/\sim = \{[\varphi]_\sim \mid \varphi \in \Phi\}$. It should now be obvious that the relation $\leq_\Gamma$, defined by

$$[\varphi]_\sim \leq_\Gamma [\psi]_\sim \quad \text{iff} \quad \Gamma \vdash \varphi \to \psi$$

is a well-defined partial order over $\mathcal{L}_\Gamma$. We have the equivalence:

$$[\varphi]_\sim = [\psi]_\sim \quad \text{iff} \quad [\varphi]_\sim \leq_\Gamma [\psi]_\sim \text{ and } [\psi]_\sim \leq_\Gamma [\varphi]_\sim.$$

Our next step is to discover that the partial order $\langle \mathcal{L}_\Gamma, \leq \rangle$ is a lattice. Define

$$[\varphi]_\sim \sqcup [\psi]_\sim = [\varphi \vee \psi]_\sim \,;$$
$$[\varphi]_\sim \sqcap [\psi]_\sim = [\varphi \wedge \psi]_\sim \,.$$

The operations $\sqcup$ and $\sqcap$ are well-defined, because the following are provable:

$$(\varphi \leftrightarrow \varphi') \to ((\psi \leftrightarrow \psi') \to ((\varphi \vee \psi) \leftrightarrow (\varphi' \vee \psi')));$$
$$(\varphi \leftrightarrow \varphi') \to ((\psi \leftrightarrow \psi') \to ((\varphi \wedge \psi) \leftrightarrow (\varphi' \wedge \psi'))).$$

The reader can easily check that $[\varphi]_\sim \sqcup [\psi]_\sim$ (resp. $[\varphi]_\sim \sqcap [\psi]_\sim$) is the supremum (resp. infimum) of the set $\{[\varphi]_\sim, [\psi]_\sim\}$. Having observed that, the next step is to see that the lattice is distributive and that $[\top]_\sim$ is the top and $[\bot]_\sim$ is the bottom. An optimistic reader might now pose the conjecture that $\mathcal{L}_\Gamma$ is a Boolean algebra with $-[\varphi]_\sim$ defined as $[\neg\varphi]_\sim$.

It would be exactly the case if we dealt with classical logic. But we do not have the excluded middle and there are (not unexpected) difficulties with this complement operation. We have $a \sqcap -a = 0$, but not necessarily $a \sqcup -a = 1$.

The best we can assert about $-a$ is that it is *the greatest element such that $a \sqcap -a = 0$*, and we can call it a *pseudo-complement*. Since negation is a special kind of implication, the above calls for a generalization.

2.4. Heyting algebras

2.4.1. DEFINITION. An element c of a lattice is called a *relative pseudo-complement* of a with respect to b, iff c is the greatest element such that $a \sqcap c \leq b$. The relative pseudo-complement, if exists, is denoted $a \Rightarrow b$, and one defines $-a$ as a special case: $-a = a \Rightarrow 0$.

For example, in our algebra $\mathcal{L}_\Gamma$, often called the *Lindenbaum algebra*, the relative pseudo-complement always exists:

$$[\varphi]_\sim \Rightarrow [\psi]_\sim = [\varphi \to \psi]_\sim.$$

We have just discovered a new type of algebra, called *Heyting algebra* or *pseudo-Boolean algebra*.

2.4.2. DEFINITION. A *Heyting algebra* is a distributive lattice H with top and bottom elements, such that the relative pseudo-complement $a \Rightarrow b$ exists for all $a, b \in H$.

In fact the word "distributive" in the definition above is redundant, see Exercise 2.9. Heyting algebras are typically taken as algebraic structures of the form $\mathcal{H} = \langle H, \sqcup, \sqcap, \Rightarrow, -, 0, 1 \rangle$, where the underlying partial order is implicit and defined as usual by the clause $a \leq b$ iff $a \sqcap b = a$ (see Lemma 2.3.2). We often ignore the notational difference between $\mathcal{H}$ and H.

2.4.3. LEMMA.

(i) *The Lindenbaum algebra $\mathcal{L}_\Gamma$ is a Heyting algebra.*

(ii) *Each Boolean algebra is a Heyting algebra with $a \Rightarrow b$ defined as $-a \sqcup b$.*

(iii) *Every finite distributive lattice is a Heyting algebra.*

PROOF. Part (i) has already been shown. For part (ii) first observe that $(-a \sqcup b) \sqcap a = (-a \sqcap a) \sqcup (b \sqcap a) = 0 \sqcup (b \sqcap a) = b \sqcap a \leq b$. Then assume that $c \sqcap a \leq b$. We have $c = c \sqcap 1 = c \sqcap (-a \sqcup a) = (c \sqcap -a) \sqcup (c \sqcap a) \leq (c \sqcap -a) \sqcup b = (c \sqcup b) \sqcap (-a \sqcup b) \leq -a \sqcup b$. Part (iii) follows from the simple fact that, for any given a and b, the set $A = \{c \mid c \sqcap a \leq b\}$ is finite, and thus it has a supremum. Because of distributivity, $\sup A$ itself belongs to A and thus $\sup A = a \Rightarrow b$. □

The following easy observations are often useful:

2.4.4. LEMMA. *In every Heyting algebra,*

(i) $a \leq b \Rightarrow c$ *is equivalent to* $a \sqcap b \leq c$;

(ii) $a \leq b$ *is equivalent to* $a \Rightarrow b = 1$.

The algebraic semantics of intuitionistic propositional logic is defined in a similar style as for classical logic (cf. Definition 2.3.10). We only need to replace Boolean algebras by Heyting algebras.

2.4.5. DEFINITION. Let $\mathcal{H} = \langle H, \sqcup, \sqcap, \Rightarrow, -, 0, 1 \rangle$ be a Heyting algebra. A *valuation* in $\mathcal{H}$ is a map $v : \mathrm{PV} \to H$. Given a valuation v in $\mathcal{H}$, we define the *value* $[\![\varphi]\!]_v$ of a formula φ with respect to v (written more precisely as $[\![\varphi]\!]_v^{\mathcal{H}}$).

$$\begin{aligned}
{[\![p]\!]_v} &= v(p), \text{ for } p \in \mathrm{PV}; \\
{[\![\bot]\!]_v} &= 0; \\
{[\![\varphi \vee \psi]\!]_v} &= [\![\varphi]\!]_v \sqcup [\![\psi]\!]_v; \\
{[\![\varphi \wedge \psi]\!]_v} &= [\![\varphi]\!]_v \sqcap [\![\psi]\!]_v; \\
{[\![\varphi \to \psi]\!]_v} &= [\![\varphi]\!]_v \Rightarrow [\![\psi]\!]_v.
\end{aligned}$$

2.4.6. NOTATION. Let $\mathcal{H}$ be a Heyting algebra. We write:

- $\mathcal{H}, v \models \varphi$, when $[\![\varphi]\!]_v = 1$;

- $\mathcal{H} \models \varphi$, when $\mathcal{H}, v \models \varphi$, for all v;

- $\mathcal{H}, v \models \Gamma$, when $\mathcal{H}, v \models \varphi$, for all $\varphi \in \Gamma$;

- $\mathcal{H} \models \Gamma$, when $\mathcal{H}, v \models \Gamma$, for all v;

- $\models \varphi$, when $\mathcal{H}, v \models \varphi$, for all $\mathcal{H}, v$;

- $\Gamma \models \varphi$, when $\mathcal{H}, v \models \Gamma$ implies $\mathcal{H}, v \models \varphi$, for all $\mathcal{H}$ and v.

If $\models \varphi$ holds, we say that φ is *intuitionistically valid* or that is an *intuitionistic tautology*. It follows from the following that the notions of a theorem and a tautology coincide for intuitionistic propositional calculus.

2.4.7. THEOREM (Completeness). *The following are equivalent:*[4]

(i) $\Gamma \vdash \varphi$;

(ii) $\Gamma \models \varphi$.

PROOF. (i) $\Rightarrow$ (ii): Let $\Gamma = \{\vartheta_1, \ldots, \vartheta_n\}$. If v is a valuation in a Heyting algebra $\mathcal{H}$, then $[\![\Gamma]\!]_v$ stands for $[\![\vartheta_1]\!]_v \sqcap \cdots \sqcap [\![\vartheta_n]\!]_v$ (which is 1 when $\Gamma = \varnothing$). By induction with respect to derivations we prove that, for all v:

$$\text{If } \Gamma \vdash \varphi \text{ then } [\![\Gamma]\!]_v \leq [\![\varphi]\!]_v.$$

---

[4]Part (i) $\Rightarrow$ (ii) is called *soundness*, while (ii) $\Rightarrow$ (i) is the proper *completeness*.

2.4. HEYTING ALGEBRAS

The statement of the theorem follows from the special case when $[\![\Gamma]\!]_v = 1$.

We proceed by cases depending on the last rule used in the proof. For instance, consider the case of ($\to$I). We need to show that $[\![\Gamma]\!]_v \leq [\![\varphi_1 \to \varphi_2]\!]_v$ follows from the induction hypothesis $[\![\Gamma]\!]_v \sqcap [\![\varphi_1]\!]_v \leq [\![\varphi_2]\!]_v$. But that is an easy consequence of $[\![\varphi_1 \to \varphi_2]\!]_v = [\![\varphi_1]\!]_v \Rightarrow [\![\varphi_2]\!]_v$. For the case of ($\vee$E) observe that $[\![\Gamma]\!]_v \leq [\![\varphi_1 \vee \varphi_2]\!]_v$ implies

$$[\![\Gamma]\!]_v = [\![\Gamma]\!]_v \sqcap ([\![\varphi_1]\!]_v \sqcup [\![\varphi_2]\!]_v) = ([\![\Gamma]\!]_v \sqcap [\![\varphi_1]\!]_v) \sqcup ([\![\Gamma]\!]_v \sqcap [\![\varphi_2]\!]_v).$$

If both components of $\sqcup$ are less than or equal to $[\![\varphi]\!]_v$ then so is $[\![\Gamma]\!]_v$.

(ii) $\Rightarrow$ (i): This follows from our construction of the Lindenbaum algebra. Indeed, suppose that $\Gamma \models \varphi$, but $\Gamma \nvdash \varphi$. Then $\varphi \not\vdash_\Gamma \top$, i.e. $[\varphi]_\sim \neq 1$ in $\mathcal{L}_\Gamma$. Define a valuation v in $\mathcal{L}_\Gamma$ by $v(p) = [p]_\sim$, and prove by induction that $[\![\psi]\!]_v = [\psi]_\sim$, for all ψ. It follows that $[\![\varphi]\!]_v \neq 1$, a contradiction. □

The completeness proof above can easily be adapted to classical logic, if we expand our natural deduction system by appropriate rules to account for the extension (we do it later in Chapter 6). The Lindenbaum algebra $\mathcal{L}_\Gamma$ is then a Boolean algebra and we conclude that classical propositional logic is sound and complete with respect to the general Boolean algebra semantics, as defined in Section 2.3. To obtain a completeness result for the two-valued semantics, one then uses Theorem 2.3.11. For an alternative approach see the proof of Theorem 6.1.10.

So far, Heyting algebras remain a nice but abstract notion. In order to actually make use of our algebraic semantics, we need concrete examples of such algebras. And these are not hard to find. The most prominent example of a Heyting algebra which is not a Boolean algebra is the family of open sets of a topological space.

An important difference between a topological space and a field of sets is that a complement of an open set is usually not open. If we want to interpret propositions as open sets, we can easily do it for conjunctions and disjunctions, using sums and intersections. But we cannot in general interpret a negation $\neg p$ as the complement of the value of p. The best we can do is to take the largest open set contained in the complement—the interior. This is exactly the Heyting algebra pseudo-complement.

2.4.8. PROPOSITION. *Let* $\mathcal{H} = \langle \mathcal{O}(\mathcal{T}), \cup, \cap, \Rightarrow, \sim, \varnothing, \mathcal{T} \rangle$, *where* $\mathcal{T}$ *is a topological space and*

- *the operations* $\cap$, $\cup$ *are set-theoretic;*
- $A \Rightarrow B = \mathrm{Int}(-A \cup B)$, *for arbitrary open sets A and B;*
- $\sim A = \mathrm{Int}(-A)$, *where* $-$ *is the set-theoretic complement.*

Then $\mathcal{H}$ is a Heyting algebra.

PROOF. Exercise 2.16. □

2.4.9. EXAMPLE. To see that Peirce's law $((p \to q) \to p) \to p$ is not intuitionistically valid, consider the algebra of open subsets of the real line. Take $v(p) = \mathbb{R} - \{0\}$ and $v(q) = \varnothing$. Then $[\![p \to q]\!]_v = \text{Int}(\{0\} \cup \varnothing) = \varnothing$, and $[\![(p \to q) \to p]\!]_v = \text{Int}(\mathbb{R} \cup (\mathbb{R} - \{0\})) = \mathbb{R}$. The value of the whole formula is thus the set $\text{Int}(\varnothing \cup (\mathbb{R} - \{0\})) = \mathbb{R} - \{0\} \neq \mathbb{R}$.

Consider now the *tertium non datur* principle in the form $p \vee \neg p$. If we take $v(p) = (0, \infty)$ then $[\![p \vee \neg p]\!]_v = \mathbb{R} - \{0\}$. Again, only one point is missing.

That the border between "yes" and "no" in the example above is so thin is not just a coincidence. By the following result, a classical tautology must always be represented in a topological space by a *dense* set, i.e. one whose complement has an empty interior.

2.4.10. THEOREM (Glivenko). *A formula φ is a classical tautology iff $\neg\neg\varphi$ is an intuitionistic tautology.*

PROOF. Exercise 2.34. □

WARNING. The above does not hold for first-order logic, cf. Example 8.2.2(i).

Intuitionistic logic, unlike classical logic, is not finite-valued. There is no single finite Heyting algebra $\mathcal{H}$ such that $\vdash \varphi$ is equivalent to $\mathcal{H} \models \varphi$. Indeed, consider the formula $\bigvee \{p_i \leftrightarrow p_j \mid i,j = 0, \ldots, n \text{ and } i \neq j\}$. (Here the symbol $\bigvee$ abbreviates the disjunction of all members of the set.) This formula is not valid in general (Exercise 2.19), although it is valid in all Heyting algebras of cardinality at most n.

A complete semantics can however be defined by an infinite Heyting algebra. Some of these are quite familiar.

2.4.11. THEOREM. *Let $\mathcal{H}$ be the algebra of all open subsets of*

- *the set $\mathbb{R}$ of reals, or*
- *the set $\mathbb{Q}$ of rationals, or*
- *any Cartesian product of the above, in particular $\mathbb{R}^2$.*

Then $\mathcal{H} \models \varphi$ iff φ is valid.

PROOF. Exercise 2.30. □

Theorem 2.4.11 may be interpreted as follows. If a formula φ is not valid intuitionistically, it is always possible to give a counterexample over the real line. The same holds for the Euclidean plane $\mathbb{R}^2$. The latter has the following nice consequence: One can always expect to produce a counterexample by drawing pictures on paper. (A similar property holds of course for classical logic. In Exercise 2.14 take $\mathcal{B} = \mathcal{P}(\mathbb{R}^2)$, the power set of the real plane.)

An alternative to one infinite algebra is the collection of all finite models.

2.5. KRIPKE SEMANTICS

2.4.12. THEOREM (Finite model property). *A formula φ of length n is valid iff it is valid in all Heyting algebras of cardinality at most 2^{2^n}.*

PROOF. Assume that φ is not valid, i.e. that $\mathcal{H}, v \not\models \varphi$, for some algebra $\mathcal{H} = \langle H, \sqcup, \sqcap, \Rightarrow, -, 0, 1 \rangle$. We show how to replace $\mathcal{H}$ by a finite algebra. Let $\varphi_1, \ldots, \varphi_m$ be all subformulas of φ. (Note that $m \leq n$.) Let $a_i = [\![\varphi_i]\!]_v$, for $i = 1, \ldots, m$, and consider the subalgebra L of $\langle H, \sqcup, \sqcap, 0, 1 \rangle$ generated by $a_1, \ldots, a_m$. Observe that L has at most 2^{2^m} elements, namely all possible unions of the 2^m possible intersections of the generators. (We get 0 if the union has no components and 1 if there is one component of length zero. The closure under $\sqcap$ follows from distributivity.) In addition, L is a Heyting algebra, because of Lemma 2.4.3(iii). But L is not necessarily a Heyting subalgebra of $\mathcal{H}$, i.e. the definition of $\Rightarrow$ in L and in $\mathcal{H}$ may be different. To distinguish the two, let us use the notation $\Rightarrow_L$ and $\Rightarrow_\mathcal{H}$, respectively.

Thus it is not necessarily the case that $[\![\psi]\!]_v^L = [\![\psi]\!]_v^\mathcal{H}$ for arbitrary ψ. However, this equation holds for all subformulas of φ, including φ itself. This is shown by induction, using the following observation: If $\psi_1 \to \psi_2$ is a subformula of φ then $[\![\psi_1]\!]_v^\mathcal{H} \Rightarrow_\mathcal{H} [\![\psi_2]\!]_v^\mathcal{H} = [\![\psi_1 \to \psi_2]\!]_v^\mathcal{H} \in L$. This implies $[\![\psi_1]\!]_v^\mathcal{H} \Rightarrow_\mathcal{H} [\![\psi_2]\!]_v^\mathcal{H} = [\![\psi_1]\!]_v^\mathcal{H} \Rightarrow_L [\![\psi_2]\!]_v^\mathcal{H}$. Details are left to the reader, and the conclusion is that $[\![\varphi]\!]_v^L \neq 1$ and thus $L, v \not\models \varphi$. □

From the finite model property it follows that intuitionistic propositional logic is decidable. The upper bound obtained this way (double exponential space) is unsatisfactory. But it can be improved down to polynomial space, see Lemma 4.2.3 and Exercise 7.13.

2.5. Kripke semantics

We now introduce another semantics of intuitionistic propositional logic. This semantics reflects the following idea. From the constructive point of view, we can assert the truth only of the propositions of which we are certain. But by learning about new facts we gain more information, and we can add new propositions to our state of knowledge. We should however not lose our knowledge. In other words, what is true now will remain true forever, but what is not recognized today as true may become true tomorrow. Because of that we have to be careful and assert *"not A"* only when it is entirely impossible that we might ever assert A.

2.5.1. DEFINITION. A *Kripke model* is a triple of the form

$$\mathcal{C} = \langle C, \leq, \Vdash \rangle,$$

where C is a non-empty set, whose elements are called *states* or *possible worlds*, $\leq$ is a partial order in C, and $\Vdash$ is a binary relation between elements of C and propositional variables. The relation $\Vdash$ (read "forces") must satisfy the following monotonicity condition.

$$\text{If } c \leq c' \text{ and } c \Vdash p \text{ then } c' \Vdash p.$$

The intuition is that elements of the model represent states of knowledge. The relation $\leq$ corresponds to extending states by gaining more knowledge, and the relation $\Vdash$ determines which propositional variables are assumed to be true in a given state. We extend this relation to provide meaning for propositional formulas as follows.

2.5.2. DEFINITION. If $\mathcal{C} = \langle C, \leq, \Vdash \rangle$ is a Kripke model, then

- $c \Vdash \varphi \vee \psi$ iff $c \Vdash \varphi$ or $c \Vdash \psi$;
- $c \Vdash \varphi \wedge \psi$ iff $c \Vdash \varphi$ and $c \Vdash \psi$;
- $c \Vdash \varphi \to \psi$ iff $c' \Vdash \psi$ for all $c' \geq c$ with $c' \Vdash \varphi$;
- $c \Vdash \bot$ does not hold.

Note that the definition above implies the following rule for negation:

- $c \Vdash \neg \varphi$ iff $c' \nVdash \varphi$, for all $c' \geq c$.

The notation $\Vdash$ can be used in various ways. Sometimes we write $\mathcal{C}, c \Vdash \varphi$ to make it clear which model is being used. We write $\mathcal{C} \Vdash \varphi$ when $c \Vdash \varphi$, for all $c \in C$. The notation $c \Vdash \Gamma$ means that $c \Vdash \varphi$ for all $\varphi \in \Gamma$, and similarly for $\mathcal{C} \Vdash \Gamma$. Finally, $\Gamma \Vdash \varphi$ that for every Kripke model $\mathcal{C}$ and every state c of $\mathcal{C}$, the condition $\mathcal{C}, c \Vdash \Gamma$ implies $\mathcal{C}, c \Vdash \varphi$.

The following generalized monotonicity follows by easy induction:

2.5.3. LEMMA. *If $c \leq c'$ and $c \Vdash \varphi$ then $c' \Vdash \varphi$.*

2.5.4. EXAMPLE. Let $C = \{c, c', c''\}$ where $c \leq c', c''$ and c', c'' are incomparable. A Kripke model $\mathcal{C} = \langle C, \leq, \Vdash \rangle$, where $c' \Vdash p$ and $c'' \Vdash q$ and $c \nVdash p, q$ can be represented by the following graph:

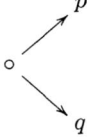

In this model we have e.g. $c \Vdash \neg\neg(p \vee q)$ and $c \Vdash (p \to q) \to q$, but $c \nVdash p \vee \neg p$.

We now want to show completeness of Kripke semantics. For this, we transform every Heyting algebra into a Kripke model.

2.5. KRIPKE SEMANTICS

2.5.5. DEFINITION. A *filter* in a Heyting algebra $\mathcal{H} = \langle H, \sqcup, \sqcap, \Rightarrow, -, 0, 1\rangle$ is a non-empty subset F of H, such that[5]

- $a, b \in F$ implies $a \sqcap b \in F$;
- $a \in F$ and $a \leq b$ implies $b \in F$.

A filter F is *proper* iff $F \neq H$. A proper filter F is *prime* iff $a \sqcup b \in F$ implies that either $a \in F$ or $b \in F$.

2.5.6. LEMMA. *Let A be any subset of a Heyting algebra $\mathcal{H}$. Then the set $F = \{a \in \mathcal{H} \mid a \geq a_1 \sqcap \ldots \sqcap a_k, \text{ for some } a_1, \ldots, a_k \in A\}$ is the least filter containing A. The filter F is proper iff $a_1 \sqcap \ldots \sqcap a_k \neq 0$, for all finite sets $\{a_1, \ldots, a_k\} \subseteq A$.*

PROOF. Left to the reader. □

2.5.7. LEMMA. *Let F be a proper filter in $\mathcal{H}$ and let $a \notin F$. There exists a prime filter G such that $F \subseteq G$ and $a \notin G$.*

PROOF. Let $\mathcal{F} = \{E \mid E \text{ is a filter and } F \subseteq E \text{ and } a \notin E\}$. One shows that the union of an arbitary chain in $\mathcal{F}$ is itself a member of $\mathcal{F}$. Thus every chain in $\mathcal{F}$ has an upper bound in $\mathcal{F}$. We can apply the Kuratowski-Zorn Lemma A.1.1 to conclude that $\mathcal{F}$ has a maximal element G with respect to inclusion. We show that G is a prime filter.

For $y \in \mathcal{H}$, let $G_y = \{x \mid x \geq g \sqcap y, \text{ for some } g \in G\}$. By Lemma 2.5.6, each G_y is a filter. We claim that if $b \sqcup c \in G$ then either G_b or G_c must belong to $\mathcal{F}$. Thus $b \in G$ or $c \in G$ by the maximality of G.

Suppose otherwise, i.e. neither G_b nor G_c is in $\mathcal{F}$. That is, $a \in G_b \cap G_c$ and there are $g_1, g_2 \in G$ with $g_1 \sqcap b \leq a$ and $g_2 \sqcap c \leq a$. Then $(g_1 \sqcap g_2) \sqcap (b \sqcup c) \in G$, and we have $a = a \sqcup a \geq (g_1 \sqcap g_2 \sqcap b) \sqcup (g_1 \sqcap g_2 \sqcap c) = (g_1 \sqcap g_2) \sqcap (b \sqcup c)$, so that $a \in G$, a contradiction. □

2.5.8. LEMMA. *Let v be a valuation in a Heyting algebra $\mathcal{H}$, where $0 \neq 1$. There is a Kripke model $\mathcal{C} = \langle C, \leq, \Vdash\rangle$, such that the conditions $\mathcal{H}, v \models \varphi$ and $\mathcal{C} \Vdash \varphi$ are equivalent for all φ.*

PROOF. We take C to be the set of all prime filters in $\mathcal{H}$. The relation $\leq$ is inclusion, and we define $F \Vdash p$ iff $v(p) \in F$. For all formulas ψ we prove

$$F \Vdash \psi \quad \text{iff} \quad [\![\psi]\!]_v \in F, \qquad (*)$$

by induction with respect to ψ. The only non-trivial case is $\psi = \psi' \to \psi''$. Suppose that $F \Vdash \psi' \to \psi''$, but $[\![\psi' \to \psi'']\!]_v = [\![\psi']\!]_v \Rightarrow [\![\psi'']\!]_v \notin F$. Take the least filter G' containing $F \cup \{[\![\psi']\!]_v\}$. Lemma 2.5.6 implies that

$$G' = \{b \mid b \geq f \sqcap [\![\psi']\!]_v \text{ for some } f \in F\},$$

---
[5]Exercises 2.11 and 2.12 explain the intuition of a filter: It is an algebraic model of a "state of knowledge", closed under logical consequence.

and we have $[\![\psi'']\!]_v \not\subseteq G'$, in particular G' is proper. Indeed, otherwise there is an $f \in F$, with $[\![\psi'']\!]_v \supseteq f \sqcap [\![\psi']\!]_v$, and thus $f \leq [\![\psi']\!]_v \Rightarrow [\![\psi'']\!]_v \in F$.

Using Lemma 2.5.7, we extend G' to a prime filter G, not containing $[\![\psi'']\!]_v$. By the induction hypothesis $G \Vdash \psi'$, because $[\![\psi']\!]_v \in G$. Since $F \Vdash \psi' \to \psi''$, it follows that $G \Vdash \psi''$. That is, $[\![\psi'']\!]_v \in G$, which is not true.

For the converse, assume that $[\![\psi' \to \psi'']\!]_v \in F$ and $F \subseteq G \Vdash \psi'$. From the induction hypothesis we have $[\![\psi']\!]_v \in G$ and since $F \subseteq G$ we obtain $[\![\psi']\!]_v \Rightarrow [\![\psi'']\!]_v \in G$. Thus $[\![\psi'']\!]_v \supseteq [\![\psi']\!]_v \sqcap ([\![\psi']\!]_v \Rightarrow [\![\psi'']\!]_v) \in G$, so that $[\![\psi'']\!]_v \in G$. Apply again the induction hypothesis to conclude $G \Vdash \psi''$.

The other cases are easy. Note that primality is essential for disjunction. Having shown (∗), assume that $\mathcal{H}, v \not\models \varphi$, i.e. $[\![\varphi]\!]_v \not\subseteq \{1\}$. The filter $\{1\}$ extends to a prime filter G such that $[\![\varphi]\!]_v \notin G$, and thus $\mathcal{C}, G \not\Vdash \varphi$. On the other hand, if $\mathcal{H}, v \models \varphi$, then $[\![\varphi]\!]_v = 1$ and 1 belongs to all filters in $\mathcal{H}$. □

2.5.9. THEOREM (Completeness). *The conditions*

$$\Gamma \vdash \varphi \qquad \Gamma \models \varphi \qquad \Gamma \Vdash \varphi$$

are equivalent to each other.

PROOF. To show that $\Gamma \vdash \varphi$ implies $\Gamma \Vdash \varphi$ we proceed by induction. As an example consider the induction step for rule (→I). Assume that we have derived $\Gamma \vdash \varphi_1 \to \varphi_2$ from $\Gamma, \varphi_1 \vdash \varphi_2$. Let $c \Vdash \Gamma$, and let $c' \geq c$ be such that $c' \Vdash \varphi_1$. By the monotonicity we have $c' \Vdash \Gamma, \varphi_1$, and from the induction hypothesis we obtain $c' \Vdash \varphi_2$ as desired. We leave the other cases to the reader. (In the case of (⊥E) observe that $\Gamma \Vdash \bot$ simply means that $c \Vdash \Gamma$ is impossible, and thus $\Gamma \Vdash \varphi$ can be safely concluded.)

By Theorem 2.4.7 it now remains to prove that $\Gamma \Vdash \varphi$ implies $\Gamma \models \varphi$. Assume the contrary. Then $\mathcal{H}, v \models \Gamma$ but $\mathcal{H}, v \not\models \varphi$, for some $\mathcal{H}, v$. From the previous lemma we have a Kripke model $\mathcal{C}$ with $\mathcal{C} \Vdash \Gamma$ and $\mathcal{C} \not\Vdash \varphi$. □

The Kripke model of Example 2.5.4 shows that the excluded middle principle is not Kripke valid. Here is another nice application of Kripke semantics.

2.5.10. PROPOSITION (Disjunction property). *If $\vdash \varphi \vee \psi$ then $\vdash \varphi$ or $\vdash \psi$.*

PROOF. Assume $\not\vdash \varphi$ and $\not\vdash \psi$. There are Kripke models $\mathcal{C}_1 = \langle C_1, \leq_1, \Vdash_1 \rangle$ and $\mathcal{C}_2 = \langle C_2, \leq_2, \Vdash_2 \rangle$ and states $c_1 \in C_1$ and $c_2 \in C_2$, such that $c_1 \not\Vdash \varphi$ and $c_2 \not\Vdash \psi$. Without loss of generality we can assume that c_1 and c_2 are least elements of $\mathcal{C}_1$ and $\mathcal{C}_2$, respectively, and that $C_1 \cap C_2 = \varnothing$. Define a new model $\mathcal{C} = \langle C_1 \cup C_2 \cup \{c_0\}, \leq, \Vdash \rangle$, where $c_0 \notin C_1 \cup C_2$ and the order is the union of $\leq_1$ and $\leq_2$, extended by c_0 taken as the least element. The relation $\Vdash$ is simply the sum of $\Vdash_1$ and $\Vdash_2$. In particular $c_0 \not\Vdash p$, for all variables p. It is easy to see that $\mathcal{C}$ is a Kripke model. In addition we have $\mathcal{C}, c_1 \Vdash \vartheta$ iff $\mathcal{C}_1, c_1 \Vdash \vartheta$, for all formulas ϑ, and a similar property holds for c_2.

Now suppose that $\vdash \varphi \vee \psi$. By soundness, we have $c_0 \Vdash \varphi \vee \psi$, and thus either $c_0 \Vdash \varphi$ or $c_0 \Vdash \psi$, by definition of $\Vdash$. Then either $c_1 \Vdash \varphi$ or $c_2 \Vdash \psi$, because of monotonicity. □

2.6. The implicational fragment

The most important logical connective is the implication. Thus, it is meaningful to study *implicational formulas*, built with only this connective. The natural deduction system NJ($\to$) for this restricted calculus consists of rules ($\to$E), ($\to$I) and the axiom scheme (Ax).

2.6.1. THEOREM. *The natural deduction system* NJ($\to$) *is complete with respect to Kripke models, i.e. if* $\to$ *is the only connective occurring in in* Γ *and* φ *then the conditions* $\Gamma \vdash \varphi$ *and* $\Gamma \Vdash \varphi$ *are equivalent.*

PROOF. The implication from left to right follows from soundness of the full natural deduction system. For the proof in the other direction, let us assume that $\Gamma \nvdash \varphi$. Define $Con(\Delta) = \{\psi \mid \Delta \vdash \psi\}$ and consider a Kripke model $\mathcal{C} = \langle C, \subseteq, \Vdash \rangle$, where

$$C = \{\Delta \mid \Gamma \subseteq \Delta, \text{ and } Con(\Delta) = \Delta\}.$$

The order is by inclusion and $\Delta \Vdash p$ holds for a propositional variable p if and only if $p \in \Delta$. By induction we show the following claim:

$$\Delta \Vdash \psi \quad \text{iff} \quad \psi \in \Delta, \tag{$*$}$$

for all implicational formulas ψ and all states Δ. The case of a variable is immediate from the definition. Let ψ be $\psi_1 \to \psi_2$ and let $\Delta \Vdash \psi$. To show that $\psi \in \Delta$, take $\Delta' = \{\vartheta \mid \Delta, \psi_1 \vdash \vartheta\}$. Then $\psi_1 \in \Delta'$ and, by the induction hypothesis, $\Delta' \Vdash \psi_1$. Thus $\Delta' \Vdash \psi_2$, because $\Delta \subseteq \Delta'$, and we get $\psi_2 \in \Delta'$, again by the induction hypothesis. Thus, $\Delta, \psi_1 \vdash \psi_2$, and by ($\to$ I) we get what we want.

Now assume $\psi_1 \to \psi_2 \in \Delta$ (in particular $\Delta \vdash \psi_1 \to \psi_2$) and take $\Delta' \supseteq \Delta$ with $\Delta' \Vdash \psi_1$. Then $\psi_1 \in \Delta'$, so $\Delta' \vdash \psi_1$. But also $\Delta' \vdash \psi_1 \to \psi_2$, because $\Delta \subseteq \Delta'$. By ($\to$ E) we can derive $\Delta' \vdash \psi_2$, which means, by the induction hypothesis, that $\Delta' \Vdash \psi_2$. This completes the proof of ($*$).

Now let $\Delta = \{\psi \mid \Gamma \vdash \psi\}$. Then $\Delta \in C$ and from ($*$) we have $\Delta \Vdash \gamma$ for all $\gamma \in \Gamma$, but $\Delta \nVdash \varphi$. □

It follows from Theorems 2.5.9 and 2.6.1 that the three conditions $\Gamma \vdash \varphi$, $\Gamma \models \varphi$, and $\Gamma \Vdash \varphi$ are equivalent for implicational formulas. The completeness theorem has a very important consequence: the *conservativity* of IPC over its implicational fragment.

2.6.2. THEOREM. *Let φ be an implicational formula, and let Γ be a set of implicational formulas. If $\Gamma \vdash_N \varphi$ then $\Gamma \vdash \varphi$ in* NJ($\to$).

2.7. Notes

The roots of constructivism in mathematics reach deeply into the 19th century, or even further into the past. Intuitionists themselves admitted for instance the inspiration of the philosophy of Immanuel Kant (1724–1804). According to Kant, such areas of mathematical cognition as time and space, are directly accessible to the *human intuition* rather than empirically "observed". Mathematics can thus be seen as a purely mental construction.

Leopold Kronecker (1823–1891) is usually cited as the first author who explicitly applied constructive ideas in mathematics. A correct definition of a number, according to Kronecker, was one which could be verified in a finite number of steps. And a proof of an existential statement should provide an explicit object witnessing that statement.

In the search for a firm basis of mathematics, Kronecker was soon joined by many others. In the second half of the 19th century mathematics was changing in a quite important way. With development of new branches, including mathematical logic, the subject of mathematical research was becoming more and more abstract and unrelated to physical experience. The mathematician's activity changed from *discovering* properties of the "real" world into *creating* an abstract one. This raised important questions concerning the foundations of mathematics. With the discovery of paradoxes, notably the well-known paradox due to Russell, these questions became really urgent.

The end of the 19th and beginning of the 20th century was a period of an intensive development and competition of ideas and trends aiming at explaining the conceptual basis of the modern mathematics. Some of these trends created the background for the philosophical school known as *intuitionism*. For instance, Poincaré emphasized the role of intuition in mathematics and Lebesgue denied the existence of objects (like number sequences) unless they were explicitly defined.

But the principles of intuitionism were first formulated in the works of the Dutch mathematician and philosopher Luitzen Egbertus Jan Brouwer (1881–1966). Beginning with his thesis of 1907, the subject was developed in subsequent works into a general and consistent exposition of a philosophy of science [56, 112, 320]. Brouwer is also the inventor of the term "intuitionism", which was originally meant to denote a philosophical approach to the foundations of mathematics, being in opposition to Hilbert's *formalism*. Here, formalism is understood as an attempt to describe mathematics within the framework of formal manipulation of symbols. By restriction to finitary methods, formalists hoped to achieve a rigorous proof of consistency of mathematics and to develop mechanical tools to verify mathematical truth. As we know, *Hilbert's programme*, as this endeavour was called, failed after Gödel's results, but formalism contributed a lot to the development of contemporary logic.

Brouwer rejected the idea that mathematics be reduced to some kind of formal game. But he also rejected the idea of mathematics as a natural science. Instead, intuitionists insisted that the human mind (the intuition), and not the external world, is the source of mathematical notions. That is, mathematical objects exist only as the creations of a mathematician, and mathematical truth is only achieved by means of proof, a mental activity. Quoting Brouwer after van Dalen [111]: *"There are no non-experienced truths"*.

Brouwer did not intend to make intuitionistic logic a part of formal logic. The intuitionistic propositonal calculus was developed between 1925 and 1930 by Kol-

2.7. NOTES

mogorov [273], Glivenko [190], and Heyting [227]. Their research in intuitionistic proof theory turned Brouwer's idealistic philosophy into a branch of mathematical logic. The interest in this logic has since then largely been motivated by its constructive character.

To explain what intuitionistic logic is about, one usually refers to the Brouwer-Heyting-Kolmogorov interpretation, as we did in Section 2.1. Only implicitly present in Brouwer's work, this explanation is due to Heyting [228, 229, 230] and Kolmogorov [274]. Although intuitively convincing and useful, the BHK interpretation causes also some misunderstanding. This is obviously due to the imprecise character of the rules and also to the numerous differences between their variants. For instance, we use the word *construction*, following Heyting [230, p. 102], Prawitz [405], and Howard [247], while other authors talk about *proofs* [189, 488, 489], *realizers* [346], *evidence* [311], etc. Kolmogorov's original paper was about *problems* and their *solutions*. Some of Heyting's informal explanations are about *fulfillment* of an *intention* expressed by a proposition [228] or about *conditions* under which a proposition can be *asserted* [230].

The way the rules are stated varies too. For instance, our interpretation of disjunction requires that a construction explicitly determines which component is taken into account. This variant is common nowadays (as it might make a difference even in the case of $\varphi \vee \varphi$), but the following would be closer to the original formulation of Heyting and Kolmogorov:

A construction of $\varphi_1 \vee \varphi_2$ is either a construction of φ_1 or a construction of φ_2.

Another possible extension of BHK (cf. [278, 462, 487]) is to require that a construction of $\varphi_1 \to \varphi_2$ should include a verification of correctness.

Various other approaches to intuitionistic semantics, most notably Kleene's realizability (see Chapter 9) are often compared to the BHK interpretation, and then obvious similarities as well as important differences are found. See for instance [15, 134, 377, 486, 488]. But a liberal understanding of the BHK rules may also lead to topological semantics [436], or even classical two-valued semantics [134].

Begining with Gödel, many authors considered various formalizations of the BHK interpretation in terms of classical logic, for instance by introducing a formal notion of "provability". In particular, one can think of *modal logic* (see [68]) as a form of provability logic. A recent development is the logic LP of Artemov [15, 16].

The natural deduction system for propositional intuitionistic logic is usually attributed to the work of Gentzen [171] but one can find a similar system in an independent paper by Jaśkowski [254]. The algebraic approach to formal semantics originated with the work of Łukasiewicz on many-valued logic. Semantic properties of intuitionistic logic were initially formulated in terms of "matrices". This includes, for instance, the solution to our Exercise 2.19 given by Gödel in 1932, and a completeness theorem and an initial version of the finite model property (Theorem 2.4.12) published by Jaśkowski in 1936. The topological semantics originated in the late 1930's with papers by Stone and Tarski. A completeness theorem (*Zweiter Hauptsatz*) in Tarski's paper [473] implies our Theorem 2.4.11. Exercise 2.30 is based on Mints and Zhang [347]. See also [47].

Heyting algebras emerged from the "closure algebras" and "Brouwerian algebras" investigated in 1944-1948 by McKinsey and Tarski,[6] who also gave the double-

---

[6] Interestingly, they initially used closed sets, rather than open sets, so that $\wedge$ was interpreted by $\cup$ and $\vee$ by $\cap$.

exponential bound of Theorem 2.4.11.

In 1956, Beth proposed a semantics ("Beth models") very similar to Kripke semantics. The latter dates from 1959, and was first designed for modal logic. The completeness of intuitionistic logic with respect to Kripke models was shown in 1963 and published in [283]. Our Lemma 2.5.8 and Exercise 2.28 are based on Fitting's [151], who attributes the translation to Beth.

Decidability of intuitionistic logic is of course a consequence of the finite model property, but the first (syntactical) proof of decidability was derived by Gentzen from his cut-elimination theorem of 1935 (see Corollary 7.3.9).

For the interested reader, let us also mention that the first proof of the independence of intuitionistic propositional connectives (Exercise 2.26) is due to Wajsberg [505], and was published in 1938. An independent solution was given next year by McKinsey [334]. Later, Prawitz [403] derived yet another solution from proof normalization. The example about $\sqrt{2}$, mentioned at the beginning of this chapter, probably appeared first in print in [253]. Since first used by Hindley in 1970, it is now obligatory in every introduction to intuitionistic logic. (In fact, there is a Gelfond-Schneider theorem which implies that $\sqrt{2}$ raised to the power $\sqrt{2}$ is irrational, but the proofs we have seen in [168, 296] do not qualify as strictly constructive.)

In this short survey we have only recalled the very basic facts about the initial history of intuitionistic logic. To learn more about the history and motivations see the books [112, 136, 230, 320, 488] and articles [111, 483, 484, 487].

For a general introduction to intuitionistic logic we recommend the article [110] of van Dalen, and also the books of Mints [346], Dragalin [134] and Fitting [151, 152]. A comprehensive study of the algebraic semantics for intuitionistic (and classical) logic is the book of Rasiowa and Sikorski [410]. See also Chapter 13 of [488, vol. II].

Although the subject of a large part of the present book is constructive logic, our approach is entirely classical, or *naive common sense*, as we perhaps should prefer to say. We do not restrict our apparatus to intuitionistic logic, accepting proofs by contradiction with no reservations. But the reader should be aware that there is a broad and active area of research, called *constructive mathematics*, aiming at re-building classical mathematics on the basis of constructive reasoning. This is far beyond the scope of this book, and the reader is referred to [38, 48, 343, 426, 488].

2.8. Exercises

2.1. (From [78].) Prove that at least one of $e + \pi$ and $e\pi$ is not algebraic.

2.2. Find constructions for the formulas in Example 2.1.3, and do not find constructions for the formulas in Example 2.1.4.

2.3. Prove Lemma 2.2.4 about weakening and substitution.

2.4. Give formal proofs of the formulas in Example 2.1.3.

2.5. Show that intuitionistic propositional logic is *consistent*, that is $\nvdash \bot$. (Note that consistency is equivalent to "There exists φ with $\nvdash \varphi$".)

2.6. Let $\langle A, \sqcup, \sqcap \rangle$ be an algebra with two binary operations (written in infix notation). Assume that the equations (i)–(iv) of Lemma 2.3.4 hold for all elements of A. Define $a \leq b$ by $a \sqcup b = b$. Prove that $\langle A, \leq \rangle$ is a lattice, where suprema and infima are given by $\sqcup$ and $\sqcap$.

2.7. Show that the family of all convex subsets of a plane is a lattice but not a distributive one.

2.8. EXERCISES

2.8. Let A be a lattice satisfying $(a \sqcup b) \sqcap c \leq (a \sqcap c) \sqcup (b \sqcap c)$, for all a, b, c. Show that A is distributive.

2.9. Assume that $a \Rightarrow b$ exists for all elements a, b of a lattice L. Prove that L is distributive. *Hint:* Use Exercise 2.8.

2.10. Consider the partial orders represented below as directed graphs where an edge from a to b means that $a < b$ (edges implied by transitivity not shown). Which of these are lattices? Heyting algebras? Boolean algebras?

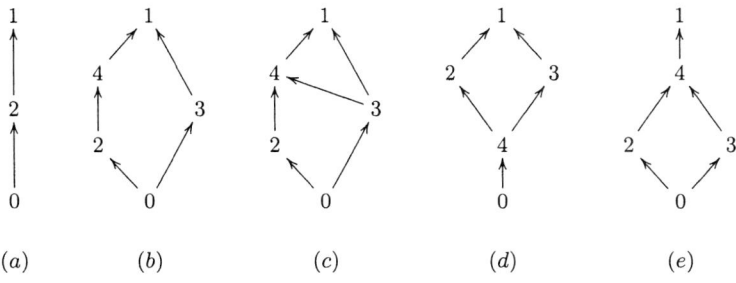

2.11. For $\Gamma \subseteq \Delta$ show that the set $F_\Delta = \{[\varphi]_\sim \mid \Delta \vdash \varphi\}$ is a filter in the Lindenbaum algebra $\mathcal{L}_\Gamma$.

2.12. Let F be a filter in a Heyting algebra $\mathcal{H}$ and let v be a valuation in $\mathcal{H}$. Define $\Gamma = \{\varphi \mid [\varphi]_v \in F\}$. Show that Γ is closed under $\vdash$, i.e. $\Gamma \vdash \varphi$ implies $\varphi \in \Gamma$.

2.13. Assume that $\mathcal{B}, v \not\models \varphi$, for some Boolean algebra $\mathcal{B}$ and some φ and v. Show that there exists a prime filter F in $\mathcal{B}$, with $[\neg\varphi]_v \in F$. Then define a binary valuation by $w(p) = 1$ iff $v(p) \in F$ and show that $[\varphi]_w = 0$.

2.14. Let $\mathcal{B}_0$ be a Boolean algebra with $0 \neq 1$, and let $\mathbb{B}$ be the two-element Boolean algebra of truth values. Show that the following three conditions are equivalent:

(i) $\mathbb{B} \models \varphi$;

(ii) $\mathcal{B}_0 \models \varphi$;

(iii) $\mathcal{B} \models \varphi$, for all Boolean algebras $\mathcal{B}$.

Do not use Stone's Theorem 2.3.9. *Hint:* Apply Exercise 2.13 to prove (i)$\Rightarrow$(iii).

2.15. For a Boolean algebra $\mathcal{B}$, let $\mathcal{Z}$ be the set of all prime filters in $\mathcal{B}$. Show that $\mathcal{B}$ is isomorphic to a subalgebra of $\mathcal{P}(\mathcal{Z})$, consisting of all sets of the form $Z_a = \{F \in \mathcal{Z} \mid a \in F\}$, for $a \in \mathcal{B}$.

2.16. Prove Proposition 2.4.8 (topological spaces regarded as Heyting algebras).

2.17. Prove that if $\not\Vdash \varphi$ then $\mathcal{C} \not\Vdash \varphi$ for a finite Kripke model $\mathcal{C}$. *Hint:* Turn an infinite counterexample into a finite one using a quotient construction.

2.18. Show that the formulas in Example 2.1.4 are not intuitionistically valid. (Use open subsets of $\mathbb{R}^2$ or construct Kripke models.)

2.19. Show that the formula $\bigvee\{p_i \leftrightarrow p_j \mid i, j = 0, \ldots, n \text{ and } i \neq j\}$ is not intuitionistically valid.

2.20. Which of the following judgements are provable?

(i) $p \to \neg p \vdash \neg p$;

(ii) $\neg p \to p \vdash p$;

(iii) $\neg p \to \neg q, p \to \neg q \vdash \neg q$;

(iv) $\neg p \to q, p \to q \vdash q$.

2.21. Which of the following formulas are intuitionistically valid?

(i) $((p \to q) \to p) \to \neg\neg p$;

(ii) $((((p \to q) \to p) \to p) \to q) \to q$;

(iii) $\neg p \vee \neg\neg p$;

(iv) $\neg p \vee \neg q \to \neg(p \wedge q)$;

(v) $(p \to p \wedge q) \vee (q \to p \wedge q)$;

(vi) $(p \to q \vee r) \to (p \to q) \vee r$;

(vii) $(p \to q \vee r) \to (p \to q) \vee (p \to r)$;

(viii) $((p \vee \neg p) \to \neg q) \to \neg q$;

(ix) $(p \to \neg p) \to \neg(\neg p \to p)$.

2.22. Which of the following equivalences are correct intuitionistically?

(i) $p \vee q \leftrightarrow (\neg p \to q)$;

(ii) $p \wedge q \leftrightarrow \neg(p \to \neg q)$;

(iii) $p \wedge q \leftrightarrow (p \leftrightarrow (p \to q))$;

(iv) $(p \to q) \leftrightarrow ((p \vee q) \leftrightarrow q)$.

2.23. Which of the following are intuitionistic tautologies?

(i) $\neg\neg(\varphi \to \psi) \to (\neg\neg\varphi \to \neg\neg\psi)$.

(ii) $\neg\neg(\varphi \wedge \psi) \to (\neg\neg\varphi \wedge \neg\neg\psi)$.

(iii) $\neg\neg(\varphi \vee \psi) \to (\neg\neg\varphi \vee \neg\neg\psi)$.

(iv) $(\neg\neg\varphi \to \neg\neg\psi) \to \neg\neg(\varphi \to \psi)$.

(v) $(\neg\neg\varphi \wedge \neg\neg\psi) \to \neg\neg(\varphi \wedge \psi)$.

(vi) $(\neg\neg\varphi \vee \neg\neg\psi) \to \neg\neg(\varphi \vee \psi)$.

2.24. Show that the following formulas are intuitionistic tautologies:

(i) $\neg\neg(\neg\neg\varphi \to \neg\neg\psi) \to \neg\neg\varphi \to \neg\neg\psi$;

(ii) $(\varphi \to \neg\psi) \to \neg\neg\varphi \to \neg\psi$.

2.25. A propositional formula is *negative* iff every propositional variable p occurs only in the form $p \to \bot$ in φ, and φ contains no occurrence of $\vee$. Show that $\neg\neg\varphi \to \varphi$ is intuitionistically valid for negative φ. *Hint:* Use Exercise 2.23.

2.8. EXERCISES

2.26. In classical logic some propositional connectives are *definable* from the others. For instance, we say that $\vee$ is definable from $\rightarrow$ and $\bot$, because $\alpha \vee \beta$ is equivalent to $(\alpha \rightarrow \bot) \rightarrow \beta$. Show that none of the connectives $\vee, \wedge, \rightarrow, \bot$ is definable from the others in propositional intuitionistic logic.

2.27. Show that $\vee, \wedge, \rightarrow$ and $\bot$ are definable from the ternary connective given by the following *Kuznetsov formula* [291]:

$$((p \vee q) \wedge \neg r) \vee (\neg p \wedge (q \leftrightarrow r)).$$

2.28. In the proof of Lemma 2.5.8 we have shown how to translate every Heyting algebra into a Kripke model. A translation in the other direction is also possible. Let $\mathcal{C} = \langle C, \leq, \Vdash \rangle$ be a Kripke model and let H be the set of all upward-closed subsets of C. Show that H is a Heyting algebra with

$$X \Rightarrow Y = \{c \mid \forall c' \in X (c \leq c' \rightarrow c' \in Y)\}.$$

Then prove that $\mathcal{C} \Vdash \varphi$ implies $H, v \models \varphi$, where $v(p) = \{c \in C \mid c \Vdash p\}$, for $p \in \mathrm{PV}$. *Hint:* Show that $[\![\varphi]\!]_v = \{c \in C \mid c \Vdash \varphi\}$ for all formulas φ.

2.29. In this exercise we fix a finite Kripke model $\mathcal{C} = \langle C, \leq, \Vdash \rangle$, with a least element c_0, and we define a certain function $\pi : (0, 1) \rightarrow C$. First we associate a "label" $\ell(w) \in C$ to every word $w \in \{0, 1\}^*$. We begin with $\ell(\varepsilon) = c_0$ and proceed by induction. Assume that $\ell(w) = c$ and that $\ell(w')$ has not yet been defined for any word w' extending w. Let c have m immediate successors in C, say $c_1, \ldots, c_m$. Then we extend the labeling by:

$$\begin{aligned}
\ell(w0^{2i}1) &= c, & \text{for } i = 0, \ldots, m; \\
\ell(w0^{2i-1}1) &= c_i, & \text{for } i = 1, \ldots, m; \\
\ell(w0^i) &= c, & \text{for } i = 1, \ldots, 2m+1.
\end{aligned}$$

This definition is illustrated, for $m = 2$, by the figure below.

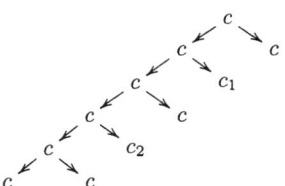

If a number $x \in (0, 1)$ is written in binary as $x = 0.b_1 b_2 \ldots b_n 1$, or equivalently as $x = 0.b_1 b_2 \ldots b_n 100 \ldots$ with an infinite sequence of zeros (we do not allow infinite sequences of 1's), then define $\pi(x) = \ell(b_1 b_2 \ldots b_n)$. Note that $\pi(x) \leq \ell(b_1 b_2 \ldots b_n 1)$, and that the inequality may happen to be strict. Now assume that x has an infinite binary representation $x = 0.b_1 b_2 b_3 \ldots$ Since the model is finite, there is $c \in C$ such that $\ell(b_1 b_2 \ldots b_n) = c$, for almost all n. In this case we take $\pi(x) = c$. Prove that π is a surjective map such that

1. If $X \subseteq C$ is upward-closed then $\pi^{-1}(X)$ is open.

2. If $A \subseteq (0, 1)$ is open then $\pi(A)$ is upward-closed;

Hint: First prove the following statements:[7]

---
[7] Note that ε_x in (i) depends on x, and that y may require an ε_y different than ε_x.

(i) $\forall x \in (0,1)\, \exists \varepsilon_x > 0\, \forall y \in (0,1)[|x-y| < \varepsilon_x \to \pi(x) \le \pi(y)]$;

(ii) $\forall x \in (0,1)\, \forall c \in C\, \forall \varepsilon > 0\, [\pi(x) \le c \to \exists y \in (0,1)[|x-y| < \varepsilon \wedge \pi(y) = c]]$.

2.30. Let $\mathcal{C}$ and $\pi : (0,1) \to \mathcal{C}$ be as in Exercise 2.29. Define a valuation v in the algebra $\mathcal{O}((0,1))$ by $v(p) = \pi^{-1}(\{c \in C \mid c \Vdash p\})$. (Observe that $v(p)$ is open by Exercise 2.29.) Prove that $[\![\varphi]\!]_v = \pi^{-1}(\{c \in C \mid c \Vdash \varphi\})$ holds for all φ.

2.31. Derive Theorem 2.4.11 (completeness for $\mathbb{R}$ etc.) from Exercises 2.17 and 2.30.

2.32. A state c in a Kripke model $\mathcal{C}$ *determines* p iff either $c \Vdash p$ or $c \Vdash \neg p$. Define a binary valuation v_c by $v_c(p) = 1$ iff $c \Vdash p$. Show that if c determines all propositional variables in φ then $[\![\varphi]\!]_{v_c} = 1$ implies $c \Vdash \varphi$. Conclude that a formula is a classical tautology if and only if it is forced in all one-element models.

2.33. Let φ be a classical tautology such that all propositional variables in φ are among $p_1, \ldots, p_n$. Show that the formula $(p_1 \vee \neg p_1) \to \cdots \to (p_n \vee \neg p_n) \to \varphi$ is intuitionistically valid. *Hint:* Use Exercise 2.32.

2.34. Prove Glivenko's Theorem 2.4.10: A double negation of a classical tautology is intuitionistically valid. *Hint:* Use Exercise 2.32.

Chapter 3

Simply typed λ-calculus

It is always a central issue in logic to determine if a formula is *valid* with respect to a certain semantics. Or, more generally, if a set of assumptions entails a formula in all models. In the "semantic tradition" of classical logic, this question is often the primary subject, and the construction of sound and complete proof systems is mainly seen as a tool for determining validity. In this case, *provability* of formulas and judgements is the only proof-related question of interest. One asks whether a proof exists, but not necessarily which proof it is.

In proof theory the perspective is different. We want to study the structure of proofs, compare various proofs, identify some of them, and distinguish between others. This is particularly important for constructive logics, where a proof (construction), not semantics, is the ultimate criterion.

It is thus natural to ask for a convenient proof notation. We can for instance write $M : \varphi$ to denote that M is a proof of φ. In presence of additional assumptions Γ, we may enhance this notation to

$$\Gamma \vdash M : \varphi.$$

Now, if M and N are proofs of $\varphi \to \psi$ and φ, respectively, then the proof of ψ obtained using ($\to$E) could be denoted by something like $@(M, N) : \psi$, or perhaps simply written as $MN : \psi$. This gives an "annotated" rule ($\to$E)

$$\frac{\Gamma \vdash M : \varphi \to \psi \quad \Gamma \vdash N : \varphi}{\Gamma \vdash MN : \psi}$$

When trying to design an annotated version of (Ax), one discovers that it is also convenient to use names for assumptions so that e.g.,

$$x : \varphi, \ y : \psi \vdash x : \varphi$$

represents the use of the first assumption. This idea also comes in handy when we want to annotate rule ($\to$I). The result of discharging an assumption x

in a proof M can be then written for example as $\natural x\, M$, or $\xi x\, M$, or... why don't we try lambda?

$$\frac{\Gamma, x : \varphi \vdash M : \psi}{\Gamma \vdash \lambda x\, M : \varphi \to \psi}$$

Yes, what we get is lambda-notation. To use the words of a famous writer in an entirely different context, the similarity is not intended and not accidental. It is unavoidable. Indeed, a proof of an implication represents a construction, and according to the BHK interpretation, a construction of an implication is a function.

However, not every lambda-term can be used as a proof notation. For instance, the self-application xx does not seem to represent any propositional proof, no matter what the assumption annotated by x is. So before we explore the analogy between proofs and terms (which will happen in Chapter 4) we must look for the appropriate subsystem of lambda-calculus.

As we said, the BHK interpretation identifies a construction of an implication with a function. In mathematics, a function f is usually defined on a certain domain A and ranges over a co-domain B. This is written as $f : A \to B$. Similarly, a construction of a formula $\varphi \to \psi$ can only be applied to designated arguments (constructions of the premise). Then the result is a construction of the conclusion, i.e. it is again of a specific *type*.

In lambda-calculus, one introduces types to describe the functional behaviour of terms. An application MN is only possible when M has a *function type* of the form $\sigma \to \tau$ and N has type σ. The result is of type τ. This is a type discipline quite like that in strictly typed programming languages.

The notion of type assignment expressing functionality of terms has been incorporated into combinatory logic and lambda-calculus almost from the very beginning, and a whole spectrum of typed calculi has been investigated since then. In this chapter we introduce the most basic formalization of the concept of type: system $\lambda_\to$.

3.1. Simply typed λ-calculus à la Curry

We begin with the *simply typed λ-calculus à la Curry*, where we deal with the same ordinary lambda-terms as in Chapter 1.

3.1.1. DEFINITION.

(i) An implicational propositional formula is called a *simple type*. The set of all simple types is denoted by $\Phi_\to$.

(ii) An *environment* is a finite set of pairs of the form $\{x_1 : \tau_1, \ldots, x_n : \tau_n\}$, where x_i are distinct λ-variables and τ_i are types. In other words, an environment is a finite partial function from variables to types.[1]

---
[1] Occasionally, we may talk about an "infinite environment," and this means that a certain finite subset is actually used.

3.1. SIMPLY TYPED λ-CALCULUS À LA CURRY

Thus, if $(x : \tau) \in \Gamma$ then we may write $\Gamma(x) = \tau$. We also have:

$$\mathrm{dom}(\Gamma) = \{x \in \Upsilon \mid (x : \tau) \in \Gamma, \text{ for some } \tau\};$$
$$\mathrm{rg}(\Gamma) = \{\tau \in \Phi_\to \mid (x : \tau) \in \Gamma, \text{ for some } x\}.$$

It is quite common in the literature to consider a variant of simply typed lambda-calculus where all types are built from a single type variable (which is then called a type *constant*). The computational properties of such a lambda-calculus are similar to those of our $\lambda_\to$. But from the "logical" point of view the restriction to one type constant is not as interesting, cf. Exercise 4.10.

NOTATION. The abbreviation $\tau^n \to \sigma$ is used for $\tau \to \cdots \to \tau \to \sigma$, where τ occurs n times. An environment $\{x_1 : \tau_1, \ldots, x_n : \tau_n\}$ is often simply written as $x_1 : \tau_1, \ldots, x_n : \tau_n$. If $\mathrm{dom}(\Gamma) \cap \mathrm{dom}(\Gamma') = \varnothing$ then we may also write Γ, Γ' for $\Gamma \cup \Gamma'$. In particular, $\Gamma, x : \tau$ stands for $\Gamma \cup \{x : \tau\}$, where it is assumed that $x \notin \mathrm{dom}(\Gamma)$. Similar conventions will be used in later chapters.

3.1.2. DEFINITION. A *judgement* is a triple, written $\Gamma \vdash M : \tau$, consisting of an environment, a lambda-term and a type. The rules in Figure 3.1 define the notion of a *derivable* judgement of system $\lambda_\to$. (One has to remember that in rules (Var) and (Abs) the variable x is not in the domain of Γ.) If $\Gamma \vdash M : \tau$ is derivable then we say that M *has type* τ *in* Γ, and we write $\Gamma \vdash_{\lambda_\to} M : \tau$ or just $\Gamma \vdash M : \tau$ (cf. Definition 2.2.1).

(Var) $\quad \Gamma, x : \tau \vdash x : \tau$

(Abs) $\quad \dfrac{\Gamma, x : \sigma \vdash M : \tau}{\Gamma \vdash (\lambda x\, M) : \sigma \to \tau}$

(App) $\quad \dfrac{\Gamma \vdash M : \sigma \to \tau \quad \Gamma \vdash N : \sigma}{\Gamma \vdash (MN) : \tau}$

FIGURE 3.1: THE SIMPLY TYPED LAMBDA-CALCULUS $\lambda_\to$

3.1.3. EXAMPLE. Let σ, τ, ρ be arbitrary types. Then:

(i) $\vdash \mathbf{I} : \sigma \to \sigma$;

(ii) $\vdash \mathbf{K} : \sigma \to \tau \to \sigma$;

(iii) $\vdash \mathbf{S} : (\sigma \to \tau \to \rho) \to (\sigma \to \tau) \to \sigma \to \rho$.

A type assignment of the form $M : \tau \to \sigma$ can of course be explained as "M is a function with the domain τ and co-domain σ". But we must understand that this understanding of a "domain" and "co-domain" is not set-theoretic. In Curry-style typed calculi, types are more adequately decribed as predicates or specifications (to be satisfied by terms) than as set-theoretical function spaces. The meaning of $f : A \to B$ in set theory is that arguments of f are exactly the elements of A, and that all values must belong to B. In contrast, $M : \tau \to \sigma$ only means that M applied to an argument of type τ must yield a result of type σ.

We conclude this section with a brief review of some of the basic properties of system $\lambda_\to$.

3.1.4. LEMMA.

(i) *If* $\Gamma \vdash M : \tau$ *then* $\text{FV}(M) \subseteq \text{dom}(\Gamma)$.

(ii) *If* $\Gamma, x{:}\tau \vdash M : \sigma$ *and* $y \notin \text{dom}(\Gamma) \cup \{x\}$ *then* $\Gamma, y{:}\tau \vdash M[x := y] : \sigma$.

PROOF. Both parts are shown by induction with respect to the length of M. As an example we treat the case of abstraction in part (ii).

Suppose that $M = \lambda z M'$ and $\sigma = \sigma'' \to \sigma'$ and that we have derived $\Gamma, x{:}\tau \vdash M : \sigma$ from $\Gamma, x{:}\tau, z{:}\sigma'' \vdash M' : \sigma'$. If $z \neq y$ then from the induction hypothesis we know that $\Gamma, y{:}\tau, z{:}\sigma'' \vdash M'[x := y] : \sigma'$, whence $\Gamma, y{:}\tau \vdash \lambda z M'[x := y] : \sigma$. Now note that $\lambda z M'[x := y] = (\lambda z M')[x := y]$.

If $z = y$ and $\Gamma, x{:}\tau, y{:}\sigma'' \vdash M' : \sigma'$ then we can choose a variable $v \notin \text{dom}(\Gamma) \cup \{x, y\}$ and obtain $\Gamma, x{:}\tau, v{:}\sigma'' \vdash M'[y := v] : \sigma'$ from the induction hypothesis. The next application of the induction hypothesis yields $\Gamma, y{:}\tau, v{:}\sigma'' \vdash M'[y := v][x := y] : \sigma'$, because the size of the term has not been increased. It follows that $\Gamma, y{:}\tau \vdash \lambda v. M'[y := v][x := y] : \sigma$. The reader will easily verify that $\lambda v. M'[y := v][x := y] = M[x := y]$. □

The following lemma is a direct consequence of the "syntax-oriented" character of the rules.

3.1.5. LEMMA (Generation lemma). *Suppose that* $\Gamma \vdash M : \tau$.

(i) *If M is a variable x then* $\Gamma(x) = \tau$.

(ii) *If M is an application PQ then* $\Gamma \vdash P : \sigma \to \tau$, *and* $\Gamma \vdash Q : \sigma$, *for some σ.*

(iii) *If M is an abstraction $\lambda x N$ and $x \notin \text{dom}(\Gamma)$ then* $\tau = \tau_1 \to \tau_2$, *where* $\Gamma, x{:}\tau_1 \vdash N : \tau_2$.

PROOF. The last step in the derivation of $\Gamma \vdash M : \tau$ is determined by the shape of M. This is all that is needed to prove the first two parts. In

3.1. SIMPLY TYPED λ-CALCULUS À LA CURRY

part (iii) the last rule must be (Abs) applied to $\Gamma, y : \tau_1 \vdash N_1 : \tau_2$, where $\lambda x N = \lambda y N_1$, and it can happen that $y \neq x$. Thus, by Lemma 1.2.20(v), we have $\Gamma, y : \tau_1 \vdash N[x := y] : \tau_2$, and then from Lemma 3.1.4(ii) we obtain $\Gamma, x : \tau_1 \vdash N[x := y][y := x] : \tau_2$. But $N[x := y][y := x] = N$, by Lemmas 1.2.20(iii) and 1.2.5(iii). □

3.1.6. LEMMA.

(i) If $\Gamma \vdash M : \sigma$ and $\Gamma(x) = \Gamma'(x)$ for all $x \in \mathrm{FV}(M)$ then $\Gamma' \vdash M : \sigma$.

(ii) If $\Gamma, x : \tau \vdash M : \sigma$ and $\Gamma \vdash N : \tau$ then $\Gamma \vdash M[x := N] : \sigma$.

PROOF. We proceed by induction with respect to the size of M. In the proof of (i) we only consider the case when $M = \lambda y M'$ and $\sigma = \sigma_1 \to \sigma_2$. If y is chosen so that $y \notin \mathrm{FV}(\Gamma) \cup \mathrm{FV}(\Gamma')$ then by Lemma 3.1.5(iii) we have $\Gamma, y : \sigma_1 \vdash M' : \sigma_2$. The induction hypothesis yields $\Gamma', y : \sigma_1 \vdash M' : \sigma_2$, which in turn gives $\Gamma' \vdash M : \sigma$.

The proof of part (ii) is also routine. Note that we use part (i) in the case of abstraction. □

The following result establishes the correctness of the type assignment system. A well-typed expression remains well-typed after reductions. In particular, no run-time error can be caused by an ill-typed function application. ("Well-typed programs do not go wrong.").

3.1.7. THEOREM (Subject reduction). *If* $\Gamma \vdash M : \sigma$ *and* $M \twoheadrightarrow_\beta N$, *then* $\Gamma \vdash N : \sigma$.

PROOF. By induction with respect to the definition of $\twoheadrightarrow_\beta$. We consider the base case when M is a redex, $M = (\lambda x P)Q$, and $N = P[x := Q]$. Without loss of generality we can assume that $x \notin \mathrm{dom}(\Gamma)$, so by the generation lemma we have $\Gamma, x : \tau \vdash P : \sigma$ and $\Gamma \vdash Q : \tau$. From Lemma 3.1.6(ii) we obtain $\Gamma \vdash P[x := Q] : \sigma$. □

3.1.8. DEFINITION. The *substitution of type τ for type variable p in type σ*, written $\sigma[p := \tau]$, is defined by:

$$\begin{aligned} p[p := \tau] &= \tau; \\ q[p := \tau] &= q, \text{ if } q \neq p; \\ (\sigma_1 \to \sigma_2)[p := \tau] &= \sigma_1[p := \tau] \to \sigma_2[p := \tau]. \end{aligned}$$

The notation $\Gamma[p := \tau]$ stands for $\{(x : \sigma[p := \tau]) \mid (x : \sigma) \in \Gamma\}$. Similar notation applies for equations, sets of equations etc.

The following shows that the type variables range over all types; this is a limited form of *polymorphism* (cf. Chapter 11).

3.1.9. PROPOSITION. *If* $\Gamma \vdash M : \sigma$, *then* $\Gamma[p := \tau] \vdash M : \sigma[p := \tau]$.

PROOF. By induction on the derivation of $\Gamma \vdash M : \sigma$. □

3.2. Type reconstruction algorithm

A term $M \in \Lambda$ is *typable* if there are Γ and σ such that $\Gamma \vdash M : \sigma$. The set of typable terms is a proper subset of the set of all λ-terms. It is thus a fundamental problem to determine exactly which terms can be assigned types in system $\lambda_\to$ and how to find these types effectively. In fact, one can consider a number of decision problems arising from the analysis of the ternary predicate "$\Gamma \vdash M : \tau$". The following definition makes sense for every type assignment system deriving judgements of this form.

3.2.1. DEFINITION.

(i) The *type checking* problem is to decide whether $\Gamma \vdash M : \tau$ holds, for a given environment Γ, a term M, and a type τ.

(ii) The *typability* problem, also called *type reconstruction* problem, is to decide if a given term M is typable.

(iii) The *type inhabitation* problem, also called *type emptiness* problem, is to decide, for a given type τ, whether there exists a closed term M, such that $\vdash M : \tau$ holds. (Then we say that τ is *non-empty*, and has the *inhabitant M*.)

The type inhabitation problem will be discussed in Chapter 4. In this section we consider typability and type checking. At first sight it might seem that determining whether a given term has a given type in a given environment could be easier than determining whether it has any type at all. This impression however is generally wrong. For many type assignment systems, typability is easily reducible to type checking. Indeed, to determine if a term M is typable, where $FV(M) = \{x_1, \ldots, x_n\}$, we may ask if

$$x_0 : p \vdash \mathbf{K}x_0(\lambda x_1 \ldots x_n. M) : p,$$

and this reduces typability to type checking. In fact, in the simply typed case, the two problems are equivalent (Exercise 3.11), although reducing the latter to the former is not as easy. But for some type assignment systems, the two problems are not equivalent: compare Proposition 13.4.3 and Theorem 13.4.4.

We now show how the typability problem can be reduced to unification[2] over the signature consisting only of the binary function symbol $\to$. Terms over this signature are identified with simple types. For every term M we define by induction

- a system of equations E_M;
- a type τ_M.

---

[2] See the Appendix for definitions related to terms and unification.

3.2. Type reconstruction algorithm

The idea is as follows: E_M has a solution iff M is typable, and τ_M is (informally) a pattern of a type for M. Type variables (unknowns) occurring in E_M are of two sorts: some of them, denoted p_x, correspond to types of free variables x of M, the other variables are auxiliary.

3.2.2. DEFINITION.

(i) If M is a variable x, then $E_M = \varnothing$ and $\tau_M = p_x$, where p_x is a fresh type variable.

(ii) Let M be an application PQ. First rename all auxiliary variables in E_Q and τ_Q so that auxiliary variables used by E_P and τ_P are distinct from those occurring in E_Q and τ_Q. Then define $\tau_M = p$, where p is a fresh type variable, and $E_M = E_P \cup E_Q \cup \{\tau_P = \tau_Q \to p\}$.

(iii) If M is an abstraction $\lambda x P$, then we define $E_M = E_P[p_x := p]$ and $\tau_M = p \to \tau_P[p_x := p]$, where p is a fresh variable.

In the definition above, it should be assumed that the renamings and the choice of "fresh" variables are made according to a certain systematic pattern, so that E_M is defined in a unique way for each M. An alternative is to think about M as a fixed alpha-representative (a pre-term) where the choice of bound and free variables is made so that no confusion is possible.

3.2.3. LEMMA.

(i) If $\Gamma \vdash M : \rho$, then there exists a solution S of E_M, such that $\rho = S(\tau_M)$ and $S(p_x) = \Gamma(x)$, for all variables $x \in \mathrm{FV}(M)$.

(ii) Let S be a solution of E_M, and let Γ be such that $\Gamma(x) = S(p_x)$, for all $x \in \mathrm{FV}(M)$. Then $\Gamma \vdash M : S(\tau_M)$.

PROOF. Induction with respect to the length of M. □

It follows that M is typable iff E_M has a solution. But E_M then has a principal solution, and this has the following consequence. (Here, $S(\Gamma)$ is the environment such that $S(\Gamma)(x) = S(\Gamma(x))$.)

3.2.4. DEFINITION. A pair (Γ, τ), consisting of an environment and a type, is a *principal pair* for M iff the following are equivalent for all Γ' and τ':

(i) $\Gamma' \vdash M : \tau'$;

(ii) $S(\Gamma) \subseteq \Gamma'$ and $S(\tau) = \tau'$, for some substitution S.

We then also say that τ is the *principal type* of M.

3.2.5. COROLLARY. *If a term M is typable, then there exists a principal pair for M. This principal pair is unique up to renaming of type variables.*

PROOF. Immediate from Lemma A.2.1. □

We conclude that a judgement $\Gamma \vdash M : \tau$ is derivable if and only if (Γ, τ) is a substitution instance of the principal pair. In this way, the principal pair provides a full characterization of all type assignments possible for M.

3.2.6. EXAMPLE.

- The principal type of **S** is $(p \to q \to r) \to (p \to q) \to p \to r$. The type $(p \to q \to p) \to (p \to q) \to p \to p$ can also be assigned to **S**, but it is not principal.

- Type **int** $= (p \to p) \to p \to p$ is the principal type of Church numerals c_n, for $n \geq 2$. For **0** and **1** the principal types are respectively $p \to q \to q$ and $(p \to q) \to p \to q$. But every Church numeral can also be assigned the type $((p \to q) \to p \to q) \to (p \to q) \to p \to q$.

3.2.7. THEOREM. *Typability and type checking in the simply typed lambda-calculus are decidable in polynomial time.*

PROOF. The system of equations E_M can be constructed in logarithmic space (and thus also in polynomial time) from M. Thus, by Lemma 3.2.3, typability reduces in logarithmic space to unification, which is decidable in polynomial time (Theorem A.5.4).

To decide if $\Gamma \vdash M : \tau$ holds, consider a signature containing (in addition to the binary symbol $\to$) all free variables occurring in Γ and τ as constant symbols. It is now enough to extend E_M to include the equations $\tau_M = \tau$ and $p_x = \Gamma(x)$, for $x \in \text{FV}(M)$. The extended system of equations has a solution if and only if $\Gamma \vdash M : \tau$. □

3.2.8. REMARK (Related problems). The typability problem is often written as "$? \vdash M : ?$", and the type inhabitation problem is abbreviated "$\vdash ? : \tau$". This notation can be used for other related problems, as one can choose to replace various parts of our ternary predicate by question marks, and choose the environment to be empty or not. A little combinatorics shows that we have a total of 12 problems. Out of these 12 problems, four are completely trivial, since the answer is always "yes":

$$? \vdash ? : ? \quad \Gamma \vdash ? : ? \quad \vdash ? : ? \quad ? \vdash ? : \tau.$$

Thus we end up with eight non-trivial problems, as follows:

(i) $? \vdash M : ?$ (typability);

(ii) $\vdash M : ?$;

(iii) $\Gamma \vdash M : ?$;

3.3. SIMPLY TYPED λ-CALCULUS À LA CHURCH

(iv) $\Gamma \vdash M : \tau$ (type checking);

(v) $\vdash M : \tau$;

(vi) $? \vdash M : \tau$;

(vii) $\vdash ? : \tau$ (inhabitation);

(viii) $\Gamma \vdash ? : \tau$;

We have already noticed that problem (i) reduces to (iv). In fact, for the simply typed lambda-calculus, all problems (i)–(vi) are equivalent to unification, and thus also to each other, with respect to logarithmic-space reductions (Exercise 3.11). Thus, all these problems are PTIME-complete. We will see in Chapter 4 that (vii) and (viii) are complete for polynomial space.

3.3. Simply typed λ-calculus à la Church

Typed lambda-calculi usually occur in two variants, called *Curry-style* and *Church-style* systems. In the previous section we have seen an example of a Curry-style system. In such calculi, types are assigned (or not) to ordinary type-free lambda-terms, according to a set of rules. In this way, one term can be assigned more than one type.

The idea of a typed calculus à la Church is different. In the "orthodox" approach, all the type information is contained in a term, as follows.[3] For each $\sigma \in \Phi_\rightarrow$, let Υ_σ be a separate denumerable set of variables. Define the sets Λ_σ of simply typed terms of type σ so that $\Upsilon_\sigma \subseteq \Lambda_\sigma$ and:

- If $M \in \Lambda_{\sigma \rightarrow \tau}$ and $N \in \Lambda_\sigma$ then $(MN) \in \Lambda_\tau$;
- If $M \in \Lambda_\tau$ and $x^\sigma \in \Upsilon_\sigma$ then $(\lambda x^\sigma M) \in \Lambda_{\sigma \rightarrow \tau}$.

The set of simply typed terms is then taken as the union of all Λ_σ.

It is sometimes convenient to think of typed lambda-terms this way, but nowadays it is more customary to define Church-style calculi in a slightly different manner. Instead of assuming that the set of variables is partitioned into disjoint sets indexed by types one uses environments to declare types of free variables as in the system à la Curry. But types of bound variables remain part of the term syntax.

3.3.1. DEFINITION.

(i) *Raw terms* of Church-style $\lambda_\rightarrow$ are defined by the following rules:

- An object variable is a raw term;

---

[3] In case the reader cannot remember which approach is named after Church and which one after Curry, we recommend *Walukiewicz's test*: Observe that the name "Church" is longer than "Curry". And Church-style typed terms are longer too.

- If M, N are raw terms then the *application* (MN) is a raw term;
- If M is a raw term, x is a variable, and σ is a type then the *abstraction* $(\lambda x{:}\sigma M)$ is a raw term.

(ii) *Free variables* of a raw term M are defined as follows.

$$\begin{aligned} \mathrm{FV}(x) &= \{x\} \\ \mathrm{FV}(\lambda x{:}\sigma P) &= \mathrm{FV}(P) - \{x\} \\ \mathrm{FV}(PQ) &= \mathrm{FV}(P) \cup \mathrm{FV}(Q) \end{aligned}$$

If $\mathrm{FV}(M) = \varnothing$ then M is called *closed*.

(iii) The variable x is considered *bound* in the term $(\lambda x{:}\sigma\, P)$. We identify raw terms which differ only in their bound variables.[4]

3.3.2. CONVENTION. We adopt similar terminology, notation, and conventions for raw terms as for untyped λ-terms, see Chapter 1, *mutatis mutandis*. In particular, we omit parentheses if this does not create confusion, and we use dot notation, so that e.g. $\lambda x{:}\tau.\, yx$ stands for $(\lambda x{:}\tau(yx))$. In addition, to enhance readability, we sometimes write x^τ instead of $x:\tau$, like for instance in $\lambda x^{p \to q \to r} \lambda y^{p \to q} \lambda z^p . xz(yz)$.

The following definition of substitution on raw terms takes into account our assumption of identifying alpha-equivalent expressions.

3.3.3. DEFINITION. The *substitution* of a raw term N for x in M, written $M[x := N]$, is defined as follows:

$$\begin{aligned} x[x := N] &= N; \\ y[x := N] &= y, \text{ if } x \neq y; \\ (PQ)[x := N] &= P[x := N]Q[x := N]; \\ (\lambda y{:}\sigma.P)[x := N] &= \lambda y{:}\sigma.P[x := N], \text{ where } x \neq y \text{ and } y \notin \mathrm{FV}(N). \end{aligned}$$

The notion of β-reduction for Church-style expressions is also similar to that for Curry-style terms. In the definition below, the notion of "compatible" applies to Church-style syntax (see the explanation following Definition 1.3.1).

3.3.4. DEFINITION. The relation $\to_\beta$ (*single step β-reduction*) is the least compatible relation on raw terms, such that

$$(\lambda x{:}\sigma P)Q \to_\beta P[x := Q]$$

The notation $\twoheadrightarrow_\beta$ and $=_\beta$ is used accordingly, cf. Remark 1.3.4.

3.3.5. DEFINITION. We say that M is a *term* of type τ in Γ, and we write $\Gamma \vdash M : \tau$, when $\Gamma \vdash M : \tau$ can be derived using the rules in Figure 3.2.

---

[4]Strictly speaking, we should proceed as in the case of λ-terms and define a notion of raw pre-terms, then define substitution and α-equivalence on these, and finally adopt the convention that by a term we always mean the α-equivalence class, see Section 1.2.

3.4. Church versus Curry typing

$$
\begin{array}{ll}
\text{(Var)} & \Gamma, x:\tau \vdash x:\tau \\[1em]
\text{(Abs)} & \dfrac{\Gamma, x:\sigma \vdash M:\tau}{\Gamma \vdash (\lambda x{:}\sigma\, M):\sigma \to \tau} \\[1em]
\text{(App)} & \dfrac{\Gamma \vdash M:\sigma \to \tau \quad \Gamma \vdash N:\sigma}{\Gamma \vdash (MN):\tau}
\end{array}
$$

FIGURE 3.2: THE SIMPLY TYPED LAMBDA-CALCULUS À LA CHURCH

3.3.6. EXAMPLE. Let σ, τ, ρ be arbitrary simple types. Then:

(i) $\vdash \lambda x^\sigma\, x : \sigma \to \sigma$;

(ii) $\vdash \lambda x^\sigma \lambda y^\tau.\, x : \sigma \to \tau \to \sigma$;

(iii) $\vdash \lambda x^{\sigma \to \tau \to \rho} \lambda y^{\sigma \to \tau} \lambda z^\sigma.\, xz(yz) : (\sigma \to \tau \to \rho) \to (\sigma \to \tau) \to \sigma \to \rho$.

As in the Curry-style calculus, we have a subject reduction theorem:

3.3.7. THEOREM. *If* $\Gamma \vdash M:\sigma$ *and* $M \twoheadrightarrow_\beta N$ *then* $\Gamma \vdash N:\sigma$.

PROOF. Similar to the proof of Theorem 3.1.7. □

It is sometimes convenient to consider also η-reduction and η-equality on typed terms.

3.3.8. DEFINITION. The relation $\to_\eta$ is the smallest compatible relation on raw terms, such that
$$\lambda x:\sigma.\, Mx \to_\eta M,$$
whenever $x \notin \mathrm{FV}(M)$.

It is not difficult to see that $\to_\eta$ preserves types, i.e. that an analogue of Theorem 3.3.7 holds for eta-reductions.

3.4. Church versus Curry typing

As mentioned at the beginning of Section 3.3, the principle of typing à la Church (at least in the orthodox way) is that types of all variables and terms are "fixed". The full type information is "built into" an expression, and a given well-formed term is correctly typed by definition. There is no issue of typability because the type of a term is simply a part of it. This corresponds to the use of types in programming languages like e.g. Pascal. In such languages, it is the responsibility of the programmer to provide proper types of all identifiers, functions, etc. In contrast, Curry-style typing resembles ML or Haskell, where a compiler or interpreter does the type inference.

Because of the difference above, $\lambda_\to$ à la Curry and other similar systems are often called *type assignment* systems, in contrast to $\lambda_\to$ à la Church and similar systems which are called *typed* systems.

Our formulation of simply typed lambda-calculus à la Church is however halfway between the Curry style and the "orthodox" Church style. Types of bound variables are "embedded" in terms, but types of free variables are declared in the environment rather than being part of syntax. A raw term becomes a true "typed Church-style term" only within an environment which determines types of its free variables. Then, unlike in Curry style (cf. Proposition 3.1.9) a type of a term is unique.

3.4.1. PROPOSITION. *If* $\Gamma \vdash M : \sigma$ *and* $\Gamma \vdash M : \tau$ *in the simply-typed λ-calculus à la Church then* $\sigma = \tau$.

PROOF. Induction with respect to M. □

3.4.2. CONVENTION. In what follows, we often refer to Church-style terms without explicitly mentioning the environment, but if not stated otherwise it is implicitly assumed that some environment is given, and types of all variables are known. By Proposition 3.4.1 we can thus assume that every term has a uniquely determined type. We then proceed as if we actually dealt with "orthodox" Church-style terms.

In order to improve readability we sometimes write types of terms as superscripts, like in $(M^{\sigma \to \tau} N^\sigma)^\tau$. This notation is not part of the syntax and is only used informally to stress that e.g. M has type $\sigma \to \tau$ in a certain fixed environment.

Similar conventions are used in the later chapters, whenever Church-style systems are discussed. Also the word *term* always refers to a well-typed expression, cf. Definition 3.3.5.

Although the simply typed λ-calculi à la Curry and Church are different, one has the feeling that essentially the same thing is going on. To a large extent this intuition is correct. A Church-style term M can in fact be seen as a linear notation for a type derivation that assigns a type to a Curry-style term. This term is the "core" or "skeleton" of M, obtained by erasing the domains of abstractions.

3.4.3. DEFINITION. The *erasing* map $|\cdot|$ from Church-style to Curry-style terms is defined as follows:

$$|x| = x;$$
$$|MN| = |M||N|;$$
$$|\lambda x{:}\sigma\, M| = \lambda x|M|.$$

3.4.4. PROPOSITION.

(i) If $M \to_\beta N$ then $|M| \to_\beta |N|$.

(ii) If $\Gamma \vdash M : \sigma$ à la Church then $\Gamma \vdash |M| : \sigma$ à la Curry.

PROOF. For (i) prove by induction on M that

$$|M[x := N]| = |M|[x := |N|] \qquad (*)$$

Then proceed by induction on the definition of $M \to_\beta N$ using $(*)$.

Part (ii) follows by induction on the derivation of $\Gamma \vdash M : \sigma$. □

Conversely, one can "lift" every Curry derivation to a Church one.

3.4.5. PROPOSITION. *For all* $M, N \in \Lambda$:

(i) If $M \to_\beta N$ and $M = |M'|$ then $M' \to_\beta N'$, for some N' such that $|N'| = N$.

(ii) If $\Gamma \vdash M : \sigma$ then there is a Church-style term M' with $|M'| = M$ and $\Gamma \vdash M' : \sigma$.

PROOF. By induction on $M \to_\beta N$ and $\Gamma \vdash M : \sigma$, respectively. □

The two propositions above allow one to "translate" various properties of Curry-style typable lambda-terms to analogous properties of Church-style typed lambda-terms, or conversely. For instance, strong normalization for one variant of $\lambda_\to$ implies strong normalization for the other (Exercise 3.15). But one has to be cautious with such proof methods (Exercise 3.19).

3.5. Normalization

In this section we show that all simply typed terms have normal forms. Even more, all such terms M are *strongly* normalizing, i.e. there exists no infinite reduction $M = M_1 \to_\beta M_2 \to_\beta \cdots$ In other words, no matter how we evaluate a well-typed expression, we must eventually reach a normal form. In programming terms: a program in a language based on the simply typed lambda-calculus can only diverge as a result of an explicit use of a programming construct such as a loop, a recursive call, or a circular data type.

Strong normalization makes certain properties easier to prove. Newman's lemma below is a good example. Another example is deciding equality of typed terms by comparing their normal forms (Section 3.7). Yet another related aspect is normalization of proofs, which will be seen at work in the chapters to come.

The results of this section hold for both Church-style and Curry-style terms. Our proofs are for Church style, but the Curry style variant can be easily derived (Exercise 3.15). In what follows, we assume that types of all variables are fixed, so that we effectively work with the "orthodox" Church style (cf. Convention 3.4.2).

3.5.1. THEOREM. *Every term of Church-style $\lambda_\to$ has a normal form.*

PROOF. We show that a certain reduction strategy is normalizing. The idea is to always reduce a redex of a most complex type available and to begin from the right if there are several candidates. To make this idea precise, define the *degree* $\delta(\tau)$ of a type τ by:

$$\begin{aligned} \delta(p) &= 0; \\ \delta(\tau \to \sigma) &= 1 + \max(\delta(\tau), \delta(\sigma)). \end{aligned}$$

The *degree* $\delta(\Delta)$ of a redex $\Delta = (\lambda x^\tau P^\rho)R$ is $\delta(\tau \to \rho)$. If a term M is not in normal form then we define

$$\boldsymbol{m}_M = (\delta_M, n_M),$$

where $\delta_M = \max\{\delta(\Delta) \mid \Delta$ is a redex in $M\}$ and n_M is the number of redex occurrences in M of degree δ_M. For $M \in \mathrm{NF}_\beta$ put $\boldsymbol{m}_M = (0,0)$. The proof is by induction on lexicographically ordered pairs $\boldsymbol{m}_M$.

If $M \in \mathrm{NF}_\beta$ the assertion is trivially true. If $M \notin \mathrm{NF}_\beta$, let Δ be the rightmost redex in M of maximal degree δ (we determine the position of a subterm by the position of its first symbol, i.e. the rightmost redex means the redex which *begins* as much to the right as possible).

Let M' be obtained from M by reducing the redex Δ. The term M' may in general have more redexes than M. But we claim that the number of redexes of degree δ in M' is smaller than in M. Indeed, the redex Δ has disappeared, and the reduction of Δ may only create new redexes of degree less than δ. To see this, note that the number of redexes can increase by either copying existing redexes or by creating new ones. The latter happens when a non-abstraction A occurs in a context AB, and it is turned into an abstraction by the reduction of Δ. This is only possible when A is a variable or $A = \Delta$. It follows that one of the following cases must hold:

1. The redex Δ has form $(\lambda x^\tau \ldots x P^\rho \ldots)(\lambda y^\rho Q^\mu)^\tau$, where $\tau = \rho \to \mu$, and it reduces to $\ldots (\lambda y^\rho Q^\mu) P^\rho \ldots$ The new redex $(\lambda y^\rho Q^\mu) P^\rho$ is of degree $\delta(\tau) < \delta$.

3.5. NORMALIZATION

2. We have $\Delta = (\lambda x^\tau \lambda y^\rho . R^\mu) P^\tau$, occurring in the context $\Delta^{\rho \to \mu} Q^\rho$. The reduction of Δ to $\lambda y^\rho R_1^\mu$, for some R_1, creates a new redex $(\lambda y^\rho R_1^\mu) Q^\rho$ of degree $\delta(\rho \to \mu) < \delta(\tau \to \rho \to \mu) = \delta$.

3. The last case is when $\Delta = (\lambda x^\tau x)(\lambda y^\rho P^\mu)$, with $\tau = \rho \to \mu$, and it occurs in the context $\Delta^\tau Q^\rho$. The reduction creates the new redex $(\lambda y^\rho P^\mu) Q^\rho$ of degree $\delta(\tau) < \delta$.

The other way to add redexes is by copying. If $\Delta = (\lambda x{:}\tau . P^\rho) Q^\tau$, and the term P contains more than one free occurrence of x, then all redexes in Q are multiplied by the reduction. But we have chosen Δ to be the rightmost redex of degree δ, and thus all redexes in Q must be of smaller degrees, because they are to the right of Δ.

Thus, in all cases $m_M > m_{M'}$, so by the induction hypothesis M' has a normal form, and then M also has a normal form. $\square$

Theorem 3.5.1 states that every typed term is normalizing. We now set out to show that every term is in fact *strongly* normalizing, i.e. that *every* reduction sequence from the term must eventually terminate. Our aim is to infer the strong property from the weak one, with the help of the conservation property of $\lambda \mathbf{I}$-terms (Corollary 1.6.7). This can be done by translating an arbitrary typed λ-term M into a $\lambda \mathbf{I}$-term $\iota(M)$, of the same type, such that $\iota(M) \in \mathrm{SN}$ implies $M \in \mathrm{SN}$.

3.5.2. DEFINITION. For every propositional variable p and every type σ, we choose a fixed λ-variable $k_{p,\sigma}$ of type $p \to \sigma \to p$. If M is a variable or an application then $\iota(M)$ is defined as:

$$\iota(x) = x, \qquad \iota(PQ) = \iota(P)\iota(Q).$$

Otherwise our term is an abstraction, say of the form

$$M = \lambda x_1^{\sigma_1} \ldots x_r^{\sigma_r} . N^{\tau_1 \to \cdots \to \tau_m \to p},$$

where $r > 0$ and N is not an abstraction. In this case we define

$$\iota(M) = \lambda x_1^{\sigma_1} \ldots x_r^{\sigma_r} y_1^{\tau_1} \ldots y_m^{\tau_m} . k_{p,\sigma_1}(\cdots (k_{p,\sigma_r}(\iota(N) y_1 \ldots y_m) x_r) \cdots) x_1.$$

Finally, we define a term $t(M)$ by replacing every $k_{p,\sigma}$ occurring in $\iota(M)$ by the appropriate version of the **K** combinator, namely $\mathbf{K}_{p,\sigma} = \lambda x^p \lambda y^\sigma . x$.

Note that $t(M) \twoheadrightarrow_{\beta\eta} M$. Note also that $\iota(M)$ is a typed $\lambda \mathbf{I}$-term, and thus it is strongly normalizing. We now want to prove that also $t(M) \in \mathrm{SN}$, by showing that reductions involving $\mathbf{K}_{p,\sigma}$ are not essential.

3.5.3. LEMMA. *If $M \notin \mathrm{SN}_\beta$ then M has an infinite reduction sequence where no redex of the form $\mathbf{K}_{p,\sigma} A$ is reduced.*

PROOF. If a redex of the form $\mathbf{K}_{p,\sigma}A$ is reduced to $\lambda y^\sigma A$, where p is a type variable, then we say that the reduction step is *of type* 1. Clearly, an infinite reduction sequence may not consist exclusively of type 1 steps. It would thus suffice to show that steps of type 1 can always be "postponed" by permuting them with other reductions. Unfortunately, this property is not true: In the context $\mathbf{K}_{p,\sigma}AB$, a reduction of type 1 creates a redex which can be reduced in the next step. These two steps cannot be permuted.

To solve this problem we first postpone reduction steps where a redex of the form $(\lambda y^\sigma A^p)B$, with $y \notin \mathrm{FV}(A)$, is reduced to A. Let us say that such reductions are *of type* 2. We write $M \to_2 M'$ to indicate that the reduction is of type 2, and we write $M \to_0 M'$ otherwise. We prove that $M \to_2 M' \to_0 M''$ implies $M \to_0 M''' \twoheadrightarrow_2 M''$ for some M'''.

Let $\Delta = (\lambda y^\sigma A^p)B^\sigma$ be the redex reduced in the step from $M \to_2 M'$. Since A cannot be an abstraction, and $y \notin \mathrm{FV}(A)$, the redex Σ reduced in the next step is not a "new" one, i.e. it is obtained from a redex Σ' in M (possibly containing Δ). It is left to the reader to check that the two reduction steps can easily be permuted. (There is a double arrow in $M''' \twoheadrightarrow_2 M''$, because Δ can be duplicated or erased by reducing Σ.)

The above allows us to postpone reduction steps of type 2, i.e. to conclude that there is an infinite reduction sequence without such steps. Using a similar argument, we can also postpone steps of type 1. Indeed, if $(\lambda x^p \lambda y^\sigma . x)A^p$ is reduced to $\lambda y^\sigma A$ then again the next redex is either "inside" or "outside" the term A. □

3.5.4. LEMMA. *Terms of the form $t(M)$ are strongly β-normalizing.*

PROOF. Suppose that $t(M) = M_0 \to_0 M_1 \to_0 M_2 \to_0 \cdots$ is an infinite reduction sequence where no redex of the form $\mathbf{K}_{p,\sigma}A$ is reduced. In this reduction sequence, the combinators $\mathbf{K}_{p,\sigma}$ behave just like variables. It follows that $\iota(M) = M_0' \to_\beta M_1' \to_\beta M_2' \to_\beta \cdots$ where M_i are obtained from M_i' by substituting $\mathbf{K}_{p,\sigma}$ for $k_{p,\sigma}$. But $\iota(M)$ is strongly normalizing. □

3.5.5. THEOREM. *The simply typed lambda-calculus has the strong normalization property: Any term is strongly normalizing.*

PROOF. If M is not β-normalizing then $t(M)$ has an infinite $\beta\eta$-reduction sequence $t(M) \twoheadrightarrow_{\beta\eta} M \to_\beta \cdots$ By Lemma 1.3.11, it must also have an infinite β-reduction sequence, contradicting Lemma 3.5.4. □

3.6. Church-Rosser property

Proving the Church-Rosser property for Church-style typed terms is not as obvious as it may seem (Exercise 3.19). Fortunately, the typed lambda-calculus is strongly normalizing, and under this additional assumption, it is enough to show the *weak Church-Rosser property*.

3.6. CHURCH-ROSSER PROPERTY

3.6.1. DEFINITION. Let $\to$ be a binary relation in a set A. Recall from Chapter 1 that $\to$ has the *Church-Rosser property* (CR) iff for all $a, b, c \in A$ such that $a \twoheadrightarrow b$ and $a \twoheadrightarrow c$ there exists $d \in A$ with $b \twoheadrightarrow d$ and $c \twoheadrightarrow d$. We say that the relation $\to$ has the *weak Church-Rosser property* (WCR) when $a \to b$ and $a \to c$ imply $b \twoheadrightarrow d$ and $c \twoheadrightarrow d$, for some d.

We also say that a binary relation $\to$ is *strongly normalizing* (SN) iff there is no infinite sequence $a_0 \to a_1 \to a_2 \to \cdots$

Clearly, CR implies WCR. The converse is not true, see Exercise 3.16. But the two properties coincide for strongly normalizing relations.

3.6.2. PROPOSITION (Newman's lemma). *Let $\to$ be a binary relation satisfying SN. If $\to$ satisfies WCR, then $\to$ satisfies CR.*

PROOF. Let $\to$ be a relation on a set A satisfying SN and WCR. As usual, a *normal form* is an $a \in A$ such that $a \not\to b$, for all $b \in A$.

Since $\to$ satisfies SN, any $a \in A$ reduces to a normal form. Call an element a *ambiguous* if a reduces to two distinct normal forms. It is easy to see that $\to$ satisfies CR if there are no ambiguous elements of A.

But for any ambiguous a there is another ambiguous a' such that $a \to a'$. Indeed, suppose $a \twoheadrightarrow b_1$ and $a \twoheadrightarrow b_2$ and let b_1, b_2 be different normal forms. Both of these reductions must make at least one step since b_1 and b_2 are distinct, so the reductions have form $a \to a_1 \twoheadrightarrow b_1$ and $a \to a_2 \twoheadrightarrow b_2$. If $a_1 = a_2$ we can choose $a' = a_1 = a_2$. If $a_1 \neq a_2$ we know by WCR that $a_1 \twoheadrightarrow b_3$ and $a_2 \twoheadrightarrow b_3$, for some b_3. We can assume that b_3 is a normal form.

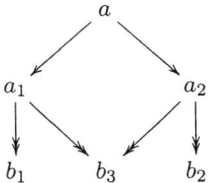

Since b_1 and b_2 are distinct, b_3 is different from b_1 or b_2 so we can choose $a' = a_1$ or $a' = a_2$. Thus, a has an infinite reduction sequence, contradicting strong normalization. Hence, there are no ambiguous terms. □

Newman's lemma is a very useful tool for proving the Church-Rosser property. We will use it many times, and here is its debut.

3.6.3. THEOREM. *The Church-style simply typed λ-calculus has the Church-Rosser property.*

PROOF. By Newman's lemma it suffices to check that the simply typed lambda-calculus has the WCR property. This is almost immediate if we observe that two different β-redexes in a lambda-term can only be disjoint, or one is a part of the other. That is, redexes never "overlap" and if we reduce one of them we can still reduce the other. A formal proof can be done by induction with respect to the definition of β-reduction. □

The subject reduction property together with the Church-Rosser property and strong normalization imply that reduction of any typed λ-term terminates in a normal form of the same type, where the normal form is independent of the particular order of reduction chosen.

3.7. Expressibility

As we saw in the preceding section, every simply typed λ-term has a normal form, and the normalization process always terminates. To verify whether two given terms of the same type are beta-equal or not, it thus suffices to reduce them to normal form and compare the results. However, this straightforward algorithm is of unexpectedly high complexity. It requires time and space proportional to the size of normal forms of the input terms. As demonstrated by Exercises 3.20–3.22, a normal form of a term of length n can (in the worst case) be of size

$$\left. 2^{2^{\cdot^{\cdot^{2}}}} \right\} \mathcal{O}(n)$$

This is a *non-elementary* function, i.e. it grows faster than any of the iterated exponentials defined by $\exp_0(n) = n$ and $\exp_{k+1}(n) = 2^{exp_k(n)}$. The following result, which we quote without proof (see [318, 456]), states that the difficulty caused by the size explosion cannot be avoided.

3.7.1. THEOREM (Statman). *The problem of deciding whether any two given Church-style terms M and N of the same type are beta-equal is of non-elementary complexity. That is, for each r, every decision procedure takes more than $\exp_r(n)$ steps on some inputs of size n.*

The strong normalization result gives a hint that the expressive power of $\lambda_\to$ should be weaker than that of the untyped lambda-calculus, i.e. that one should not be able to represent all recursive functions by simply typed λ-terms. On the other hand, Theorem 3.7.1 might lead one to expect that the class of definable total functions should still be quite rich.

As we shall now see, the latter expectation is not quite correct, but first the notion of a definable function must be made precise.

3.7.2. DEFINITION. Let $\mathbf{int} = (p \to p) \to (p \to p)$, where p is an arbitrary type variable. A function $f : \mathbb{N}^k \to \mathbb{N}$ is $\lambda_\to$-*definable* if there is an $F \in \Lambda$ with $\vdash F : \mathbf{int}^k \to \mathbf{int}$, such that

$$F\mathbf{c}_{n_1}\ldots\mathbf{c}_{n_k} =_\beta \mathbf{c}_{f(n_1,\ldots,n_k)}$$

for all $n_1, \ldots, n_k \in \mathbb{N}$.

3.7.3. DEFINITION. The class of *extended polynomials* is the smallest class of functions over $\mathbb{N}$ which is closed under compositions and contains the constant functions 0 and 1, projections, addition, multiplication, and the conditional function

$$cond(n_1, n_2, n_3) = \begin{cases} n_2, & \text{if } n_1 = 0; \\ n_3, & \text{otherwise.} \end{cases}$$

3.7.4. THEOREM (Schwichtenberg). *The $\lambda_\rightarrow$-definable functions are exactly the extended polynomials.*

PROOF. Exercises 3.24–3.26. □

3.8. Notes

Types are often seen as a method to avoid paradoxes occurring in the type-free world as a result of various forms of self-application. Undoubtedly, paradoxes gave the impulse for the creation of various type theories at the beginning of the 20th century. But, as pointed out in [165, 257], it is natural in mathematics to classify or stratify objects into categories or "types", and that occurred well before the paradoxes were discovered.

The history of formal type theory begins with Russell. The work of Chwistek, Ramsey, Hilbert, and others contributed to the development of the subject. An influential presentation of the simple theory of types was given in Church's paper [73] of 1940. For this purpose Church introduced the simply typed lambda-calculus, the core language of his type theory.

A typed version of combinatory logic was proposed by Curry a few years earlier, in the the 1934 paper [100], although Curry must have had the idea already in 1928, see [439, 440]. Curry's full "theory of functionality" turned out later to be contradictory [104] but it was readily corrected [105]. Soon types became a standard concept in the theory of combinators and in lambda-calculus.

Later types turned out to be useful in programming languages. Just like the type-free λ-calculus provides a foundation of untyped programming languages, various typed λ-calculi provide a foundation of programming languages with types. In particular, the design of languages like ML [387] motivated the research on type checking and typability problems. But, as noted in [237, pp. 103–104], the main ideas of a type reconstruction algorithm can be traced as far back as the 1920's.[5] The unification-based principal type inference algorithm, implicit in Newman's 1943 paper [362], was first described by Hindley [234] in 1969 (see [233, 237, 238] for historical notes). The PTIME-completeness of typability in simple types was shown in 1988 by Tyszkiewicz. Hindley's algorithm was later reinvented by Milner [342] for the language ML (but for ML it is no longer polynomial [259, 263]).

If we look at the typability problem from the point of view of the Curry-Howard isomorphism, then we may restate it by asking whether a given "proof skeleton" can actually be turned into a correct proof by inserting the missing formulas. Interestingly, such questions (like the "skeleton instantiation" problem of [500]) are indeed motivated by proof-theoretic research.

---

[5] The good old Polish school again...

Our normalization proof follows Turing's unpublished note from the 1930's [166]. This method was later rediscovered by several other authors, including Prawitz [403]. The first published normalization proof can be found in [107] (see [166] for discussion). The translation from "weak" to strong normalization is based on Klop's [272], here modified using ideas of Gandy [167]. A similar technique was earlier used by Nederpelt [359]. Variations of this approach have been invented (some independently from the original) by numerous authors, see e.g. [184, 202, 261, 264, 265, 312, 508, 507]. Paper [448] gives a survey of some of these results, see also [199].

A widely used normalization proof method, the *computability method*, is due to Tait. We will see it for the first time in Chapter 5. Another approach (first used by Lévy and van Daalen in the late 1970's) is based on inductive characterizations of strongly normalizable terms [65, 120, 255]. Other proofs are based on different forms of explicit induction, or semantical completeness [87, 167, 398, 406, 501]. Complexity bounds for the length of reductions are given in [433].

Various extensions of the simply typed lambda calculus can be proved strongly normalizable and we will do a number of such proofs in the chapters to follow. See also Chapter 7 for a related subject: cut-elimination.

Newman's lemma dates from 1942. A special case of it was known to Thue as early as in 1910 [458]. The proof given here, a variant of Huet's proof in [249], is taken from [31]. Theorem 3.7.4 is from [431] and remains true if type **int** is replaced by $\mathbf{int}_\sigma = (\sigma \to \sigma) \to \sigma \to \sigma$, for any fixed σ, see [305]. If one does not insist that σ is fixed, more functions become $\lambda_\to$-definable [153]. For instance the predecessor function is definable by a term of type $\mathbf{int}_\sigma \to \mathbf{int}_\tau$, for suitable σ and τ. But various natural functions, like e.g. subtraction, remain undefinable even this way. To obtain a more interesting class of definable functions, one has to extend $\lambda_\to$ by some form of iteration or recursion, cf. Chapter 10.

The books [237, 241] and the survey [32] are recommended for further reading on simply typed lambda-calculus and combinatory logic. Papers [165, 257] give the historical perspective. For applications in programming languages we suggest [349, 393]. In our short exposition we have entirely omitted semantical issues and extensions of simply typed λ-calculus, such as PCF. See [8, 211, 285, 349] for more on these subjects.

3.9. Exercises

3.1. Show that the following λ-terms are not typable in $\lambda_\to$ à la Curry.

$$\mathbf{KI}\Omega, \quad \mathbf{Y}, \quad \lambda xy.y(x\mathbf{K})(x\mathbf{I}), \quad 2\mathbf{K}.$$

3.2. Does the Curry style $\lambda_\to$ have the *subject conversion* property: *If* $\Gamma \vdash M : \sigma$ *and* $M =_\beta N$, *then* $\Gamma \vdash N : \sigma$? What if we assume in addition that N is typable?

3.3. Show an example of an untypable λI-term M, such that $M \to_\beta N$, for some typable term N.

3.4. Show an example of a typable closed λI-term M, and a type τ such that $\not\vdash M : \tau$, but $M \to_\beta N$, for some term N with $\vdash N : \tau$.

3.5. Assume an environment Γ consisting of type assumptions of the form $(x_p : p)$. Define terms t_τ such that $\Gamma \vdash t_\tau : \sigma$ holds if and only if $\tau = \sigma$.

3.6. How long (in the worst case) is the shortest type of a Curry-style term of length n?

3.9. EXERCISES

3.7. Show that the general unification problem can be reduced to the case of a signature consisting exclusively of the binary function symbol $\rightarrow$.

3.8. Show that the unification problem reduces to solving a single equation, provided the signature contains a binary function symbol.

3.9. Show that the unification problem reduces in logarithmic space to the typability problem.

3.10. Show that problems (vii) and (viii) of Remark 3.2.8 reduce to each other in logarithmic space.

3.11. Prove that the unification problem and problems (i)–(vi) of Remark 3.2.8 reduce to each other in logarithmic space.

3.12. What is wrong with the following reduction of problem (vi) to problem (i): *To answer* $? \vdash M : \tau$ *ask the question* $? \vdash \lambda yz.y(zM)(zt_\tau) : ?$

3.13. Prove the following *converse principal type theorem*: If τ is a non-empty type, then there exists a closed term M such that τ is the principal type of M. (In fact, if N is a closed term of type τ, then we can require M to be beta-reducible to N.) *Hint*: Use the technique of Exercise 3.5.

3.14. Let $\Gamma \vdash M[x := N] : \sigma$ in Church-style, and let $x \in FV(M)$. Show that $\Gamma, x : \tau \vdash M : \sigma$, for some τ with $\Gamma \vdash N : \tau$. Does it hold for Curry style?

3.15. Show that strong normalization for $(\lambda_\rightarrow)$ à la Curry implies strong normalization for $(\lambda_\rightarrow)$ à la Church, and conversely.

3.16. Show that the weak Church-Rosser property does not in general imply the Church-Rosser property.

3.17. Let M_1 and M_2 be Church-style normal forms of the same type, and let $|M_1| = |M_2|$. Show that $M_1 = M_2$.

3.18. Let M_1 and M_2 be Church-style terms of the same type, and assume that $|M_1| =_\beta |M_2|$. Show that $M_1 =_\beta M_2$. Does $|M_1| = |M_2|$ imply $M_1 = M_2$?

3.19. Can you derive Church-Rosser property for Church-style terms from the Church-Rosser property for untyped terms?

3.20. How long is the normal form of the term $M = 22 \cdots 2xy$ with n occurrences of 2? How long (including types) is a Church-style term M' such that $|M'| = M$?

3.21. (Based on [153].) This exercise uses the notation introduced in the proof of Theorem 3.5.1. In addition we define the *depth* $d(M)$ of M as:

$$\begin{aligned} d(x) &= 0; \\ d(MN) &= 1 + \max(d(M), d(N)); \\ d(\lambda x{:}\sigma\, M) &= 1 + d(M). \end{aligned}$$

Let $\delta = \delta(M)$ and let $d(M) = d$. Show that M reduces in at most 2^d steps to some term M_1 such that $\delta(M_1) < \delta$ and $d(M_1) \leq 2^d$.

3.22. Use Exercise 3.21 to prove that if $\delta(M) = \delta$ and $d(M) = d$ then the normal form of a term M is of depth at most $\exp_\delta(d)$, and can be obtained in at most $\exp_{\delta+1}(d)$ reduction steps.

3.23. How long (in the worst case) is the normal form of a Curry-style typable term of length n?

3.24. Show that all the extended polynomials are $\lambda_\rightarrow$-definable.

3.25. Let $\Gamma = \{f : p \to p,\ a : \mathbf{int},\ b : \mathbf{int},\ x_1 : p,\ \ldots,\ x_r : p\}$, and let $\Gamma \vdash M : p$. Prove that there exists a polynomial $P(m,n)$ and a number $i \in \{1,\ldots,r\}$ such that $M[a := \mathbf{c}_m, b := \mathbf{c}_n] =_\beta f^{P(m,n)} x_i$ holds for all $m, n \neq 0$.

3.26. Prove that all functions definable in simply typed lambda-calculus are extended polynomials (Theorem 3.7.4). *Hint:* Use Exercise 3.25.

Chapter 4

The Curry-Howard isomorphism

Having discussed intuitionistic logic in Chapter 2 and typed lambda-calculus in Chapter 3, we now concentrate on the relationship between the two—the Curry-Howard isomorphism. We have already discovered this amazing analogy between logic and computation in Chapter 3, where we readily accepted lambda terms as an appropriate notation for proofs. Indeed, the Brouwer-Heyting-Kolmogorov interpretation, in particular the functional understanding of constructive implication, strongly supports this choice.

In this chapter we have a closer look at the correspondence between proofs and terms. We soon discover that, depending on the choice of a specific proof formalism, it can be more or less adequate to call it an "isomorphism". In most cases, lambda-terms can be seen as a refinement of proofs rather than an isomorphic image. On the other hand, we also discover that the analogy goes further than the equivalence between non-empty types and provable implicational formulas. It goes further "in width", because it extends to a correspondence between various logical connectives and data types. But it also goes further "in depth". There is a fundamental relationship between term reduction (computation) and proof normalization. This aspect of the formulas-as-types correspondence is perhaps the most important one.

4.1. Proofs and terms

As we have discovered in Chapter 3, the type assignment rules of $\lambda_\to$ can be seen as annotated versions of the natural deduction system $NJ(\to)$. The following fact is an immediate consequence of this similarity.

4.1.1. PROPOSITION (Curry-Howard isomorphism).

(i) *If* $\Gamma \vdash M : \varphi$ *in* $\lambda_\to$ *then* $\text{rg}(\Gamma) \vdash \varphi$ *in* $\text{IPC}(\to)$.

(ii) *If* $\Delta \vdash \varphi$ *in* $\text{IPC}(\to)$ *then* $\Gamma \vdash M : \varphi$ *in* $\lambda_\to$, *for some M and some Γ with* $\text{rg}(\Gamma) = \Delta$.

In particular an implicational formula is an intuitionistic theorem if and only if it is an inhabited type.

PROOF. Easy induction with respect to derivations. In part (ii) it is convenient to choose $\Gamma = \{(x_\varphi : \varphi) \mid \varphi \in \Delta\}$, where x_φ are distinct variables. □

The correspondence between logic and lambda-calculus makes it possible to translate results concerning one of these systems into results about the other. For instance we immediately derive that

- There is no combinator of type $((p \to q) \to p) \to p$;
- Formula $((((p \to q) \to p) \to p) \to q) \to q$ is an intuitionistic tautology,

because we know that Peirce's law (Example 2.1.4(i)) is not intuitionistically valid, and because we can construct the term

$$\lambda x^{(((p \to q) \to p) \to p) \to q}. x(\lambda y^{(p \to q) \to p}. y(\lambda z^p. x(\lambda u^{(p \to q) \to p} z))).$$

APPLICATIONS OF NORMALIZATION. Many properties regarding unprovability can be difficult to obtain directly. In Chapter 2 we used semantics for this purpose. The Curry-Howard isomorphism suggests an alternative approach to such problems, which is based on normalization. Here is an example.

4.1.2. PROPOSITION. *The implicational logic* IPC($\to$) *is consistent, i.e. there are underivable judgements.*

PROOF. Assume that $\vdash p$, where p is a propositional variable. Then $\vdash M : p$ for some closed term M, which by Theorem 3.5.1 can be assumed normal. So $M = \lambda y_1 \ldots y_m. x M_1 \ldots M_n$, by Remark 1.3.7, and because of the type of M, we must have $m = 0$. But then x is free in M, a contradiction. □

The reader will perhaps find this result entirely obvious. Indeed, the consistency of intuitionistic propositional calculus is an immediate consequence of soundness. However, other logical calculi and theories are not always so simple. A semantical consistency proof (construction of a model) may then require concepts and tools more sophisticated than normalization. Proposition 4.1.2 demonstrates an important proof method on a simple example.

In the next section (and also in Section 8.8) we will see another application of the Curry-Howard isomorphism and normalization. Questions concerning decidability and complexity of a logical system can easier be resolved if one can identify provability with the existence of normal inhabitants. Other applications of proof normalization are various conservativity results (cf. Remark 7.3.7), program extraction procedures (like in the proof of Theorem 12.7.11), and syntactic proofs of various metamathematical properties (disjunction property, existence property, independence of connectives, etc.)

4.2. Type inhabitation

We now prove that the inhabitation problem for the simply typed lambda-calculus is PSPACE-complete. In particular it is decidable. An immediate consequence is that provability in IPC($\to$) is also decidable and PSPACE-complete. Compare this to the semantic proof obtained in Chapter 2 as a consequence of Theorem 2.4.12.

First we prove that our problem is in PSPACE. It follows from normalization that an inhabited type must have an inhabitant in normal form. Thus it suffices to consider terms in normal form. In fact, it is convenient to make a further restriction.

4.2.1. DEFINITION. We define the notion of a Church-style term in η-long normal form (or just long normal form) as follows:

- If x is a variable of type $\tau_1 \to \cdots \to \tau_n \to p$, and $M_1^{\tau_1}, \ldots, M_n^{\tau_n}$ are in η-long normal form, then $(xM_1 \ldots M_n)^p$ is in η-long normal form.

- If M^σ is in η-long normal form then so is $(\lambda x{:}\tau.\, M)^{\tau\to\sigma}$.

Equivalently, a long normal form is a term of shape $\lambda x_1^{\tau_1} \ldots x_n^{\tau_n}.\, y\vec{N}$, where $\vec{N}$ are normal forms. Informally speaking, a long normal form is a normal form where all function variables are "fully applied" (all arguments are provided).

4.2.2. LEMMA. *For every Church-style term M^σ in β-normal form, there exists a term L^σ in η-long normal form, such that $L^\sigma \twoheadrightarrow_\eta M^\sigma$.*

PROOF. Easy induction with the help of Remark 1.3.7. □

4.2.3. LEMMA. *There is a polynomial space algorithm to determine whether a given type τ is inhabited in a given environment Γ.*

PROOF. If a type is inhabited then, by Proposition 4.1.1 and Lemma 4.2.2, it is inhabited by a long normal form. To determine whether there exists a long normal form M, satisfying $\Gamma \vdash M : \tau$, we proceed as follows:

- If $\tau = \tau_1 \to \tau_2$, then M must be an abstraction, $M = \lambda x{:}\tau_1.\, M'$. Thus, we look for an M' satisfying $\Gamma, x{:}\tau_1 \vdash M' : \tau_2$.

- If τ is a type variable, then M is an application of a variable to a sequence of terms. We nondeterministically choose a variable z, declared in Γ to be of type $\tau_1 \to \cdots \to \tau_n \to \tau$. (If there is no such variable, we reject.) If $n = 0$ then we accept. If $n > 0$, we recursively answer all the questions whether τ_i are inhabited in Γ.

This nondeterministic recursive (or alternating) procedure is repeated as long as there are new questions of the form $\Gamma \vdash ? : \tau$. We can assume that in a successful run of the procedure no such question is raised again (otherwise we can consider a shorter run). Note also that if there are two variables in Γ, say x and y, declared to be of the same type σ, then each term M can be replaced with $M[y := x]$ with no change of type. This means that a type τ is inhabited in Γ iff it is inhabited in $\Gamma - \{y : \sigma\}$, and it suffices to consider contexts with all declared types being different. At each step of our procedure, the environment Γ either stays the same or it expands. And it can remain unchanged for at most as many steps as there are different possible formulas to the right of $\vdash$. Thus the depth of recursion does not exceed the squared number of subformulas of types in Γ, τ, where $\Gamma \vdash ? : \tau$ is the initially posed question. It follows that the space needed to implement our algorithm is polynomial: The stack of procedure activations is of polynomial height, and each record on the stack is of polynomial size. □

The reader who has noticed the game flavour of the algorithm used in the proof will see more of that in Section 4.6. Now we have to show PSPACE-hardness. We define a reduction from the satisfiability problem for classical second-order propositional formulas (or "quantified Boolean formulas"), see Section A.5.

Assume that a quantified Boolean formula Φ is given. Without loss of generality we may assume that the negation symbol $\neg$ does not occur in Φ, except in the context $\neg p$, where p is a propositional variable.

Assume that all bound variables of Φ are different and that no variable occurs both free and bound. For each propositional variable p, occurring in Φ (free or bound), let α_p and $\alpha_{\neg p}$ be fresh type variables. Also, for each subformula φ of Φ, let α_φ be a fresh type variable. We construct an environment Γ_Φ from the following types:

- $(\alpha_p \to \alpha_\psi) \to (\alpha_{\neg p} \to \alpha_\psi) \to \alpha_{\forall p \psi}$, for each subformula $\forall p\, \psi$;
- $(\alpha_p \to \alpha_\psi) \to \alpha_{\exists p \psi}$, $(\alpha_{\neg p} \to \alpha_\psi) \to \alpha_{\exists p \psi}$, for each subformula $\exists p\, \psi$;
- $\alpha_\psi \to \alpha_\vartheta \to \alpha_{\psi \wedge \vartheta}$, for each subformula $\psi \wedge \vartheta$;
- $\alpha_\psi \to \alpha_{\psi \vee \vartheta}$ and $\alpha_\vartheta \to \alpha_{\psi \vee \vartheta}$, for each subformula $\psi \vee \vartheta$.

If v is a zero-one valuation of propositional variables, then Γ_v is Γ_Φ extended with the type variables

- α_p, when $v(p) = 1$;
- $\alpha_{\neg p}$, when $v(p) = 0$.

The following lemma is proved by a routine induction with respect to the length of formulas. Details are left to the reader.

4.3. Not an exact isomorphism

4.2.4. LEMMA. *For every subformula φ of Φ, and every valuation v, defined on the free variables of φ, the type α_φ is inhabited in Γ_v iff $[\![\varphi]\!]_v = 1$. Thus, to verify if Φ is a tautology, one checks if α_Φ is inhabited in Γ_Φ.*

Observe that the number of subformulas of Φ does not exceed the length of Φ. Thus the reduction from Φ to Γ_Φ and α_Φ can be performed in logarithmic space (provided we represent a variable α_ψ using a binary number, rather than ψ itself). This implies the PSPACE-hardness, which together with Lemma 4.2.3 gives the main result of this section.

4.2.5. THEOREM. *The inhabitation problem for simply typed lambda-calculus is complete for polynomial space.* □

An immediate consequence of the above (and Theorem 2.6.2) is that the intuitionistic propositional logic (with all the connectives) is also PSPACE-hard. In Chapter 7 we will define a polynomial space algorithm for arbitrary formulas, thus proving PSPACE-completeness of the full system (Exercise 7.13).

4.3. Not an exact isomorphism

We have just demonstrated that the formulas-as-types analogy can indeed be useful. However, the reader may find Proposition 4.1.1 a little unsatisfactory. If we talk about an "isomorphism" then perhaps the statement of the proposition should have the form of an equivalence? The concluding sentence is indeed of this form, but it only holds on a fairly high level: We must abstract from the proofs and only ask about conclusions. (Or, equivalently, we abstract from terms and only ask which types are non-empty.) To support the idea of an "isomorphism," we would certainly prefer an exact, bijective correspondence between proofs and terms.

Unfortunately, we cannot improve Proposition 4.1.1 in this respect, at least not for free. While it is correct to say that lambda-terms are essentially annotated proofs, the problem is that some proofs can be annotated in more than one way. For instance, the proof

$$\frac{\dfrac{p \vdash p}{p \vdash p \to p}}{\vdash p \to p \to p}$$

can be annotated as either $\lambda x^p \lambda y^p\, x$ or $\lambda x^p \lambda y^p\, y$. And the three judgements

$$f : p \to p \to q,\ x : p \vdash fxx : q;$$
$$f : p \to p \to q,\ x : p,\ y : p \vdash fxy : q;$$
$$f : p \to p \to q,\ x : p,\ y : p,\ z : p \vdash fxx : q$$

(and many others) correspond to the same derivation:

$$\frac{p \to p \to q, p \vdash p \to p \to q \quad p \to p \to q, p \vdash p}{\frac{p \to p \to q, p \vdash p \to q \qquad\qquad p \to p \to q, p \vdash p}{p \to p \to q, p \vdash q}}$$

As we can thus see, the difference between terms and proofs is that the former carry more information than the latter. The reason is that in logic, the primary isssue is usually to determine provability of a formula, and in this respect, certain issues are irrelevant. It does not matter whether we use the same assumption twice or if we use two different assumptions about the same formula. In fact, many of us would have objections against the very idea of "two different assumptions about φ." Indeed, in classical or intuitionistic logic, once φ becomes part of our knowledge, it can be asserted as many times as we want. We don't make any distinction between repeated statements of φ, it is just repeating the same information. On the contrary, in lambda-calculus we can have many variables of the same type φ, and this corresponds to making a difference between various assumptions about the same formula.

But when studying proofs as such (rather than mere provability) it may actually be useful to maintain exact bookkeeping of assumptions. A further discovery is that, from a purely logical point of view, it can make a lot of sense to see whether an assumption is used or not in an argument, or how many times it has to be used. In *resource-conscious* logics, a major example of which is *linear logic*, these issues play a fundamental role. If we realize all this, we conclude that our natural deduction proofs are simply too schematic and perhaps a more informative proof system may turn our Curry-Howard correspondence into a real isomorphism.

In fact, our formulation of natural deduction is not very convenient for this specific purpose. It can be more informative to write natural deduction proofs in the "traditional" way. We use this occasion to explain it informally. In this style, quite common in the literature on proof theory, one does not associate a set of assumptions (the environment) to every node of a derivation. Instead, one writes all the assumptions at the top, and puts in brackets (or crosses out) those assumptions that have been discharged by the implication introduction rule. To indicate which assumptions are eliminated by which proof steps, one can enhance the notation by using labels:

$$\frac{\begin{array}{c}[\varphi]^{(i)}\\ \vdots \\ \psi\end{array}}{\varphi \to \psi} {\scriptstyle (i)}$$

In general, in an $\to$-introduction step, we may discharge zero, one, or more occurrences of the assumption. A label referring to an arrow-introduction proof step is then attached to all discharged assumptions.

4.4. Proof normalization

The arrow elimination is represented as usual:

$$\frac{\varphi \to \psi \quad \varphi}{\psi}$$

and a proof is a tree with all assumptions at the top and the conclusion as a root. It is now apparent that the "traditional" natural deduction is more fine-grained than our simplified notation (although often less convenient in use). Indeed, for instance

$$\cfrac{\cfrac{[p]^{(1)}}{p \to p}\,(1)}{\vdash p \to p \to p}\,(2) \quad \text{and} \quad \cfrac{\cfrac{[p]^{(1)}}{p \to p}\,(2)}{\vdash p \to p \to p}\,(1)$$

are two different proofs of $p \to p \to p$, corresponding respectively to terms $\lambda x^p \lambda y^p\, y$ and $\lambda x^p \lambda y^p\, x$. However, this additional information is only available about discharged assumptions, and not about the free ones. Indeed, our other example, proving q from $p \to p \to q$ and p, would be written as

$$\cfrac{\cfrac{p \to p \to q \quad p}{p \to q} \quad p}{q}$$

which of course still gives no clue about the possible identification of the two p's, until we discharge one or both of them. Thus, the "traditional" natural deduction allows us to claim an isomorphism[1] for closed terms and proofs with no free assumptions (Exercise 4.8), but not for arbitrary proofs and terms. To extend the isomorphism to arbitrary terms we need a proof system with labeled free assumptions. One can design such a proof system, only to discover that the resulting proofs, up to syntactic sugar, are... another representation of lambda-terms.

4.4. Proof normalization

The correspondence between proofs and lambda-terms suggests that reduction of lambda-terms should be related to a meaningful operation on proofs. Indeed, *proof normalization* is a central issue in proof theory and has been studied by logicians independently. No surprise. Once we agree that proofs are as important in logic as the formulas they derive, it becomes a basic question whether two given proofs of the same formula should be considered essentially identical, or whether they are truly different. In particular it is desirable that an arbitrary proof can be identified with a *normal proof* of

---

[1] To be precise, we also need alpha-conversion on discharge labels.

a certain simple shape. A natural way to identify two proofs is to rewrite one into another by means of a sequence of proof reductions.[2] As an example, consider the following natural deduction proof:

$$\cfrac{\cfrac{p \vdash p}{\vdash p \to p}\,(\to\!I) \qquad \cfrac{\cfrac{\cfrac{p \to p,\ q \vdash p \to p}{p \to p \vdash q \to p \to p}\,(\to\!I)}{\vdash (p \to p) \to q \to p \to p}\,(\to\!I)}{\vdash q \to p \to p}}{\vdash q \to p \to p}\,(\to\!E)$$

This proof demonstrates that $q \to p \to p$ is derivable. However, it does so in a somehow inefficient way. By $\to$-introduction we derive the formula $(p \to p) \to q \to p \to p$. Since we can *prove* the assumption $p \to p$, we apply the $\to$-elimination rule to obtain the conclusion. That is, we introduce the implication only to immediately eliminate it. Such a pair consisting of an introduction step followed by an elimination step (applied to the formula just introduced) is called a *detour*. Here is a template of a detour.

$$\cfrac{\Gamma \vdash \varphi \qquad \cfrac{\Gamma, \varphi \vdash \psi}{\Gamma \vdash \varphi \to \psi}\,(\to\!I)}{\Gamma \vdash \psi}\,(\to\!E)$$

This resembles of course a beta-redex $(\lambda x{:}\varphi.M^\psi)N^\varphi$ which is reduced by replacing every occurrence of x with a copy of N. In a similar way, to get rid of a detour, one replaces every use of the assumption φ in the right-hand side proof by a copy of the left-hand side proof. This operation turns our proof of $q \to p \to p$ into a simplified one:

$$\cfrac{\cfrac{\cfrac{p \vdash p}{\vdash p \to p}\,(\to\!I)}{\vdash q \to p \to p}\,(\to\!I)}{}$$

Our example becomes more transparent if we use the "traditional" notation for natural deduction proofs. Our initial proof is then written as

$$\cfrac{\cfrac{[p]^{(3)}}{p \to p}\,(3) \qquad \cfrac{\cfrac{[p \to p]^{(1)}}{q \to p \to p}\,(2)}{(p \to p) \to q \to p \to p}\,(1)}{q \to p \to p}$$

---

[2]This was first discovered by Gentzen, who designed *sequent calculus* especially for this purpose (see Chapter 7).

4.4. PROOF NORMALIZATION

and reduces to

$$\frac{\dfrac{[p]^{(3)}}{p \to p}\,(3)}{q \to p \to p}\,(2)$$

as a result of replacing the assumption $[p \to p]^{(1)}$ by the proof of $p \to p$. In general, we have the following reduction rule:

$$\begin{array}{c}(*)\\ \vdots\\ \psi\end{array}\quad \dfrac{[\psi]^{(i)}\\ \vdots\\ \varphi\qquad \psi \to \varphi}{\varphi}\,(i) \quad \Longrightarrow \quad \begin{array}{c}(*)\\ \vdots\\ \psi\\ \vdots\\ \varphi\end{array}$$

which should be read as follows. Replace each occurrence of the assumption $[\psi]^{(i)}$ in the proof of φ by a separate copy of the proof of ψ marked by $(*)$. Now we have an explicit analogy with a substitution $M[x := N]$.

The process of eliminating proof detours is called *proof normalization*, and a proof tree with no detours is said to be in *normal form*, quite like a term without redexes. An important application of the Curry-Howard isomorphism is that every proof can be turned into a normal form, because every typed term is normalizing. By the strong normalization theorem for $\lambda_\to$ (Theorem 3.5.5), the order of eliminating detours is also irrelevant.

4.4.1. PROPOSITION. *Each provable judgement of* NJ($\to$) *has a normal proof.*

PROOF. Suppose that $\Gamma \vdash \varphi$ is derivable. Then by Proposition 4.1.1(ii) we have $\Delta \vdash M : \varphi$ where $\mathrm{rg}(\Delta) = \Gamma$. By Theorem 3.5.5 we can assume that M is in normal form. We show that $\Gamma \vdash \varphi$ has a normal proof by easy induction with respect to M, using Remark 1.3.7. $\square$

Another form of detour, corresponding to eta-reduction, occurs when an arrow-elimination step is followed by an immediate re-introduction of the same implication. Abusing the formalism slightly, we can represent this situation as

$$\dfrac{\dfrac{\Gamma \vdash \varphi \to \psi \quad \dfrac{}{\Gamma, \varphi \vdash \varphi}\,(\mathrm{Ax})}{\Gamma, \varphi \vdash \psi}\,(\to \mathrm{E})}{\Gamma \vdash \varphi \to \psi}\,(\mathrm{I} \to)$$

Again, the traditional notation is more illustrative, so we use it to show the eta-rule for proofs:

$$\frac{\varphi \to \psi \quad [\varphi]^{(i)}}{\dfrac{\psi}{\varphi \to \psi}\,(i)} \begin{array}{c}(*)\\ \vdots\end{array} \quad \Longrightarrow \quad \begin{array}{c}(*)\\ \vdots\\ \varphi \to \psi\end{array}$$

where it is assumed that the assumption number i does not occur in the proof represented by $(*)$.

4.5. Sums and products

The correspondence between derivations in NJ($\to$) and $\lambda_\to$ can be extended to the whole system of intuitionistic propositional logic if the simply typed λ-calculus is appropriately extended. The BHK interpretation of Chapter 2 gives an explicit guidance on how this extension should be made. First, a construction of a conjunction is a *pair*. That is, the formula $\varphi \wedge \psi$ corresponds to a product type, and could equally well be written as $\varphi \times \psi$. A canonical element of this type is a pair $\langle M, N \rangle$, where M is of type φ and N is of type ψ. The creation of a pair corresponds to $\wedge$-introduction. The two $\wedge$-elimination rules correspond to the two projections.

In type-free λ-calculus, pairs and projections were definable, but this is no longer the case in $\lambda_\to$. Indeed, as we have seen in Chapter 2 (Exercise 2.26), conjunction is not definable by any implicational formula. This means we have to add pairs and projections to the term language.

In the same spirit, $\varphi \vee \psi$ is a *disjoint sum* (or *variant*) type, also called *co-product*. An object of a variant type consists of data of either type φ or ψ together with a flag indicating the variant chosen.

It remains to observe that $\bot$, the formula with no construction, should correspond to the *empty type*.

4.5.1. WARNING. We use the logical symbols $\wedge$ and $\vee$ to denote also the corresponding types. Similar symbols are sometimes used to denote *intersection* and *union* types, which have quite a different meaning (see e.g. [26]), and which correspond to set-theoretic intersections and unions rather than to products and co-products.

4.5.2. DEFINITION.

(i) The syntax of *extended* Church-style simply typed lambda-calculus is defined by the type assignment rules of Figure 4.1. As in Chapter 3, this should be understood so that a "raw" expression becomes a term (in a given environment) only if it can be assigned a type according to these rules. The superscript in $\mathbf{in}_i^{\varphi \vee \psi}(M)$, which ensures type uniqueness, is often omitted for simplicity, and we write just $\mathbf{in}_i(M)$.

4.5. Sums and Products

(ii) The free variables in a (raw) term are defined as usual, with the following clauses added to Definition 3.3.1(ii):

- $FV(\langle M, N \rangle) = FV(M) \cup FV(N)$;
- $FV(\pi_i(M)) = FV(\mathbf{in}_1(M)) = FV(\mathbf{in}_2(M)) = FV(\varepsilon_\varphi(M)) = FV(M)$;
- $FV(\mathbf{case}\ M\ \mathbf{of}\ [x]P\ \mathbf{or}\ [y]Q) = FV(M) \cup (FV(P) - \{x\}) \cup (FV(Q) - \{y\})$.

That is, a variable in brackets is bound in the corresponding branch of a **case**-expression. Of course, the particular choice of bound variables does not matter; that is, we consider terms up to an (appropriately generalized) alpha-conversion. Substitution respects the bindings, so that for instance (**case** M **of** $[x]P$ **or** $[y]Q)[z := N]$ is defined as **case** $M[z := N]$ **of** $[x]P[z := N]$ **or** $[y]Q[z := N]$, for $x, y \notin FV(N)$.

(iii) The *beta-reduction* relation on extended terms is defined as the smallest compatible relation $\to_\beta$ extending the ordinary beta-reduction by:

$$\pi_1(\langle M_1, M_2 \rangle) \to_\beta M_1;$$
$$\pi_2(\langle M_1, M_2 \rangle) \to_\beta M_2;$$

$\mathbf{case}\ \mathbf{in}_1^{\varphi \vee \psi}(N)\ \mathbf{of}\ [x]P\ \mathbf{or}\ [y]Q \to_\beta P[x := N]$;
$\mathbf{case}\ \mathbf{in}_2^{\varphi \vee \psi}(N)\ \mathbf{of}\ [x]P\ \mathbf{or}\ [y]Q \to_\beta Q[y := N]$.

The intuitive meaning of pairs $\langle M, N \rangle$ and projections $\pi_i(M)$ in the rules of Figure 4.1 should be obvious. Also, the reduction rules (a projection applied to a pair returns a component) are as expected. As far as disjunction is concerned, a term of the form $\mathbf{in}_1^{\varphi \vee \psi}(M)$ or of the form $\mathbf{in}_2^{\varphi \vee \psi}(N)$ represents the conversion of $M : \varphi$ (resp. $N : \psi$) into an element of the disjoint sum. These are the canonical elements of $\varphi \vee \psi$. The actual contents of a given $M : \varphi \vee \psi$ is examined by a **case**-statement **case** M **of** $[x]P$ **or** $[y]Q$. Depending on the result, either the first branch or the second one is executed.

Finally, the expression $\varepsilon_\varphi(M)$ is a *miracle of type* φ. If $M : \bot$ then M is "the thing which is not". Once we have achieved the impossible, we should be able to do whatever we wish.

4.5.3. PROPOSITION (Curry-Howard isomorphism).

(i) *If* $\Gamma \vdash M : \varphi$ *then* $\mathrm{rg}(\Gamma) \vdash \varphi$.
(ii) *If* $\Delta \vdash \varphi$, *then there are M and Γ with* $\Gamma \vdash M : \varphi$ *and* $\mathrm{rg}(\Gamma) = \Delta$.

There is a general pattern in the above rules, related to the duality between introduction and elimination in natural deduction. Each data type (function space, product, sum) comes with its own *constructors*. The constructors are operators *creating* canonical objects of the particular types and the destructors *use* these objects in computation. It may happen that

$$\frac{}{\Gamma, x:\psi \vdash x:\psi}$$

$$\frac{\Gamma, x:\varphi \vdash M:\psi}{\Gamma \vdash (\lambda x{:}\varphi\, M):\varphi \to \psi} \qquad \frac{\Gamma \vdash M:\varphi \to \psi \quad \Gamma \vdash N:\varphi}{\Gamma \vdash (MN):\psi}$$

$$\frac{\Gamma \vdash M:\psi \quad \Gamma \vdash N:\varphi}{\Gamma \vdash \langle M,N\rangle : \psi \wedge \varphi} \qquad \frac{\Gamma \vdash M:\psi \wedge \varphi}{\Gamma \vdash \pi_1(M):\psi} \qquad \frac{\Gamma \vdash M:\psi \wedge \varphi}{\Gamma \vdash \pi_2(M):\varphi}$$

$$\frac{\Gamma \vdash M:\varphi}{\Gamma \vdash \mathbf{in}_1^{\varphi \vee \psi}(M):\psi \vee \varphi} \qquad \frac{\Gamma \vdash M:\psi}{\Gamma \vdash \mathbf{in}_2^{\varphi \vee \psi}(M):\psi \vee \varphi}$$

$$\frac{\Gamma \vdash L:\varphi \vee \psi \quad \Gamma, x:\varphi \vdash M:\rho \quad \Gamma, y:\psi \vdash N:\rho}{\Gamma \vdash (\mathbf{case}\ L\ \mathbf{of}\ [x]M\ \mathbf{or}\ [y]N):\rho}$$

$$\frac{\Gamma \vdash M:\bot}{\Gamma \vdash \varepsilon_\psi(M):\psi}$$

FIGURE 4.1: EXTENDED SIMPLY TYPED LAMBDA-CALCULUS

a destructor decomposes a complex object and returns a simpler object of a component type. This is how a projection works, but notice that the result of applying a **case**-statement to an object of a sum type may be of an arbitrary type. However there is always an analogy between a proof detour and a beta-redex. The introduction-elimination detour corresponds to a constructor directly passed to a destructor. A projection applied to a pair, an abstraction given an argument, or a **case**-dispatch on an explicitly indicated variant—all these expressions are ready for evaluation and may be subject to a reduction step.

We omit the proof of the following theorem, because it is a corollary of stronger results we consider later in the book. See Exercises 6.16 and 11.29.

4.5.4. THEOREM. *The extended simply typed lambda-calculus is strongly normalizing.*

A consequence of the above is the Church-Rosser property, which can be obtained with help of Newman's lemma, after an easy verification of WCR (cf. Section 3.6).

4.6. Prover-skeptic dialogues

A SMALL SUMMARY. Various other aspects and components of proofs also correspond to computational phenomena. For instance, the natural deduction axiom (a free assumption) is reflected by the use of a variable in a term. The provability problem translates to the inhabitation problem. The subject reduction property means that removing detours in a proof yields a proof of the same formula, and the Church-Rosser theorem states that the order of proof normalization is immaterial. Here is a little glossary:

logic	lambda-calculus
formula	type
propositional variable	type variable
connective	type constructor
implication	function space
conjunction	product
disjunction	disjoint sum
absurdity	empty type
proof	term
assumption	object variable
introduction	constructor
elimination	destructor
proof detour	redex
normalization	reduction
normal proof	normal form
provability	inhabitation

In the remainder of this book the concepts corresponding to one another under the isomorphism are often used interchangeably. Systems to be introduced in the following chapters will be considered from these two points of view: as typed lambda-calculi and as logics. And our glossary will expand.

4.6. Prover-skeptic dialogues

Recall the BHK-interpretation from Chapter 2. We can imagine the development of a construction for a formula as a dialogue between a *prover*, who is to produce the construction, and a *skeptic*, who doubts that it exists.

Prover 1: I assert that $((p \to p) \to (q \to r \to q) \to s) \to s$ holds.

Skeptic 1: Really? Let us suppose that I provide you with a construction of $(p \to p) \to (q \to r \to q) \to s$. Can you then give me one for s?

Prover 2: I got it by applying your construction to a construction of $p \to p$ and then by applying the result to a construction of $q \to r \to q$.

Skeptic 2: Hmm... you are making two new assertions. I doubt the first one the most. Are you sure you have a construction of $p \to p$? Suppose I give you a construction of p, can you give me one back for p?

Prover 3: I can just use the same construction!

A *dialogue* starts with the prover making an *assertion*. Based on the prover's assertion, the skeptic presents a list of *offers* to the prover and then presents a *challenge*. The offers do not consist of *actual* constructions; rather, the skeptic is telling the prover to continue under the assumption that she has such constructions. The prover must meet the challenge using these offers as well as previous offers. In doing so, the prover is allowed to introduce new assertions. The skeptic then reacts to each of these assertions, etc.

Here is another dialogue starting from the same assertion (the first three steps are identical to the preceding dialogue):

Skeptic 2': Hmm... you are making two new assertions. I doubt the second one the most. Are you sure you have a construction of $q \to r \to q$? If I give you constructions of q and r, can you give me one for q?

Prover 3': I can use the one you just gave me!

If the prover, in a given step, introduces several new assertions, then the skeptic may challenge any one of them—but only one, and always one from the latest prover step. Conversely, the prover must always respond to the latest challenge, but she may use any previous offer that the skeptic has made, not just those from the latest step.

A dialogue *ends* when the player who is up cannot respond, in which case the other player *wins;* here we think of the dialogues as a sort of game or debate. The skeptic has no response, if the prover's last step did not introduce new assertions to address the skeptic's last challenge. The prover has no response, if the skeptic's last step introduced a challenge that cannot be addressed using any of the skeptic's offers.

A dialogue can continue infinitely, which is regarded as a win for the skeptic. For example, the prover starts by asserting $(a \to b) \to (b \to a) \to a$, the skeptic offers $a \to b$ and $b \to a$ and challenges the prover to produce a construction of a. She can do this by applying the second offer to an alleged construction of b, which the skeptic will then challenge the prover to construct. Now the prover can respond by applying the first offer to an alleged construction of a, etc.

A *prover strategy* is a technique for arguing against the skeptic. When the skeptic has a choice—when the prover introduces more than one new assertion—the prover has to anticipate all the different choices the skeptic can make and prepare a reaction to each of them. Such a strategy can be represented as a tree where every path from the root is a dialogue, and where

4.6. Prover-skeptic dialogues

every node labeled with a prover step has as many children as the number of assertions introduced by the prover in her step; each of these children is labeled with a skeptic step challenging an assertion. For instance, the above two dialogues can be collected into the prover strategy in Figure 4.2. A prover strategy is *winning* if all its dialogues are won by the prover.

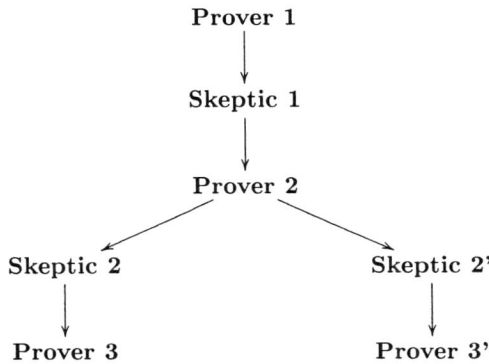

FIGURE 4.2: PROVER STRATEGY

One can extract the construction for a formula from a winning prover strategy. Indeed, if we accept simply typed λ-terms of type φ as constructions of φ, then the construction is obtained by viewing the skeptic's steps as lambda-abstractions and the prover's steps as applications, as outlined in Figure 4.3.

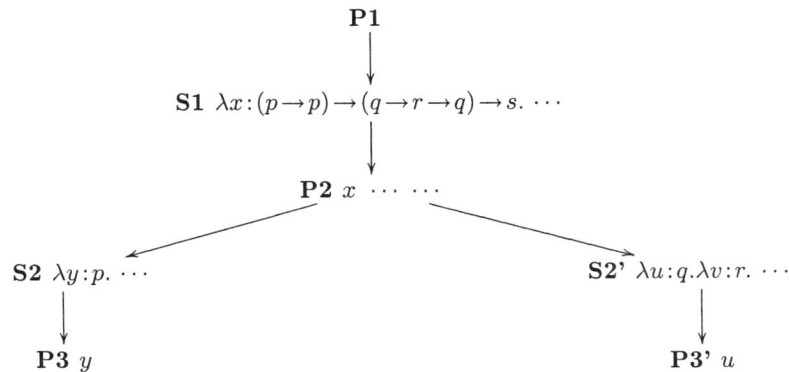

FIGURE 4.3: CONSTRUCTIONS FROM STRATEGIES

In fact, we can think of the extracted λ-term, which is an η-long normal form, as a convenient representation of the strategy. And the construction of long forms in the Ben-Yelles algorithm (see Lemma 4.2.3) can be seen as the search for a winning prover strategy.

We now define these concepts rigorously for minimal implicational logic. For technical reasons it is convenient to consider dialogues (about some formula φ) relative to a collection Γ of offers provided in advance.

WARNING. The dialogues considered in this chapter are somewhat different than the ones originating with Lorenzen—see Chapter 6.

4.6.1. DEFINITION. A *dialogue* over (Γ, φ) is a possibly infinite sequence $(\Sigma_1, \alpha_1), (\Pi_1, \beta_1), (\Sigma_2, \alpha_2), (\Pi_2, \beta_2) \ldots$, where α_i, β_j are propositional variables and Σ_i, Π_j are finite or empty sequences of formulas, satisfying:

(i) $\Sigma_1 = \varphi$ and $\alpha_1 = \top$. (Prover begins.)[3]

(ii) $\alpha_{i+1} = \beta_i$. (Prover meets the challenge of the preceding step.)

(iii) $\Sigma_{i+1} \to \alpha_{i+1} \in \Gamma \cup \Pi_1 \cup \ldots \cup \Pi_i$. (Prover uses an available offer.)[4]

(iv) $\Pi_i \to \beta_i \in \Sigma_i$. (Skeptic challenges a formula from the preceding step.)

A *winning prover strategy for* (Γ, φ) (or just "for φ" when Γ is empty) is a labeled tree where:

- Every branch is a dialogue over (Γ, φ).

- Every node labeled with a prover step with n assertions has n children labeled with distinct steps.

- Every node labeled with a skeptic step has one child.

- All branches are finite and end in a leaf labeled with a prover step with no assertions.

A prover step (Σ, α) means meeting challenge α by introducing new assertions Σ. A skeptic step (Π, β) means a challenge to prove β with permission to use offers Π. An assertion (e.g. the initial formula) may be atomic; in this case, the prover will be challenged to prove it without any additional offers.

4.6.2. THEOREM. *There is a winning prover strategy for φ iff φ is a theorem in minimal implicational logic*

---

[3]The choice $\alpha_1 = \top$ in this case is an arbitrary convention.
[4]$(\psi_1, \ldots, \psi_n) \to \varphi$ means $\psi_1 \to \cdots \to \psi_n \to \varphi$. For $n = 0$, this boils down to just φ.

4.6. PROVER-SKEPTIC DIALOGUES

PROOF. For the direction from left to right, let D be a winning prover strategy for φ. For each skeptic node (Π, β), we consider the accumulated set of offers Γ, i.e. the union of all offers made in the dialogue from the root to this node. For any subtree D' with root (Π, β) that has accumulated offers Γ, we now show that $\Gamma \vdash \beta$ in minimal implicational logic. The form of D' must be ($n \geq 0$):

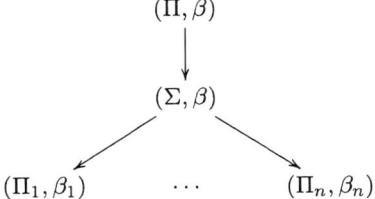

Each child has accumulated offers $\Gamma \cup \Pi_i$ (viewing Π_i as the set of its elements). By the induction hypothesis $\Gamma, \Pi_i \vdash \beta_i$, so $\Gamma \vdash \Pi_i \to \beta_i$. Since $\Sigma = \Pi_1 \to \beta_1, \ldots, \Pi_n \to \beta_n$, also $\Gamma \vdash (\Pi_1 \to \beta_1) \to \cdots \to (\Pi_n \to \beta_n) \to \beta$ (an equivalent offer must have been stated) so $\Gamma \vdash \beta$ by *modus ponens*.

In particular, D has root label $((\varphi), \mathsf{T})$ and single child labeled (Π, β), where $\varphi = \Pi \to \beta$, so $\Pi \vdash \beta$, i.e. $\vdash \varphi$.

For the opposite direction, assume $\vdash \varphi$. There is a closed λ-term of type φ. By normalization and Lemma 4.2.2 we can assume that the term is in η-long normal form, so it suffices to show that if

$$x_1 : \psi_1, \ldots, x_n : \psi_n \vdash M : \varphi$$

for η-long normal form M, then there is a winning prover strategy for $(\{\psi_1, \ldots, \psi_n\}, \varphi)$. We proceed by induction on the size of M. Let

$$M = \lambda x_{n+1}^{\psi_{n+1}} \ldots x_{n+m}^{\psi_{n+m}} . x_k \, N_1 \ldots N_l,$$

where $1 \leq k \leq n + m$. Then

$$x_1 : \psi_1, \ldots, x_{n+m} : \psi_{n+m} \vdash N_i : \rho_i$$

for each i, where $\psi_k = \rho_1 \to \ldots \rho_l \to \alpha$ and $\varphi = \psi_{n+1} \to \ldots \to \psi_{n+m} \to \alpha$. By the induction hypothesis, there is a winning prover strategy for each $(\{\psi_1, \ldots, \psi_{n+m}\}, \rho_i)$. It must have form

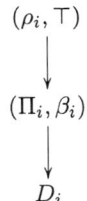

Then we have the following strategy for $(\{\psi_1, \ldots, \psi_n\}, \varphi)$:

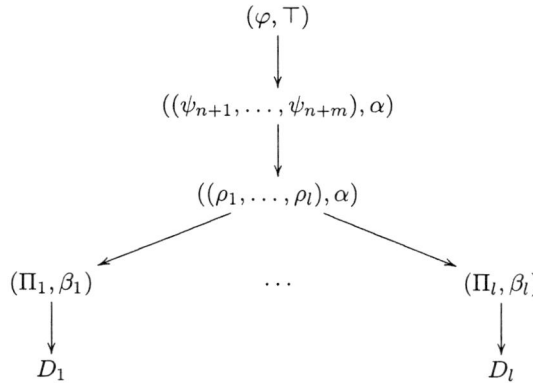

This concludes the proof. □

In the direction from left to right we could construct λ-terms of type β rather than prove $\Gamma \vdash \beta$, and it is not difficult to see that these are actually η-long normal forms. Thus we can state the following identification:

$$\text{Winning prover strategy} \quad \sim \quad \eta\text{-long normal form.}$$

More precisely, in a term

$$\lambda x_1^{\psi_1} \ldots x_n^{\psi_n}.\, x_i\, M_1 \ldots M_m,$$

the abstractions correspond to offers made by the skeptic, and the type of $x_i M_1 \ldots M_m$ is the challenge by the skeptic. The variable x_i represents the prover's choice of offer, and the terms $M_1, \ldots, M_m$ represent the continued dialogue in each branch of the winning prover strategy.

4.7. Prover-skeptic dialogues with absurdity

What was said above concerns minimal logic. What happens when we add absurdity and the rule *ex falso sequitur quodlibet*? The BHK interpretation states that "there is no construction of ⊥." If the skeptic actually makes ⊥ as an offer, or makes an offer from which ⊥ can be inferred, he is creating false assumptions. In this case the prover should be permitted to abort the current challenge—the skeptic is not playing by the book. The prover can provoke such circumstances by making assertions that will force the skeptic to provide contradictory information, as in the following example.

4.7. PROVER-SKEPTIC DIALOGUES WITH ABSURDITY

Prover 1: I assert that $p \to \neg p \to q$ holds!

Skeptic 1: Really? Suppose I give you constructions of p and $\neg p$, can you give me a construction of q?

Prover 2: I do not believe you can provide such constructions. By applying your second offer to a construction of p, I get absurdity, and that's impossible!

Skeptic 2: Sure, but I doubt you have a construction of p.

Prover 3: I can just use the one you gave me earlier on. I win!

4.7.1. DEFINITION. Dialogues and strategies are as in Definition 4.6.1 except that α, β now range over all atomic propositions, i.e. not only propositional variables, but also $\bot$. (This will be the convention for the rest of this section.) Moreover, condition (ii) is replaced by:

(ii) $\alpha_{i+1} = \beta_i$ (prover meets challenge of preceding step) or $\alpha_{i+1} = \bot$ (prover aborts preceding challenge).

Note that if the prover has to make new assertions to derive absurdity, then the skeptic will challenge these, so it is not the dialogue as such that is aborted, only the latest challenge.

In order to represent strategies as terms, we use the extended simply typed λ-calculus from Section 4.5 without pairs and sums, but with ε, here denoted $\lambda\varepsilon$-calculus. The following reduction rules are needed.

4.7.2. DEFINITION. The relation $\to_\varepsilon$ is the smallest compatible relation containing the following rules:

$$\varepsilon_{\varphi \to \psi}(M)N \to_\varepsilon \varepsilon_\psi(M) \qquad (\varepsilon_1)$$

$$\varepsilon_\psi(\varepsilon_\bot(M)) \to_\varepsilon \varepsilon_\psi(M) \qquad (\varepsilon_2)$$

The term $\varepsilon_\varphi(M)$ expresses the fact that the prover has inferred absurdity from a skeptic offer, and that she is aborting the attempt to meet the skeptic's challenge. We can view this as though the prover has produced an imaginary construction of φ. The reduction rule (ε_1) captures the fact that rather than producing an imaginary construction of an implication $\varphi \to \psi$ and then applying this construction to a construction of φ to obtain a construction of ψ, we might as well "abort" the argument and produce an imaginary construction of ψ directly.

In order to relate the revised prover-skeptic games to provability in intuitionistic, implicational logic we need some preparations.

4.7.3. THEOREM. *The reduction $\to_{\beta\varepsilon}$ is normalizing.*

PROOF. First reduce a term to β-normal form. From the point of view of β-reductions, there is no difference between $\varepsilon_\varphi(M)$ and $x^{\bot \to \varphi}M$, where x is a fresh variable, so by the normalization theorem, such a β-normal form exists. Then reduce to ε-normal form, which is possible since each reduction decreases the term size. Finally note that ε-reduction steps cannot create β-redexes, so the resulting term is still in β-normal form. □

4.7.4. DEFINITION. The set of η-long normal forms of $\lambda\varepsilon$-calculus is the smallest set closed under the following rules (where $m \geq 0$):

- If y has type $\tau_1 \to \cdots \to \tau_n \to \alpha$ and M_i is an η-long normal form of type τ_i, then the following term is also an η-long normal form:

$$\lambda y_1^{\sigma_1} \ldots y_m^{\sigma_m}. y M_1 \ldots M_n.$$

- If y has type $\tau_1 \to \cdots \to \tau_n \to \bot$ and M_i is an η-long normal form of type τ_i, then the following term is also an η-long normal form:

$$\lambda y_1^{\sigma_1} \ldots y_m^{\sigma_m}. \varepsilon_\alpha(y M_1 \ldots M_n).$$

4.7.5. LEMMA. *For every simply typed $\lambda\varepsilon$-term, there is another simply typed $\lambda\varepsilon$-term of the same type in η-long normal form.*

PROOF. Given a term of type $\varphi = \sigma_1 \to \cdots \to \sigma_m \to \alpha$, we proceed by induction on the term. We can assume it is in $\beta\varepsilon$-normal form, i.e.

$$\lambda y_1^{\sigma_1} \ldots y_k^{\sigma_k}. \varepsilon_\psi(y M_1 \ldots M_n) \quad \text{or} \quad \lambda y_1^{\sigma_1} \ldots y_k^{\sigma_k}. y M_1 \ldots M_n,$$

where $k \leq m$ and $\psi = \sigma_{k+1} \to \cdots \to \sigma_m \to \alpha$. By the induction hypothesis, we can let N_i be an η-long normal form of M_i. Then the desired term is, respectively:

$$\lambda y_1^{\sigma_1} \ldots y_m^{\sigma_m}. \varepsilon_\alpha(y N_1 \ldots N_n) \quad \text{or} \quad \lambda y_1^{\sigma_1} \ldots y_m^{\sigma_m}. y N_1 \ldots N_n \, y_{k+1} \ldots y_m. \quad \square$$

4.7.6. THEOREM. *There is a winning prover strategy for φ iff φ is a theorem in intuitionistic implicational logic.*

PROOF. Similar to the proof of Theorem 4.6.2. □

4.8. Notes

Surprisingly, the ancient Greeks probably did not know about the Curry-Howard isomorphism. But some authors trace the roots of the idea back to Schönfinkel, Brouwer, Heyting, or even Frege and Peano.

Although it is hard to deny that the proof interpretation of Brouwer-Heyting-Kolmogorov is fundamental for the propositions-as-types correspondence, we would not go so far as to claim that the BHK interpretation and Curry-Howard isomorphism are the same. The formulas-as-types paradigm is one possible way of understanding BHK. And e.g. realizability (see Chapter 9) can be seen as another.

4.8. NOTES

FORMULAS-AS-TYPES. To our knowledge, the first explicit statement that may be considered the beginning of the subject occurs in Curry's [102], as footnote 28:

Note the similarity of the postulates for F and those for P. If in any of the former postulates we change F to P and drop the combinator we have the corresponding postulate for P.

Here P stands for implication and F for a "functionality" combinator, such that $FABf$ essentially means as much as $f : A \to B$. In fact, Curry must have realized the similarity several years before. As pointed out by Hindley [237], a remark on page 588 of [100] is a hint of that. In the same paper, and also in [101], there is another hint. Properties of $\supset$ are named PB, PC, PW and PK, after the combinators B, C, W and K. From [439] we know that Curry used this notation as early as 1930. Years later, Carew Meredith independently observed the analogy between proofs and combinators, cf. [237, 339].

The correspondence was made precise (for typed combinatory logic) by Curry and Feys in [107, Chapter 9] by means of two theorems. Theorem 1 (p. 313) states that inhabited types are provable, and Theorem 2 (p. 314) states the converse.

Before Howard, various authors, e.g. Kreisel, Goodman, and Scott, contributed to the understanding of the relationship between formulas and types. In the paper of Läuchli [298], preceded by the quarter-page abstract [297], formulas are assigned sets and provability corresponds to non-emptiness. Stenlund [459], who probably first used the word "isomorphism", points out the inspiration coming from the *Dialectica* interpretation, so that also Gödel should be counted as one of the fathers of the paradigm. However, all these works are relevant to only one aspect of what we call Curry-Howard isomorphism: the essentially "static" correspondence between provable theorems and inhabited types.

REDUCTION AS COMPUTATION. The other important aspect is the relationship between proof normalization and computation (term reduction). The paper often credited for the discovery of this is Tait's [466]. For instance, Howard himself declares in [247] that he was inspired by Tait. Instead of *proving* that an expression of type "number" is equal to a certain number, Tait *reduces* the expression to normal form. Thus, in a sense, term reduction is used instead of proof normalization. But the statement of the propositions-as-types paradigm can hardly be considered explicit in [466]. Also Curry and Feys, who used cut-elimination in [107] to prove the normalization theorem for simply typed terms, seem not to pay real attention to the relationship between proof reduction and computation. An explicit statement of the "dynamic" aspect of the isomorphism did not occur in the literature until 1969, when the famous paper of Howard began to be "privately circulated".

Howard did not publish his manuscript [247] until 1980 when it became part of Curry's Festschrift [441]. By then, the paper was already well-known, and the idea of formulas-as-types had gained considerable popularity. (As early as 1969, Martin-Löf applied Howard's approach in a strong normalization proof.)

The dozen pages of [247] explain the fundamentals of the proofs-as-terms paradigm on the level of implicational logic, with explicit emphasis on the relationship between normalization of terms and normalization of proofs. The approach is then extended to other propositional connnectives, and finally a term language for Heyting Arithmetic (first-order intuitionistic arithmetic) is introduced and discussed.

PROOFS AS OBJECTS. While Curry and Howard are given fair credit for their discovery, the contribution of Nicolaas Govert de Bruijn is not always properly recognized. De Bruijn not only made an independent statement of the correspondence between proofs and objects, but he was also the first to make practical use of it. His "mathematical language" Automath [58, 60, 57, 360], developed in Eindhoven from 1967 onward (at the same time when Howard's paper was written), was a tool for writing mathematical proofs in a precise way so that they could be verified by a computer. Automath became an inspiration for the whole area of logical frameworks and proof assistants. Many fundamental concepts were for the first time used in Automath. Dependent types, with implication defined as a special case of universal quantifier is an outstanding example. The idea of proofs as objects is exploited from the very beginning. Propositions are represented as sets ("categories") of their proofs, and the question of provability becomes the question of non-emptiness.

Scott appreciated the importance of de Bruijn's work very early. The influential paper [437], which appeared in the same volume as [58], applied the concepts and ideas from Automath (designed for classical proof encoding) in an intuitionistic "calculus of constructions".

In the early 1970's also Per Martin-Löf began his effort to develop a constructive type theory, aimed at providing a foundational basis for constructive mathematics. After overcoming some initial difficulties (the first version was inconsistent—this is now called *Girard's paradox*) Martin-Löf's type theory [325, 326, 327] turned out to be very successful. It is often said to be as important for constructivism as axiomatic set theory is for classical mathematics. In particular, Martin-Löf's work was essential for our present understanding of the Curry-Howard isomorphism as the *identification* between propositions and types. Not so unexpectedly, Martin-Löf's theory can be seen as a prototypic programming language [328, 365, 366] and a programming logic in one.

As mentioned above, Automath preceded many systems introduced later with the aim of mechanizing the process of proof design and proof verification. The design of such systems is often more or less related to the "proofs-as-objects" paradigm, see [33]. Among these, let us name a few.

The acronym LCF (for "Logic of Computable Functions") denotes several versions of a system originating in 1972 from a proof-checker designed by Milner. Currently, the most important descendants of LCF are system HOL (for "Higher-Order Logic") and Isabelle, see [197, 364].

Another line of research, inspired by Martin-Löf's type theory, was undertaken by Constable and his co-workers also in the early 1970's. The "Program Refinement Logic" (or "Proof Refinement Logic") PRL gave rise to the system Nuprl, currently one of the most well-known and actively evolving systems for proof development and software verification [81, 82].

The systems of the 1970's can all be recognized as inspiration for the Formel project initiated in France at the beginning of the 1980's. An important theoretical tool in this research was the Calculus of Constructions proposed by Coquand and Huet in mid 1980's and then developed and extended by Luo, Paulin-Mohring and others. The Calculus of Constructions, to which we return in Chapter 14, provided the formalism on which the proof-assistants Coq [44] and Lego are based.

PROOFS INTO PROGRAMS. *Witness extraction* is the problem of extracting from a proof of $\exists x P(x)$ a specific value a such that $P(a)$ holds. Similarly, one can consider the problem of extracting from a proof of $\forall x \exists y P(x,y)$ a function f such that $\forall x P(x, f(x))$; this is called *program extraction*. For instance, a constructive proof of a specification, say of the form

$$\forall x \ \exists y. \ ordered(y) \wedge permutation(x,y)$$

should implicitly contain a sorting algorithm. The idea that a proof of a specification might be turned into a program fulfilling this specification attracted researchers in theoretical computer science already in the 1960's, and various techniques were applied in this direction, see e.g. [321]. Witness extraction methods based on resolution proofs developed in the late 1960's contributed to the birth of what is now called logic programming. For instance, Constable [79] and then Marini and Miglioli [322] discussed the idea of extracting an algorithm from a proof with the help of Kleene's realizability approach. Constable's PRL and Nuprl were partly motivated by this direction of work. A few years later, Goto and Sato investigated a program extraction procedure based on the *Dialectica* interpretation (see Chapter 10). The Japanese school provided the background on which Hayashi invented the system PX [217].

The Curry-Howard isomorphism identifies a proof (of a specification) and the program (meeting that specification). Thus the program extraction process simply amounts to deleting all "computationally irrelevant" contents from the proof. In Chapter 12 we will see an example of a direct program extraction procedure in the proof of Girard's result: If $\forall x \exists y P(x,y)$ is provable in second-order arithmetic, then a function f satisfying $\forall x P(x, f(x))$ is definable in the polymorphic λ-calculus.

Issues related to Curry-Howard style program extraction occur in the literature since mid 1970's. See [25, 43, 80, 191, 280, 288, 382, 386] as a small sample of that category. Although the issue of practical software development based on these ideas is still a matter of the future, the research goes on (see e.g. the system MINLOG of [430]) and the last word is certainly not said yet in this area.

MISCELLANEA. To make a few more specific comments, the decidability of propositional intuitionistic logic was already known to Gentzen in 1935 and we will discuss his (syntactic) proof in Chapter 7. The algorithm we used in the proof of Lemma 4.2.3, implicit in in Wajsberg's [505], is often referred to as the *Ben-Yelles algorithm*, for Choukri-Bey Ben-Yelles, who defined a procedure to *count* the number of inhabitants of a type (see [237]). A polynomial space algorithm (based on a Kripke model construction) was also given by Ladner [292], and the PSPACE-completeness is from Statman [455]. There is also a semantical PSPACE-hardness proof by Švejdar [463].

Normalization for natural deduction proofs with all the standard propositional connectives was first shown by Prawitz [403], and variants of this proof occur in newer books like [113, 489]. Strong normalization proofs can be found e.g. in [398] and [160], and for especially short ones the reader is referred to [122, 255]. The latter paper only covers the case of implication and disjunction, but that is quite enough, see Exercise 4.16.

We are concerned in this book with the correspondence between logics and lambda-calculi. However, there is another correspondence: between logics and dialogue games. The "dialogical" approach to logic originates with Lorenzen [315], and works by Lorenz [314], Felscher [147] and others contributed to the subject (see notes to Chapters 6 and 7). However, the games considered in this chapter are somewhat different than those of Lorenzen. We return to the latter type in Chapter 7. There are various other game-related approaches to logic, of which we only mention Girard's new concept of *ludics* [186, 187].

If we look at the formulas-as-types paradigm from the point of view of categorical semantics, we discover yet another side of the same correspondence. Objects and morphisms correspond to types and terms, and also to formulas and proofs.

This analogy was first discovered by Lawvere [300] and Lambek. See [18, 294] for more on that subject.

Various other forms of the correspondence have been discovered in recent years, and many old ideas turned out to be relevant. For instance, Ajdukiewicz's and Lambek's grammars, used in the analysis of natural languages [40, 64, 293, 354], have a close relationship to type assignment, as first observed by Curry [107, p. 274]. Papers [16, 123, 127, 128, 287, 372, 373, 497] discuss various other faces of the Curry-Howard correspondence, related to modal logic, illative combinatory logic, intersection types, low-level programming, etc. etc. Among these, linear logic (see the notes to Chapter 5) takes a distinguished place.

The historical article [247], as well as the crucial section 9E of [107] are reprinted in the collection [206]. Other articles included in [206] cover various aspects of the Curry-Howard correspondence. Hindley's book [237] contains a detailed study of the paradigm on the simply typed level. In particular, the exact correspondence between closed terms and assumption-free proofs (cf. Exercise 4.8) is explicitly formulated there as Theorem 6B7(iii). The reader will enjoy the explanation in [22] on how the isomorphism extends to various data types and how the "inversion principle" works on these. For a further and deeper study of the Curry-Howard interpretation of inductive types and families, see [139, 140]. A textbook presentation of the Curry-Howard isomorphism can be found in [189], and a recent monograph [444] takes a serious look at the subject. Among various survey articles of the Curry-Howard isomorphism we recommend [85, 161, 238, 487, 504].

4.9. Exercises

4.1. (From [510].) Are the following formulas intuitionistically valid?

  (i) $((p \to q) \to r) \to (p \to r) \to r$?

  (ii) $((((p \to q) \to r) \to (p \to r) \to r) \to q) \to q$?

Hint: Are these types inhabited?

4.2. Show that **K** and **S** are the only closed normal forms of types $p \to q \to p$ and $(p \to q \to r) \to (p \to q) \to p \to r$, respectively.

4.3. For $n \in \mathbb{N} \cup \{\aleph_0\}$, show examples of types with exactly n normal inhabitants.

4.4. Run the Ben-Yelles algorithm on the inputs:

  (i) $(p \to p) \to p$;

  (ii) $((p \to q) \to q) \to (p \to q)$.

4.5. (Ben-Yelles) Show that it is decidable whether a type has a finite or an infinite number of different normal inhabitants. *Hint:* Modify the proof of Lemma 4.2.3.

4.6. Observe that the proof of Lemma 4.2.3 makes it possible to actually construct an inhabitant of a given type if it exists. Show that the size of this inhabitant is at most exponential in the size of the given type. Then show that the exponential bound is best possible. That is, for every n define a type τ_n of length $\mathcal{O}(n)$ such that every normal inhabitant of τ_n is of length at least of order 2^n. (If you find that too easy, delete the word "normal".)

4.7. Recall that the implication introduction rule in our natural deduction system

$$\frac{\Gamma, \varphi \vdash \psi}{\Gamma \vdash \varphi \to \psi} \ (\text{I} \to)$$

is understood so that Γ, φ abbreviates $\Gamma \cup \{\varphi\}$. It may be the case that $\varphi \in \Gamma$. In the "traditional" proof notation this is expressed by the possibility of not discharging the assumption φ. What will change if we instead choose the *complete discharge convention:* When introducing an implication $\varphi \to \psi$, discharge *all* occurrences of the assumption φ?

4.8. Define precisely a natural deduction calculus with labeled discharges (in the style of Section 4.3). Establish a bijective correspondence between closed λ-terms and proofs in your system without free assumptions.

4.9. Let $\varphi = \tau_1 \to \cdots \to \tau_n \to p$ be an implicational formula and let p be the only type variable occurring in φ. Prove that φ is an inhabited type if and only if at least one of the τ_i's is *not* inhabited.

4.10. (Statman) Let p be the only propositional variable, and let $\to$ be the only connective occurring in a classical propositional tautology φ. Show that φ is intuitionistically valid.

4.11. (Prucnal, Dekkers [126]) A proof rule of the form

$$\frac{\tau_1, \ldots, \tau_n}{\tau}$$

is *sound* for IPC($\to$) iff for every substitution S with $S(\tau_1), \ldots S(\tau_n)$ all valid, also $S(\tau)$ must be valid. Prove that if such a rule is sound then the implication $\tau_1 \to \cdots \to \tau_n \to \tau$ is valid.

4.12. Write an inhabitant of type $(p \to \neg p) \to (\neg p \to p) \to \bot$.

4.13. Find inhabitants for the formulas in Example 2.1.3 in the extended calculus of Section 4.5.

4.14. Consider again Exercises 2.21–2.24. Give inhabitants for all the tautologies.

4.15. Prove that the simply typed lambda-calculus extended with product types, pairs and projections (no disjunction types or $\bot$) is strongly normalizing.

4.16. Assume that the simply typed lambda-calculus extended with disjunction types, injections and **case**-statements has the strong normalization property. Prove the SN property for the whole system of Section 4.5 (Theorem 4.5.4).

4.17. How should the notion of η-reduction be generalized for sums and products?

4.18. Design a Church-style calculus with $\wedge$ and $\vee$ and show that subject reduction property holds for that calculus for both beta- and eta-reductions

4.19. Define a Curry-style variant of lambda-calculus with $\wedge$ and show that the subject reduction property does not hold for η-reductions.

4.20. Explain the difference between the above two results.

Chapter 5

Proofs as combinators

In the preceding chapters we have been concerned with the Curry-Howard isomorphism in the shape of a correspondence between natural deduction proofs and terms of λ-calculus. As we have seen, the isomorphism connects many concepts from the two worlds. In this chapter we present a related correspondence between two other fundamental concepts: *Hilbert style* proofs and terms of *combinatory logic*.

One rather inconvenient aspect of λ-calculi is the role played by bound variables, especially in connection with substitution. Similar problems always occur when one deals with the notion of free and bound variables. In order to avoid these problems, Schönfinkel and Curry introduced combinatory logic, a system without bound variables, that is equivalent to λ-calculus in many respects. Instead of using λ-abstraction, this calculus deals only with applications of constants.

From the standpoint of the Curry-Howard isomorphism, this change corresponds to rejecting the implication introduction rule, and replacing it by axioms. The result is a Hilbert style proof system, the most traditional way to express the notion of a formal proof.

The fact, discovered by Schönfinkel and Curry, that combinatory logic is capable of expressing λ-abstraction, translates directly to a very basic property of Hilbert style provability—the *deduction theorem*, discovered by Herbrand. Thus, the Curry-Howard isomorphism again makes a bridge between two independently developed mathematical worlds.

5.1. Hilbert style proofs

Natural deduction was not historically the first approach to the notion of a formal proof, nor it is generally the most commonly used one. Many classical presentations of mathematical logic, like [337, 352, 410], used the more traditional notion of a Hilbert style proof. Usually a Hilbert style proof system consists of a set of formulas, called *logical axioms*, and only a few proof rules. Moreover, rather than tree-structured derivations, formal proofs in Hilbert style systems are traditionally defined as sequences of formulas.

The systems we consider in the present chapter (like most systems for propositonal logics) have only one proof rule, called *modus ponens* or the *detachment rule*:
$$\frac{\varphi \quad \varphi \to \psi}{\psi}$$
Thus, a Hilbert style proof system can be identified with its set of logical axioms. The following definition applies to all such propositional systems.

5.1.1. DEFINITION. A *proof sequence* (or a *Hilbert style proof*) of a formula φ is a finite sequence of formulas $\psi_1, \psi_2, \ldots, \psi_n$, such that $\psi_n = \varphi$, and such that for all $i = 1, \ldots, n$,

- either ψ_i is an axiom, or
- there are $j, \ell < i$ such that $\psi_j = \psi_\ell \to \psi_i$ (i.e. ψ_i is obtained from ψ_j and ψ_ℓ using *modus ponens*).

If such a proof exists, the formula φ is called a *theorem*.

5.1.2. DEFINITION. Our system for the implicational fragment of propositional intuitionistic logic is given by two *axiom schemes*. That is, all formulas that fit the patterns below are considered to be logical axioms:

(A1) $\varphi \to \psi \to \varphi$;

(A2) $(\varphi \to \psi \to \vartheta) \to (\varphi \to \psi) \to \varphi \to \vartheta$.

Note that there are in fact infinitely many axioms. This does not bother us as long as it can be effectively decided whether or not a formula is an axiom.

5.1.3. EXAMPLE. Take any formula φ. Here is a proof of $\varphi \to \varphi$:

1. $(\varphi \to (\psi \to \varphi) \to \varphi) \to (\varphi \to \psi \to \varphi) \to \varphi \to \varphi$ (axiom A2);

2. $\varphi \to (\psi \to \varphi) \to \varphi$ (axiom A1);

3. $(\varphi \to \psi \to \varphi) \to \varphi \to \varphi$ (*modus ponens*: detach 2 from 1);

4. $\varphi \to \psi \to \varphi$ (axiom A1);

5. $\varphi \to \varphi$ (*modus ponens*: detach 4 from 3).

In order to use our proof system to derive a formula from some additional assumptions, e.g. to derive a propositional variable q from p and $p \to q$, we must include the assumptions in the set of axioms. Of course, p and $p \to q$ are not axiom *schemes*, and should not be considered logical axioms. Such additonal assumptions are sometimes called *non-logical axioms*.

5.1. HILBERT STYLE PROOFS

5.1.4. EXAMPLE. Here is a proof of ϑ from the set of non-logical axioms $\{\varphi, \varphi \to \psi, \psi \to \vartheta\}$:

1. $\varphi \to \psi$ (non-logical axiom);

2. φ (non-logical axiom);

3. ψ (*modus ponens*: detach 2 from 1);

4. $\psi \to \vartheta$ (non-logical axiom);

5. ϑ (*modus ponens*: detach 3 from 4).

The definition of a proof as a sequence of formulas is simple and mathematically elegant. Because of this simplicity it is often convenient when provability of formulas is the issue. However, if one is concerned with proofs themselves, the sequence model becomes less adequate.

Indeed, as the reader might have noticed, our examples of Hilbert style proofs are annotated. Each formula is accompanied with a justification for its occurrence in the proof. Without annotations, a proof sequence becomes ambiguous. Indeed, there may be more than one way to justify an occurrence of a formula in a proof sequence. For instance, a formula can be seen as a result of applying *modus ponens* in two or more different ways. Also, an axiom occurring in a proof sequence may at the same time be considered a legitimate result of applying *modus ponens*.

Another inconvenience of the formalization of proofs as sequences is that a proof sequence may contain redundant formulas, i.e. intermediate formulas which are never used (as a premise of *modus ponens*).

Note that the two problems mentioned above are overlapping. A formula which is seen as redundant under one interpretation of a proof sequence is not necessarily so under another interpretation, see Exercise 5.2.

As proofs, not theorems, are our main concern, we prefer to consider a formalization which is free of both ambiguity and redundancy. A formal definition of an annotated sequence might be an option, but we prefer to make the underlying tree structure of proofs explicit. From this point of view, *modus ponens* can be seen as a variant of the implication elimination rule ($\to$E) of Figure 2.1.

5.1.5. DEFINITION. We write $\Gamma \vdash_H \varphi$ iff the judgement "$\Gamma \vdash \varphi$" is derivable using the rules in Figure 5.1. The appropriate labeled tree is then a formal proof of the judgement.

5.1.6. EXAMPLE. Here is a proof of the judgement $\varphi, \varphi \to \psi, \psi \to \vartheta \vdash_H \vartheta$ (cf. Example 5.1.4):

$$
\begin{array}{ll}
\text{(Id)} & \Gamma, \varphi \vdash \varphi \\[1em]
\text{(Ax)} & \Gamma \vdash \varphi, \text{ if } \varphi \text{ is a logical axiom} \\[1em]
\text{(MP)} & \dfrac{\Gamma_1 \vdash \alpha \quad \Gamma_2 \vdash \alpha \to \beta}{\Gamma_1, \Gamma_2 \vdash \beta}
\end{array}
$$

FIGURE 5.1: TREE VARIANT OF HILBERT STYLE SYSTEM

$$
\dfrac{\dfrac{\varphi \vdash \varphi, \quad \varphi \to \psi \vdash \varphi \to \psi}{\varphi, \varphi \to \psi \vdash \psi} \quad \psi \to \vartheta \vdash \psi \to \vartheta}{\varphi, \varphi \to \psi, \psi \to \vartheta \vdash \vartheta}
$$

The reader may wonder why we have chosen to present the *modus ponens* rule in a "multiplicative" form, rather than as the ordinary arrow-elimination:

$$
\text{(MP')} \quad \dfrac{\Gamma \vdash \alpha \quad \Gamma \vdash \alpha \to \beta}{\Gamma \vdash \beta}
$$

The reason is that the "multiplicative" variant may be more informative, if we wish to exhibit the exact set of formulas in Γ that were used to prove each of the two premises. This formulation will be useful in Section 5.6. Of course, as long as ordinary intuitionistic logic is concerned, the choice of the rule is not essential. The reader may easily check that the two rules are equivalent in terms of provable judgements.

5.1.7. PROPOSITION. *A formula φ has a Hilbert style proof from a set of non-logical axioms Γ if and only if $\Gamma \vdash_H \varphi$.*

PROOF. Left to the reader. □

In conclusion, Hilbert style differs from natural deduction mainly in that implication introduction is replaced by the logical axioms. Moreover, the following theorem states that implication introduction can be regained.

5.1.8. THEOREM (Deduction theorem). $\Gamma, \varphi \vdash_H \psi$ *iff* $\Gamma \vdash_H \varphi \to \psi$.

PROOF. The proof from right to left is easy. If $\Gamma \vdash_H \varphi \to \psi$, then of course $\Gamma, \varphi \vdash_H \varphi \to \psi$. Since we also have a proof of $\Gamma, \varphi \vdash_H \varphi$, an application of (MP) gives a proof of $\Gamma, \varphi \vdash_H \psi$.

5.1. HILBERT STYLE PROOFS

For the direction from left to right, we proceed by induction with respect to the size of the proof of $\Gamma, \varphi \vdash_H \psi$. There are three cases.

First of all, it may happen that the assumption φ is not used in the proof. Then we are lucky, because in fact we have $\Gamma \vdash_H \psi$. In this case we obtain $\Gamma \vdash_H \varphi \to \psi$ by detaching ψ from the axiom (A1) in the form $\psi \to \varphi \to \psi$.

The second case is when $\psi = \varphi$ and the proof consists of a single application of rule (Id). Now we are lucky too, because we have already worked out Example 5.1.3, and we know that $\Gamma \vdash_H \varphi \to \varphi$.

Finally, assume that ψ was obtained by an application of *modus ponens*. This means that, for a certain formula ϑ, we have $\Gamma, \varphi \vdash_H \vartheta \to \psi$ and $\Gamma, \varphi \vdash_H \vartheta$, and the corresponding proofs are simpler. From the induction hypothesis, we have both $\Gamma \vdash_H \varphi \to \vartheta \to \psi$ and $\Gamma \vdash_H \varphi \to \vartheta$. To obtain a proof of $\Gamma \vdash \varphi \to \psi$, we must apply (MP) twice to the axiom (A2) in the form $(\varphi \to (\vartheta \to \psi)) \to (\varphi \to \vartheta) \to \varphi \to \psi$. □

It is often much easier to find out that a given formula is provable by applying the deduction theorem, than by writing an actual proof. For instance, it is readily seen from Example 5.1.4 that $(\varphi \to \psi) \to (\psi \to \vartheta) \to \varphi \to \vartheta$ is provable. (Try Exercise 5.1 for comparison.)

5.1.9. PROPOSITION. $\Gamma \vdash_N \varphi$ iff $\Gamma \vdash_H \varphi$.

PROOF. The right to left part is an easy induction. The converse is also easy, using the deduction theorem and conservativity of NJ over NJ($\to$) (Theorem 2.6.2). □

We conclude that our Hilbert style system is complete.

5.1.10. THEOREM. $\Gamma \vdash_H \varphi$ iff $\Gamma \models \varphi$.

PROOF. Immediate from Proposition 5.1.9 and the completeness of the implicational fragment of natural deduction. □

By adding new logical axioms to handle connectives other than implication, one obtains a complete Hilbert style proof system for the full propositional intuitionistic logic. One possible choice of axioms is as follows.

(A3) $\bot \to \varphi$;

(D1) $\varphi \to \varphi \vee \psi$;

(D2) $\psi \to \varphi \vee \psi$;

(D3) $(\varphi \to \vartheta) \to (\psi \to \vartheta) \to \varphi \vee \psi \to \vartheta$;

(C1) $\varphi \wedge \psi \to \varphi$;

(C2) $\varphi \wedge \psi \to \psi$;

(C3) $(\vartheta \to \varphi) \to (\vartheta \to \psi) \to \vartheta \to \varphi \wedge \psi$.

Note that we have not included axioms for negation. This is because we only use $\neg \alpha$ as an abbreviation for $\alpha \to \bot$.

Also note that the system discussed above adds new axioms to the basic system, but no new rules. The usefulness of Hilbert style systems is exactly this: the properties of logics or theories are expressed in terms of lists of axioms, which are often more concise and comprehensible than proof rules.

5.1.11. REMARK. The reader has learned from the previous chapters that the Curry-Howard isomorphism is about identifying proofs with algorithms. This applies to various notions of *formal proofs* (Hilbert style, natural deduction, etc.), but our book also contains *meta-level proofs*, written in English in order to convince the reader about various facts.

Perhaps we should be prepared to "taste our own medicine" and see if, also at the meta-level, proofs may describe computations. It is not always the case that they do, as not all our proofs are constructive. But often the computational contents of a proof is quite evident. Let us take the proof of the deduction theorem as an example. The "hard" part is the implication:

If $\Gamma, \varphi \vdash_H \psi$ *then* $\Gamma \vdash_H \varphi \to \psi$.

According to the BHK interpretation (see Chapter 2), a proof (construction) of an implication $\sigma \to \tau$ is a method for turning constructions of σ into constructions of τ. If we agree for simplicity, that a construction of a judgement $\Gamma \vdash_H \phi$ is simply a Hilbert-style derivation, then a proof of the deduction theorem does indeed contain such a method fairly explicitly. Given a proof of ψ from Γ, φ we actually *construct* a new proof of $\varphi \to \psi$ from Γ.

This is also the case for many other proofs. We encourage the reader to unravel the algorithms implicit in the meta-level proofs in the rest of this book, whenever possible. (A good place to start with is Proposition 5.1.9.)

5.2. Combinatory logic

What should be the Curry-Howard equivalent of the Hilbert style? Of course, *modus ponens* corresponds to application. Furthermore, the logical axioms can be represented by typed constants, e.g.

$\vdash A1 : \varphi \to \psi \to \varphi$.

That is, terms are *combined* from basic constants by means of applications, and that is why we call them *combinators*. The calculus of combinators is traditionally called *combinatory logic*.

5.2. COMBINATORY LOGIC

As mentioned in Chapter 1, the original systems of combinatory logic and λ-calculus that aimed at providing a foundation of mathematics and logic turned out to be inconsistent, due to the presence of arbitrary fixed-points. Nevertheless, one may distinguish a useful subsystem of the original system of combinators dealing only with pure functions, and this is the system that we call *combinatory logic* (CL) below.

As with λ-calculus, combinatory logic comes in both typed and untyped variants. In the rest of this section we shall be concerned only with the latter. The objects of study in combinatory logic are the *combinatory terms*.

5.2.1. DEFINITION. The set $\mathcal{C}$ of *combinatory terms* is defined as follows:

- All object-variables are in $\mathcal{C}$, that is, $\Upsilon \subseteq \mathcal{C}$.

- The constants K and S are in $\mathcal{C}$.

- If G and H are in $\mathcal{C}$ then also the *application* (GH) is in $\mathcal{C}$.

The notational conventions are similar as for lambda terms. Note the difference between symbols K and S used for the combinatory constants, and symbols **K** and **S** denoting the λ-terms $\lambda xy.\, x$ and $\lambda xyz.\, xz(yz)$, respectively.

5.2.2. DEFINITION. The reduction relation $\to_w$ on combinatory terms, called *weak reduction*, is the least relation such that:

- $\mathsf{K}FG \to_w F$;

- $\mathsf{S}FGH \to_w FH(GH)$;

- If $F \to_w F'$ then $FG \to_w F'G$ and $GF \to_w GF'$.

The symbol $\twoheadrightarrow_w$ denotes the smallest reflexive and transitive relation containing $\to_w$, and $=_w$ denotes the least equivalence relation containing $\to_w$, called *weak equality*.

A combinatory term F such that $F \not\to_w G$, for all combinatory terms G, is said to be in *w-normal form*, or simply *normal form*. A combinatory term F is *normalizing* iff there exists a normal form G such that $F \twoheadrightarrow_w G$. A term F is *strongly normalizing* iff there is no infinite reduction sequence $F \to_w F_1 \to_w F_2 \to_w \cdots$

5.2.3. EXAMPLE.

- Let I = SKK. Then $\mathsf{I}F \to_w \mathsf{K}F(\mathsf{K}F) \to_w F$, for all F.

- The term SII(SII) reduces to itself in five steps.

- Let W = SS(KI). Then $\mathsf{W}FG \twoheadrightarrow_w FGG$, for all F, G.

- Let B = S(KS)K. Then $\mathsf{B}FGH \twoheadrightarrow_w F(GH)$, for all F, G, H.

- Let $C = S(BBS)(KK)$. Then $CFGH \twoheadrightarrow_w FHG$, for all F, G, H.

- Terms K, S, KS, SK, I, B, W are in w-normal form, but C is not.

5.2.4. DEFINITION. For $F \in \mathcal{C}$ define the set $FV(F)$ of *free variables* of F:

$$\begin{aligned}
FV(x) &= \{x\}; \\
FV(HG) &= FV(H) \cup FV(G); \\
FV(S) &= \emptyset; \\
FV(K) &= \emptyset.
\end{aligned}$$

For $F, G \in \mathcal{C}$ and $x \in \Upsilon$ define substitution of G for x in F by:

$$\begin{aligned}
x[x := G] &= G; \\
y[x := G] &= y, \quad \text{if } x \neq y; \\
(HE)[x := G] &= H[x := G]E[x := G]; \\
S[x := G] &= S; \\
K[x := G] &= K.
\end{aligned}$$

The following is similar to the Church-Rosser property for λ-calculus.

5.2.5. THEOREM (Church-Rosser property). *If $F \twoheadrightarrow_w F_1$ and $F \twoheadrightarrow_w F_2$ then $F_1 \twoheadrightarrow_w G$ and $F_2 \twoheadrightarrow_w G$, for some G.*

PROOF. Exercise 5.8. □

One can now infer consequences similar to Corollary 1.4.6, and conclude the consistency of CL by observing that $K \neq S$.

5.3. Typed combinators

We now return to the idea of typed combinators (Curry style).

5.3.1. DEFINITION. Type assignment rules for combinatory terms are given in Figure 5.2. As for lambda-terms, the judgement $\Gamma \vdash M : \tau$ is read as "*the term M has type τ in the environment Γ.*"

This defines simply typed combinatory terms à la Curry. Alternatively, one can consider combinatory terms à la Church. Instead of two constants K and S one introduces infinitely many constants, denoted $K_{\sigma,\tau}$, $S_{\sigma,\tau,\rho}$, for each possible choice of types σ, τ and ρ. An orthodox variant of the Church style would also split the set of variables into a family of disjoint sets of variables of each type.

To distinguish between the typing relation for simply typed combinatory logic and the one for simply typed λ-calculus, we shall use $\vdash_C$ and $\vdash_\Lambda$ for the two, respectively.

5.3. Typed combinators

$$\Gamma, x : \tau \vdash x : \tau$$

$$\Gamma \vdash \mathsf{K} : \sigma \to \tau \to \sigma$$

$$\Gamma \vdash \mathsf{S} : (\sigma \to \tau \to \rho) \to (\sigma \to \tau) \to \sigma \to \rho$$

$$\frac{\Gamma \vdash M : \sigma \to \tau \quad \Gamma \vdash N : \sigma}{\Gamma \vdash (MN) : \tau}$$

FIGURE 5.2: SIMPLY TYPED COMBINATORY LOGIC

Following the path of Chapter 3 we could derive properties like a generation lemma, subject reduction and so on. In some cases, e.g. the proof of subject reduction, the proof is simpler, since reduction does not involve any substitutions, in contrast to the case of β-reduction. Moreover, one can also prove an equivalence result analogous to Propositions 3.4.4–3.4.5. For the sake of brevity, we shall not do so.

However, we cannot resist inserting here a proof of strong normalization for typed combinators. This proof follows the standard method called the *computability* or *reducibility* method, or the method of *logical relations*. In the case of simply typed CL, the proof is particularly simple, so we can enjoy the technique without having to cope with too many details.

5.3.2. DEFINITION. Let $SN_\mathcal{C}$ be the set of all strongly normalizing combinatory terms. For each τ, we define the set $[\![\tau]\!] \subseteq \mathcal{C}$ of terms *computable* for τ:

- $[\![p]\!] = SN_\mathcal{C}$, for any variable p;
- $[\![\sigma \to \rho]\!] = \{G \in \mathcal{C} \mid \forall H(H \in [\![\sigma]\!] \to GH \in [\![\rho]\!])\}$.

Computable terms are also called *reducible* or *stable* in type τ.

5.3.3. LEMMA. *For all τ:*

(i) $[\![\tau]\!] \subseteq SN_\mathcal{C}$;

(ii) *If $H_1, \ldots, H_k \in SN_\mathcal{C}$ then $xH_1 \ldots H_k \in [\![\tau]\!]$. (In particular all variables are computable in all types.)*

PROOF. Induction with respect to τ. If τ is a variable, then the result is obvious. Suppose that $\tau = \sigma \to \rho$, and let $G \in [\![\sigma \to \rho]\!]$. If x is a variable, then $x \in [\![\sigma]\!]$, from the induction hypothesis (ii), so we have $Gx \in [\![\rho]\!]$. From the induction hypothesis (i), it follows that $Gx \in SN_{\mathcal{C}}$, so $G \in SN_{\mathcal{C}}$ must also hold. We have shown (i).

For part (ii) we must show that $xH_1 \ldots H_k P \in [\![\rho]\!]$, for each $P \in [\![\sigma]\!]$. But $P \in SN_{\mathcal{C}}$, by the induction hypothesis (i), so it suffices to apply the induction hypothesis (ii) for ρ. □

5.3.4. LEMMA. *For all τ:*

(i) *If $GQ(PQ)R_1 \ldots R_k \in [\![\tau]\!]$ then $SGPQR_1 \ldots R_k \in [\![\tau]\!]$;*

(ii) *If $PR_1 \ldots R_k \in [\![\tau]\!]$ and $Q \in SN_{\mathcal{C}}$ then $KPQR_1 \ldots R_k \in [\![\tau]\!]$.*

PROOF. (i) Induction with respect to τ. Let τ be a variable. We show that the term $H = SGPQR_1 \ldots R_k$ is in $SN_{\mathcal{C}}$, i.e. there is no infinite reduction

$$H = H_0 \to H_1 \to H_2 \to \cdots$$

First suppose that $H_i = SG^i P^i Q^i R_1^i \ldots R_k^i$ for all i, i.e. all the reductions are "internal". This cannot happen because all the terms $G, P, Q, R_1, \ldots, R_k$ are in $SN_{\mathcal{C}}$. Indeed, all are subterms of $GQ(PQ)R_1 \ldots R_k \in [\![\tau]\!] \subseteq SN_{\mathcal{C}}$.

It follows that the head occurrence of the combinator S must eventually get reduced, and we have $H_{n+1} = G^n Q^n (P^n Q^n) R_1^n \ldots R_k^n$, for some n. In addition, $GQ(PQ)R_1 \ldots R_k \twoheadrightarrow H_{n+1}$, and we know that $GQ(PQ)R_1 \ldots R_k$ is strongly normalizing, so the reduction must terminate.

Let $\tau = \sigma \to \rho$. For $H \in [\![\sigma]\!]$, we must show that $SGPQR_1 \ldots R_k H \in [\![\rho]\!]$. We know that $GQ(PQ)R_1 \ldots R_k H \in [\![\rho]\!]$ and we apply induction for ρ.

(ii) The proof is similar to (i). We leave it to the reader, who should notice that the additional assumption that $Q \in SN_{\mathcal{C}}$ is necessary. Otherwise we may have an infinite sequence of "internal" reductions

$$KPQR_1 \ldots R_k \to KPQ'R_1 \ldots R_k \to KPQ''R_1 \ldots R_k \to \cdots$$

and thus $KPQR_1 \ldots R_k \notin SN_{\mathcal{C}}$. □

5.3.5. LEMMA. *If $\Gamma \vdash G : \tau$, for some Γ and τ, then $G \in [\![\tau]\!]$. That is, each typable term is computable.*

PROOF. Induction with respect to G. We know that variables are computable by Lemma 5.3.3(ii). To show that $K \in [\![\sigma \to \rho \to \sigma]\!]$ it sufices to prove $KPQ \in [\![\sigma]\!]$, for each $P \in [\![\sigma]\!]$ and $Q \in [\![\rho]\!]$. This follows from Lemma 5.3.4(ii), for $k = 0$. Similarly, in order to prove that S is computable we need to show that $SPQR$ is computable for P, Q, R computable (in appropriate types). But then $PR(QR)$ is computable, as it is obtained by application from computable terms. Thus $SPQR$ is computable by Lemma 5.3.4(i). Finally, for $G = G_1 G_2$, we use the induction hypothesis for G_1 and G_2. □

5.4. Combinators versus lambda terms

5.3.6. THEOREM. *Each typable term of $\mathcal{C}$ is strongly normalizing.*

PROOF. Immediate from Lemmas 5.3.3(i) and 5.3.5. □

The following is a version of the Curry-Howard isomorphism for Hilbert style proof systems and combinatory logics.

5.3.7. PROPOSITION.

(i) *If* $\Gamma \vdash_\mathcal{C} F : \varphi$ *then* $\mathrm{rg}(\Gamma) \vdash_H \varphi$;

(ii) *If* $\Gamma \vdash_H \varphi$ *then there exists* $F \in \mathcal{C}$ *such that* $\Delta \vdash_\mathcal{C} F : \varphi$, *for some* Δ *with* $\mathrm{rg}(\Delta) = \Gamma$.

PROOF. (i) By induction on the derivation of $\Gamma \vdash_\mathcal{C} M : \varphi$. (ii) By induction on the Hilbert style proof of $\Gamma \vdash_H \varphi$. □

The Curry-Howard isomorphism for Hilbert style proofs and combinatory terms can be summarized by the following pairs of equivalent notions:

logic	combinators
modus ponens	application
assumption	variable
axiom schemes	constants K and S

Of course, if we prefer the Church style approach, we would rather replace the last line by:

| specific logical axioms | constants $\mathsf{K}_{\sigma,\tau}$, $\mathsf{S}_{\sigma,\tau,\rho}$ |

5.4. Combinators versus lambda terms

The deduction theorem makes our Hilbert style proof system as strong as natural deduction. An immediate consequence of Proposition 5.1.9 is that the same types are inhabited in simply typed lambda calculus and simply typed combinatory logic.

5.4.1. PROPOSITION. *The following conditions are equivalent:*

(i) *There exists a term* $M \in \Lambda$ *such that* $\Gamma \vdash_\Lambda M : \tau$;

(ii) *There exists a term* $H \in \mathcal{C}$ *such that* $\Gamma \vdash_\mathcal{C} H : \tau$.

A careful analysis of the proof of the deduction theorem 5.1.8 reveals that the relation between Λ and $\mathcal{C}$ is deeper than stated in Proposition 5.4.1. Not only are the same types inhabited, but there is also a correspondence between the inhabitants. In fact, this correspondence extends to the untyped calculi. Let us begin with a translation

$$(\ )_\Lambda : \mathcal{C} \to \Lambda,$$

which is easy to define. We just identify K and S with the corresponding lambda terms:

5.4.2. DEFINITION.

- $(x)_\Lambda = x$, for $x \in \Upsilon$;
- $(\mathsf{K})_\Lambda = \lambda xy.x$;
- $(\mathsf{S})_\Lambda = \lambda xyz.xz(yz)$;
- $(FG)_\Lambda = (F)_\Lambda (G)_\Lambda$.

5.4.3. PROPOSITION. *If* $F \to_w G$ *then* $(F)_\Lambda \twoheadrightarrow_\beta^+ (G)_\Lambda$.

PROOF. By induction with respect to $F \to_w G$. □

5.4.4. COROLLARY.

(i) *If* $F \twoheadrightarrow_w G$ *then* $(F)_\Lambda \twoheadrightarrow_\beta (G)_\Lambda$;

(ii) *If* $F =_w G$ *then* $(F)_\Lambda =_\beta (G)_\Lambda$.

PROOF. Immediate from Proposition 5.4.3. □

Part (ii) of Corollary 5.4.4 states that the translation $(\ )_\Lambda$ can be seen as a mapping from $\mathcal{C}/{=_w}$ to $\Lambda/{=_\beta}$ (i.e. a mapping from CL to λ-calculus understood as equational theories).

It is less obvious how to find a translation backward, because we have to define lambda abstraction without bound variables. The solution can be found in the proof of the deduction theorem. As we noted in Remark 5.1.11, this meta-proof defines a certain function on the object level proofs. These are represented by certain combinators, so we actually have a function, called *combinatory abstraction*, that turns some combinators into other combinators. It is a remarkable fact that this function works for untyped combinators as well. To define it, one can just replace every detachment from (A1) by an application of K, and every use of (A2) by S. Also, Example 5.1.3 is nothing else (cf. Example 5.2.3) but the definition of I in terms of S and K.

5.4. COMBINATORS VERSUS LAMBDA TERMS

5.4.5. DEFINITION. For each $F \in \mathcal{C}$ and $x \in \Upsilon$ we define the term $\lambda^* x\, F \in \mathcal{C}$.

- $\lambda^* x\, x = \mathsf{I}$;
- $\lambda^* x\, F = \mathsf{K} F$, if $x \notin \mathrm{FV}(F)$;
- $\lambda^* x\, FG = \mathsf{S}(\lambda^* x\, F)(\lambda^* x\, G)$, otherwise.

Notational conventions for λ^* are similar to those for λ. The following shows that combinatory abstraction behaves as expected.

5.4.6. PROPOSITION.

(i) $(\lambda^* x\, F)G \twoheadrightarrow_w F[x := G]$;

(ii) $(\lambda^* x\, F)_\Lambda \twoheadrightarrow_\beta \lambda x\, (F)_\Lambda$.

PROOF. Left to the reader. □

Another formulation of Proposition 5.4.6(i) is as follows.

5.4.7. COROLLARY. *For all x and F, there exists H such that for all G:*

$$HG \twoheadrightarrow_w F[x := G].$$

This property is sometimes called "combinatory completeness". It can be explained as follows. A term F (possibly containing a variable x) defines a mapping $f : \mathcal{C} \to \mathcal{C}$ such that $f(G) = F[x := G]$. Combinatory completeness states that every such function can actually be represented by a combinator H so that $HG = f(G)$, for each G.

Using the operator $\lambda^* x$, we can define a translation $(\ )_\mathcal{C} : \Lambda \to \mathcal{C}$.

5.4.8. DEFINITION.

- $(x)_\mathcal{C} = x$, for $x \in \Upsilon$;
- $(MN)_\mathcal{C} = (M)_\mathcal{C}(N)_\mathcal{C}$;
- $(\lambda x M)_\mathcal{C} = \lambda^* x\, (M)_\mathcal{C}$.

5.4.9. PROPOSITION. *For all $M \in \Lambda$, we have $((M)_\mathcal{C})_\Lambda \twoheadrightarrow_\beta M$.*

PROOF. Induction with respect to M, using Proposition 5.4.6(ii). □

In particular, $((M)_\mathcal{C})_\Lambda =_\beta M$, which means that the mapping from $\mathcal{C}/_{=_w}$ to $\Lambda/_{=_\beta}$ induced by $(\ )_\Lambda$ is surjective. The reader may now feel a temptation to say that $(\ )_\Lambda$ behaves as a "retraction" of $\mathcal{C}/_{=_w}$ onto $\Lambda/_{=_\beta}$. But it is better to resist such temptations (cf. the discussion following Example 5.5.1).

The following property is sometimes expressed by saying that $\mathbf{K}$ and $\mathbf{S}$ make a *basis* for untyped λ-calculus.

5.4.10. COROLLARY. *Every closed lambda term M is beta-equal to a term obtained from $\mathbf{K}$ and $\mathbf{S}$ solely by application.*

PROOF. The desired term is $((M)_C)_\Lambda$. □

See Exercises 5.15 and 5.16 for more about the relationship between sets of axioms and bases for the λ-calculus.

It is not difficult to see that $(\ )_\Lambda$ preserves types. In addition, the same is true for the translation $(\ )_C$, but this requires the following lemma (which can be seen as an annotated variant of the deduction theorem):

5.4.11. LEMMA. *Let $\Gamma, x:\rho \vdash_C F:\tau$. Then $\Gamma \vdash_C \lambda^* x\, F : \rho \to \tau$.*

PROOF. By induction on F. □

5.4.12. PROPOSITION.

(i) *If $\Gamma \vdash_C F:\tau$ then $\Gamma \vdash_\Lambda (F)_\Lambda : \tau$.*

(ii) *If $\Gamma \vdash_\Lambda M:\tau$ then $\Gamma \vdash_C (M)_C : \tau$.*

PROOF. (i) By induction on the derivation of $\Gamma \vdash_C F:\tau$. (ii) By induction on the derivation of $\Gamma \vdash_\Lambda M:\tau$, using Lemma 5.4.11. □

5.4.13. REMARK. Note that part (i) of the above gives an alternative proof that the simply typed version of the calculus of combinators has the strong normalization property (Theorem 5.3.6). Indeed, the result follows from the strong normalization of simply typed λ-calculus, using Proposition 5.4.3.

5.5. Extensionality

Unfortunately, the translation $(\ )_C : \Lambda \to \mathcal{C}$ is not as good as $(\ )_\Lambda$.

5.5.1. EXAMPLE.

(i) The composition $((\ )_\Lambda)_C$ is not identity, e.g. $((\mathsf{K})_\Lambda)_C = \mathsf{S}(\mathsf{KK})\mathsf{I} \neq_w \mathsf{K}$.

(ii) An equality $M =_\beta N$ does not in general imply $(M)_C =_w (N)_C$. For instance, $\lambda x\, \mathbf{KI}x \twoheadrightarrow_\beta \lambda x\, \mathbf{I}$, but $(\lambda x\, \mathbf{KI}x)_C = \mathsf{S}(\mathsf{K}(\mathsf{S}(\mathsf{KK})\mathsf{II}))\mathsf{I} =_w \mathsf{S}(\mathsf{K}(\mathsf{KI}))\mathsf{I} \neq_w \mathsf{KI} = (\lambda x.\mathbf{I})_C$.

Part (ii) of Example 5.5.1 states that the translation $(\ )_C$ does not induce a properly defined mapping from $\Lambda/_{=_\beta}$ to $\mathcal{C}/_{=_w}$. This is quite disappointing. The problem is that the *weak extensionality* principle

$$\frac{M = N}{\lambda x\, M = \lambda x\, N} \tag{ξ}$$

5.5. EXTENSIONALITY

does not hold for λ^* and $=_w$. Indeed, we have

$$\lambda^* x \, \mathsf{K} \mathsf{I} x = \mathsf{S}(\mathsf{K}(\mathsf{K}\mathsf{I}))\mathsf{I} \neq_w \mathsf{K}\mathsf{I} = \lambda^* x \, \mathsf{I},$$

while of course $\mathsf{K}\mathsf{I}x \to_w \mathsf{I}$.

In a sense, combinatory logic has too many normal forms; the evaluation of a combinatory term stops too early. (That is why the weak reduction is called "weak".) It seems that everything should go smoothly, if we could evaluate "incomplete applications" like $\mathsf{S}GH$ or $\mathsf{K}G$.

Indeed, this is the case. The correspondence between Λ and $\mathcal{C}$ is much more elegant in the *extensional* versions of these calculi. Recall the rule (ext) discussed in Section 1.3. In a similar way we can add rule (ext) to the equational theory of combinatory terms.

5.5.2. DEFINITION. Let $=_{ext}$ be the least equivalence on $\mathcal{C}$, such that:

- If $G =_w H$ then $G =_{ext} H$;

- If $Gx =_{ext} Hx$, and $x \notin \mathrm{FV}(G) \cup \mathrm{FV}(H)$, then $G =_{ext} H$;

- If $G =_{ext} G'$ then $GH =_{ext} G'H$ and $HG =_{ext} HG'$.

Part (iv) of the following proposition is of course a weaker version of Proposition 5.4.9. But we included it to "complete the quartet."

5.5.3. PROPOSITION.

(i) *For all $G, H \in \mathcal{C}$, the conditions $G =_{ext} H$ and $(G)_\Lambda =_{\beta\eta} (H)_\Lambda$ are equivalent.*

(ii) *For all $M, N \in \Lambda$, the conditions $M =_{\beta\eta} N$ and $(M)_\mathcal{C} =_{ext} (N)_\mathcal{C}$ are equivalent.*

(iii) *The identity $((G)_\Lambda)_\mathcal{C} =_{ext} G$ holds for all $G \in \mathcal{C}$.*

(iv) *The equation $((M)_\mathcal{C})_\Lambda =_{\beta\eta} M$ holds for all $M \in \Lambda$.*

PROOF. First prove the left-to-right implications in parts (i) and (ii). The latter requires showing the following statements first:

- If $G =_{ext} H$ then $\lambda^* x \, G =_{ext} \lambda^* x \, H$ (rule ξ for extensional equality).

- The equation $(N[x:=M])_\mathcal{C} = (N)_\mathcal{C}[x:=(M)_\mathcal{C}]$ holds for all $M, N \in \Lambda$.

The details are quite routine and we leave them to the reader.

In order to prove part (iii) one proceeds by induction with respect to G. The cases of a variable and application are easy. For the two constants, it is enough to observe that $(\mathbf{K})_C xy = (\lambda^* xy.x)xy =_w x =_w \mathsf{K}xy$ and $(\mathbf{S})_C xyz = (\lambda^* xyz.xz(yz))xyz =_w xz(yz) =_w \mathsf{S}xyz$.

Now, with help of part (iii) and Proposition 5.4.9, one can show the other implications in parts (i) and (ii). For instance, if $(G)_\Lambda =_{\beta\eta} (H)_\Lambda$ then $G =_{ext} ((G)_\Lambda)_C) =_{ext} ((H)_\Lambda)_C) =_{ext} H$. □

We conclude that in the extensional case we have indeed an isomorphism between lambda calculus and combinatory logic. Another way to obtain the equivalence (see [31, 241, 459]) is to replace weak reduction by *strong reduction*, such that the induced equality $=_s$ satisfies rule (ξ) and $((M)_\Lambda)_C =_s M$ holds for all terms.[1]

REMARK. A consequence of Proposition 5.5.3 is that many results can be translated from one extensional calculus to the other. For instance, it suffices to know that extensional equality is undecidable for lambda terms, to conclude the same for combinatory terms. This is because we have:

$$M =_{\beta\eta} N \quad \text{iff} \quad (M)_C =_{ext} (N)_C.$$

One would expect that there should be a similar proof technique for weak vs. β-reductions, so that e.g. the undecidability of $=_\beta$ could be used to infer the undecidability of weak equality. Unfortunately, the translations $(\ )_\Lambda$ and $(\ )_C$ are useless for this purpose, and surprisingly, no "good" translation seems to be known. There is however a translation $(\ )_a$, preserving strong normalization (each way). This translation uses another combinatory abstraction

$$\lambda^a x\, G = \mathsf{K}(\lambda^* x\, G)G^\bullet,$$

where $G^\bullet$ is G with x replaced by a new constant (or a fresh variable). See Akama [7] for details.

5.6. Relevance and linearity

Neither intuitionistic nor classical logic have any objections against the axiom scheme $\varphi \to \psi \to \varphi$, which expresses the following rule of reasoning: "an unnecessary assumption can be forgotten". This rule is however dubious when we are interested in the *relevance* of assumptions with respect to the conclusion. Just like *no* use of an assumption may be regarded as a dubious phenomenon, *multiple* use of an assumption may also raise important doubts. Therefore, logicians and philosophers study various *substructural* logics in which restrictions are made concerning the use of assumptions.[2]

---

[1] The terminology used here is confusing. Indeed, $=_w$ implies $=_s$, but not conversely, and thus the weak equality is actually strictly stronger than strong equality.

[2] The name "substructural" is related to the restricted use of the *structural rules* of sequent calculus (cf. Chapter 7).

5.6. Relevance and linearity

With the Curry-Howard isomorphism at hand, we can easily identify the corresponding fragments of (the implicational fragment of) intuitionistic propositional logic, by characterizing lambda terms with respect to the number of occurrences of bound variables within their scopes. For instance, the λI-terms, as defined in Section 1.7, correspond to reasoning where each assumption is used *at least once*, but all assumptions are reusable.

5.6.1. DEFINITION.

(i) *Affine terms* (also called **BCK**-*terms*) are defined as follows:

- Every variable is an affine term;
- An application MN is an affine term iff both M and N are affine terms, and $\mathrm{FV}(M) \cap \mathrm{FV}(N) = \varnothing$;
- An abstraction $\lambda x\, M$ is an affine term iff M is an affine term.

(ii) A term is called *linear* (or a **BCI**-*term) iff it is both a λI-term and an affine term.

The affine terms, where each variable occurs at most once, represent the idea of *disposable* assumptions that are thrown away after use, and cannot be used again. A strict control over all assumptions, with each one being used *exactly once*, is maintained in proofs corresponding to linear terms. The three classes of lambda terms correspond to three fragments of intuitionistic implicational logic, defined below by Hilbert style logical axioms.

5.6.2. DEFINITION.

(i) The axioms of the *relevant* propositional logic $\mathbf{R}_\to$ are all instances of the following schemes:

$(A_S)\ (\varphi \to \psi \to \vartheta) \to (\varphi \to \psi) \to \varphi \to \vartheta$;

$(A_B)\ (\psi \to \vartheta) \to (\varphi \to \psi) \to \varphi \to \vartheta$;

$(A_C)\ (\varphi \to \psi \to \vartheta) \to \psi \to \varphi \to \vartheta$;

$(A_I)\ \varphi \to \varphi$.

(ii) The axioms of the **BCK** propositional logic are all instances of the axiom schemes (A_B) and (A_C) and

$(A_K)\ \varphi \to \psi \to \varphi$.

(iii) The axioms of the **BCI** propositional logic are all instances of the axiom schemes (A_B) and (A_C) and (A_I).

(iv) The notions of a proof sequence and a theorem (for any of the three logics) are as in Definition 5.1.1.

WARNING. The expression "linear logic" denotes a system which is a strict extension of the **BCI** logic. Linear logic differs from our **BCI** logic in that it involves more logical connectives, and it is *conservative* over **BCI** logic, i.e. all purely implicational theorems of linear logic are **BCI** theorems. But the other connectives are what makes linear logic so interesting and important, so we would rather not use this name for the (quite weak) implicational fragment. Also, the expression "affine logic" stands for logical systems that are richer than our **BCK** logic.

Of course the axioms (A_K) and (A_S) are exactly our axioms (A1) and (A2) of the full intuitionistic logic. The other axioms can also be seen as types of combinators. For $B = S(KS)K$ and $C = S(BBS)(KK)$, one can check that

- $\vdash B : (\psi \to \vartheta) \to (\varphi \to \psi) \to \varphi \to \vartheta$;

- $\vdash C : (\varphi \to \psi \to \vartheta) \to \psi \to \varphi \to \vartheta$,

holds for all φ, ψ and ϑ. We have $BFGH \twoheadrightarrow_w F(GH)$ and $CFGH \twoheadrightarrow_w FHG$, for all F, G, H. The corresponding closed lambda terms are:

- $B = \lambda xyz.x(yz) =_\beta (B)_\Lambda$;

- $C = \lambda xyz.xzy =_\beta (C)_\Lambda$.

Clearly, our three logics correspond to fragments of $\mathcal{C}$ generated by the appropriate choices of combinators, hence the abbreviations **BCK** and **BCI**.

We still have to justify the correspondence between Definition 5.6.2 and the restrictions on the use of variables in lambda terms. First we have the obvious part:

5.6.3. LEMMA.

(i) *The combinators* **S**, **B**, **C**, **I** *are* λI-*terms, and so are all terms obtained from* **S**, **B**, **C**, **I** *by applications;*

(ii) *The combinators* **B**, **C**, **K** *are affine terms, and so are all terms obtained from* **B**, **C**, **K** *by applications;*

(iii) *The combinators* **B**, **C**, **I** *are linear terms, and so are all terms obtained from* **B**, **C**, **I** *by applications.*

Thus, the translation $(\ )_\Lambda$ maps the appropriate fragments of $\mathcal{C}$ into the appropriate fragments of Λ. But the inverse translation $(\ )_\mathcal{C}$ cannot be used anymore, as it requires S and K to be available. We need a new combinatory abstraction operator.

5.6. RELEVANCE AND LINEARITY

5.6.4. DEFINITION.

(i) For each term in $\mathcal{C}$ and each $x \in \Upsilon$ we define the term $\lambda^\circ x\, M \in \mathcal{C}$.

- $\lambda^\circ x\, x = \mathsf{I}$;
- $\lambda^\circ x\, F = \mathsf{K}F$, whenever $x \notin \mathrm{FV}(F)$;
- $\lambda^\circ x\, FG = \mathsf{S}(\lambda^\circ x\, F)(\lambda^\circ x\, G)$, if $x \in \mathrm{FV}(F) \cap \mathrm{FV}(G)$;
- $\lambda^\circ x\, FG = \mathsf{C}(\lambda^\circ x\, F)G$, if $x \in \mathrm{FV}(F)$ and $x \notin \mathrm{FV}(G)$;
- $\lambda^\circ x\, FG = \mathsf{B}F(\lambda^\circ x\, G)$, if $x \notin \mathrm{FV}(F)$ and $x \in \mathrm{FV}(G)$.

(ii) Now define a translation $[\ ]_\mathcal{C} : \Lambda \to \mathcal{C}$, as follows:

- $[x]_\mathcal{C} = x$, for $x \in \Upsilon$;
- $[MN]_\mathcal{C} = [M]_\mathcal{C}[N]_\mathcal{C}$;
- $[\lambda x\, M]_\mathcal{C} = \lambda^\circ x\, [M]_\mathcal{C}$.

The translation $[\ ]_\mathcal{C} : \Lambda \to \mathcal{C}$ has all the good properties of $(\ )_\mathcal{C}$. That is, Propositions 5.4.6, 5.4.9 and 5.4.12 remain true. (For the proof note first that $(\lambda^\circ x\, F)_\Lambda =_\beta (\lambda^* x\, F)_\Lambda$.) In addition we have:

5.6.5. PROPOSITION.

(i) *If M is a $\lambda\mathbf{I}$-term, then $[M]_\mathcal{C}$ is built solely from variables and the combinators S, B, C, and I.*

(ii) *If M is an affine term then $[M]_\mathcal{C}$ is weakly equal to a term built solely from variables and the combinators B, C, and K.*

(iii) *If M is a linear term then $[M]_\mathcal{C}$ is built solely from variables and the combinators B, C, and I.*

PROOF. Left to the reader. For part (ii) one has to do Exercise 5.17. □

We can conclude with the following summary:

5.6.6. THEOREM.

(i) *A formula φ is a theorem of $\mathbf{R}_\to$ if and only if it is a type of a closed $\lambda\mathbf{I}$-term;*

(ii) *A formula φ is a theorem of $\mathbf{BCK}$-logic if and only if it is a type of a closed affine term.*

(iii) *A formula φ is a theorem of $\mathbf{BCI}$-logic if and only if it is a type of a closed linear term.*

PROOF. We have already observed that a formula is a theorem of any of these three logics iff it is a type of a combinatory term built from the appropriate choice of basic combinators. It remains to show that the latter is equivalent to being a type of an appropriate closed lambda term. This follows from Lemma 5.6.3, Proposition 5.6.5, and the appropriate modification of Proposition 5.4.12. □

Theorem 5.6.6 allows us to draw the following table:

logic	lambda-calculus
Relevant logic	λI-terms
BCK-logic	affine terms
BCI-logic	linear terms

WARNING. Our "naive" Hilbert style formalization of substructural reasoning does not properly reflect the intuition concerning the use of assumptions in substructural logics. In addition, in all the three cases, the deduction theorem fails. For instance, according to Definition 5.6.2, the judgement $\alpha, \alpha \to \alpha \to \beta \vdash \beta$, has a legal **BCI**-proof, while $\alpha \to \alpha \to \beta \vdash \alpha \to \beta$ is not a correct **BCI** (nor even affine) judgement. Also, $\alpha, \beta \vdash \alpha$ has a correct relevant proof but $\alpha \vdash \beta \to \alpha$ has not. The first example also shows that the correspondence in the above table does not apply to affine and linear terms with free variables (the proof corresponds to a non-affine term yxx).

The tree variant (Figure 5.1) of Hilbert style (appropriately adjusted) can provide a more adequate formulation of our three logics. Each of them needs a different adjustment though (Exercises 5.19–5.20).

5.7. Notes

Combinatory logic was invented by Schönfinkel and Curry in the 1920's shortly before Church introduced the λ-calculus. To be more precise, Schönfinkel lectured on his system as early as 1920, but the paper [427] was published only in 1924. Schönfinkel introduced the combinators S (the *Verschmelzungsfunktion*) and K (the *Konstantfunktion*), but he used the letter C to denote the latter.

Both Schönfinkel and Curry aimed at the same goal: to build the foundations of logic on a formal system in which logical formulas could be handled in a *variable-free* manner. Curry, who began his research in the mid 1920's, initially did not know about Schönfinkel's work. Curry's first paper [97] appeared in 1929, and initiated a systematic development of combinatory logic. In 1934 Curry published his first work [100] about types in CL. Most of the theory of combinators has been developed in the books [107, 108].

In more recent times, the idea of a combinator, understood as a function defined by an equation, has proven to be useful in practical applications, namely in compiler construction. The reason is that the evaluation of an expression becomes more efficient after a translation into a combinator form, using a procedure resembling very much our combinatory abstraction λ°.

Curry's effort to build the foundations of mathematics with illative combinatory logic was continued by Bunder and others [62, 127]. Certain systems of illative combinatory logic were shown complete with respect to propositional and first-order logic, by means of translations inspired by the Curry-Howard isomorphism. Provability of a formula is represented under such translations by the existence of an inhabitant of an appropriate type.

The strong normalization proof given in Section 5.3 is based on an idea due to Tait, who first used the computability method in [467] to show normalization for Gödel's system **T** (see Chapter 10). This approach (and its generalizations) was later successfully used to prove normalization and other properties of various typed systems, see [161, 378, 397, 443, 454] and Chapters 10–11. In particular, the closely related idea of *logical relations* [349] turned out to be a successful tool in semantics.

The oldest reference to a logic in which multiple use of assumptions was taken into account is a work of Fitch from 1936, and the idea of relevant logics dates back to early 50's [237]. Incidentally, Church's lambda-calculus of [74] was essentially the λI-calculus. Propositions-as-types interpretations of substructural calculi began with C.A. Meredith, and were investigated by Helman, Rezus, and others.

These ideas have found an ultimate exposition in the *linear logic*, introduced in a classic paper by Girard [183] and studied by many for almost twenty years now. Linear logic provides a framework which highlights practically all aspects of the Curry-Howard isomorphism discused in this book, and which adds various new ones. (The basic concept of a *proof net* is the principal example.) The reader interested in linear logic is referred to the books [144, 189, 188, 485][3] and papers [4, 95, 178, 183] for the basic material. There are also various introductory papers available from the Web. See [17, 118, 319] for the geometry of interaction and its applications for optimal reductions, and consult [186, 187] to find out about *ludics*.

As a survey on combinatory logic, we recommend [241] and Chapters 7 and 9 of [31]. The books [107, 108] are a little older, but still very useful. See [238, 440] for the early history of combinators. The classical references for relevant and related logics are [9] and [10]. A more concise introduction can be found in [137]. For the theory of BCK-logics the reader is referred to the book [130] and research papers [158, 235, 236, 242, 275, 479]. For applications of combinators in functional programming see e.g. [149, 391].

5.8. Exercises

5.1. Write down a Hilbert style proof of $(p \to q) \to (q \to r) \to p \to r$.

5.2. Let $\Gamma = \{\varphi, \varphi \to \psi, \gamma, \gamma \to \psi, \psi \to \psi \to \vartheta\}$. Consider the proof sequence

$$\varphi, \varphi \to \psi, \gamma, \gamma \to \psi, \psi \to \psi \to \vartheta, \psi, \psi, \psi \to \vartheta, \vartheta$$

using the set of non-logical axioms Γ. How many different annotations (in the style of Examples 5.1.3 and 5.1.4) can you associate with this sequence? How many such annotations are possible if φ, ψ, ϑ are logical axioms?

5.3. Let $\vdash'_H$ denote provability in a variant of Hilbert style system where the "multiplicative" rule (MP) is replaced by its "additive" version:

$$(MP') \quad \frac{\Gamma \vdash \varphi \quad \Gamma \vdash \varphi \to \psi}{\Gamma \vdash \psi}$$

---
[3] Reading [485] requires a little care, because of the non-standard notation.

Show that $\Gamma \vdash'_H \varphi$ and $\Gamma \vdash_H \varphi$ are equivalent for all Γ and φ.

5.4. Prove that the Hilbert-style proof system with axioms (A1)–(A3), (C1)–(C3) and (D1)–(D3) of Section 5.1 is complete for propositional intuitionistic logic.

5.5. (Hindley, Meredith [240]) Consider a Hilbert-style proof system with only two axioms (where the variables p and q are fixed):

(A1) $p \to q \to p$;

(A2) $(p \to q \to r) \to (p \to q) \to p \to r$,

and with *modus ponens* replaced by the following rule of *condensed detachment*:

$$\frac{\varphi \quad \psi \to \vartheta}{S_{\varphi',\psi}(\vartheta)}$$

where φ' is φ with all variables replaced by fresh ones, and $S_{\varphi',\psi}$ denotes the most general unifier of φ' and ψ. Write $\vdash_{cd} \varphi$ if an implicational formula φ is provable in this system.

(i) Prove that $\vdash_{cd} \varphi$ holds iff φ is a principal type of a closed combinatory term.

(ii) Show that the system is complete for IPC($\to$).

Hint: Prove the converse to Proposition 5.4.12(ii), and use Exercise 3.13.

5.6. Find a combinator $2 \in \mathcal{C}$ such that $2FA \twoheadrightarrow_w F(FA)$, for all F and A in $\mathcal{C}$.

5.7. Is there a combinator M such that $M \to_w M$?

5.8. Prove the Church-Rosser property for weak reduction (Theorem 5.2.5) using the Tait & Martin-Löf method from Chapter 1. Note that the proof for combinatory logic is somewhat simpler due to the fact that non-overlapping redexes remain non-overlapping during reduction of other redexes.

5.9. Prove an analogue of the subject reduction theorem (Theorem 3.1.7) for typable combinatory terms.

5.10. Prove the strong normalization theorem for the simply typed lambda-calculus, by means of Tait's computability technique, adapting the proof of Theorem 5.3.6.

5.11. Describe a notion of reduction for Hilbert style proofs, corresponding to weak reduction on combinatory terms.

5.12. Here is an alternative definition of combinatory abstraction:

- $\lambda^\bullet x \, x = \mathsf{I}$;
- $\lambda^\bullet x \, F = \mathsf{K}F$, if F is a constant (K or S), or a variable $y \neq x$;
- $\lambda^\bullet x \, FG = \mathsf{S}(\lambda^\bullet x \, F)(\lambda^\bullet x \, G)$, otherwise.

Show that the combinatory abstraction operator $\lambda^\bullet$ has a property analogous to Proposition 5.4.6. Define a translation from Λ to $\mathcal{C}$ based on $\lambda^\bullet$, and investigate its properties.

5.13. Consider the following variant of the calculus of combinators: there are typed constants $\mathsf{K}_{\sigma,\tau}$, and $\mathsf{S}_{\tau,\tau,\tau}$, with typing and reduction rules as usual, and in addition, there are constants $\mathsf{I}_\tau : \tau \to \tau$ with the reduction rule $\mathsf{I}_\tau F \to F$. (The identity combinator cannot now be defined as SKK because not all typed forms of S are available.) By embedding into Church style combinatory logic, show that this variant satisfies subject reduction and strong normalization properties.

5.8. EXERCISES

5.14. (Hindley) Give an example of a term $G \in \mathcal{C}$, in w-normal form, such that $(G)_\Lambda$ does not have a β-normal form. *Hint:* When is $\lambda^* x F$ in w-normal form?

5.15. Let A be a set of simply typed terms such that their types make a complete set of axiom schemes for the implicational fragment of the intuitionistic propositional logic. More precisely, assume that the Hilbert style system obtained by taking each of these types as an axiom (together with modus ponens as the single rule) leads to the same set of provable formulas as our system in Section 5.1. Prove that A is a basis for the untyped lambda calculus in the sense of Corollary 5.4.10.

5.16. (Based on [55].) Let $\mathbf{S}° = \lambda ixyz.i(i(x(iz)(i(y(iz)))))$. Show that the terms K, I and $\mathbf{S}°$ form a basis for lambda calculus in the sense of Corollary 5.4.10, but their types (whatever choice is taken) do not make a complete Hilbert style axiom system for the implicational intuitionistic logic. *Hint:* One cannot derive the formula $(p \to p \to q) \to p \to q$.

5.17. Define I by means of K, B and C.

5.18. Prove or disprove: If a formula φ is both a **BCK**-theorem and a relevant theorem, then it is a **BCI**-theorem.

5.19. Let the symbol $\vdash_R$ stand for provability in a system consisting of rule (MP) of Figure 5.1 and the premise-free rules:

(Id-R) $\varphi \vdash \varphi$ (Ax-R) $\vdash \varphi$, if φ is a logical axiom of $\mathbf{R}_\to$.

Prove that $\vdash_R$ satisfies the deduction theorem and:

(i) If M is a λI-term with $\mathrm{FV}(M) = \mathrm{dom}(\Gamma)$ and $\Gamma \vdash M : \varphi$ then $\mathrm{rg}(\Gamma) \vdash_R \varphi$.

(ii) If $\Gamma \vdash_R \varphi$, then there exists a λI-term M, such that $\Delta \vdash M : \varphi$, for some Δ, with $\mathrm{dom}(\Delta) = \mathrm{FV}(M)$ and $\mathrm{rg}(\Delta) = \Gamma$.

5.20. An adequate formalism for the **BCK** and **BCI** logics should be able to control multiple use of assumptions. A judgement $\Gamma \vdash \varphi$, where the context Γ is a set, does not tell us how many times an assumption $\alpha \in \Gamma$ is actually used in the proof. Using multisets (see the Appendix) of assumptions rather than sets, define Hilbert-style proof systems for **BCK** and **BCI** logics satisfying the deduction theorem and properties analogous to (i) and (ii) in Exercise 5.19.

5.21. Prove that every affine term is typable, and that affine terms satisfy *subject conversion*, i.e. that $\Gamma \vdash M : \sigma$ and $M =_\beta N$ implies $\Gamma \vdash N : \sigma$, for affine M, N.

5.22. Prove that two affine terms with the same principal pair are $\beta\eta$-equal.

5.23. (From [449].) A lambda-term M is *user-friendly* iff every application PQ occurring in M is such that either P or Q is an affine term. Prove that user-friendly terms are strongly normalizing.

Chapter 6

Classical logic and control operators

In the previous chapters we have encountered the Curry-Howard isomorphism in various incarnations. Each of these states a correspondence between some calculus for computation and a system of formal logic. Until now these systems of formal logic have been intuitionistic; for instance, in none of them have we been able to derive the principle of double negation elimination $\neg\neg\varphi \to \varphi$ that one finds in classical logics.

This is by no means a coincidence. Until around 1990 it was believed by many that "there is no Curry-Howard isomorphism for classical logic." However, at that time Timothy Griffin discovered that the principle of double negation elimination corresponds to the typing of a *control operator* known from programming language theory. The idea was soon generalized and refined by Chet Murthy.

This chapter presents the discovery and its consequences, e.g. the result that Kolmogorov's double negation embedding of classical logic into intuitionistic logic corresponds to a *continuation passing style translation*.

6.1. Classical propositional logic

Classical logic differs from intuitionistic logic by embracing these principles:

(i) *Tertium non datur (law of the excluded middle):* $\varphi \vee \neg\varphi$. Either φ holds, or φ does not hold.

(ii) *Double negation elimination:* $\neg\neg\varphi \to \varphi$. If it is not the case that φ does not hold, then φ holds.

(iii) *Peirce's law:* $((\varphi \to \psi) \to \varphi) \to \varphi$.

(iv) *Reductio ad absurdum:* $(\neg\varphi \to \varphi) \to \varphi$. If the assumption that φ does not hold implies φ, then φ must hold.

A natural deduction system for classical propositional calculus (CPC) is obtained by replacing the *ex falso* rule in intuitionistic logic (see Figure 2.1) with a rule for double negation elimination. We study the implicational fragment with absurdity; negation is defined using absurdity as usual.

6.1.1. DEFINITION. The natural deduction presentation $NK(\to, \bot)$, of classical propositional calculus for implication and absurdity, is defined by the axiom and rules in Figure 6.1. We write $\Gamma \vdash_N \varphi$ when $\Gamma \vdash \varphi$ can be derived in this system (we sometimes leave out "N"). We use the same terminology, notations, and conventions as in Chapter 2.

6.1.2. EXAMPLE.

(i) The following is a derivation of the formula $(\varphi \to \psi) \to (\neg\varphi \to \psi) \to \psi$ ("proof by cases"). Let $\Gamma = \{\varphi \to \psi, \neg\varphi \to \psi, \neg\psi\}$.

$$
\cfrac{
  \cfrac{
    \Gamma \vdash \neg\psi \qquad
    \cfrac{
      \cfrac{
        \Gamma \vdash \neg\varphi \to \psi \qquad
        \cfrac{
          \cfrac{
            \Gamma, \varphi \vdash \neg\psi \qquad
            \cfrac{
              \Gamma, \varphi \vdash \varphi \to \psi \qquad \Gamma, \varphi \vdash \varphi
            }{\Gamma, \varphi \vdash \psi}
          }{\Gamma, \varphi \vdash \bot}
        }{\Gamma \vdash \neg\varphi}
      }{\Gamma \vdash \psi}
    }{}
  }{
    \cfrac{
      \cfrac{
        \cfrac{\Gamma \vdash \bot}{\varphi \to \psi, \neg\varphi \to \psi \vdash \psi}
      }{\varphi \to \psi \vdash (\neg\varphi \to \psi) \to \psi}
    }{\vdash (\varphi \to \psi) \to (\neg\varphi \to \psi) \to \psi}
  }
}{}
$$

(ii) The following is a derivation of Peirce's law. Let $\Gamma = \{(\varphi \to \psi) \to \varphi\}$.

$$
\cfrac{
  \cfrac{
    \Gamma, \neg\varphi \vdash \neg\varphi \qquad
    \cfrac{
      \Gamma, \neg\varphi \vdash (\varphi \to \psi) \to \varphi \qquad
      \cfrac{
        \cfrac{
          \cfrac{
            \Gamma, \neg\varphi, \varphi, \neg\psi \vdash \neg\varphi \qquad \Gamma, \neg\varphi, \varphi, \neg\psi \vdash \varphi
          }{\Gamma, \neg\varphi, \varphi, \neg\psi \vdash \bot}
        }{\Gamma, \neg\varphi, \varphi \vdash \psi}
      }{\Gamma, \neg\varphi \vdash \varphi \to \psi}
    }{\Gamma, \neg\varphi \vdash \varphi}
  }{
    \cfrac{\Gamma, \neg\varphi \vdash \bot}{\cfrac{\Gamma \vdash \varphi}{\vdash ((\varphi \to \psi) \to \varphi) \to \varphi}}
  }
}{}
$$

6.1. CLASSICAL PROPOSITIONAL LOGIC

$$\Gamma, \varphi \vdash \varphi \;\; (\text{Ax})$$

$$\frac{\Gamma, \varphi \vdash \psi}{\Gamma \vdash \varphi \to \psi} \; (\to\text{I}) \qquad \frac{\Gamma \vdash \varphi \to \psi \quad \Gamma \vdash \varphi}{\Gamma \vdash \psi} \; (\to\text{E})$$

$$\frac{\Gamma, \varphi \to \bot \vdash \bot}{\Gamma \vdash \varphi} \; (\neg\text{E})$$

FIGURE 6.1: CLASSICAL PROPOSITIONAL CALCULUS.

6.1.3. EXAMPLE. *Ex falso* is a derived rule in NK($\to, \bot$). That is, if $\Gamma \vdash \bot$ then also $\Gamma \vdash \varphi$. Indeed, assume we have a derivation of $\Gamma \vdash \bot$. By adding $\neg \varphi$ to all environments in this derivation, we obtain a derivation of $\Gamma, \neg \varphi \vdash \bot$. Then $\Gamma \vdash \varphi$ by the ($\neg$E) rule.

6.1.4. COROLLARY. *If $\Gamma \vdash \varphi$ in NJ($\to, \bot$), then $\Gamma \vdash \varphi$ in NK($\to, \bot$).*

PROOF. By induction on the derivation of $\Gamma \vdash \varphi$, using Example 6.1.3. □

6.1.5. REMARK. An alternative to the rule ($\neg$E) of Figure 6.1 are the following two rules, namely *ex falso* and a form of Peirce's law.

$$\frac{\Gamma \vdash \bot}{\Gamma \vdash \varphi} (\bot\text{E}) \qquad \frac{\Gamma, \varphi \to \psi \vdash \varphi}{\Gamma \vdash \varphi} (\text{P}).$$

Using the rules ($\to$E), ($\to$I), and (Ax), the two above rules can be derived from ($\neg$E)—see Examples 6.1.2 and 6.1.3—and vice versa.

Moreover, we can replace the rule (P) by the special case (P$\bot$), which is very similar to the formula (iv) mentioned in the introduction:

$$\frac{\Gamma, \varphi \to \bot \vdash \varphi}{\Gamma \vdash \varphi} (\text{P}\bot).$$

Using *ex falso*, (P) can be derived from (P$\bot$).

In the case of Hilbert-style systems, there are several ways to obtain classical logic from intuitionistic logic. The following is one possibility.

6.1.6. DEFINITION. We consider the same terminology and notation as in Chapter 5. The language of propositions is extended with $\bot$, and we use the abbreviation $\neg \varphi = \varphi \to \bot$ again. The Hilbert-style system for classical logic is obtained by adding the following axiom scheme:

$$(A3_\bot) \quad ((\varphi \to \bot) \to \bot) \to \varphi$$

to the system with $(A1)$ and $(A2)$. We write $\Gamma \vdash_H \varphi$ if $\Gamma \vdash \varphi$ can be derived in the system from Definition 5.1.2 with these three axioms.

The following is an immediate consequence of Theorem 5.1.8, because classical provability can be seen as intuitionistic provability with some additional axioms.

6.1.7. THEOREM (Deduction theorem). $\Gamma, \varphi \vdash_H \psi$ iff $\Gamma \vdash_H \varphi \to \psi$.

6.1.8. PROPOSITION. $\Gamma \vdash_N \varphi$ iff $\Gamma \vdash_H \varphi$.

PROOF. The only new issue compared to the intuitionistic case is to show that one can express $(\neg E)$ using $(A3_\bot)$ and vice versa. For the former, assume that we have a Hilbert-style derivation of $\Gamma, \neg \varphi \vdash \bot$. Then by the deduction theorem, $(A3_\bot)$, and *modus ponens* we obtain a derivation of $\Gamma \vdash \varphi$. Conversely, the natural deduction derivation

$$\frac{\dfrac{\dfrac{\Gamma, \neg\neg\varphi, \neg\varphi \vdash \neg\neg\varphi \quad \Gamma, \neg\neg\varphi, \neg\varphi \vdash \neg\varphi}{\Gamma, \neg\neg\varphi, \neg\varphi \vdash \bot}}{\Gamma, \neg\neg\varphi \vdash \varphi}}{\Gamma \vdash \neg\neg\varphi \to \varphi}$$

shows that one can express $(A3_\bot)$ using $(\neg E)$. □

In view of the above proposition it is natural to talk about "classical propositional calculus" independently of the chosen formalism.

6.1.9. NOTATION. The phrase "theorem of $\text{CPC}(\to, \bot)$" means a formula that can be proved in $\text{NK}(\to, \bot)$, or in the Hilbert-style system of Definition 6.1.6, or in any other equivalent system. Similar terminology will be employed for other sets of connectives, e.g. including $\wedge$ and $\vee$.

6.1.10. THEOREM (Completeness). *Every theorem of* $\text{CPC}(\to, \bot)$ *is a classical tautology, and vice versa.*

PROOF (Kalmár, 1935). We show that $\vdash_H \varphi$ iff φ is a tautology. From left to right, show that the three axioms are tautologies, and show the following property: if $\varphi \to \psi$ is a tautology and φ is a tautology, so is ψ.

For the direction from right to left, we proceed as follows. Given a valuation v, let

$$\psi' = \begin{cases} \psi & \text{if } [\![\psi]\!]_v = 1 \\ \neg\psi & \text{if } [\![\psi]\!]_v = 0. \end{cases}$$

6.1. CLASSICAL PROPOSITIONAL LOGIC   131

We show the following claim: for any valuation v and formula φ,
$$p'_1, \ldots, p'_n \vdash \varphi', \qquad (*)$$
where $p_1, \ldots, p_n$ include all the propositional variables in φ. The proof of this result is by induction on φ. The only interesting case is when $\varphi = \varphi_1 \to \varphi_2$. We consider three subcases.

CASE 1: $[\![\varphi_1]\!]_v = 0$. Then $[\![\varphi_1 \to \varphi_2]\!]_v = 1$. By the induction hypothesis,
$$p'_1, \ldots, p'_n \vdash \neg \varphi_1.$$
Now prove $\vdash \neg\varphi_1 \to (\varphi_1 \to \varphi_2)$ and use *modus ponens* to obtain:
$$p'_1, \ldots, p'_n \vdash \varphi_1 \to \varphi_2.$$
CASE 2: $[\![\varphi_2]\!]_v = 1$. Then $[\![\varphi_1 \to \varphi_2]\!]_v = 1$. By the induction hypothesis, $p'_1, \ldots, p'_n \vdash \varphi_2$. Then use axiom (A1) and *modus ponens* to obtain:
$$p'_1, \ldots, p'_n \vdash \varphi_1 \to \varphi_2.$$
CASE 3: $[\![\varphi_1]\!]_v = 1$ and $[\![\varphi_2]\!] = 0$. Then $[\![\varphi_1 \to \varphi_2]\!]_v = 0$. By the induction hypothesis,
$$p'_1, \ldots, p'_n \vdash \varphi_1 \qquad p'_1, \ldots, p'_n \vdash \neg\varphi_2.$$
Now prove $p'_1, \ldots, p'_n \vdash \varphi_1 \to \neg\varphi_2 \to \neg(\varphi_1 \to \varphi_2)$. Then by *modus ponens*,
$$p'_1, \ldots, p'_n \vdash \neg(\varphi_1 \to \varphi_2).$$
This concludes the proof of $(*)$. Now let φ be a tautology with propositional variables $p_1, \ldots, p_n$. We show for all $i \in \{0, \ldots, n\}$ and each valuation v that
$$p'_1, \ldots, p'_i \vdash \varphi.$$
The desired result is then the special case $i = 0$.

The proof is by induction on $n - i$. If $i = n$, the claim follows from $(*)$. If $i < n$, let v be any valuation. Consider the two different valuations arising from v by assigning 0 and 1 to p_{i+1}. We have by the induction hypothesis that $p'_1, \ldots, p'_i, p_{i+1} \vdash \varphi$ and $p'_1, \ldots, p'_i, \neg p_{i+1} \vdash \varphi$, and then also $p'_1, \ldots, p'_i \vdash \varphi$ using Example 6.1.2.   □

Since every theorem is a classical tautology, consistency follows, i.e. $\not\vdash \bot$. In fact, our logic is *maximally* consistent in the following sense.

6.1.11. PROPOSITION. *Let φ be a formula which is not a tautology. Consider the system Z obtained from Definition 6.1.6 by adding φ as an axiom scheme. The system Z is inconsistent.*

PROOF. Let $p_1, \ldots, p_n$ be the propositional letters in φ. Since φ is not a tautology there is a valuation v such that $[\![\varphi]\!]_v = 0$. Then let

$$\psi_i = \begin{cases} p_i \to p_i & \text{if } v(p_i) = 1 \\ \neg(p_i \to p_i) & \text{if } v(p_i) = 0. \end{cases}$$

Let ψ be obtained from φ by replacing all p_i by ψ_i. We have $\vdash \psi$ in Z. However, $[\![\psi]\!]_u = 0$ for all valuations u, so $[\![\neg\psi]\!]_u = 1$. By completeness, $\vdash \neg\psi$ in classical logic, and hence also in Z. Thus $\vdash \bot$ in Z. □

Thus, with classical logic the addition of new axiom schemes must come to an end; otherwise, the system becomes inconsistent.

6.2. The $\lambda\mu$-calculus

We now introduce the $\lambda\mu$-calculus, which bears the same relationship to classical propositional logic as the simply typed λ-calculus bears to minimal propositional logic.[1] We consider a formulation in the Church style and use the notation, terminology, and conventions from Chapter 3.

6.2.1. DEFINITION.

(i) Let A be a denumerably infinite set of *addresses* ranged over by a, let V be a denumerably infinite set of variables ranged over by x, and let σ range over $\Phi(\to, \bot)$. The set of raw $\lambda\mu$-terms is defined as follows:

$$M ::= x \mid (MM) \mid (\lambda x{:}\sigma M) \mid ([a]M) \mid (\mu a{:}\neg\sigma M).$$

We write $\varepsilon_\varphi(M) = \mu{:}\neg\varphi M$, when $a \notin \mathrm{FV}(M)$.

(ii) $\mathrm{FV}(M)$ denotes the set of variables and addresses occurring free in M, defined as usual, taking into account that $(\mu a{:}\neg\sigma M)$ binds a in M. Terms differing only in bound variables and addresses are identified.

(iii) $M[x := N]$ denotes the result of substituting the term N for free occurrences of the variable x in M, defined in the usual way.

(iv) An environment Γ is a set of pairs of form $x : \tau$ and $a : \neg\sigma$, such that $x : \tau_1 \in \Gamma$ and $x : \tau_2 \in \Gamma$ implies $\tau_1 = \tau_2$, and similarly for addresses.

(v) We write $\Gamma \vdash M : \sigma$ when this follows from the rules in Figure 6.2. Following Convention 3.4.2, $\lambda\mu$-terms are always assumed typed.

6.2.2. EXAMPLE. Peirce's law $((p \to q) \to p) \to p$ is inhabited by the term

$$\lambda x{:}(p \to q) \to p.\mu a{:}\neg p.[a](x(\lambda z{:}p.\mu b{:}\neg q.[a]z)).$$

---

[1] The $\lambda\mu$-calculus should not be confused with μ-calculus, which is an entirely different system that is not discussed in this book—see [14].

6.2. The λμ-calculus

$$\Gamma, x:\tau \vdash x:\tau$$

$$\frac{\Gamma, x:\sigma \vdash M:\tau}{\Gamma \vdash (\lambda x{:}\sigma M) : \sigma \to \tau} \qquad \frac{\Gamma \vdash M:\sigma \to \tau \quad \Gamma \vdash N:\sigma}{\Gamma \vdash (MN):\tau}$$

$$\frac{\Gamma, a:\neg\sigma \vdash M:\bot}{\Gamma \vdash (\mu a{:}\neg\sigma M):\sigma} \qquad \frac{\Gamma, a:\neg\sigma \vdash M:\sigma}{\Gamma, a:\neg\sigma \vdash ([a]M):\bot}$$

FIGURE 6.2: λμ-CALCULUS.

REMARK. Terms of the form $[a]M$ and $\mu a{:}\neg\sigma.M$ are called *address application* and *abstraction*, respectively. The address application rule (lower right corner) has no counterpart in the logical system, which does not distinguish between variables and addresses. Also, the axiom $\Gamma, x : \tau \vdash x : \tau$ has no companion for addresses, since these are not terms. But we have $\Gamma, a{:}\neg\sigma \vdash \lambda x{:}\sigma.[a]x : \neg\sigma$.

The following properties are proved similarly to Lemmas 3.1.4–3.1.6.

6.2.3. LEMMA.

(i) *If* $\Gamma \vdash M : \tau$ *then* $\mathrm{FV}(M) \subseteq \mathrm{dom}(\Gamma)$.

(ii) *If* $\Gamma, x:\tau \vdash M : \sigma$ *and* $y \notin \mathrm{dom}(\Gamma) \cup \{x\}$ *then* $\Gamma, y:\tau \vdash M[x := y] : \sigma$.

6.2.4. LEMMA (Generation lemma). *Suppose that* $\Gamma \vdash M : \tau$. *Then:*

(i) *If M is a variable x then* $\Gamma(x) = \tau$;

(ii) *If M is an application PQ then* $\Gamma \vdash P : \sigma \to \tau$, *and* $\Gamma \vdash Q : \sigma$, *for some σ;*

(iii) *If M is an abstraction $\lambda x{:}\tau_1.P$ and $x \notin \mathrm{dom}(\Gamma)$ then $\tau = \tau_1 \to \tau_2$, where* $\Gamma, x:\tau_1 \vdash P : \tau_2$.

(iv) *If M is an application $[a]P$ then $\tau = \bot$ and for some σ it holds that* $\Gamma(a) = \neg\sigma$ *and* $\Gamma \vdash P : \sigma$;

(v) *If M is an abstraction $\mu a{:}\neg\sigma.P$ and $a \notin \mathrm{dom}(\Gamma)$ then $\tau = \sigma$ and* $\Gamma, a:\neg\sigma \vdash P : \bot$.

6.2.5. LEMMA.

(i) *If $\Gamma \vdash M : \sigma$ and $\Gamma(x) = \Gamma'(x)$ for all $x \in \mathrm{FV}(M)$ then $\Gamma' \vdash M : \sigma$.*

(ii) *If $\Gamma, x : \tau \vdash M : \sigma$ and $\Gamma \vdash N : \tau$ then $\Gamma \vdash M[x := N] : \sigma$.*

Inhabitation in $\lambda\mu$-calculus corresponds to provability in $\mathrm{NK}(\to, \bot)$.

6.2.6. PROPOSITION (Curry-Howard isomorphism).

(i) *If $\Gamma \vdash M : \varphi$ in $\lambda\mu$ then $\mathrm{rg}(\Gamma) \vdash \varphi$ in $\mathrm{CPC}(\to, \bot)$.*

(ii) *If $\Gamma \vdash \varphi$ in $\mathrm{CPC}(\to, \bot)$, then there are Δ, M with $\Delta \vdash M : \varphi$ in $\lambda\mu$ and $\mathrm{rg}(\Delta) = \Gamma$.*

PROOF. For part (i) we show by induction on M that $\mathrm{rg}(\Gamma) \vdash \varphi$ in $\mathrm{NK}(\to, \bot)$. For part (ii) we show by induction on the natural deduction derivation of $\Gamma \vdash \varphi$ that $\mathrm{lf}(\Gamma) \vdash M : \varphi$ for some M, where $\mathrm{lf}(\Gamma) = \{(x_\varphi : \varphi) \mid \varphi \in \Gamma\}$ and x_φ is a unique variable for each type φ. The only new case compared to Proposition 4.1.1 is when the derivation ends with ($\neg$E)

$$\frac{\Gamma, \neg\varphi \vdash \bot}{\Gamma \vdash \varphi}.$$

We consider two subcases.
CASE 1: $\neg\varphi \in \Gamma$. By the induction hypothesis there is M with $\mathrm{lf}(\Gamma) \vdash M : \bot$. Then $\mathrm{lf}(\Gamma), a : \neg\varphi \vdash M : \bot$, so $\mathrm{lf}(\Gamma) \vdash \mu a : \neg\varphi.M : \varphi$.
CASE 2: $\neg\varphi \notin \Gamma$. By the induction hypothesis there is an M with

$$\mathrm{lf}(\Gamma), x_{\neg\varphi} : \neg\varphi \vdash M : \bot.$$

Then $\mathrm{lf}(\Gamma), a : \neg\varphi, x_{\neg\varphi} : \neg\varphi \vdash M : \bot$. Moreover, $\mathrm{lf}(\Gamma), a : \neg\varphi \vdash \lambda y : \varphi.[a]y : \neg\varphi$, so $\mathrm{lf}(\Gamma), a : \neg\varphi \vdash M[x_{\neg\varphi} := \lambda y : \varphi.[a]y] : \bot$ by Lemma 6.2.5(ii). Then finally $\mathrm{lf}(\Gamma) \vdash \mu a : \neg\varphi.M[x_{\neg\varphi} := \lambda y : \varphi.[a]y] : \varphi$, as required. □

Computationally, μ-abstraction and address application can be seen as a *channel* device. In the term $\mu a : \neg\sigma. \cdots [a]M \cdots$, the address a is a channel that values can be transmitted along, and the subterm $[a]M$ represents transmittal of M to $\mu a : \neg\sigma. \cdots$. Roughly speaking, the value of such a term should be M. Such devices are usually known as *jumps* or *control operators*.

Next we introduce reduction on $\lambda\mu$-terms to capture this idea. For this end it is useful to introduce $\lambda\mu$-*contexts*.

6.2.7. DEFINITION.

(i) The set of (*applicative*) $\lambda\mu$-*contexts* is defined by:

$$C ::= [\,] \mid CM \mid [a]C.$$

6.2. THE λμ-CALCULUS

(ii) The λμ-term $C[M]$ is defined by:

$$\begin{aligned} {[\,]}[M] &= M; \\ (CN)[M] &= C[M]N; \\ ([a]\,C)[M] &= [a]\,C[M]. \end{aligned}$$

The λμ-context $C[D]$ is defined as $C[M]$ (replace M by D everywhere).

(iii) The set $\mathrm{FV}(C)$ is defined by:

$$\begin{aligned} \mathrm{FV}([\,]) &= \varnothing; \\ \mathrm{FV}(CM) &= \mathrm{FV}(M) \cup \mathrm{FV}(C); \\ \mathrm{FV}([a]\,C) &= \mathrm{FV}(C) \cup \{a\}. \end{aligned}$$

(iv) For a λμ-context C and an address a, define $M[a := C]$ as follows (where $y, b \notin \mathrm{FV}(C)$ and $b \neq a$):

$$\begin{aligned} x[a:=C] &= x; \\ (\lambda y\!:\!\sigma.M)[a:=C] &= \lambda y\!:\!\sigma.M[a:=C]; \\ (M_1\,M_2)[a:=C] &= M_1[a:=C]\,M_2[a:=C]; \\ (\mu b\!:\!\neg\sigma.M)[a:=C] &= \mu b\!:\!\neg\sigma.M[a:=C]; \\ ([a]M)[a:=C] &= C[M[a:=C]]; \\ ([b]M)[a:=C] &= [b]M[a:=C]. \end{aligned}$$

(v) Let $\to_\beta, \to_{\eta_\mu}, \to_{\beta_\mu}, \to_\zeta$ denote the least compatible relations satisfying

$$\begin{aligned} (\lambda x\!:\!\varphi.M)N &\to_\beta M[x := N]; \\ \mu a\!:\!\neg\varphi.[a]M &\to_{\eta_\mu} M \text{ if } a \notin \mathrm{FV}(M); \\ [b](\mu a\!:\!\neg\varphi.M) &\to_{\beta_\mu} M[a := [b]([\,])]; \\ (\mu a\!:\!\neg(\varphi\to\psi).M)N &\to_\zeta \mu b\!:\!\neg\psi.M[a := [b]([\,]N)] \text{ if } a \neq b \notin \mathrm{FV}(MN) \end{aligned}$$

By $\to_\mu$ we denote the union of these relations and adopt the usual terminology μ-redex, etc.

6.2.8. EXAMPLE. Let M, N be terms of type $p \to p$ and $q \to q$, respectively. The term $\mu a\!:\!\neg(p \to p).[a]((\varepsilon_{(q\to q)\to(p\to p)}([a]M))\,N)$ reduces to M. Indeed, by using all three reduction rules for μ:

$$\begin{aligned} &\mu a\!:\!\neg(p \to p).[a]((\varepsilon_{(q\to q)\to(p\to p)}([a]M))\,N) \\ \to_\zeta\ &\mu a\!:\!\neg(p \to p).[a](\varepsilon_{p\to p}([a]M)) \\ \to_{\beta_\mu}\ &\mu a\!:\!\neg(p \to p).[a]M \\ \to_{\eta_\mu}\ &M. \end{aligned}$$

In fact, in the example we only use the following special case of (β_μ) and (ζ):

$$\begin{aligned} {[b](\varepsilon_\varphi(L))} &\to L; \\ (\varepsilon_{\varphi\to\psi}(L))J &\to \varepsilon_\psi(L). \end{aligned}$$

By iteration of the latter, we see that for a context C with no free addresses, $C[\varepsilon_\rho(L)] \twoheadrightarrow_\mu \varepsilon_\sigma(L)$ for appropriate σ, so $\mu a\!:\!\neg\sigma.[a]C[\varepsilon_\rho([a]M)] \twoheadrightarrow_\mu M$, provided neither C nor M contains free addresses.

The rule (η_μ) connects transmittal and receipt of a result along a channel. The rule (β_μ) is a form of "jump optimization." For a term $[b](\mu a\!:\!\neg\varphi.M)$, suppose that inside M some term L is transmitted to a. It will be received at μa, and then transmitted to b. It would be simpler to transmit L directly to b. This is obtained by finding all places in M where we transmit to a, and in each place transmit to b instead. That is exactly what the term $M[a := [b]\,[\,]]$ means. The rule (ζ) states that rather than receiving an abstraction and then applying it to some argument, we might as well move the application down to the transmitted term, and receive the result instead.

Proof-theoretically (η_μ) concerns a proof in which we apply an assumption $\neg\varphi$ to a proof of φ that does not make use of the assumption, and then infer φ using proof by contradiction. Instead the proof of φ can be used directly. The rule (ζ) concerns the situation in which a proof of $\varphi \to \psi$ is constructed by assuming $\neg(\varphi \to \psi)$ and deriving $\bot$, i.e. the proof is $\mu a\!:\!\neg(\varphi \to \psi).M$. Then this proof is applied to a proof N of φ to obtain a proof of ψ. In practice, a proof by contradiction of $\varphi \to \psi$ is often constructed by assuming φ and $\neg\psi$, and deriving a contradiction. In fact, this latter proof can be obtained from $\mu a\!:\!\neg(\varphi \to \psi).M$ thus:

$$\lambda x\!:\!\varphi.\mu b\!:\!\neg\psi.M[a := [b]\,([\,]\,x)]. \qquad (*)$$

Since we have a proof N of φ available, we can substitute this proof for x, arriving at the right hand side of (ζ). Reducing $\mu a\!:\!\neg(\varphi \to \psi).M$ to $(*)$ can be considered an "eager" variant of (ζ), which we return to in Section 6.5.

6.2.9. EXAMPLE. The programming language Lisp [200] has control operators **catch** and **throw**. Evaluation of a term[2]

$$\textbf{catch } a \textbf{ in } M \qquad (*)$$

proceeds first with M. If the latter yields some result, then this also becomes the overall result of the term $(*)$. This is known as a *normal return*. If, however, we encounter the term

$$\textbf{throw } N \textbf{ to } a$$

then evaluation of M is aborted. Instead the result of N becomes the overall result of $(*)$. This is an *exceptional return*. Thus, $1 + \textbf{catch } a \textbf{ in } (10 + 41)$ gives 52, whereas $1 + \textbf{catch } a \textbf{ in } (10 + \textbf{throw } 41 \textbf{ to } a)$ results in 42.

---

[2]This example, though referring to Lisp, does not use official Lisp syntax.

6.3. SUBJECT REDUCTION, CONFLUENCE, STRONG NORMALIZATION

The following shows how **catch** and **throw** can be expressed with μ:

$$\begin{aligned}\textbf{catch } a \textbf{ in } M &= \mu a{:}\neg\sigma.[a]M; \\ \textbf{throw } M \textbf{ to } a &= \varepsilon_\rho([a]M)\end{aligned}$$

(where M has type σ, and ρ depends on the context). The conclusion of Example 6.2.8 shows that **catch** a **in** $C[\textbf{throw } M \textbf{ to } a]$ indeed reduces to M, provided there are no free addresses in M and C.

6.3. Subject reduction, confluence, strong normalization

We now present three of the main properties of μ-reduction: subject reduction, confluence, and strong normalization.

6.3.1. LEMMA. *If* $\Gamma, a : \neg\varphi \vdash M : \psi$ *and* $\Gamma, z : \varphi \vdash C[z] : \bot$, *then also* $\Gamma \vdash M[a := C] : \psi$, *provided* $z \notin \mathrm{FV}(C)$.

PROOF. First show, by induction on C, that if $\Gamma, z : \varphi \vdash C[z] : \psi$, where $z \notin \mathrm{FV}(C)$, and $\Gamma \vdash L : \varphi$, then $\Gamma \vdash C[L] : \psi$. Using this, then prove the lemma by induction on M. □

6.3.2. THEOREM (Subject reduction). *If* $M \to_\mu N$ *and* $\Gamma \vdash M : \varphi$, *then* $\Gamma \vdash N : \varphi$.

PROOF. Induction on $M \to_\mu N$ using Lemmas 6.2.5 and 6.3.1. □

We prove strong normalization by translating $\lambda\mu$-terms to simply typed λ-terms in such a way that the translation respects reductions. For the proof it is convenient to assume that types of all variables and addresses are fixed, i.e. that we deal with "orthodox Church-style" version of $\lambda\mu$. In contrast, our translation is to Curry-style $\lambda_\to$, because it is simpler this way. But we assume that the simple types include $\bot$ (we can use some fixed type variable).

6.3.3. NOTATION. For every address a of type $\neg(\sigma_1 \to \cdots \to \sigma_n \to \alpha)$, we fix a sequence of fresh variables $\vec{y}^a = y_1^a, \ldots, y_n^a$, one for each argument.

6.3.4. DEFINITION (Ong). To a $\lambda\mu$-term M we assign a pure λ-term $\underline{M}$:

$$\begin{aligned}\underline{x} &= x \\ \underline{\lambda x{:}\sigma.M} &= \lambda x\,\underline{M} \\ \underline{MN} &= \underline{M}\,\underline{N} \\ \underline{\mu a{:}\neg\sigma.M} &= \lambda \vec{y}^a\,\underline{M} \\ \underline{[a]M} &= \underline{M}\,\vec{y}^a\end{aligned}$$

For a type σ, define $\underline{\sigma}$ by replacing all type variables in σ by $\bot$. The following two lemmas follow by routine induction.

6.3.5. LEMMA. *If M is a $\lambda\mu$-term of type σ then $\Gamma \vdash \underline{M} : \underline{\sigma}$ in the Curry-style $\lambda_\to$, where Γ consists of the declarations:*

- $x : \underline{\tau}$, *for each free variable x of M of type τ;*
- $y_1^a : \underline{\tau_1}, \ldots, y_n^a : \underline{\tau_n}$, *for each free address a of type $\neg(\tau_1 \to \cdots \to \tau_n \to \alpha)$.*

6.3.6. LEMMA. *The following equations hold (note that the type of a in (iii) is different than the type of c) :*

(i) $\underline{P[x := Q]} = \underline{P}[x := \underline{Q}]$.

(ii) $\underline{P[a := [c][\,]]} = \underline{P}[\vec{y}^a := \vec{y}^c]$.

(iii) $\underline{P[a := [c]([\,]Q)]} = \underline{P}[\vec{y}^a := Q, \vec{y}^c]$.

6.3.7. LEMMA. *If $M \to_\mu N$ in $\lambda\mu$, then $\underline{M} \twoheadrightarrow_{\beta\eta} \underline{N}$. Moreover, if $M \to_{\beta\zeta} N$ then $\underline{M} \to_\beta \underline{N}$.*

PROOF. Induction on the definition of $M \to_\mu N$. Consider, as an example, the case where

$$M = (\mu a {:} \neg(\sigma \to \tau).P)\, Q \to \mu b {:} \neg \tau.P[a := [b]([\,]Q)] = N.$$

By the preceding lemma, we have

$$\underline{M} = (\lambda x \vec{y}.\, \underline{P})Q \to_\beta \lambda \vec{y}.\, \underline{P}[x := \underline{Q}] = \lambda \vec{z}.\, \underline{P}[x, \vec{y} := \underline{Q}, \vec{z}]$$
$$= \lambda \vec{z}.\, \underline{P[a := [b]([\,]Q)]} = \underline{\mu b {:} \neg \tau.P[a := [b]([\,]Q)]} = \underline{N}. \qquad \Box$$

6.3.8. THEOREM. *Each $\lambda\mu$-term is strongly normalizing.*

PROOF. For the sake of a contradiction assume that M is a $\lambda\mu$-term with an infinite μ-reduction

$$M \to_\mu M' \to_\mu M'' \to_\mu \cdots \qquad (*)$$

By the preceding proposition,

$$\underline{M} \twoheadrightarrow_{\beta\eta} \underline{M'} \twoheadrightarrow_{\beta\eta} \underline{M''} \twoheadrightarrow_{\beta\eta} \cdots \qquad (+)$$

Since every $\beta_\mu \eta_\mu$-reduction decreases the number of occurrences of μ in the term, there are no infinite $\beta_\mu \eta_\mu$-reductions. Hence, the reduction $(*)$ must have infinitely many $\beta\zeta$-reductions, and the preceding proposition then implies that there are infinitely many non-empty steps in the reduction $(+)$. By Proposition 6.3.5 we then have a simply typed λ-term $\underline{M}$ with an infinite $\beta\eta$-reduction. By Lemma 1.3.11, this contradicts strong normalization of simply typed λ-calculus. $\qquad \Box$

6.3. SUBJECT REDUCTION, CONFLUENCE, STRONG NORMALIZATION

We finally show that the relation $\to_\mu$ is confluent, i.e. that it satisfies the Church-Rosser property. We use Newman's lemma.

6.3.9. THEOREM. *The relation $\to_\mu$ is weakly Church-Rosser.*

PROOF. We show that if $M_1 \to_\mu M_2$ and $M_1 \to_\mu M_3$ then there is M_4 with $M_2 \twoheadrightarrow_\mu M_4$ and $M_3 \twoheadrightarrow_\mu M_4$. We proceed by induction on the derivation of $M_1 \to_\mu M_2$. Consider first the following cases (where $a \notin \text{FV}(R)$):

CASE 1:
$$M_1 = (\mu a{:}\neg(\varphi \to \psi).[a]R)\, Q \longrightarrow M_2 = R\, Q$$
$$\downarrow$$
$$M_3 = \mu b{:}\neg\psi.([a]R)[a := [b]([]Q)]$$

CASE 2:
$$M_1 = \mu a{:}\neg\varphi.[a]\mu b{:}\neg\varphi.R \longrightarrow M_2 = \mu b{:}\neg\varphi.R$$
$$\downarrow$$
$$M_3 = \mu a{:}\neg\varphi.R[b := [a][\,]]$$

CASE 3:
$$M_1 = [b]\, \mu a{:}\neg\varphi.[a]R \longrightarrow M_2 = [b]\, R$$
$$\downarrow$$
$$M_3 = ([a]R)[a := [b][\,]]$$

In the remaining cases use this property: If $M \to_\mu N$, then

(i) $M[a := [b]([\,]L)] \twoheadrightarrow_\mu N[a := [b]([\,]L)]$.

(ii) $L[a := [b]([\,]M)] \twoheadrightarrow_\mu L[a := [b]([\,]N)]$.

(iii) $M[x := L] \to_\mu N[x := L]$.

(iv) $L[x := M] \twoheadrightarrow_\mu L[x := N]$. □

6.3.10. REMARK. The Cases 1–3 in the above proof cover all *critical pairs*. A critical pair occurs when two redexes *overlap* in such a way that reducing either of them requires and "consumes" the same "resource" (a syntactic construct). For instance, the term M_1 in Case 1 is a ζ-redex, and the subterm $\mu a{:}\neg(\varphi \to \psi).[a]R$ is also an η_μ-redex. Each of these redexes "consumes" $\mu a{:}\neg(\varphi \to \psi)\ldots$ thus disallowing the reduction of the other.

6.3.11. THEOREM (Church-Rosser). *If $M_1 \twoheadrightarrow_\mu M_2$ and $M_1 \twoheadrightarrow_\mu M_3$, then there is an M_4 with $M_2 \twoheadrightarrow_\mu M_4$ and $M_3 \twoheadrightarrow_\mu M_4$.*

PROOF. By Newman's lemma, strong normalization, and the weak Church-Rosser property. □

6.4. Logical embedding and CPS translation

Interpretations of classical logic within intuitionistic logic, known as *logical embeddings*, have been studied since the 1920s. In this section we study one such embedding, known as the *Kolmogorov double negation translation*. It has an interesting computational interpretation.

6.4.1. DEFINITION (Kolmogorov translation). Define a translation as follows, where α ranges over atomic propositions:

$$k(\alpha) = \neg\neg\alpha$$
$$k(\varphi \to \psi) = \neg\neg(k(\varphi) \to k(\psi)).$$

Moreover, $k(\Gamma) = \{k(\varphi) \mid \varphi \in \Gamma\}$.

6.4.2. REMARK. Clearly $k(\varphi) = \neg\neg\varphi^*$, where φ^* is defined by:

$$\alpha^* = \alpha$$
$$(\varphi \to \psi)^* = \neg\neg\varphi^* \to \neg\neg\psi^*.$$

6.4.3. THEOREM.

(i) $\vdash \varphi \to k(\varphi)$ and $\vdash k(\varphi) \to \varphi$ in $\mathrm{CPC}(\to, \bot)$.

(ii) $\vdash \varphi$ in $\mathrm{CPC}(\to, \bot)$ iff $\vdash k(\varphi)$ in $\mathrm{IPC}(\to, \bot)$.

PROOF. For (i) show by induction on φ that for any valuation v, we have $[\![\varphi]\!]_v = [\![k(\varphi)]\!]_v$, and use completeness.

The right-to-left direction in (ii) follows from (i). For the left-to-right direction, we show by induction on the derivation of $\Gamma \vdash \varphi$ in $\mathrm{NK}(\to, \bot)$ that $k(\Gamma) \vdash k(\varphi)$ in $\mathrm{NJ}(\to, \bot)$. The case (Ax) is trivial, and for ($\to$I) use Example 2.1.3(iv). For ($\to$E) use Remark 6.4.2 and Exercise 2.24(i). Finally, assume the derivation ends with

$$\frac{\Gamma, \neg\varphi \vdash \bot}{\Gamma \vdash \varphi}.$$

By the induction hypothesis $k(\Gamma), \neg\neg(k(\varphi) \to \neg\neg\bot) \vdash \neg\neg\bot$, so Example 2.1.3(iv) yields $k(\Gamma), k(\varphi) \to \bot \vdash \bot$. Now use Example 2.1.3(v) and Remark 6.4.2. □

The proof of part (ii) induces a translation of classical proofs to intuitionistic proofs. In terms of $\lambda\mu$ this amounts to a translation that eliminates occurrences of μ. The remainder of the section explains the computational significance of the translation.

Consider the term[3]

$$(22 + 19) + 1.$$

---

[3] To reduce clutter we temporarily adopt Curry style and admit numbers and infix addition as $\lambda\mu$-terms.

6.4. LOGICAL EMBEDDING AND CPS TRANSLATION

In order to calculate the result, we must reduce $22 + 19$, because the surrounding part of the term (the context $[\,] + 1$) requires a number before the addition with 1 can be calculated. The context forces the evaluation of $22 + 19$, and once it has been reduced to 41, the resulting term $41 + 1$ can be reduced to 42.

In *continuation passing style* (CPS), this idea is turned upside down. Every subterm M becomes parameterized by a function, a *continuation*, that, when provided with the value of M, returns the overall result of the term. For example, the term 22 becomes $\lambda l.l\, 22$, which is parameterized by a continuation l, and since 22 is a number, i.e. no computation is required to calculate the value of 22, the continuation is applied directly to 22.

The composite term $22 + 19$ becomes the more complicated term:

$$\lambda h.\underline{22}(\lambda u.\underline{19}(\lambda v.h(u+v)))$$

in which $\underline{22}$ and $\underline{19}$ denote the CPS versions $\lambda l.l\, 22$ and $\lambda i.i\, 19$ of 22 and 19, respectively, and in which the structure $\lambda h.E'(\lambda u.F'(\lambda v.h(u+v)))$ is the CPS version of $E + F$ (when E' and F' are CPS versions of E and F, respectively). This term is parameterized by a continuation h. Given such a continuation, evaluation of the term first produces the value 22, and it then applies the continuation $\lambda u. \ldots$ to it, leading to the production of 19 and the application of the continuation $\lambda v. \ldots$ to this value. Finally, the overall continuation h is applied to the sum 41.

The term $(22 + 19) + 1$ becomes:

$$\lambda k.[\lambda h.\underline{22}(\lambda u.\underline{19}(\lambda v.h(u+v)))](\lambda m.\underline{1}(\lambda n.k(m+n))).$$

In this term h becomes the continuation $\lambda m. \ldots$ corresponding to the context $[\,] + 1$. Indeed, the result of applying the continuation $\lambda m. \ldots$ to 41 is that 1 is produced and the continuation $\lambda n. \ldots$ applied to it, leading finally to the overall value 42 and the overall continuation k being applied to it.

If $(22 + 19) + 1$ is the full term to be expressed in CPS we must apply the above CPS term to the *top-level continuation* $\lambda e.e$. Then k will become bound to $\lambda e.e$, indicating that the number 42 is returned without any further processing. The reader should check that we indeed get 42 by reducing

$$\{\lambda k.[\lambda h.\underline{22}(\lambda u.\underline{19}(\lambda v.h(u+v)))](\lambda m.\underline{1}(\lambda n.k(m+n)))\}(\lambda e.e).$$

This may seem like an awfully complicated way to add a few numbers, but writing programs in CPS has many advantages—see Section 6.8.

CPS is a certain style that terms can be written in. We can translate arbitrary terms into this style with the *CPS translation*

$$\begin{array}{rcl}\underline{n} & = & \lambda l.l\, n \\ \underline{M+N} & = & \lambda k.\underline{M}(\lambda m.\underline{N}(\lambda n.k(m+n))).\end{array}$$

Then $(22 + 19) + 1$ ($\lambda e.e$) is our CPS term reducing to 42.

How do we CPS translate abstractions and applications? Like a number, an abstraction $\lambda x.M$ is a value in the sense that no further evaluation is required before a redex $(\lambda x.M) N$ can be reduced. However, inside M there can be computations and these should be transformed as well. Thus,

$$\underline{\lambda x.M} = \lambda k.k(\lambda x.\underline{M})$$

is natural. As for an application MN, we expect that M be reduced to a value before something can be done to the application. Hence the translation must have form $\underline{MN} = \lambda k.\underline{M}(\lambda m \ldots m\underline{N} \ldots)$. The term $m\underline{N}$, where m gets bound to a value of $\underline{M}$, may again require reduction to reach a value, so $m\ \underline{N}$ will expect a continuation that can be applied once the value has been reached. Thus we should apply $m\ \underline{N}$ to the continuation k:

$$\underline{MN} = \lambda k.\underline{M}\ (\lambda m.m\ \underline{N}\ k).$$

Since $(\lambda x.M)N \twoheadrightarrow \lambda k.\underline{M}[x := \underline{N}]k$, i.e. x becomes bound to a translated term that expects a continuation, the natural translation of x is

$$\underline{x} = \lambda k.x\ k.$$

With control operators we can manipulate the context. For instance, in

$$1 + \mu a.[a](20 + \varepsilon([a]22)) \qquad (*)$$

the subterm $[a]\ 22$ skips part of the context, namely $[a](20 + \varepsilon([\,]))$. Thus, the translation of $\mu a \ldots$ should introduce an explicit continuation k_a corresponding to the context $1 + [\,]$ of this term, and the translation of $[a]22$ must discard the continuation corresponding to its context and instead use k_a. What continuation should the translation of M be applied to in the translation of $\mu a.M$? We expect that inside M there will be a result transmitted to $\mu a.M$, so M as such will not return any value. Thus, the continuation that M is applied to will be discarded, so any choice will do, e.g. $\lambda v.v$. Using a as the name for k_a, we have:

$$\begin{aligned}\underline{\mu a.M} &= \lambda a.\underline{M}(\lambda v.v) \\ \underline{[a]M} &= \lambda d.\underline{M}a.\end{aligned}$$

With these definitions, the CPS version of $(*)$, applied to top-level continuation $\lambda e.e$, indeed reduces to 23.

How about typing? If we consider an overall term O of type **int**, and a subterm N of type **int**, then $\underline{N}$ becomes parameterized by a continuation which, when supplied with the value of N, will return the overall result of O. Thus the continuation must have type **int** $\to$ **int**. Since $\underline{N}$ is parameterized by this continuation, and since it will apply the continuation to the value

6.4. LOGICAL EMBEDDING AND CPS TRANSLATION

of N, the term $\underline{N}$ will have type $(\text{int} \to \text{int}) \to \text{int}$. If N had had type **bool** instead, the type for $\underline{N}$ would be $(\text{bool} \to \text{int}) \to \text{int}$. In general, if N has type α, then $\underline{N}$ has type $t(\alpha)$, where

$$t(\alpha) \;=\; (\alpha \to \text{int}) \to \text{int}$$

In particular, Q has type $(\text{int} \to \text{int}) \to \text{int}$, and $\underline{Q}(\lambda v.v)$ has type **int**.

What happens if the subterm N is a function of type $\sigma \to \tau$? Inside the function, the production of a result of type τ involves computations just like the overall expression, and these are done in CPS as well. Similarly, the function is applied to an argument in CPS, so the type of $\underline{N}$ will be

$$t(\sigma \to \tau) \;=\; [[t(\sigma) \to t(\tau)] \to \text{int}] \to \text{int}.$$

Thus, the Kolmogorov translation k can be considered a special case of the CPS translation t on types. The special case arises by taking $\bot$ as the type of the overall term, and the implicit translation of classical to intuitionistic proofs in the proof of Theorem 6.4.3 is a CPS translation on terms. The following summarizes our considerations.

6.4.4. DEFINITION. Define a translation from $\lambda\mu$-terms to λ-terms as follows (recall the definition of φ^* from Remark 6.4.2):

$$
\begin{aligned}
\underline{x^\varphi} &= \lambda h\!:\!\neg\varphi^*.\, x^{k(\varphi)}\, h \\
\underline{\lambda x\!:\!\varphi.M^\psi} &= \lambda h\!:\!\neg(k(\varphi) \to k(\psi)).\, h(\lambda x\!:\!k(\varphi).\, \underline{M}^{k(\psi)}) \\
\underline{MN} &= \lambda h\!:\!\neg\varphi^*.\, \underline{M}^{\neg\neg(k(\psi)\to k(\varphi))}(\lambda m\!:\!k(\psi) \to k(\varphi).\, m\underline{N}^{k(\psi)}\, h) \\
\underline{\mu a\!:\!\neg\varphi.M^\bot} &= \lambda a\!:\!\neg\varphi^*.\, \underline{M}^{\neg\neg\bot}(\lambda v\!:\!\bot.v) \\
\underline{[a^{\neg\varphi}]M^\varphi} &= \lambda d\!:\!\neg\bot.\, \underline{M}^{k(\varphi)}\, a^{\neg\varphi^*}.
\end{aligned}
$$

Let $\underline{\Gamma} = \{x\!:\!k(\varphi) \mid x : \varphi \in \Gamma\} \cup \{a\!:\!\neg\varphi^* \mid a : \neg\varphi \in \Gamma\}$.[4]

The translation respects equality provided we use a slight variant of $\lambda\mu$.

6.4.5. DEFINITION. The set of *restricted* $\lambda\mu$-terms is defined as follows:

$$M ::= x \mid MM \mid \lambda x\!:\!\sigma.M \mid \mu a\!:\!\neg\sigma.[b]M.$$

Restricted $\lambda\mu$-terms enforce two mild requirements compared to general $\lambda\mu$-terms. The first is that an address application $[b]\, M$ always must be accompanied by a μ-abstraction μa (where a, b may be identical). When programming with μ we want this anyway—either because a is the address we are transmitting to, or because μa is actually the type coercion ε permitting the application to be placed in a context. The second requirement is that a μ always needs an address application. Again, we want this, because we have to simulate exceptional returns by normal returns. Thus, the unlimited

---
[4]Thus, the handling of negations in Γ is slightly optimized compared to Theorem 6.4.3.

syntax of $\lambda\mu$-terms is perhaps too free; anyway, by adopting a fixed address b_0 of type $\neg\bot$, any $\lambda\mu$-term can be translated into a restricted $\lambda\mu$-term with the same type.[5]

We then have the following extension of Theorem 6.4.3(ii).

6.4.6. THEOREM.

(i) *If $\Gamma \vdash M : \varphi$ in $\lambda\mu$-calculus, then $\underline{\Gamma} \vdash \underline{M} : k(\varphi)$ in simply typed λ-calculus.*

(ii) *If $M =_\mu N$ then $\underline{M} =_\beta \underline{N}$, for any restricted M, N.*

PROOF. Part (i) is by induction on M. For (ii), confluence implies that $M \twoheadrightarrow_\mu L$ and $N \twoheadrightarrow_\mu L$, for some $\lambda\mu$-term L. Thus, it suffices to show (ii) where $M \twoheadrightarrow_\mu N$. To this end, note that the set of restricted $\lambda\mu$-terms is closed under reduction, and use the following auxiliary properties:

$$\begin{array}{ll}
\lambda k : \varphi . \underline{P}\, k & =_\beta \ \underline{P} \\
\underline{P[x := Q]} & =_\beta \ \underline{P}[x := \underline{Q}] \\
\underline{P[a := [b][\,]]} & =_\beta \ \underline{P}[a := \underline{b}] \\
\underline{P[a := [b]([\,]Q)]} & =_\beta \ \underline{P}[a := \lambda m^{k(\psi) \to k(\varphi)} . m\, \underline{Q}\, \underline{b}]
\end{array}$$

The first follows from the observation that $\underline{P}$ is always an abstraction. The remaining ones are proved by induction on P using the first property. □

6.4.7. REMARK. The restriction on M, N in (ii) is essential. Leaving out type information again: $[b]\mu a . x =_\mu x$, but $\lambda d . (\lambda a . (\lambda h . x\, h) \lambda v . v)\, b \neq_\beta \lambda h . x\, h$.

6.5. Classical prover-skeptic dialogues

In the classical version, the prover is allowed to play tricks on the skeptic.

Prover 1: I assert that $((p \to q) \to p) \to p$ holds.

Skeptic 1: I don't believe you. Suppose that I give you a construction of $(p \to q) \to p$, can you give me a construction of p?

Prover 2: Yes, I have it. I got it by inserting a construction of $p \to q$ into the construction I just got from you!

Skeptic 2: You're lying, suppose I give you a construction of p, can you give me a construction of q?

Prover 3: Ah, your construction of p is actually what you requested in your first move, then we were done already there.

---

[5] Indeed, the restricted syntax with address b_0 is the syntax of $\lambda\mu$ in some presentations.

6.5. CLASSICAL PROVER-SKEPTIC DIALOGUES

Here the prover met the challenge p raised in the first skeptic move by asserting $p \to q$ in her following move. In order to challenge this formula, the skeptic provides an offer p meeting his own previous challenge. His own attack becomes the defense—the prover has pulled a *Catch-22* trick.[6]

Although we are only concerned with dialogues for implication and absurdity, consider, as another example, the one for *tertium non datur*.

Prover 1: I assert that $q \vee \neg q$ holds.

Skeptic 1: Yeah, right. Can you give me a construction of one of the two?

Prover 2: Yes, I have it. It is $\neg q$, i.e. $q \to \bot$.

Skeptic 2: You're lying, suppose I give you a construction of q, can you give me a construction of $\bot$? Ha—got you!

Prover 3: Ah, this construction of q was what you requested in your first step. I should have chosen q, not $\neg q$. So we were done already there.

Here the prover plays a Catch-22 trick in its purest form: she asserts $q \vee \neg q$ by claiming $\neg q$. If the skeptic challenges this assertion by producing a construction of q, then $q \vee \neg q$ still holds, though with a different construction.

To adapt the definition of dialogues to permit Catch-22 tricks, we allow the prover to respond to an earlier challenge instead of the most recent one. That is, we consider dialogues (about some formula φ) relative not only to a collection Γ of offers provided in advance, but also a collection Δ of challenges made in advance. Winning prover strategies remain as in Definition 4.6.1, *mutatis mutandis*.

6.5.1. DEFINITION. A *dialogue* over $(\Gamma, \varphi, \Delta)$ is a possibly infinite sequence $(\Sigma_1, \alpha_1), (\Pi_1, \beta_1), (\Sigma_2, \alpha_2), (\Pi_2, \beta_2) \ldots$, where Σ_i, Π_j are finite (maybe empty) sequences of formulas and α_i, β_j are propositional variables, satisfying:

(i) $\Sigma_1 = \varphi$ (Prover begins) and $\alpha_1 = \top$.

(ii) $\alpha_{i+1} \in \Delta \cup \{\beta_1, \ldots, \beta_i, \bot\}$ (Prover meets a challenge or aborts).

(iii) $\Sigma_{i+1} \to \alpha_{i+1} \in \Gamma \cup \Pi_1 \cup \ldots \cup \Pi_i$ (Prover uses an available offer).

(iv) $\Pi_i \to \beta_i \in \Sigma_i$ (Skeptic challenges a formula from a prover move).

In Chapter 4 we represented winning prover strategies for intuitionistic dialogues by simply typed λ-terms in η-long normal form. Each λ-abstracted variable labels an offer from the skeptic, and an application of this variable

---

[6] From the book by Joseph Heller [219]. As explained in [289], "The 'catch' in Catch-22 involves a mysterious Air Force regulation which asserts that a man is considered insane if he willingly continues to fly dangerous combat missions, but that if he makes the necessary formal request to be relieved of such missions, the very act of making the request proves that he is sane and therefore ineligible to be relieved."

indicates a use of this particular offer by the prover. What we need now is an abstraction operator to label each of the skeptic's challenges, and a corresponding variable reference mechanism to indicate the prover's choice of challenge—in the intuitionistic setting this was not necessary because the prover always responded to the challenge of the preceding step.

We can use μ and address applications! To represent a challenge σ by the skeptic we use the term $\mu a \colon \neg \sigma.M$, and when representing the continued dialogue inside M we indicate the choice of challenge σ by the term $[a]N$. Thus, a skeptic step followed by a prover step is represented by

$$\lambda y_1 \colon \sigma_1 \ldots y_m \colon \sigma_m.\, \mu a \colon \neg \alpha.[b](y\, M_1 \ldots M_n),$$

where $y_1, \ldots, y_m$ are the skeptic's offers, a is the skeptic's challenge, b is the previous challenge now being met by the prover, y is the offer used by the prover, and $M_1, \ldots, M_n$ are the continued dialogues over the new assertions introduced by the prover when using this offer. For example, the winning prover strategy for $((p \to q) \to p) \to p$ is represented by the term

$$\lambda x \colon (p \to q) \to p.\, \mu a \colon \neg p.[a](x(\lambda z \colon p.\, \mu b \colon \neg q.[a]z)).$$

We introduce the notion of an η-long form for $\lambda\mu$-terms, show that every $\lambda\mu$-term can be transformed to this form, and then show that they represent winning prover strategies.

6.5.2. DEFINITION. The set of η-long forms of $\lambda\mu$-calculus is the smallest set closed under the following rules (where α, β are atomic).

- If y has type $\tau_1 \to \cdots \to \tau_n \to \beta$, and a, b are (possibly equal) addresses of type $\neg \alpha, \neg \beta$, and $M_1, \ldots, M_n$ are η-long forms of type $\tau_1, \ldots, \tau_n$, then the following is an η-long form:

$$\lambda y_1 \colon \sigma_1 \ldots y_m \colon \sigma_m.\, \mu a \colon \neg \alpha.[b](y\, M_1 \ldots M_n).$$

- If y has type $\tau_1 \to \cdots \to \tau_n \to \bot$, and a is an address of type $\neg \alpha$, and $M_1, \ldots, M_n$ are η-long forms of type $\tau_1, \ldots, \tau_n$, then the following is an η-long form:

$$\lambda y_1 \colon \sigma_1 \ldots y_m \colon \sigma_m.\, \mu a \colon \neg \alpha.y\, M_1 \ldots M_n.$$

WARNING. An η-long form is not necessarily in μ-normal form.

To show that any $\lambda\mu$-term can be transformed to an η-long form, we introduce a strongly normalizing extension of μ-reduction, called $\hat{\mu}$-reduction, and show that any $\hat{\mu}$-normal form can be transformed to an η-long form.

6.5. CLASSICAL PROVER-SKEPTIC DIALOGUES   147

6.5.3. DEFINITION. Let $\to_{\hat\mu}$ be the smallest compatible relation closed under (β), (η_μ), (β_μ) and the following rules:

$$\begin{array}{lll}
\lambda x\!:\!\varphi.M\, x & \to\; M \text{ if } x \notin \mathrm{FV}(M) & (\eta) \\
\mu b\!:\!\neg\bot.M & \to\; M[b := [\,]] & (\mu_\bot) \\
{[b^{\neg\bot}]}\, M & \to\; M & ([\,]_\bot) \\
\mu a\!:\!\neg(\varphi \to \psi).M & \to\; \lambda x\!:\!\varphi.\mu b\!:\!\neg\psi.M[a := [b]\,([\,]\,x)] & (\zeta')
\end{array}$$

6.5.4. THEOREM. *No $\lambda\mu$-term has an infinite $\hat\mu$-reduction.*

PROOF. The proof is similar to the proof of Theorem 6.3.8. First extend Lemma 6.3.6 with the following result, which is proved by induction on M: If a is an address of type $\neg\bot$, then

$$\underline{M[a := [\,]]} = \underline{M}.$$

Using this result and Lemma 6.3.6, we can extend Lemma 6.3.7 to state: If $M \to_{\hat\mu} N$, then $\underline{M} \twoheadrightarrow_{\beta\eta} \underline{N}$. Moreover, if $M \to_{\beta\eta} N$, then $\underline{M} \twoheadrightarrow_{\beta\eta} \underline{N}$. The proof is by induction on the derivation of $M \to_{\hat\mu} N$. For instance, let

$$M = \mu a\!:\!\neg(\sigma \to \tau).P \;\to_{\zeta'}\; \lambda u\!:\!\sigma.\mu b\!:\!\neg\tau.P[a := [b]([\,]u)] = N.$$

Then $\underline{M} = \lambda \vec{y}^a \underline{P} = \lambda u \lambda \vec{y}^b.\underline{P}[\vec{y}^a := u, \vec{y}^b] = \underline{N}$.

Finally, we must prove that there are no infinite $\beta_\mu \eta_\mu \zeta' [\,]_\bot \mu_\bot$-reductions. Let $|\tau|$ denote the length of τ, and define the *weight* of an address application or μ-abstraction thus: $w([a^\tau]M) = w(\mu a\!:\!\tau.M) = |\tau|$. Then define the weight $s(M)$ of a term M as the sum of all weights of occurrences of address applications and μ-abstractions in M. Now note that each $\beta_\mu \eta_\mu \zeta' [\,]_\bot \mu_\bot$-reduction step decreases $s(M)$. □

6.5.5. THEOREM (Subject reduction). *Let $M \to_{\hat\mu} N$. If $\Gamma \vdash M\!:\!\varphi$ then also $\Gamma \vdash N\!:\!\varphi$.*

PROOF. By induction on the derivation of $M \to_{\hat\mu} N$ using Lemma 6.3.1. □

We now show how any $\lambda\mu$-term in $\hat\mu$-normal form can be transformed to η-long form. We need a small twist to the idea.

6.5.6. DEFINITION. A $\lambda\mu$-term M is *atomic* iff all addresses occurring in M (free or bound) have atomic types; that is, if $[a^{\neg\tau}]N$ or $\mu a\!:\!\neg\tau.N$ is a subterm of N then τ must be atomic. Every closed $\hat\mu$-normal form is atomic.

6.5.7. LEMMA. *Let $\Gamma \vdash M : \varphi$, where M is an atomic $\hat\mu$-normal form. There is an η-long form M' with $\Gamma \vdash M' : \varphi$ and $M =_{\hat\mu} M'$.*

PROOF. By induction on the size of M. Let

$$M = \lambda x_1 : \sigma_1 \ldots x_m : \sigma_m . E N_1 \ldots N_k,$$

where $k, m \geq 0$ and E is neither an application nor an abstraction. By the induction hypothesis, there are η-long forms N_i' with $N_i =_{\hat{\mu}} N_i'$. We consider each possible form of E.

CASE 1: $E = x$. Let $\tau_1 \to \cdots \to \tau_n \to \alpha$ be x's type, and $\ell = \max\{n - k, 0\}$. Then the desired η-long form is

$$\lambda x_1 : \sigma_1 \ldots x_m : \sigma_m . \lambda y_1 : \tau_{k+1} \ldots y_\ell : \tau_n . \mu a : \neg \alpha . [a](x\, N_1' \ldots N_k'\, y_1 \ldots y_\ell).$$

CASE 2: $E = [a]L$. For typing reasons, $k = 0$. Since M is atomic, L is not an abstraction. Thus, $L = O\, O_1 \ldots O_j$, where $j \geq 0$ and O is neither an application nor an abstraction. Moreover, O does not have form $[b]\, S$ (for typing reasons, this would imply $j = 0$, but then a has type $\neg \bot$, contradicting the fact that $[a]L$ is in $[\,]_\bot$-normal form). Finally, O does not have form $\mu c : \neg \delta . S$ (then either $j = 0$, contradicting the fact that $[a]\, L$ is in β_μ-normal form, or $j > 0$, so $\delta = \delta_1 \to \delta_2$, contradicting the fact that L is in ζ'-normal form). Thus, $O = x$. By the induction hypothesis, there are η-long forms O_i' with $O_i =_{\hat{\mu}} O_i'$. Then the desired η-long form is:

$$M = \lambda x_1 : \sigma_1 \ldots x_m : \sigma_m . \mu b : \neg \bot . [a](x\, O_1' \ldots O_j').$$

CASE 3: $E = \mu a : \neg \alpha . L$. Then $k = 0$ (otherwise α is an arrow type, but then N is not in ζ'-normal form). Since L has type $\bot$, it is not an abstraction, i.e. $L = O\, O_1 \ldots O_j$, where $j \geq 0$ and O is neither an application nor an abstraction. Moreover, O does not have form $\mu c : : \neg \delta . P$ (if $j = 0$, then $\delta = \bot$, contradicting the fact that M is in $\mu_\bot$-normal form, and if $j > 0$, then $\delta = \delta_1 \to \delta_2$ contradicting the fact that M is in ζ'-normal form). So O must be a variable x or an address application $[b]P$. In the former case, the induction hypothesis, yields η-long forms O_i' with $O_i =_{\hat{\mu}} O_i'$. Then the desired η-long form is:

$$M = \lambda x_1 : \sigma_1 \ldots x_m : \sigma_m . \mu a : \neg \alpha . x\, O_1' \ldots O_j'.$$

In the latter case, $j = 0$ for typing reasons, and as in Case 2, P must have form $x\, P_1 \ldots P_i$. By the induction hypothesis, there are η-long forms P_i' with $P_i =_{\hat{\mu}} P_i'$. Then the desired η-long form is:

$$M = \lambda x_1 : \sigma_1 \ldots x_m : \sigma_m . \mu a : \neg \alpha . [b](x P_1' \ldots P_k'). \qquad \square$$

6.5.8. COROLLARY. *For every closed $\lambda\mu$-term M of type φ, there is another closed $\lambda\mu$-term M' of type φ such that M is in η-long form and $M =_{\hat{\mu}} M'$.*

PROOF. By strong normalization, subject reduction, and Lemma 6.5.7. $\square$

6.5. CLASSICAL PROVER-SKEPTIC DIALOGUES

6.5.9. THEOREM. *There is a winning prover strategy for φ iff φ is a classical theorem.*

PROOF. For the direction from left to right, let D be a winning prover strategy for φ. For each skeptic node (Π, β), we consider the accumulated set of offers and challenges (Γ, Δ), i.e. the union of all offers and challenges, respectively, made in the dialogue from the root to this node. For any subtree D' of D with accumulated offers and challenges (Γ, Δ) in the root, we now show that $\Gamma, \neg \Delta \vdash \bot$ in classical logic. The form of D' must be $(n \geq 0)$:

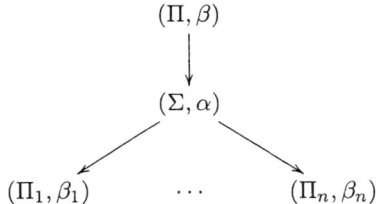

Each child has offers and challenges $(\Gamma \cup \Pi_i, \Delta \cup \{\beta_i\})$. By the induction hypothesis, $\Gamma, \Pi_i, \neg \Delta, \neg \beta_i \vdash \bot$, and then we have $\Gamma, \neg \Delta, \vdash \Pi_i \to \beta_i$. Since $\Sigma = \Pi_1 \to \beta_1, \ldots, \Pi_n \to \beta_n$, also $\Gamma \vdash (\Pi_1 \to \beta_1) \to \cdots \to (\Pi_n \to \beta_n) \to \alpha$ (the formula must have been stated as an offer), so $\Gamma, \neg \Delta \vdash \alpha$ by *modus ponens*. Now, either $\alpha = \bot$ in which case we are done, or $\alpha \in \Delta$ in which case $\Gamma, \neg \Delta \vdash \bot$.

In particular, D has the root label $(\varphi, \top)$ and the single child labeled (Π, β), where $\varphi = \Pi \to \beta$, so $\Pi, \neg \beta \vdash \bot$, i.e. $\vdash \varphi$.

For the opposite direction, assume $\vdash \varphi$. There is a closed $\lambda\mu$-term of type φ. By Corollary 6.5.8, we can assume that the term is in η-long form, so it suffices to show that if, for an η-long form M,

$$x_1 : \psi_1, \ldots, x_n : \psi_n, a_1 : \neg\alpha_1, \ldots, a_p : \neg\alpha_p \vdash M : \varphi,$$

then there is a winning prover strategy for $(\{\psi_1, \ldots, \psi_n\}, \varphi, \{\alpha_1, \ldots, \alpha_p\})$. We proceed by induction on the size of M. Let

$$M = \lambda x_{n+1}^{\psi_{n+1}} \ldots x_{n+m}^{\psi_{n+m}} . \mu a_{p+1}^{\neg \alpha_{p+1}} . [a_r] \, (x_k \, N_1 \ldots N_l),$$

where $1 \leq k \leq n+m$ and $1 \leq r \leq p+1$ (the case with no $[a_r]$ is similar). Then

$$x_1 : \psi_1, \ldots, x_{n+m} : \psi_{n+m}, a_1 : \neg\alpha_1, \ldots, a_{p+1} : \neg\alpha_{p+1} \vdash N_i : \rho_i$$

for all i, where $\psi_k = \rho_1 \to \cdots \to \rho_l \to \alpha_r$ and $\varphi = \psi_{n+1} \to \cdots \to \psi_{n+m} \to \alpha_{p+1}$. By the induction hypothesis, there is a winning prover strategy for each

$(\{\psi_1, \ldots, \psi_{n+m}\}, \rho_i, \{\alpha_1, \ldots, \alpha_{p+1}\})$. It must have form

$$(\rho_i, \top)$$
$$\downarrow$$
$$(\Pi_i, \beta_i)$$
$$\downarrow$$
$$D_i$$

Then we have the following strategy for $(\{\psi_1, \ldots, \psi_n\}, \varphi, \{\alpha_1, \ldots, \alpha_p\})$:

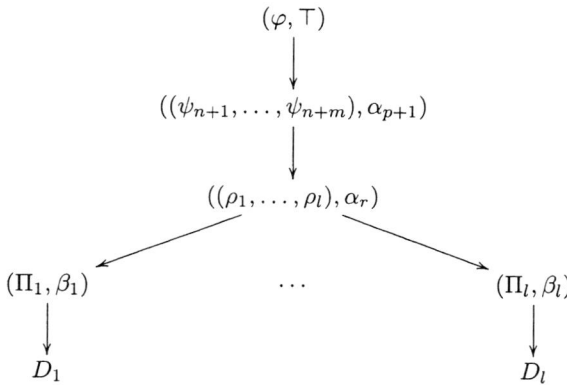

This concludes the proof. □

6.6. The pure implicational fragment

We have considered intuitionistic logic with and without absurdity, i.e. the systems NJ($\to, \bot$) and NJ($\to$), and we have considered classical logic with absurdity NK($\to, \bot$). Classical logic without absurdity is missing.

6.6.1. DEFINITION. We consider formulas in $\Phi(\to)$. The rules for NK($\to$) are as in Figure 6.1 with ($\neg$E) replaced by (P) from Remark 6.1.5.

Figure 6.3 relates different logical systems. An arrow from one system to another labeled with a rule indicates that adding the rule to the former system yields the theorems of the latter system (in the upper right direction one must also add $\bot$ to the language). Steps in the upper right direction are conservative. For the intuitionistic systems, this follows from Theorem 2.6.2. For the classical systems, the result is Theorem 6.6.7 below, which is the most important property of NK($\to$). We prove it by translating η-long forms of $\lambda\mu$ to $\lambda\xi$, a λ-calculus corresponding to NK($\to$).

6.6. THE PURE IMPLICATIONAL FRAGMENT

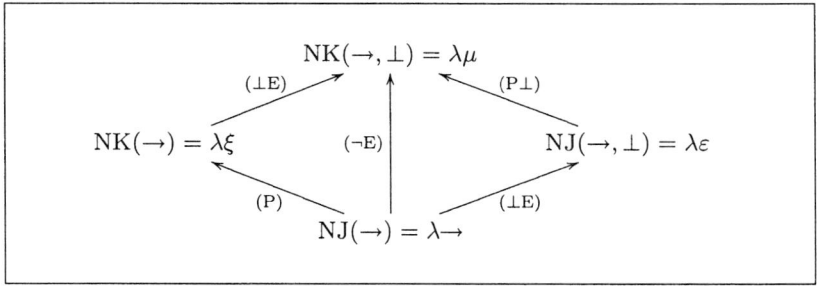

FIGURE 6.3: SOME LOGICAL SYSTEMS AND λ-CALCULI.

6.6.2. DEFINITION. Let φ range over $\Phi(\to)$ and define $\lambda\xi$-terms by:

$$M ::= x \mid (MM) \mid (\lambda x{:}\varphi M) \mid ([a]M) \mid (\xi a{:}\varphi \to \varphi\, M).$$

Let $\lambda\xi$ be the lambda-calculus obtained by replacing the rules for $[a]\,M$ and $\mu a{:}\neg\varphi.M$ in Figure 6.2 by the following rules

$$\frac{\Gamma,\ a{:}\sigma \to \tau \vdash M : \sigma}{\Gamma,\ a{:}\sigma \to \tau \vdash [a]M : \tau} \qquad \frac{\Gamma,\ a{:}\sigma \to \tau \vdash N : \sigma}{\Gamma \vdash \xi a{:}\sigma \to \tau.N : \sigma}.$$

WARNING. The system $\lambda\xi$ has no common name in the literature.

6.6.3. PROPOSITION.

(i) *If* $\Gamma \vdash M : \varphi$ *in* $\lambda\xi$, *then* $\mathrm{rg}(\Gamma) \vdash \varphi$ *in* $\mathrm{NK}(\to)$.

(ii) *If* $\Gamma \vdash \varphi$, *then there are* Δ, M *with* $\Delta \vdash M : \varphi$ *and* $\mathrm{rg}(\Delta) = \Gamma$.

PROOF. Similar to the proof of Proposition 6.2.6. □

The following shows that only terms of a certain form need to be considered in the translation from $\lambda\mu$ to $\lambda\xi$.

6.6.4. LEMMA. *Assume* $\Gamma \vdash M{:}\tau$ *in* $\lambda\mu$, *where* $\tau \in \Phi(\to)$ *and* $\sigma \in \Phi(\to)$ *for all* $(x : \sigma) \in \Gamma$. *If* M *is in* η-*long form, then* M *must have form*

$$\lambda y_1{:}\sigma_1 \ldots y_k{:}\sigma_k.\, \mu a{:}\neg\alpha.[b](y\, M_1 \ldots M_l). \tag{$*$}$$

PROOF. The problem is to show that M does not have form:

$$\lambda y_1{:}\sigma_1 \ldots y_k{:}\sigma_k.\, \mu a{:}\neg\alpha.y\, M_1 \ldots M_l.$$

If M has this form, then y has type $\rho_1 \to \cdots \to \rho_l \to \bot$, but then $\bot$ occurs in a type in Γ (if y is free) or in σ (if y is bound). □

152   CHAPTER 6. CLASSICAL LOGIC AND CONTROL OPERATORS

6.6.5. DEFINITION. Assume $\Gamma \vdash M : \tau$ in $\lambda\mu$, where $\tau \in \Phi(\to)$, $\sigma \in \Phi(\to)$ for all $(x : \sigma) \in \Gamma$, and M is in η-long form. Let $(c : \neg\gamma) \in \Gamma$.

(i) Define the set $\mathrm{img}_c(M)$ of *image types for c* by:

$$\mathrm{img}_c(M) = \{\delta \mid M \text{ has a subterm of form } \mu d{:}\neg\delta.[c]N\}.$$

(ii) Let $\mathrm{con}_c(M)$ be the environment defined by:

$$\mathrm{con}_c(M) = \{c_\delta : \gamma \to \delta \mid \delta \in \mathrm{img}_c(M)\}$$

where c_δ is a fresh variable for each $\delta \in \mathrm{img}_c(M)$.

6.6.6. LEMMA. *Let $\Gamma, a_1 : \neg\alpha_1, \ldots, a_m : \neg\alpha_m \vdash M : \tau$ in $\lambda\mu$, where M is in η-long form, $\tau \in \Phi(\to)$, $\sigma \in \Phi(\to)$ for all $(x : \sigma) \in \Gamma$, and Γ contains no addresses. Then there exists $\underline{M}$ such that in $\lambda\xi$:*

$$\Gamma, \mathrm{con}_{a_1}(M), \ldots, \mathrm{con}_{a_m}(M) \vdash \underline{M} : \tau$$

PROOF. By induction on the size of M. According to Lemma 6.6.4,

$$M = \lambda y_1{:}\sigma_1 \ldots y_k{:}\sigma_k.\, \mu a{:}\neg\alpha.[b](y\, M_1 \ldots M_l).$$

Here y has type $\rho_1 \to \cdots \to \rho_l \to \beta$. Whether y is a free or bound variable, its type does not contain $\bot$, so $\rho_i \in \Phi(\to)$. Since $\tau = \sigma_1 \to \cdots \to \sigma_k \to \alpha$, also $\sigma_j \in \Phi(\to)$. By the induction hypothesis we therefore have:

$$\Gamma, y_1{:}\sigma_1, \ldots, y_k{:}\sigma_k, \mathrm{con}_{a_1}(M_i), \ldots, \mathrm{con}_{a_m}(M_i), \mathrm{con}_a(M_i) \vdash \underline{M_i} : \rho_i.$$

Let $\Delta = \mathrm{con}_{a_1}(M) \cup \ldots \cup \mathrm{con}_{a_m}(M)$. Then $\mathrm{con}_{a_1}(M_i) \cup \ldots \cup \mathrm{con}_{a_m}(M_i) \subseteq \Delta$. Also, let $\{a_{\delta_1}{:}\alpha{\to}\delta_1 \ldots a_{\delta_p}{:}\alpha{\to}\delta_p\} = \mathrm{con}_a(M_1) \cup \ldots \cup \mathrm{con}_a(M_l)$. We have:

$$\Gamma, y_1{:}\sigma_1, \ldots, y_k{:}\sigma_k, \Delta, a_{\delta_1}{:}\alpha{\to}\delta_1, \ldots, a_{\delta_p}{:}\alpha{\to}\delta_p \vdash \underline{M_i} : \rho_i.$$

If $a \neq b$, then $\alpha \in \mathrm{img}_b(M)$, and $(b_\alpha : \beta \to \alpha) \in \Delta$ for some address b_α. In this case we define

$$\underline{M} = \lambda y_1{:}\sigma_1 \ldots y_k{:}\sigma_k.\, \xi a_{\delta_1}{:}\alpha{\to}\delta_1 \ldots \xi a_{\delta_p}{:}\alpha{\to}\delta_p.[b_\alpha](y\, \underline{M_1} \ldots \underline{M_l}).$$

If $a = b$, let $\hat{a}$ be a fresh address, and define

$$\underline{M} = \lambda y_1{:}\sigma_1 \ldots y_k{:}\sigma_k.\, \xi a_{\delta_1}{:}\alpha{\to}\delta_1 \ldots \xi a_{\delta_p}{:}\alpha{\to}\delta_p.\xi \hat{a}{:}\alpha{\to}\alpha.[\hat{a}](y\, \underline{M_1} \ldots \underline{M_l}).$$

The reader will easily check that $\Gamma, \Delta \vdash \underline{M} : \tau$ in both cases. □

6.6.7. THEOREM (Conservativity). *Let $\varphi \in \Phi(\to)$. Then $\vdash \varphi$ in $\mathrm{NK}(\to, \bot)$ iff $\vdash \varphi$ in $\mathrm{NK}(\to)$.*

PROOF. For the direction from left to right, assume $\vdash \varphi$ in $NK(\rightarrow, \bot)$. By Proposition 6.2.6 there is a closed $\lambda\mu$-term M of type φ. By Corollary 6.5.8 we can assume that M is in η-long form. Then by the preceding lemma, there is a term $\underline{M}$ with type φ in $\lambda\xi$, so Proposition 6.6.3 implies $\vdash \varphi$ in $NK(\rightarrow)$.

The other direction follows from Remark 6.1.5. □

6.6.8. COROLLARY. *A formula* $\varphi \in \Phi(\rightarrow)$ *is a classical tautology iff* $\vdash \varphi$ *in* $NK(\rightarrow)$.

PROOF. By conservativity and Theorem 6.1.10. □

The dialogues for $NK(\rightarrow)$ permit Catch-22 tricks, but not abort tricks.

6.6.9. DEFINITION. Dialogues and strategies are defined as in Definition 6.5.1, with rule (ii) replaced by

(ii) $\alpha_{i+1} \in \Delta \cup \{\beta_1, \ldots, \beta_i\}$ (Prover meets challenge).

6.6.10. THEOREM. *Let* $\varphi \in \Phi(\rightarrow)$. *There is a winning prover strategy for* φ *in the above sense iff* $\vdash \varphi$ *in* $NK(\rightarrow)$.

PROOF. Assume that there is a winning prover strategy for φ in the above type of dialogue. Then the same strategy is a winning prover strategy in the classical type of dialogue (Definition 6.5.1), and then by Theorem 6.5.9 $\vdash \varphi$ in $NK(\rightarrow, \bot)$, so by conservativity $\vdash \varphi$ in $NK(\rightarrow)$.

Conversely, assume $\vdash \varphi$ in $NK(\rightarrow)$. Then also $\vdash \varphi$ in $NK(\rightarrow, \bot)$, so there is a closed $\lambda\mu$-term M of type φ, and we may assume that M has the shape described in Lemma 6.6.4. We can then proceed as in the proof of Theorem 6.5.9, noticing that there will be no prover steps using the abort trick due to the fact that M has the special shape. □

6.7. Conjunction and disjunction

Until now we have been concerned with the implicational fragment of classical propositional calculus (with and without absurdity). Now we now add conjunction and disjunction. One approach is to add to the language the two connectives $\wedge$ and $\vee$ and adopt the rules of Section 2.2.

6.7.1. DEFINITION (Classical propositional calculus). We consider the full set $\Phi(\rightarrow, \bot, \vee, \wedge)$ of propositional formulas. Figure 6.4 contains the natural deduction rules for $NK(\rightarrow, \bot, \vee, \wedge)$.

6.7.2. THEOREM. *A formula in* $\Phi(\rightarrow, \bot, \vee, \wedge)$ *is a theorem of* $NK(\rightarrow, \bot, \vee, \wedge)$ *iff it is a classical tautology.*

PROOF. Similar to the proof of Theorem 6.1.10. □

$$\Gamma, \varphi \vdash \varphi \ (\text{Ax})$$

$$\frac{\Gamma, \varphi \vdash \psi}{\Gamma \vdash \varphi \to \psi}(\to\text{I}) \qquad \frac{\Gamma \vdash \varphi \to \psi \quad \Gamma \vdash \varphi}{\Gamma \vdash \psi}(\to\text{E})$$

$$\frac{\Gamma \vdash \varphi \quad \Gamma \vdash \psi}{\Gamma \vdash \varphi \wedge \psi} \ (\wedge\text{I}) \qquad \frac{\Gamma \vdash \varphi \wedge \psi}{\Gamma \vdash \varphi}(\wedge\text{E})\frac{\Gamma \vdash \varphi \wedge \psi}{\Gamma \vdash \psi}$$

$$\frac{\Gamma \vdash \varphi}{\Gamma \vdash \varphi \vee \psi} \ (\vee\text{I}) \ \frac{\Gamma \vdash \psi}{\Gamma \vdash \varphi \vee \psi} \qquad \frac{\Gamma, \varphi \vdash \vartheta \quad \Gamma, \psi \vdash \vartheta \quad \Gamma \vdash \varphi \vee \psi}{\Gamma \vdash \vartheta} \ (\vee\text{E})$$

$$\frac{\Gamma, \neg\varphi \vdash \bot}{\Gamma \vdash \varphi}(\neg\text{E})$$

FIGURE 6.4: FULL CLASSICAL PROPOSITIONAL CALCULUS.

The Kolmogorov translation is extended to the full system as follows.

6.7.3. DEFINITION.

$$\begin{aligned} k(\alpha) &= \neg\neg\alpha, \text{ for atomic } \alpha; \\ k(\varphi \circ \psi) &= \neg\neg(k(\varphi) \circ k(\psi)), \text{ for } \circ \in \{\to, \vee, \wedge\}. \end{aligned}$$

Theorem 6.4.3 holds for the extended translation.

6.7.4. THEOREM.

(i) $\vdash \varphi \to k(\varphi)$ and $\vdash k(\varphi) \to \varphi$ in CPC$(\to, \bot, \vee, \wedge)$.

(ii) $\vdash \varphi$ in CPC$(\to, \bot, \vee, \wedge)$ iff $\vdash k(\varphi)$ in IPC$(\to, \bot, \vee, \wedge)$.

PROOF. Similar to the proof of Theorem 6.4.3. In the right-to-left direction in (ii), the new introduction rules are easy. For $(\wedge\text{E})$ use Example 2.23(ii).

Now suppose the derivation ends with

$$\frac{\Gamma, \varphi \vdash \rho \quad \Gamma, \psi \vdash \rho \quad \Gamma \vdash \varphi \vee \psi}{\Gamma \vdash \rho}.$$

By the induction hypothesis we have $k(\varphi) \vee k(\psi), k(\Gamma) \vdash k(\rho)$, so also $k(\Gamma) \vdash \neg\neg(k(\varphi) \vee k(\psi)) \to k(\rho)$ by Exercise 2.24(ii). Now use the induction hypothesis and *modus ponens*. □

6.7. CONJUNCTION AND DISJUNCTION

The extended system is conservative over the implicational fragment with absurdity.

6.7.5. PROPOSITION. *Let $\varphi \in \Phi(\to, \bot)$. Then $\vdash \varphi$ in $\mathrm{CPC}(\to, \bot, \vee, \wedge)$ iff $\vdash \varphi$ in $\mathrm{CPC}(\to, \bot)$.*

PROOF. The right-to-left direction is trivial. As for the other direction, if $\vdash \varphi$ in $\mathrm{NK}(\to, \bot, \vee, \wedge)$, then φ is a tautology by the preceding theorem, so Theorem 6.1.10 implies $\vdash \varphi$ in $\mathrm{NK}(\to, \bot)$. □

However, we can get the whole system without adding rules for conjunction and disjunction.

6.7.6. DEFINITION. Let $\varphi \wedge \psi$ and $\varphi \vee \psi$ abbreviate the following formulas:

$$\varphi \wedge \psi = (\varphi \to \psi \to \bot) \to \bot;$$
$$\varphi \vee \psi = (\varphi \to \bot) \to (\psi \to \bot) \to \bot.$$

WARNING. A widespread definition of $\varphi \vee \psi$ is $\varphi \vee \psi = \neg \varphi \to \psi$. The abbreviations in Definition 6.7.6 arise as a special case of a general scheme for representing data types in second-order typed λ-calculi, as we shall see.

6.7.7. COROLLARY. *Let $\psi \in \Phi(\to, \bot, \vee, \wedge)$ abbreviate $\varphi \in \Phi(\to, \bot)$. Then $\vdash \varphi$ in $\mathrm{NK}(\to, \bot)$ iff $\vdash \psi$ in $\mathrm{NK}(\to, \bot, \vee, \wedge)$.*

PROOF. By completeness in both systems. □

We can also define pairs and sums as derived constructions in $\lambda\mu$. This is not possible in the simply typed λ-calculus, because one cannot define conjunction and disjunction in intuitionistic propositional calculus.

6.7.8. EXAMPLE. Let P_1, P_2 have types σ_1, σ_2, respectively, let Q have type $(\sigma_1 \to \sigma_2 \to \bot) \to \bot$, and let R have type $(\sigma_1 \to \bot) \to (\sigma_2 \to \bot) \to \bot$. Finally, let S_1, S_2 both have type ρ. Define the following abbreviations.

$$\begin{aligned}
\langle P, Q \rangle^{\sigma \wedge \tau} &= \lambda z : \sigma \to \tau \to \bot.\, zPQ; \\
\pi_i(Q) &= \mu a {:} \neg \sigma_i.\, Q\, (\lambda x_1 : \sigma_1. \lambda x_2 : \sigma_2.\, [a]x_i); \\
\mathrm{in}_i^{\sigma_1 \vee \sigma_2}(P_i) &= \lambda y_1 : \sigma_1 \to \bot.\, \lambda y_2 : \sigma_2 \to \bot.\, y_i\, P_i; \\
\mathbf{case}\ R^{\sigma \vee \tau}\ \mathbf{of}\ [x]S\ \mathbf{or}\ [y]T &= \mu a {:} \neg \rho.\, R\, (\lambda x : \sigma.[a]S)\, (\lambda y : \tau.[a]T),
\end{aligned}$$

where z, a, y_1, y_2 are fresh. Then indeed

$$\pi_1(\langle P_1, P_2 \rangle^{\sigma_1 \wedge \sigma_2}) \twoheadrightarrow_\mu P_1$$

and

$$\mathbf{case}\ (\mathrm{in}_2^{\sigma_1 \vee \sigma_2}(P_2))\ \mathbf{of}\ [\,x_1^{\sigma_1}]S_1\ \mathbf{or}\ [\,x_2^{\sigma_2}]S_2 \twoheadrightarrow_\mu S_2[x := P_2].$$

6.8. Notes

Prawitz [403] is a classic work on normalization in intuitionistic and classical natural deduction. Later contributions include, for instance, [122, 196, 438, 454]; the second and third study systems based on Peirce's law. Curry's book [106] contains an elaborate study of many different systems with and without rules for negation, with and without Peirce's law, etc. Embeddings of classical logic into intuitionistic logic were studied since the 1920s by, among others, Kolmogorov [273], Glivenko [190], Gentzen [169] and Gödel [193]—see also [303].

Control operators occur in many programming languages. Obvious examples include the *exception mechanism* of e.g. C++ and Java, or **catch** and **throw** from Lisp [200]. Such constructions were formalized by Felleisen and co-workers [145, 146] as calculi with the $\mathcal{C}$-operator (*control*) and $\mathcal{A}$-operator (*abort*). In hindsight one can partially recognize reduction rules for these in Prawitz' proof normalization rules. Felleisen also studied CPS translations of control operators, extending classic results for the pure (i.e. without control operators) λ-calculus by Plotkin [396] and Reynolds [414]. The translations by Felleisen, Plotkin, and Reynolds all applied to untyped calculi only. Meyer and Wand [341] studied translations for typed λ-calculi (without control operators).

CPS and CPS translations are used for various purposes. For instance, CPS is used as the intermediate language for some compilers (see e.g. [11]), and it is well-suited for this purpose because the control flow is very explicit in a CPS term, compared to an arbitrary λ-term. Also, CPS is used for specifying the semantics of programming languages; the use of continuations makes it easy to model backtracking, control operators, etc. As a last example, CPS translation is used as a pre-phase to some automatic program optimization techniques (see e.g. [77, 363]) because translation into CPS enables optimization techniques to improve the program more.

The correspondence between classical natural deduction and double-negation embeddings on the one hand and control operators and CPS translation on the other hand was presented by Griffin [201] in a paper that opened up a new direction for research on the Curry-Howard isomorphism. We do not attempt to survey all relevant work that has followed, but merely provide an incomplete list of works in which the reader can find additional references.

Murthy [356] extended Griffin's study and gave another solution than Griffin's to the problem of preserving types under reduction. He changes the type system, rather than changing the reduction rules. Other type systems based on classical logic for control operators have followed, see e.g. [135, 205, 215, 216]. There are also lines of work attempting to type control operators in type systems corresponding to *constructive* logics—see e.g. [121, 358, 424]

Whereas Griffin considered the problem of typing Felleisen's $\mathcal{C}$-operator, Parigot developed a new type of natural deduction system for classical logic [380] and introduced $\lambda\mu$ as the corresponding logic of terms [381]. His formulations all use two sets of assumptions, one for variables and one for addresses, and the addresses are always written to the right of $\vdash$, thus obtaining a variant of natural deduction with multiple conclusions. Moreover, in some formulations [383, 384] the two rules for μ and address application are collected in a single rule, and the terms are then what we have called *restricted* $\lambda\mu$-terms. Systems with similar aims and properties as $\lambda\mu$, but based on the usual type of natural deduction for classical logic, were developed by Rezus [416, 417] ($\lambda\gamma$) and Rehof and Sørensen [413] (λ_Δ).

The relation between $\lambda\mu$ on the one hand and the calculi studied by Felleisen and Griffin on the other hand was addressed by de Groote [204] who also introduced several elegant CPS translations of $\lambda\mu$ [203, 207] which were used to establish normalization results; the CPS translation in this chapter is inspired by de Groote. The paper [13] studies the implicational fragment of classical propositional logic and its relation to the calculi of Parigot and Felleisen. More about λ-calculi related to classical logic can be found in [24, 25, 43, 91, 117, 243, 286, 287, 376, 423, 490]. Many of these papers establish subject reduction, confluence, and strong normalization of the system under consideration. Our proof of the latter is due to Ong. Another short proof can be found in [122].

Girard [185], in an independent line of work, introduces a classical system **LC** with a translation from **LC** into intuitionistic logic, and Murthy [357] relates this system to Felleisen's control operators and CPS translations. The papers [115, 116] show that Parigot's free deduction and $\lambda\mu$ and Girard's **LC** are related to corresponding embeddings of certain classical sequent calculi into linear logic (see the notes for Chapter 5). See also [369, 370, 371].

Filinsky [150] discovered that values and continuations can be seen as categorically dual concepts. Categorical semantics for control operators and continuations are considered in e.g. [244, 375, 442, 461, 476].

For notes about dialogues, see the next chapter.

6.9. Exercises

In all the exercises of this chapter we consider formulas in $\Phi(\to, \bot)$ and take negation as defined by $\neg\varphi = \varphi \to \bot$, except when explicitly stated otherwise.

6.1. Show that rules (P) and ($\neg$E) are derived rules in the system consisting of rules ($\to$E), ($\to$I), (Ax), (P$\bot$), and ($\bot$E) (see Definition 6.1.1 and Remark 6.1.5).

6.2. We consider some variants of the system in Figure 6.1

(i) Show that the ($\neg$E) rule can be replaced by the rule

$$\frac{\Gamma \vdash \neg\neg\varphi}{\Gamma \vdash \varphi} \ (\neg E_1)$$

without affecting the set of derivable judgments.

(ii) Show that the ($\neg$E) rule can be replaced by

$$\Gamma \vdash \neg\neg\varphi \to \varphi \quad (\neg E_2)$$

without affecting the set of derivable judgments.

6.3. In many presentations of Hilbert-style systems for classical logic, negation is taken as primitive; that is, one extends the language with propositions of form $\neg\varphi$ instead of introducing $\bot$ and the abbreviation $\neg\varphi = \varphi \to \bot$. An example of such a system is obtained by adding to the axioms (A1) and (A2) the single axiom:

$$(A5_\neg) \ (\neg\varphi \to \neg\psi) \to (\neg\varphi \to \psi) \to \varphi.$$

Let H_1 denote the Hilbert-style system for formulas in $\Phi(\to, \bot)$ with axioms (A1), (A2), (A3$_\bot$), and let H_2 denote the Hilbert-system for formulas in $\Phi(\to, \neg)$ with axioms (A1), (A2), (A5$_\neg$). Show the following properties:

(i) If $\Gamma \vdash \varphi$ in H_2 then $\Gamma \vdash \varphi'$ in H_1, where φ' is the formula obtained by replacing every subformula of the form $\neg \psi$ by $\psi \to \bot$.

(ii) If $\Gamma \vdash \varphi$ in H_1 then $\Gamma \vdash \varphi'$ in H_2, where φ' is the formula obtained by replacing every occurrence of $\bot$ by $\neg(p \to p)$, where p is some arbitrary propositional symbol.

6.4. Let H_2 be the same system as in the preceding exercise, and let H_3 denote the Hilbert-system for formulas consisting of implication and negation with axioms (A1), (A2) and the following two axioms:

$$(A3_\neg)\ \neg\neg\varphi \to \varphi$$
$$(A4_\neg)\ (\varphi \to \psi) \to (\varphi \to \neg\psi) \to \neg\varphi.$$

Show that $\Gamma \vdash \varphi$ in H_2 iff $\Gamma \vdash \varphi$ in H_3. Also show that this equivalence no longer holds, if $(A4_\neg)$ is dropped from H_3.

6.5. (Mendelson) In this exercise we show that our Hilbert-style formalization with axioms (A1), (A2), and $(A3_\bot)$ is *minimal* in the sense that we cannot leave out any of the axioms without reducing the set of provable formulas.

(i) Show that $(A3_\bot)$ does not follow from *modus ponens* and (A1),(A2). *Hint:* Let φ^* denote the formula obtained by replacing all occurrences of $\bot$ in φ by $\top$, where $\top = p \to p$, for some propositional variable p. Show that ψ^* is a tautology for every formula following from (A1),(A2) and *modus ponens*. Then find an instance ψ of $(A3_\bot)$ where ψ^* is not a tautology.

(ii) Show that (A1) cannot be derived from (A2), $(A3_\bot)$, and *modus ponens*. *Hint:* Define a binary operator $\Rightarrow$ over the set $\mathbf{3} = \{0,1,2\}$, so that for all $a, b, c \in \mathbf{3}$,

- $(a \Rightarrow (b \Rightarrow c)) \Rightarrow (a \Rightarrow b) \Rightarrow (a \Rightarrow c) = 0$ and $((a \Rightarrow 1) \Rightarrow 1) \Rightarrow a = 0$;
- If $a \Rightarrow b = 0$ and $a = 0$ then also $b = 0$;
- But $a \Rightarrow (b \Rightarrow a)$ can be non-zero.

(iii) Show that (A2) does not follow from (A1), $(A3_\bot)$, and *modus ponens*.

6.6. Construct $\lambda\mu$-terms of the following types:

(i) $(p \to q) \to (\neg p \to q) \to q$.

(ii) $(p \lor q \to p) \lor (p \lor q \to q)$.

(iii) $(p \to (q \lor r)) \to ((p \to q) \lor r)$.

6.7. The proofs of Theorem 6.1.10 and Proposition 6.2.6 provide a way to construct a closed $\lambda\mu$-term of type φ from any valid formula φ. Write up this algorithm explicitly. What is (the $\hat\mu$-normal form of) the term you get for Peirce's law?

6.8. Suppose we extend Definition 6.2.7(i) to allow also the clause

$$\cdots \mid MC.$$

Show that the following rule (considered alone)

$$\mu a{:}\neg\varphi.C[[a]M] \to M \text{ if } a \notin \mathrm{FV}(M)$$

does not satisfy Church-Rosser property.

6.9. EXERCISES

6.9. Show that **catch** a **in** $C[\text{throw } M \text{ to } a] \twoheadrightarrow_{\hat\mu} M$, provided there are no free addresses in M, i.e. if we consider $\hat\mu$-reduction instead of μ-reduction, the restriction that there be no free addresses in C can be lifted (see Example 6.2.9). Show that this property remains true even if we extend Definition 6.2.7(i) to allow also the clause
$$\cdots \mid \mu b{:}\neg\tau.C.$$

6.10. (Ong) Show that adding the eta rule
$$\lambda x{:}\varphi.Mx \to M, \text{ if } x \notin \text{FV}(M)$$
to $\lambda\mu$ results in the failure of the Church-Rosser property for μ-reduction.

6.11. Show that replacing ζ by ζ' results in the failure of the Church-Rosser property for μ-reduction.

6.12. For each of the following formulas, draw a tree representing a winning prover strategy:

(i) $(\neg p \to \neg q) \to (\neg p \to q) \to p$.

(ii) $(p \to q) \to (p \to \neg q) \to \neg p$.

6.13. Does $\hat\mu$-reduction satisfy the Church-Rosser property?

6.14. Show in the direction from left to right in the proof of Theorem 6.5.9 how to construct η-long forms, rather than classical proofs.

6.15. Prove that $\vdash p \vee \neg p$ in $\text{NK}(\to, \bot, \wedge, \vee)$.

6.16. Prove Theorem 4.5.4 (stating that the reductions for the simply typed λ-calculus with pairs and injections are strongly normalizing). *Hint:* Use the translation in Example 6.7.8. Moreover, prove that the theorem remains correct even if (ε_1) and (ε_2) are added as reductions in the extended λ-calculus (see Definition 4.7.2).

6.17. The *Gödel-Gentzen* translation is defined as follows:

$$\begin{array}{rcl} g(\alpha) & = & \neg\neg\alpha \\ g(\varphi \to \psi) & = & (g(\varphi) \to g(\psi)). \end{array} \quad \text{for any atomic formula } \alpha;$$

Show that Theorem 6.4.3 remains true if we replace the Kolmogorov translation with the Gödel-Gentzen translation. *Hint:* Use Exercise 2.25.

6.18. Define $\text{NK}(\to, \vee, \wedge)$ as the fragment of $\text{NK}(\to, \bot, \vee, \wedge)$ obtained by omitting the constant $\bot$ and replacing ($\neg$E) by (P). Prove in $\text{NK}(\to, \vee, \wedge)$:

(i) $\varphi \vee (\varphi \to \psi)$.

(ii) $(\varphi \to (\psi \vee \delta)) \to ((\varphi \to \psi) \vee \delta)$.

The formula (i) bears the same relationship to *tertium non datur* $\varphi \vee \neg\varphi$ as Peirce's law $((\varphi \to \psi) \to \varphi) \to \varphi$ bears to its special case $(\neg\varphi \to \varphi) \to \varphi$.

6.19. (Selinger) Here is a fairy tale: The evil king calls the poor shepherd and gives him these orders: "You must bring me the philosopher's stone, or you have to find a way to turn the philosopher's stone into gold. If you don't, your head will be taken off tomorrow!" What can the poor shepherd do to save his life?

Chapter 7

Sequent calculus

In the preceding chapters we have seen two different types of proof systems: natural deduction and Hilbert style. Each of these has its advantages. For instance, the Hilbert-style formalism is a mathematically simple model, whereas natural deduction proofs are closer in spirit to the proofs actually developed in mathematical practice.

A third kind of proof formalism, the *sequent calculus,* was introduced in the 1930's by Gerhard Gentzen who also introduced natural deduction. Despite similar syntax (at least in some expositions), sequent calculus and natural deduction are quite different and serve different purposes. While natural deduction highlights the most fundamental properties of each connective by its introduction and elimination rule, sequent calculus makes a more adequate tool for actual proof construction; the rules (read bottom-up) define ways to replace complex formulas by simpler ones, finally reducing the initial task to verifying axioms. Instead of introduction and elimination rules, there are only introduction rules. Some of these rules introduce connectives to the right of $\vdash$ in a judgement—these rules are similar to the introduction rules from natural deduction. But there are also rules introducing connectives to the left of $\vdash$ in a judgement. These rules replace the elimination rules of natural deduction.

In sequent calculus we also meet other new rules. Multiple use of the same assumption (see Section 5.6) is managed by an explicit *contraction* rule in the system. Similarly, assumptions that are not used at all are managed by a *weakening* rule. Finally, whereas detours in natural deduction are formed by certain pairs of introduction and elimination rules, there is a specific *Cut* rule in sequent calculus which represent detours. Normalization then amounts to eliminating applications of this rule.

7.1. Gentzen's sequent calculus LK

Gentzen introduced sequent calculi for both classical and intuitionistic logic; in this section we present a classical system. Since we shall be concerned extensively with the relationship to natural deduction, which is typically based on absurdity (as opposed to negation), it is convenient to consider sequent calculi based on absurdity, even though negation is the most natural approach for the latter systems.

7.1.1. DEFINITION (The system LK). We consider the set $\Phi = \Phi(\to, \land, \lor, \bot)$ of formulas. As usual, $\neg \varphi = \varphi \to \bot$.

(i) A *sequent* is a pair (Γ, Δ), where Γ, Δ are finite sequences of formulas, called the *antecedent* and *succedent*, respectively. Such a pair is also written $\Gamma \vdash \Delta$.

(ii) A *derivation* of $\Gamma \vdash \Delta$ is a tree labeled with sequents, where the root is $\Gamma \vdash \Delta$, and where every node and its children match the sequents below and above the line, respectively, of some rule in Figure 7.1. Moreover, the leaves in the tree must be axioms.

(iii) We write $\Gamma \vdash_{LK}^{+} \Sigma$ if $\Gamma \vdash \Sigma$ has a derivation, and we write $\Gamma \vdash_{LK} \Sigma$ if $\Gamma \vdash \Sigma$ has a derivation which does not use the (Cut) rule.

7.1.2. REMARK. The intuitive meaning of $\Gamma \vdash \Sigma$ is that the conjunction of all formulas in Γ implies the disjunction of all formulas in Σ, i.e., that $\varphi_1 \land \ldots \land \varphi_n$ implies $\psi_1 \lor \ldots \lor \psi_m$, where $\Gamma = \varphi_1, \ldots, \varphi_n$ and $\Sigma = \psi_1, \ldots, \psi_m$.

We begin with some examples which illustrate the rules and which allow us to compare the rules to natural deduction.

7.1.3. EXAMPLE. Here is a derivation of $\vdash (p \to q) \to (q \to r) \to (p \to r)$:

$$\cfrac{\cfrac{p \vdash p \quad q \vdash q}{\cfrac{p, p \to q \vdash q}{p \to q, p \vdash q}(LX)}(L\to) \quad \cfrac{\cfrac{\cfrac{\cfrac{r \vdash r}{r, p \to q \vdash r}(LW)}{p \to q, r \vdash r}(LX)}{\cfrac{p \to q, r, p \vdash r}{p \to q, p, r \vdash r}(LX)}(LW)}{\cfrac{\cfrac{\cfrac{\cfrac{p \to q, p, q \to r \vdash r}{p \to q, q \to r, p \vdash r}(LX)}{p \to q, q \to r \vdash p \to r}(R\to)}{p \to q \vdash (q \to r) \to (p \to r)}(R\to)}{\vdash (p \to q) \to (q \to r) \to (p \to r)}(R\to)}}(L\to)$$

7.1. Gentzen's sequent calculus LK

Axioms:

$$\bot \vdash \quad (L\bot) \qquad \varphi \vdash \varphi \quad (Ax)$$

Structural Rules:

$$\frac{\Gamma \vdash \Sigma}{\Gamma, \varphi \vdash \Sigma} \ (LW) \qquad \frac{\Gamma \vdash \Pi}{\Gamma \vdash \varphi, \Pi} \ (RW)$$

$$\frac{\Gamma, \varphi, \psi, \Gamma' \vdash \Sigma}{\Gamma, \psi, \varphi, \Gamma' \vdash \Sigma} \ (LX) \qquad \frac{\Gamma \vdash \Delta, \varphi, \psi, \Delta'}{\Gamma \vdash \Delta, \psi, \varphi, \Delta'} \ (RX)$$

$$\frac{\Gamma, \varphi, \varphi \vdash \Sigma}{\Gamma, \varphi \vdash \Sigma} \ (LC) \qquad \frac{\Gamma \vdash \varphi, \varphi, \Delta}{\Gamma \vdash \varphi, \Delta} \ (RC)$$

Logical Rules:

$$\frac{\Gamma, \varphi \vdash \Sigma}{\Gamma, \varphi \wedge \psi \vdash \Sigma} \ (L\wedge) \frac{\Gamma, \psi \vdash \Sigma}{\Gamma, \varphi \wedge \psi \vdash \Sigma} \qquad \frac{\Gamma \vdash \varphi, \Delta \quad \Gamma \vdash \psi, \Delta}{\Gamma \vdash \varphi \wedge \psi, \Delta} \ (R\wedge)$$

$$\frac{\Gamma, \varphi \vdash \Sigma \quad \Gamma, \psi \vdash \Sigma}{\Gamma, \varphi \vee \psi \vdash \Sigma} \ (L\vee) \qquad \frac{\Gamma \vdash \varphi, \Delta}{\Gamma \vdash \varphi \vee \psi, \Delta} \ (R\vee) \frac{\Gamma \vdash \psi, \Delta}{\Gamma \vdash \varphi \vee \psi, \Delta}$$

$$\frac{\Gamma \vdash \varphi, \Delta \quad \Gamma, \psi \vdash \Sigma}{\Gamma, \varphi \to \psi \vdash \Delta, \Sigma} \ (L\to) \qquad \frac{\Gamma, \varphi \vdash \psi, \Delta}{\Gamma \vdash \varphi \to \psi, \Delta} \ (R\to)$$

Cut Rule:

$$\frac{\Gamma \vdash \varphi, \Delta \quad \Gamma, \varphi \vdash \Sigma}{\Gamma \vdash \Delta, \Sigma} \ (Cut)$$

FIGURE 7.1: CLASSICAL SEQUENT CALCULUS LK.

The derivation makes use of the *structural rules*. The left *weakening* rule (LW) introduces additional assumptions, and the left *exchange* rule (LX) shifts a formula right, so that the other rules can apply to it. In natural deduction the latter rule is implicit, since left hand sides are *sets*, i.e. there is no order of the elements. Moreover, the left weakening rule is implicit because the axiom in natural deduction has form $\Gamma, \varphi \vdash \varphi$, i.e. the left hand side need not be φ alone. The structural right rules are redundant in natural deduction, because the right hand side always has exactly one formula.

The right rule (R→) is identical to the introduction rule from natural deduction if we restrict it to singleton right hand sides and consider the left hand side as a set. This similarity applies to all the logical right rules. In contrast, the left rule (L→) does not look like the elimination rule for implication in natural deduction.

7.1.4. EXAMPLE. For an example involving an empty right hand side, note that $\bot \vdash \varphi$ using (L$\bot$) and (RW). Thus, if $\Gamma \vdash \bot$, then also $\Gamma \vdash \varphi$. In other words, *ex falso* is a derived rule.

7.1.5. EXAMPLE. The following is a proof of Peirce's law:

$$\cfrac{\cfrac{\cfrac{\cfrac{\cfrac{\cfrac{p \vdash p}{p \vdash q, p} \text{(RW)}}{\vdash p \to q, p} \text{(R}\to\text{)} \qquad p \vdash p}{(p \to q) \to p \vdash p, p} \text{(L}\to\text{)}}{(p \to q) \to p \vdash p} \text{(RC)}}{\vdash ((p \to q) \to p) \to p} \text{(R}\to\text{)}}$$

Here we see two formulas at the right hand side bringing right weakening and right *contraction* into action. In natural deduction such steps are implicit (for instance, $\sigma \vdash \tau$ is the same as $\sigma, \sigma \vdash \tau$, so we may conclude $\sigma \vdash \sigma \to \tau$).

There are other formulations of sequent calculus (see Exercises 7.5–7.7). A main variation is obtained by taking *negation*, as opposed to absurdity, as primitive. In this case one replaces the axiom (L$\bot$) by the rules

$$\cfrac{\Gamma \vdash \varphi, \Delta}{\Gamma, \neg\varphi \vdash \Delta} \text{(L}\neg\text{)} \qquad \cfrac{\Gamma, \varphi \vdash \Delta}{\Gamma \vdash \neg\varphi, \Delta} \text{(R}\neg\text{)}$$

The resulting system has a left and a right rule for every connective, and is therefore more symmetric than the formulation with (L$\bot$). It is also more symmetric than any natural deduction system for classical logic.

A common variation, which emphasizes the intuition in Remark 7.1.2, is to replace (L∧) and (R∨) by:

$$\cfrac{\Gamma, \varphi, \psi \vdash \Sigma}{\Gamma, \varphi \wedge \psi \vdash \Sigma} \text{(L}'\wedge\text{)} \qquad \cfrac{\Gamma \vdash \varphi, \psi, \Delta}{\Gamma \vdash \varphi \vee \psi, \Delta} \text{(R}'\vee\text{)}$$

7.2. Fragments of LK versus natural deduction

Another idea is to adopt this variant of (L→) and (Cut):

$$\frac{\Gamma \vdash \varphi, \Sigma \quad \Gamma, \psi \vdash \Sigma}{\Gamma, \varphi \to \psi \vdash \Sigma} \, (L'{\to}) \qquad \frac{\Gamma \vdash \varphi, \Sigma \quad \Gamma, \varphi \vdash \Sigma}{\Gamma \vdash \Sigma} \, (\text{Cut}')$$

Whereas (L→) copies the left hand side Γ from the hypotheses to the conclusion, but collects the right hand sides Δ and Σ, this variant copies both left and right hand sides from the hypotheses to the conclusion.

These variations on conjunction, disjunction, implication, and cut reduce the need for contractions (see the end of Section 7.3), but complicate the formulation of the intuitionistic fragment, which we turn to next.

7.2. Fragments of LK versus natural deduction

The *intuitionistic sequent calculus* is obtained from the classical system by a syntactic restriction: there must be *at most one* formula in any succedent.

7.2.1. DEFINITION. A sequent $\Gamma \vdash \Sigma$ is called

- *intuitionistic*, if Σ has at most one formula.
- *minimal*, if Σ has exactly one formula.
- *of Peirce type*, if Σ has at least one formula.

We write $\Gamma \vdash^+_{LJ} \varphi$ ($\Gamma \vdash^+_{LM} \varphi$, $\Gamma \vdash^+_{LP} \varphi$) if $\Gamma \vdash \varphi$ has a derivation using only intuitionistic (minimal, Peirce type) sequents. We omit "+" if there is a derivation without (Cut). We also talk of the "systems" LJ, LM, and LP, meaning LK restricted to the appropriate sequents. (LP is not standard in the literature.)

Some of the rules of LK are not used in some of the fragments. For instance, in LJ, succedents (right hand sides) have at most one formula; this means that in Figure 7.1, Σ has exactly one formula, and Δ, Π are empty. As a consequence, derivations may use (L⊥) and (RW), but cannot use (RC) or (RX). This is summarized in Figure 7.2.

System	RHS	Σ	Δ	Π	(L⊥)	(RC),(RX)	(RW)
LM	$= 1$	$= 1$	$= 0$	$= 0$	×	×	×
LJ	≤ 1	≤ 1	$= 0$	$= 0$	✓	×	✓
LP	≥ 1	≥ 1	≥ 0	≥ 1	×	✓	✓
LK	≥ 0	≥ 0	≥ 0	≥ 0	✓	✓	✓

FIGURE 7.2: APPLICABILITY OF RULES IN FRAGMENTS OF LK.

We now relate natural deduction and sequent calculi (see Figure 7.3).

7.2.2. NOTATION. We use the following shorthands related to derivability in natural deduction:

- $\vdash_C$ for $NK(\to, \bot, \wedge, \vee)$ (see Definition 6.7.1).
- $\vdash_I$ for $NJ(\to, \bot, \wedge, \vee)$ (see Definition 2.2.1).
- $\vdash_P$ for $NK(\to, \wedge, \vee)$ (see Exercise 6.18).
- $\vdash_M$ for $NJ(\to, \wedge, \vee)$ (see Notation 2.2.2).

In the rest of this section it is convenient to assume that the two latter of the above systems contain $\bot$ in the language, though the systems have no inference rules for $\bot$, i.e. neither *ex falso*, nor ($\neg$E). In other words, we can think of $\bot$ as some arbitrary, fixed propositional variable in these systems.

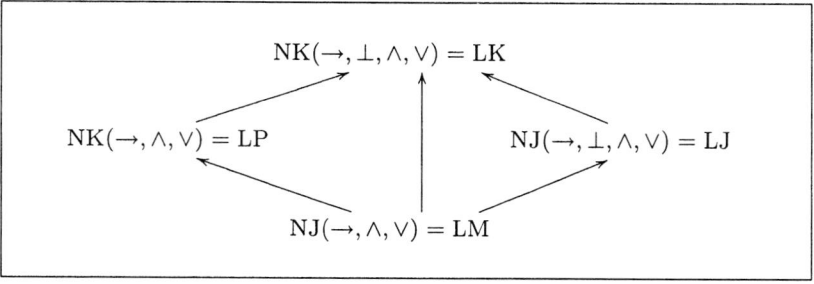

FIGURE 7.3: FRAGMENTS OF LK.

7.2.3. PROPOSITION. *Let Γ be finite (sequence or set as appropriate).*

(i) $\Gamma \vdash_M \varphi$ *implies* $\Gamma \vdash^+_{LM} \varphi$.

(ii) $\Gamma \vdash_I \varphi$ *implies* $\Gamma \vdash^+_{LJ} \varphi$.

(iii) $\Gamma \vdash_C \varphi$ *implies* $\Gamma \vdash^+_{LK} \varphi$.

(iv) $\Gamma \vdash_P \varphi$ *implies* $\Gamma \vdash^+_{LP} \varphi$.

PROOF. For (i), we construct, by induction on the derivation of $\Gamma \vdash_M \varphi$, a derivation of $\Gamma \vdash^+_{LK} \varphi$ in which all sequents are minimal. The cases where the derivation consists of (Ax) or end in an introduction rule are easy. We consider the case of implication elimination; disjunction and conjunction are similar. So assume the derivation ends with

$$\frac{\Gamma \vdash \psi \to \varphi \quad \Gamma \vdash \psi}{\Gamma \vdash \varphi}$$

7.2. FRAGMENTS OF LK VERSUS NATURAL DEDUCTION

Then we have the following derivation in LM (where the dots indicate the part of the proof obtained by the induction hypothesis):

$$\cfrac{\cfrac{\vdots}{\Gamma \vdash \psi \to \varphi} \quad \cfrac{\cfrac{\vdots}{\Gamma \vdash \psi} \quad \cfrac{\varphi \vdash \varphi}{\Gamma, \varphi \vdash \varphi}}{\Gamma, \psi \to \varphi \vdash \varphi}(L\to)}{\Gamma \vdash \varphi}(\text{Cut})$$

where the double lines indicate a number of applications of structural rules.

For (ii), the only new case is when the derivation ends with

$$\cfrac{\Gamma \vdash \bot}{\Gamma \vdash \varphi}$$

Then we have the derivation in LJ:

$$\cfrac{\cfrac{\vdots}{\Gamma \vdash \bot} \quad \cfrac{\bot \vdash}{\Gamma, \bot \vdash \varphi}}{\Gamma \vdash \varphi}(\text{Cut})$$

and the derivation contains only intuitionistic sequents.

For (iii), the new situation is when the derivation ends with

$$\cfrac{\Gamma, \neg\varphi \vdash \bot}{\Gamma \vdash \varphi}$$

Then we have the derivation in LK:

$$\cfrac{\cfrac{\cfrac{\vdots}{\Gamma, \neg\varphi \vdash \bot}}{\Gamma \vdash \neg\neg\varphi}(R\to) \quad \cfrac{\cfrac{\cfrac{\varphi \vdash \varphi}{\Gamma, \varphi \vdash \bot, \varphi}}{\Gamma \vdash \neg\varphi, \varphi}(R\to) \quad \cfrac{\bot \vdash}{\Gamma, \bot \vdash}}{\Gamma, \neg\neg\varphi \vdash \varphi}(L\to)}{\Gamma \vdash \varphi}(\text{Cut})$$

For (iv), the new case, compared to (i), is when the derivation ends with

$$\cfrac{\Gamma, \varphi \to \psi \vdash \varphi}{\Gamma \vdash \varphi}$$

In this case we have the derivation in LP:

$$\cfrac{\cfrac{\cfrac{\vdots}{\Gamma, (\varphi \to \psi) \vdash \varphi}}{\Gamma \vdash (\varphi \to \psi) \to \varphi}(R\to) \quad \cfrac{\cfrac{\cfrac{\varphi \vdash \varphi}{\Gamma, \varphi \vdash \psi, \varphi}}{\Gamma \vdash \varphi \to \psi, \varphi}(R\to) \quad \cfrac{\varphi \vdash \varphi}{\Gamma, \varphi \vdash \varphi}}{\cfrac{\Gamma, (\varphi \to \psi) \to \varphi \vdash \varphi, \varphi}{\Gamma, (\varphi \to \psi) \to \varphi \vdash \varphi}(\text{RC})}(L\to)}{\Gamma \vdash \varphi}(\text{Cut})$$

and the derivation uses only Peirce-type sequents. □

For the inverse translation, how do we handle multiple formulas on the right hand side in LK and LP? Remark 7.1.2 suggests to view the right hand side as a disjunction, and in LJ and LK we can interpret empty right hand sides as the empty disjunction, i.e. ⊥. (Another idea is to interpret formulas on the right hand side as negations on the left hand side—see Exercise 7.17). Disjunction is associative in all our logics, so the following is well-defined.

7.2.4. PROPOSITION. *For $\Delta = \varphi_1, \ldots, \varphi_n$, let $\bigvee \Delta = \varphi_1 \vee \ldots \vee \varphi_n$ if $n > 0$, and $\bigvee \Delta = \bot$ if $n = 0$. In particular, if $\Delta = \varphi$, then $\bigvee \Delta = \varphi$. Then for finite Γ and Δ (sequences or sets as appropriate):*

(i) $\Gamma \vdash^+_{LM} \Delta$ *implies* $\Gamma \vdash_M \bigvee \Delta$.

(ii) $\Gamma \vdash^+_{LJ} \Delta$ *implies* $\Gamma \vdash_I \bigvee \Delta$.

(iii) $\Gamma \vdash^+_{LP} \Delta$ *implies* $\Gamma \vdash_P \bigvee \Delta$.

(iv) $\Gamma \vdash^+_{LK} \Delta$ *implies* $\Gamma \vdash_C \bigvee \Delta$.

PROOF. For (i) proceed by induction on the derivation of $\Gamma \vdash^+_{LM} \Delta$. The cases involving right rules and (Ax) are easy, as are the structural rules and (Cut). For the left rules we consider implication; conjunction and disjunction are similar. So assume the derivation ends with

$$\frac{\Gamma \vdash \psi_1 \quad \Gamma, \psi_2 \vdash \varphi}{\Gamma, \psi_1 \to \psi_2 \vdash \varphi}$$

By the induction hypothesis $\Gamma \vdash_M \psi_1$, so $\Gamma, \psi_1 \to \psi_2 \vdash_M \psi_2$. By the induction hypothesis again, $\Gamma, \psi_1 \to \psi_2, \psi_2 \vdash_M \varphi$, so also $\Gamma, \psi_1 \to \psi_2 \vdash_M \varphi$.

For (ii) we show by induction on derivations that if $\Gamma \vdash^+_{LJ} \Sigma$, then it follows that (1) Σ singleton implies $\Gamma \vdash_I \Sigma$, and (2) Σ empty implies $\Gamma \vdash_I \bot$. The cases under (1) are identical to the cases considered in (i) with the addition now of the case (RW) which is easy. The cases for (2) include the left rules, (Cut) and (L⊥). The two former are handled exactly as the left rules under (i) (with the general φ replaced by ⊥), and the case (L⊥) is easy.

For (iii), proceed by induction on the derivation of $\Gamma \vdash^+_{LP} \Delta$. The axiom (Ax) is trivial, and (L⊥) is not used. The structural and logical left rules are trivial, except (L→), which is dealt with as follows. Assume the derivation ends with

$$\frac{\Gamma \vdash \varphi, \Delta \quad \Gamma, \psi \vdash \Sigma}{\Gamma, \varphi \to \psi \vdash \Delta, \Sigma} \; (L\to)$$

The case where Δ is empty is easy, so assume that Δ is not empty. By the induction hypothesis $\Gamma, \varphi \to \psi \vdash \varphi \vee \bigvee \Delta$ and $\Gamma, \varphi \to \psi \vdash \psi \to \bigvee \Sigma$. Then $\Gamma, \varphi \to \psi \vdash \bigvee \Delta \vee \bigvee \Sigma$, so $\Gamma, \varphi \to \psi \vdash \bigvee(\Delta, \Sigma)$.

7.3. GENTZEN'S HAUPTSATZ   169

The case of (Cut) is similar, so it remains to consider the right rules. This boils down to mimicking the effect of right rules inside disjunctions, using basic properties of disjunction and Exercise 6.18; for instance, to mimic right contraction, we use the theorem $\varphi \vee \varphi \to \varphi$ and to mimic right implication, we use $(\varphi \to (\psi \vee \delta)) \to ((\varphi \to \psi) \vee \delta)$.

For (iv) proceed by induction on the derivation of $\Gamma \vdash^+_{LK} \Delta$. The only new cases are (RW) when Π is empty, and (L$\bot$), both of which are easy. □

7.2.5. COROLLARY. $\Gamma \vdash \varphi$ in LK (resp. LJ) iff $\Gamma \vdash \varphi$ in NK (resp. NJ).

7.3. Gentzen's Hauptsatz

Recall from Chapter 4 that some natural deduction derivations contain certain detours that can be eliminated. Roughly speaking, whenever an introduction rule is immediately followed by the corresponding elimination rule, the proof has a detour that can be eliminated. When we translate natural deduction proofs to sequent calculus proofs, we can thus expect that the latter will contain some kind of redundancy. In fact, in sequent calculus this redundancy amounts to applications of the (Cut) rule, so getting rid of redundancy becomes *cut elimination*.

7.3.1. EXAMPLE. Consider a derivation of the following form:

$$
\begin{array}{cc}
(1) & (2) \qquad (3) \\
\vdots & \vdots \qquad \vdots \\
(\text{R}\to)\dfrac{\Gamma, \varphi \vdash \psi}{\Gamma \vdash \varphi \to \psi} \quad \dfrac{\Gamma \vdash \varphi \quad \Gamma, \psi \vdash \vartheta}{\Gamma, \varphi \to \psi \vdash \vartheta}(\text{L}\to) \\
\hline
\Gamma \vdash \vartheta
\end{array}(\text{Cut})
$$

We can eliminate this cut at the cost of introducing two new ones. This makes sense, because the new cut formulas are simpler. The new proof is:

$$
\begin{array}{cc}
(2) \qquad (1) & (3) \\
\vdots \qquad \vdots & \vdots \\
(\text{Cut})\dfrac{\Gamma \vdash \varphi \quad \Gamma, \varphi \vdash \psi}{\Gamma \vdash \psi} & \Gamma, \psi \vdash \vartheta \\
\hline
\Gamma \vdash \vartheta
\end{array}(\text{Cut})
$$

Note that in our example the cut formula $\varphi \to \psi$ was introduced just before the cut by the rules (R$\to$) and (L$\to$).

Gentzen's Hauptsatz states that we can always eliminate the redundancy. If one tries to prove this by induction on derivations, one runs into difficulties with the contraction rule (see Exercise 7.8). A common way to overcome this difficulty is to consider a technically more convenient form of (Cut):

$$\frac{\Gamma \vdash \Delta \quad \Gamma' \vdash \Delta'}{\Gamma, (\Gamma' - \varphi) \vdash (\Delta - \varphi), \Delta'} \text{ (Mix)}$$

where $\Sigma - \varphi$ denotes the result of removing all occurrences of φ from Σ, and where we require that $\varphi \in \Delta$ and $\varphi \in \Gamma'$. One then proves that all instances of (Mix) can be eliminated. The restriction that $\varphi \in \Delta$ and $\varphi \in \Gamma'$ is to ensure that the proof works also for the fragments of LK. Below we implicitly adopt this idea, though we still consider the system with (Cut).

In the remainder of this section, let L denote one of the systems LK, LJ, LM, or LP.

7.3.2. DEFINITION.

(i) The *degree* of a cut is the size of the formula φ eliminated in the cut.

(ii) The degree of a derivation is the maximal degree of any cut in the derivation (and 0 if there are no cuts).

(iii) We write $\Gamma \vdash_L^d \Delta$ if there is a derivation of $\Gamma \vdash \Delta$ of degree at most d.

Note that $\Gamma \vdash_L \Delta$ iff $\Gamma \vdash_L^0 \Delta$, and $\Gamma \vdash_L^+ \Delta$ iff $\Gamma \vdash_L^d \Delta$, for some d.

7.3.3. LEMMA. *Let* $d(\varphi) = d+1$, $\varphi \in \Delta$, *and* $\varphi \in \Gamma'$. *If*

$$\Gamma \vdash_L^d \Delta \text{ and } \Gamma' \vdash_L^d \Delta',$$

then

$$\Gamma, (\Gamma' - \varphi) \vdash_L^d (\Delta - \varphi), \Delta'.$$

PROOF. We first consider the case for LK. We proceed by induction on the sum of the heights of the derivations D and D' of $\Gamma \vdash^d \Delta$ and $\Gamma' \vdash^d \Delta'$, respectively. We consider the different shapes of D and D'.
CASE 1: D consists of

$$(\text{Ax}) \ \psi \vdash \psi.$$

Since $\varphi \in \Delta$, we must have $\varphi = \psi$. We are to show that $\varphi, \Gamma' - \varphi \vdash \Delta'$. This follows from the derivation D' using structural rules.
CASE 2: D' consists of (Ax). Similar to Case 1.
CASE 3: D consists of (L$\bot$). This is not possible, since $\varphi \in \Delta$.
CASE 4: D' ends with (L$\bot$). Then $\varphi = \bot$, since $\varphi \in \Gamma'$. We are to show that $\Gamma \vdash \Delta - \bot$. This follows from $\Gamma \vdash \Delta$ by induction on its derivation.
CASE 5: D ends with a logical left rule. Consider the case of the implication:

$$(\text{L}\rightarrow) \frac{\Gamma_1 \vdash \rho, \Delta_1 \quad \Gamma_1, \psi \vdash \Delta_2}{\Gamma_1, \rho \rightarrow \psi \vdash \Delta_1, \Delta_2}$$

7.3. GENTZEN'S HAUPTSATZ

If $\varphi = \rho$, then by the induction hypothesis, $\Gamma_1, (\Gamma' - \varphi) \vdash (\Delta_1 - \varphi), \Delta'$. Using structural rules, we easily obtain $\Gamma_1, \rho \to \psi, (\Gamma' - \varphi) \vdash (\Delta_1, \Delta_2) - \varphi, \Delta'$. Now assume $\varphi \neq \rho$. If $\varphi \in \Delta_1$ and $\varphi \in \Delta_2$, then by the induction hypothesis

$$\Gamma_1, (\Gamma' - \varphi) \vdash \rho, (\Delta_1 - \varphi), \Delta' \qquad \Gamma_1, \psi, (\Gamma' - \varphi) \vdash (\Delta_2 - \varphi), \Delta'.$$

If $\varphi \notin \Delta_1$, we infer the left part directly from $\Gamma_1 \vdash \rho, \Delta_1$. Similarly if $\varphi \notin \Delta_2$. Using some structural rules together with the left rule for implication, we then obtain $\Gamma_1, \rho \to \psi, (\Gamma' - \varphi) \vdash ((\Delta_1, \Delta_2) - \varphi), \Delta'$.

CASE 6: D' ends with a logical right rule. Similar to Case 5.
CASE 7: D or D' ends with a structural rule or (Cut). Similar to Case 5.
CASE 8: D ends with logical right rule or D' ends with a logical left rule, but φ is not introduced to the right and left, respectively. Similar to Case 5.
CASE 9: The only remaining case is when D ends with a right rule, and D' ends with a left rule and both rules introduce φ, to the right and left, respectively. We consider the case of implication $\varphi = \rho \to \psi$. So assume D and D' end with

$$(\text{R}\to)\frac{\Gamma, \rho \vdash \psi, \Delta_1}{\Gamma \vdash \rho \to \psi, \Delta_1} \qquad (\text{L}\to)\frac{\Gamma'_1 \vdash \rho, \Delta'_1 \quad \Gamma'_1, \psi \vdash \Delta'_2}{\Gamma'_1, \rho \to \psi \vdash \Delta'_1, \Delta'_2}$$

If $\varphi \in \Delta_1$, then by the induction hypothesis (applied above the line in (R$\to$) and below the line in (L$\to$)):

$$\Gamma, \rho, (\Gamma'_1 - \varphi) \vdash \psi, (\Delta_1 - \varphi), \Delta'_1, \Delta'_2 \tag{7.1}$$

If $\varphi \notin \Delta_1$ then (7.1) is inferred directly from $\Gamma, \rho \vdash \psi, \Delta_1$. If $\varphi \in \Gamma'_1$, then by the induction hypothesis (applied below the line in (R$\to$) and above the line in (L$\to$)):

$$\Gamma, (\Gamma'_1 - \varphi) \vdash (\Delta_1 - \varphi), \rho, \Delta'_1 \tag{7.2}$$
$$\Gamma, (\Gamma'_1 - \varphi), \psi \vdash (\Delta_1 - \varphi), \Delta'_2 \tag{7.3}$$

If $\varphi \notin \Gamma'_1$ then (7.2)-(7.3) follow directly.
Combining (7.1) and (7.2) with (Cut), and using structural rules:

$$\Gamma, (\Gamma'_1 - \varphi) \vdash \psi, (\Delta_1 - \varphi), \Delta'_1, \Delta'_2.$$

Combining this and (7.3) with (Cut) using structural rules, we get

$$\Gamma, (\Gamma'_1 - \varphi) \vdash (\Delta_1 - \varphi), \Delta'_1, \Delta'_2.$$

The degrees of the new cuts (involving ρ and ψ) are at most d.

By careful inspection one sees that if the original derivation uses only minimal, intuitionistic, or Peirce-type sequents, respectively, then the same can be made to hold for the new derivation, so the lemma holds for LM, LJ, and LP. □

7.3.4. THEOREM (Gentzen, 1935). *If* $\Gamma \vdash^+_L \Delta$ *then* $\Gamma \vdash_L \Delta$.

PROOF. We show for any d that $\Gamma \vdash^{d+1} \Delta$ implies $\Gamma \vdash^d \Delta$. The proof is by induction on the derivation of $\Gamma \vdash^{d+1} \Delta$. In all other cases than (Cut) we apply the induction hypothesis directly to the premises. In the case of

$$(\text{Cut}) \; \frac{\Gamma \vdash \varphi, \Delta \quad \Gamma, \varphi \vdash \Sigma}{\Gamma \vdash \Delta, \Sigma},$$

first apply the induction hypothesis to arrive at derivations of

$$\Gamma \vdash^d \varphi, \Delta \qquad \Gamma, \varphi \vdash^d \Sigma.$$

If $d(\varphi) < d+1$ one can use (Cut) directly; if $d(\varphi) = d+1$, the preceding lemma yields

$$\Gamma, (\Gamma - \varphi) \vdash^d (\Delta - \varphi), \Sigma.$$

Using structural rules we then get $\Gamma \vdash^d \Delta, \Sigma$. □

This result is a "weak" normalization result. Our reduction strategy selects cuts of maximal degree, not arbitrary cuts. In the next section we prove a strong normalization result.

The disjunction property (Proposition 2.5.10) was proved using Kripke semantics. Now we prove it using syntactic means.

7.3.5. COROLLARY. *If* $\vdash \varphi \vee \psi$ *in* IPC *then either* $\vdash \varphi$ *or* $\vdash \psi$ *in* IPC.

PROOF. Take a cut-free proof. Since the antecedent is empty, the proof must end with a right rule. This rule can be (R$\vee$) or (RW); the latter is impossible, because the premise of (RW) would have empty antecedent and succedent, and no rule can have such a conclusion, nor does it match any axiom. □

7.3.6. COROLLARY (Subformula property). *Let* $\Gamma \vdash^+_L \Delta$. *There is a derivation of this sequent in which each formula is a subformula of a formula occurring in* Γ, Δ.

PROOF. Take a cut-free derivation and show by induction that all formulas in it are subformulas of formulas occurring in Γ, Δ. □

7.3.7. REMARK. The subformula property implies that a cut-free proof of a sequent can only mention connectives occurring in that sequent. For instance, the formula $((p \wedge q) \to r) \leftrightarrow (p \to (q \to r))$ is provable in a system containing only rules for implication and conjunction. Similarly, LK and LJ are *conservative* over LP and LM, respectively.

7.3.8. COROLLARY. *Let* φ *not contain* $\bot$. *If* $\vdash^+_{LK} \varphi$ *then* $\vdash^+_{LP} \varphi$, *and if* $\vdash^+_{LJ} \varphi$ *then* $\vdash^+_{LM} \varphi$.

7.3. Gentzen's Hauptsatz

PROOF. A cut-free derivation in LK or LJ cannot use (L$\bot$), since $\bot$ does not occur in the derivation. □

Similarly to the Ben-Yelles algorithm searching for natural deduction proofs in η-long normal form, we can develop an algorithm that searches for cut-free sequent calculus proofs. As we reconstruct a proof by building the tree upward, the search space at each step is limited to subformulas of the formulas occurring at the present stage. This process cannot continue indefinitely, as the number of available formulas is bounded, and we will eventually repeat already considered sequents.

7.3.9. COROLLARY. *It is decidable whether, for any φ, it holds that $\vdash^+_L \varphi$.*

PROOF. First note that a proof of minimal size of a sequent $\Gamma \vdash \Delta$ cannot contain that sequent elsewhere than in the root. Indeed, the appropriate subtree would yield another proof of $\Gamma \vdash \Delta$ of a smaller size. More generally, a proof of minimal size cannot contain the same sequent twice on the path from the root to a leaf.

We say that a sequent $\Gamma \vdash \Delta$ is *bounded* if each formula occurs at most twice in Γ and at most twice in Δ. A derivation is *bounded* if all sequents occurring in it are bounded. For a sequence Θ of formulas, define Θ_1 to be the sequence obtained by listing each formula in Θ only once, according to some fixed order. It is not difficult to prove that if $\Gamma \vdash_L \Delta$, then $\Gamma_1 \vdash_L \Delta_1$ using a bounded cut-free derivation.

Now suppose that $\vdash_L \varphi$ and consider a bounded derivation of $\vdash_L \varphi$ such that the size of the derivation is minimal among all bounded derivations of this sequent. There is only a finite number, say m, of bounded sequents built of subformulas of φ. It follows that the height of our derivation cannot exceed m, as otherwise a certain sequent must occur in a proof of itself. Deciding if $\vdash_L \varphi$ thus reduces to verifying if there exists a proof of height not exceeding m. □

As mentioned earlier, the structural rules are useful for highlighting reasoning principles that are left implicit in natural deduction. However, in many cases these rules actually cause undesirable noise. For instance, the proof of cut-elimination requires the (Mix) rule to overcome problems with contraction. And in the above decidability proof, the notion of bounded sequents is needed to neutralize the same contractions. Even worse, the proof does not give the polynomial space bound one might expect for the intuitionistic case, because antecedents may both grow (by contraction) and shrink (by logical rules and weakening) when reading the rules bottom-up.

Fortunately, the trouble can be minimized (see Exercise 7.9–7.13). A brutal solution is to adopt a variant with sequents of form $\Gamma \vdash \Delta$, where Γ and Δ are sets, and take the axioms in generalized form $\Gamma, \varphi \vdash \varphi, \Delta$ and $\Gamma, \bot \vdash \Delta$.

This eliminates the need for structural rules altogether at the expense of building them into the notation.

A point between the two extremes (sequences and sets) is to use multisets, in which case the exchange rules are omitted. For both multisets and sequences, the weakening rules can be omitted by taking the generalized axioms. Contraction can be made redundant in sequence and multiset formulations in several ways. The simplest is to modify all the logical rules so that they retain the main formula (read bottom-up). For instance, the left implication rule can be taken in the form:

$$\frac{\Gamma, \varphi \to \psi \vdash \varphi, \Sigma \quad \Gamma, \varphi \to \psi, \psi \vdash \Sigma}{\Gamma, \varphi \to \psi \vdash \Sigma}$$

This type of system is known as G3. (The terminology originates with Kleene who used G1 for a system similar to Figure 7.1 and G2 for a system with (Mix) instead of (Cut).) For the intuitionistic system one only retains the main formula in the logical left rules. With generalized axioms and without weakening, antecedents cannot shrink, so this formulation can be used to obtain the polynomial space bound (sequents can be taken as sets to avoid multiple occurrences of the same formula), see Exercise 7.13.

As a more subtle approach to elimination of contraction, by taking the variations of rules for conjunction, disjunction, and implication mentioned in Section 7.1, one can actually omit the contraction rules without retaining the principal formula in the logical rules, at least in the classical case. In the intuitionistic case, there is a small fly in the ointment; one must retain the principal formula in the left hypothesis of left implication:

$$\frac{\Gamma, \varphi \to \psi \vdash \varphi \quad \Gamma, \psi \vdash \sigma}{\Gamma, \varphi \to \psi \vdash \sigma}$$

These contraction-free systems are also called G3.

7.4. Cut elimination versus normalization

Natural deduction proofs correspond to λ-terms with types, and Hilbert-style proofs correspond to combinators with types. What do sequent calculus proofs correspond to? One possibility is to assign λ-terms to sequent calculus proofs. That is, to devise an alternative version of simply typed λ-calculus, —with the same term language, but with different typing rules—which is to sequent calculus what the traditional formulation of simply typed λ-calculus is to natural deduction. A main corollary of this approach is that Gentzen's Hauptsatz can be inferred from weak normalization in natural deduction.

We now review this idea. In what follows we shall only discuss LM (not LJ, LP, and LK), and we shall be concerned exclusively with the fragment dealing with implication. In order to eliminate the noise resulting from

7.4. CUT ELIMINATION VERSUS NORMALIZATION

contractions we take a cousin of the G3 variant of LM. Weakening is left out since we take the axiom in the generalized form. And exchange is omitted since we consider sets. Note that there is little incentive to consider multisets here; we can have multiple copies of the same assumption φ by giving it different names as in $x : \varphi, y : \varphi$.

7.4.1. DEFINITION. The sets of types, terms, and environments of $\lambda^{\text{sc}}_\rightarrow$ are as for $\lambda_\rightarrow$ à la Church (Definition 3.3.1). For an environment Γ, we use the notation $\Gamma + (y : \sigma)$ for $\Gamma \cup \{y : \sigma\}$ provided $y \notin \text{dom}(\Gamma)$ or $\Gamma(y) = \sigma$. The typing rules are in Figure 7.4. We write $\vdash^+_{\text{SC}}$ and $\vdash_{\text{SC}}$ for derivability in this system with and without cut, respectively.

$$\Gamma, x : \varphi \vdash x : \varphi \ (\text{Ax})$$

$$\frac{\Gamma \vdash N : \varphi \quad \Gamma, x : \psi \vdash M : \sigma}{\Gamma + (y : \varphi \rightarrow \psi) \vdash M[x := yN] : \sigma} (\text{L}\rightarrow) \quad \frac{\Gamma, x : \varphi \vdash M : \psi}{\Gamma \vdash \lambda x{:}\varphi\, M : \varphi \rightarrow \psi} (\text{R}\rightarrow)$$

$$\frac{\Gamma \vdash N : \varphi \quad \Gamma, x : \varphi \vdash M : \sigma}{\Gamma \vdash M[x := N] : \sigma} (\text{Cut})$$

FIGURE 7.4: SEQUENT CALCULUS STYLE $\lambda_\rightarrow$.

7.4.2. PROPOSITION.
   (i) If $\Gamma \vdash^+_{\text{SC}} M : \varphi$ then $\text{rg}(\Gamma) \vdash^+_{\text{LM}} \varphi$.
   (ii) If $\Gamma \vdash^+_{\text{LM}} \varphi$ then $\Gamma' \vdash^+_{\text{SC}} M : \varphi$ for some $M \in \Lambda$ and Γ' with $\text{rg}(\Gamma') = \Gamma$.

The above system assigns types to a certain class of λ-terms, namely the terms that receive types by the usual simply typed λ-calculus à la Church, as is easy to show by "lifting" the translations from the proofs of Propositions 7.2.3 and 7.2.4 to handle terms instead of just derivations.

7.4.3. PROPOSITION. $\Gamma \vdash^+_{\text{SC}} M : \varphi$ iff $\Gamma \vdash_{\lambda_\rightarrow} M : \varphi$ à la Church.

Recall that simply typed λ-terms in normal form correspond to normal derivations in natural deduction. Normal λ-terms in $\lambda^{\text{sc}}_\rightarrow$ correspond to cut-free proofs.

7.4.4. PROPOSITION.
   (i) If $\Gamma \vdash_{\lambda_\rightarrow} L : \varphi$ à la Church and $L \in \text{NF}_\beta$, then $\Gamma \vdash_{\text{SC}} L : \varphi$.
   (ii) If $\Gamma \vdash_{\text{SC}} L : \varphi$, then $L \in \text{NF}_\beta$.

PROOF. For (i), use Exercise 1.5 (modified appropriately to deal with raw terms à la Church) and proceed by induction on the size of L. For (ii), use induction on the derivation of $\Gamma \vdash_{\sf SC} L : \varphi$. □

7.4.5. COROLLARY. $\Gamma \vdash_{\sf SC} M : \varphi$ iff $\Gamma \vdash_{\lambda_\to} M : \varphi$ à la Church and $M \in {\rm NF}_\beta$.

Thus simply typed λ-terms in normal form correspond to both normal proofs in natural deduction and to cut-free sequent calculus proofs. We therefore have the correspondence

$$\text{Cut-free proofs} \quad \sim \quad \text{Normal deductions}$$

This correspondence gives another proof of cut elimination: if $\vdash_{\sf LM}^+ \varphi$, then for some M we have $\vdash_{\sf SC}^+ M : \varphi$, so $\vdash_{\lambda_\to} M : \varphi$. By normalization there is a normal N with $\vdash_{\lambda_\to} N : \varphi$, so $\vdash_{\sf SC} N : \varphi$, and then $\vdash_{\sf LM} \varphi$.

Nevertheless, the correspondence between cut-free proofs and normal deductions is not perfect. For instance, a deduction may use the cut rule even if the corresponding λ-term is in normal form: the substitution $N[x := yM]$ may delete the term M which may contain redexes. In this case we just know that there is another typing that does not use the cut rule. Also, the map from cut-free proofs to normal deductions is not injective: several cut-free derivations may be mapped to the same normal derivation. One way of realizing this is to see that the same rules of $\lambda_\to^{\sf sc}$ applied in different order generate the same λ-term.

Thus, a fine-grained analysis of derivations is not facilitated at the level of terms. It is natural to remedy this unsatisfactory situation by replacing the meta-notation $M[x := N]$ by an *explicit substitution,* i.e. a term formation operator $M\langle x{=}N\rangle$ (where x is bound in M but not N), and introduce reduction rules for this operator, reflecting the cut elimination steps. Similarly, one may introduce explicit substitutions for the left rules.

Calculi for explicit substitutions have been studied, with and without types, motivated by a desire to provide a model that is closer to actual implementations than is the case with ordinary λ-calculus. Indeed, with the traditional formulation of β-reduction, $(\lambda xP)Q \to_\beta P[x := Q]$, the argument Q may be proliferated, leading to size explosion, and the different copies of Q may each need reduction, leading to unnecessarily duplicated work. Both of these problems may be overcome in an explicit substitution calculus, which therefore corresponds more closely to the implementation of functional programming languages.

The system $\lambda{\bf x}$ due to Rose and Bloo is one of the simplest such calculi. Unlike most related calculi it uses names for variables, as opposed to de Bruijn notation [59]. It comes in a type-free as well as a simply typed variant; the latter is obtained from simply typed λ-calculus by adding the

7.4. CUT ELIMINATION VERSUS NORMALIZATION

new term operator $M\langle x = N\rangle$ with the typing rule

$$\frac{\Gamma \vdash N : \varphi \quad \Gamma, x : \varphi \vdash M : \sigma}{\Gamma \vdash M\langle x = N\rangle : \sigma}.$$

One then considers the two notions of reduction $\to_x$ and $\to_{\beta_x}$ defined by:

$$
\begin{array}{lll}
(\lambda x.M)N & \to_{\beta_x} & M\langle x = N\rangle \\
x\langle x = N\rangle & \to_x & N \\
L\langle x = N\rangle & \to_x & L & (x \notin \mathrm{FV}(L)) \\
(\lambda y.M)\langle x = N\rangle & \to_x & \lambda y.M\langle x = N\rangle & (x \neq y,\ y \notin \mathrm{FV}(N)) \\
(ML)\langle x = N\rangle & \to_x & (M\langle x = N\rangle)(L\langle x = N\rangle)
\end{array}
$$

The first of these reductions is a variant of β-reduction which delays the substitution of N for x. The others propagate and perform the substitutions. The third rule is often referred to as *garbage collection*, since N may be regarded as garbage when it is not referred in L.

Reductions that happen in one (big) step in λ-calculus, require more steps with explicit substitutions:

$$
\begin{array}{ll}
(\lambda x \lambda y.xy)(\lambda z.z)(\lambda u.u) & \to_{\beta_x} \quad ((\lambda y.xy)\langle x = \lambda z.z\rangle)(\lambda u.u) \\
& \to_x \quad (\lambda y.(xy)\langle x = \lambda z.z\rangle)(\lambda u.u) \\
& \to_x \quad (\lambda y.x\langle x = \lambda z.z\rangle\, y\langle x = \lambda z.z\rangle)(\lambda u.u) \\
& \to_x \quad (\lambda y.x\langle x = \lambda z.z\rangle\, y)(\lambda u.u) \\
& \to_x \quad (\lambda y.(\lambda z.z)\, y)(\lambda u.u) \\
& \to_{\beta_x} \quad (\lambda y.z\langle z = y\rangle)(\lambda u.u) \\
& \to_x \quad (\lambda y.y)(\lambda u.u) \\
& \to_{\beta_x} \quad y\langle y = \lambda u.u\rangle \\
& \to_x \quad \lambda u.u
\end{array}
$$

For a typed explicit substitution calculus, one is interested in proving strong normalization for all typable terms. An obvious idea is to attempt to infer this property from strong normalization of the corresponding calculus without explicit substitutions, i.e. in the present case the simply typed λ-calculus. Even in a type-free setting one can consider this type of reduction, by showing that strong normalization of any λ-term (without explicit substitutions) with respect to ordinary β-reduction implies strong normalization of the same term with respect to $x\beta_x$-reduction. This latter property is known as *preservation of strong normalization (PSN)*.

One can show that $\to_x$ is confluent and strongly normalizing (see [50]) so we may let $\mathrm{nf}_x(M)$ denote the unique normal form of M with respect to $\to_x$. Then $M \to_x N$ implies $\mathrm{nf}_x(M) = \mathrm{nf}_x(N)$, and one can show that $M \to_{\beta_x} N$ implies $\mathrm{nf}_x(M) \twoheadrightarrow_\beta \mathrm{nf}_x(N)$, a result known as the *projection property*. To infer strong normalization of $x\beta_x$-reduction from strong normalization of β-reduction it would suffice to establish that $\mathrm{nf}_x(M) \twoheadrightarrow_\beta \mathrm{nf}_x(N)$ must happen in at least one step. However, this is not generally the case, as the example $M = x\langle y{=}(\lambda z.L)P\rangle$ and $N = x\langle y{=}L\langle z{=}P\rangle\rangle$ shows. The problem is that we may have reductions in garbage. However, by analyzing more carefully what can happen inside garbage, the proof can be made to work—see [51]. In particular, $x\beta_x$-reduction on typable terms is strongly normalizing.

One is easily tempted to adopt further reductions allowing for various substitutions to permute. For instance, when proving the Church-Rosser property of $x\beta_x$-reduction, the case $(\lambda x P)Q\langle y{=}L\rangle$ which reduces to both $P\langle x{=}Q\rangle\langle y{=}L\rangle$ and $P\langle y{=}L\rangle\langle x{=}Q\langle y{=}L\rangle\rangle$ may inspire us to consider the following rules

$$M\langle x{=}N\rangle\langle y{=}L\rangle \;\to\; M\langle y{=}L\rangle\langle x{=}N\langle y{=}L\rangle\rangle \quad (x \notin \mathrm{FV}(L))$$
$$M\langle x{=}N\rangle\langle y{=}L\rangle \;\to\; M\langle x{=}N\langle y{=}L\rangle\rangle \quad (x \notin \mathrm{FV}(L),\, y \notin \mathrm{FV}(M)).$$

The second one follows from the first and the garbage collection rule. Thus, the second one is weaker than the first. The first rule obviously destroys strong normalization since an infinite reduction can be obtained just by alternating the order of substitutions. However, even the second rule breaks PSN. Indeed, let $W = (\lambda l.L)L$, for some term L with $l \notin \mathrm{FV}(L)$, and define $O = (\lambda w.(\lambda z.x)W)W$, where $w \notin \mathrm{FV}(W)$. Then O has an infinite reduction, using $x\beta_x$-reductions together with the first rule above. Indeed, let α_0 be short for $\langle w{=}W\rangle$ and α_{m+1} be short for $\langle l{=}L\alpha_m\rangle$. Then

$$O \twoheadrightarrow x\langle z{=}L\alpha_0\alpha_1\rangle.$$

Moreover,
$$L\alpha_0\alpha_m \twoheadrightarrow L\langle w{=}L\alpha_m\alpha_{m+1}\rangle$$
and
$$L\alpha_{n+1}\alpha_m \twoheadrightarrow L\langle l{=}L\alpha_n\alpha_m\rangle.$$

Using these building blocks it is easy to construct an infinite reduction. But by taking L to be any typable term in simply typed λ-calculus, it is easy to see that O is typable and therefore strongly normalizing. Thus, adding composition of substitutions breaks PSN.

We do not go into a detailed study of $\lambda\mathbf{x}$ or other explicit substitution calculi here. What we wish to draw attention to is merely that the rule for typing the explicit substitution in $\lambda\mathbf{x}$ is an annotated version of the cut rule. Thus we may state the slogan

Cut rule $\sim$ Explicit substitution

Yet, the system $\lambda\mathbf{x}$ does not correspond exactly to sequent calculus, because the former has an elimination rule for implication, rather than the left rule. That is, the system $\lambda\mathbf{x}$ is obtained by adding explicit substitution to simply typed λ-calculus or, equivalently, by adding the cut rule to natural deduction. A λ-calculus corresponding closer to the rules of the implicational fragment of LM is as follows.

7.4. CUT ELIMINATION VERSUS NORMALIZATION

7.4.6. DEFINITION. Define raw terms for sequent calculus by the grammar:

$$M ::= x \mid M\langle x = xM \rangle \mid (\lambda x{:}\varphi M) \mid M\langle x = M \rangle,$$

where x ranges over variables. A term of shape $M\langle x = M \rangle$ is called a *cut*. We have two different substitution operators, corresponding to (L→) and (Cut). No confusion should result from using the same syntax for both, since xM is not a term. Free variables are defined as usual adding the clauses

$$\begin{aligned} \mathrm{FV}(M\langle x = yN \rangle) &= (\mathrm{FV}(M) - \{x\}) \cup \mathrm{FV}(N) \cup \{y\}; \\ \mathrm{FV}(M\langle x = N \rangle) &= (\mathrm{FV}(M) - \{x\}) \cup \mathrm{FV}(N). \end{aligned}$$

We adopt the usual conventions, mutatis mutandis. This includes the identification of α-equivalent terms.

$$\Gamma, x : \varphi \vdash x : \varphi \ (\mathrm{Ax})$$

$$\frac{\Gamma \vdash N : \varphi \quad \Gamma, x : \psi \vdash M : \sigma}{\Gamma + (y : \varphi \to \psi) \vdash M\langle x = yN \rangle : \sigma} \ (\mathrm{L}\to) \qquad \frac{\Gamma, x : \varphi \vdash M : \psi}{\Gamma \vdash \lambda x{:}\varphi.M : \varphi \to \psi} \ (\mathrm{R}\to)$$

$$\frac{\Gamma \vdash N : \varphi \quad \Gamma, x : \varphi \vdash M : \sigma}{\Gamma \vdash M\langle x = N \rangle : \sigma} \ (\mathrm{Cut})$$

FIGURE 7.5: SEQUENT CALCULUS AS A TYPE SYSTEM.

Typing $\vdash_{\mathrm{sc}}^+$ is defined by the rules in Figure 7.5. Recall that the notation $\Gamma, x{:}\varphi$ is defined only if $x \notin \mathrm{dom}(\Gamma)$. In particular x is never free in N in the context $M\langle x = N \rangle$ or $M\langle x = yN \rangle$. Again $\vdash_{\mathrm{sc}}$ indicates derivability without application of (Cut).

7.4.7. DEFINITION. Define $\to_{\mathrm{sc}}$ as the compatible closure of the below rules, where the variables x, y, z, u are all different, where $z \notin \mathrm{FV}(N)$ in the third and fourth rule, and where $u \notin \mathrm{FV}(M) \cup \mathrm{FV}(L)$ in the sixth rule.

$$\begin{aligned} x\langle x = N \rangle &\to N \\ y\langle x = N \rangle &\to y \\ (\lambda z{:}\varphi.M)\langle x = N \rangle &\to \lambda z{:}\varphi.M\langle x = N \rangle \\ M\langle z = yL \rangle \langle x = N \rangle &\to M\langle x = N \rangle \langle z = yL\langle x = N \rangle \rangle \\ M\langle z = xL \rangle \langle x = y \rangle &\to M\langle x = y \rangle \langle z = yL\langle x = y \rangle \rangle \\ M\langle z = xL \rangle \langle x = N\langle u = yJ \rangle \rangle &\to M\langle z = xL \rangle \langle x = N \rangle \langle u = yJ \rangle \\ M\langle z = xL \rangle \langle x = \lambda y{:}\varphi.N \rangle &\to M\langle x = \lambda y{:}\varphi.N \rangle \langle z = N\langle y = L\langle x = \lambda y{:}\varphi.N \rangle \rangle \rangle \end{aligned}$$

The last rule shows how β-reduction is implemented in sequent calculus; the other rules manipulate substitutions.

With this presentation of typing and reduction for sequent calculus at hand, the development in 7.4.3–7.4.5 can be restated using translations between sequent calculus and simply typed λ-calculus.

Here we concentrate on cut-elimination. The above reduction rules are similar to the transformations used to prove the Hauptsatz (for the implicational fragment of LM), though with some differences caused by the absence of contractions in the present system. Thus, the technique used to prove the Hauptsatz should carry over to the term reduction rules; however, we are now aiming for *strong* cut-elimination.

We first observe that the reductions preserve typing.

7.4.8. THEOREM (Subject reduction). *If* $\Gamma \vdash M : \sigma$ *and* $M \to_{\mathsf{SC}} N$, *then* $\Gamma \vdash N : \sigma$.

PROOF. Routine induction. $\square$

7.4.9. LEMMA. *Let* $\Gamma \vdash Q : \tau$ *and* $\Gamma, v{:}\tau \vdash P : \sigma$. *If* P *and* Q *are strongly normalizing, so is* $P\langle v = Q\rangle$.

PROOF. By induction on lexicographically ordered tuples (d, n, t, m, s), where

- d is the size of τ;
- n is the length of the longest reduction from Q;
- t is the size of Q;
- m is the length of the longest reduction from P;
- s is the size of P.

It suffices to show that whenever $P\langle v = Q\rangle \to_{\mathsf{SC}} R$, we have $R \in \mathsf{SN}_{\mathsf{SC}}$. If the reduction is inside P, then m decreases and d, n, t are unaffected; similarly if the reduction is inside Q. Now consider the different ways $P\langle v = Q\rangle$ can be a redex. Suppose, for example, that

$$M\langle z = xL\rangle\langle x = \lambda y{:}\varphi.N\rangle \to M\langle x = \lambda y{:}\varphi.N\rangle\langle z = N\langle y = L\langle x = \lambda y{:}\varphi.N\rangle\rangle\rangle.$$

By the induction hypothesis (with smaller m, or the same m but smaller s), $M\langle x = \lambda y{:}\varphi.N\rangle$ and $L\langle x = \lambda y{:}\varphi.N\rangle$ are both in $\mathsf{SN}_{\mathsf{SC}}$. By the induction hypothesis (with smaller d), $N\langle y = L\langle x = \lambda y{:}\varphi.N\rangle\rangle \in \mathsf{SN}_{\mathsf{SC}}$ and (with smaller d again) the term $M\langle x = \lambda y{:}\varphi.N\rangle\langle z = N\langle y = L\langle x = \lambda y{:}\varphi.N\rangle\rangle\rangle$ is in $\mathsf{SN}_{\mathsf{SC}}$. $\square$

7.4.10. THEOREM. *If* $\Gamma \vdash^+_{\mathsf{SC}} M : \varphi$, *then* $M \in \mathsf{SN}_{\mathsf{SC}}$.

PROOF. By induction on M. For (Cut) use the preceding lemma. $\square$

7.5. Lorenzen dialogues

In Chapters 4 and 6 we considered several variants of dialogue games. They are all asymmetric in the roles of the proponent and opponent: the opponent constantly attacks and the proponent constantly defends. We now consider a more symmetric form of the dialogue game in which both players can attack and defend. Whereas we mapped the strategies of the previous dialogue games to natural deduction derivations (or corresponding terms in long form), we map the new strategies to sequent calculus derivations.

The proponent starts by asserting some formula, and the players then take alternating turns. However, in each move the current player may now either attack a formula stated by the other player in a previous move, or defend a formula attacked by the other player in a previous move.

Proponent 1: I assert that $((p \to p) \to q) \to q$ holds!

Opponent 1: I attack that claim. If I state $(p \to p) \to q$, you have to show that the conclusion q follows.

Proponent 2: Oh yeah? Well, I attack *your* claim. If I state $p \to p$ can you show that your conclusion q holds?

Opponent 2: Sure, I can show that q holds.

Proponent 3: Right, q holds—that's what you wanted me to show.

The proponent begins by asserting $((p \to p) \to q) \to q$, and the opponent immediately attacks this formula. An implication $\sigma \to \tau$ is attacked by stating σ and is defended by stating τ. In the above dialogue, the proponent does not immediately defend the formula; she may only state a propositional variable if it has previously been stated by the opponent. Thus she instead makes a counter-attack on the opponent's formula. The opponent responds with a defense, and the proponent can then defend herself against the attack. Now the opponent has nothing further to add; he must always refer to the formula stated by the proponent in the preceding move. In contrast, the proponent may refer to any previous opponent attack or defense; in particular, the same formula may be defended twice.

Another possibility for the opponent in his second step is to launch yet another counter-attack:

Opponent 2: Bah! I grant you p, now show the conclusion p.

Proponent 3: Well, you stated p yourself, so sure it holds!

A dialogue comes to an end when the player to take the turn has nothing further to add, i.e. has no possible move, and the other player then wins. The proponent has a winning strategy for a formula if, regardless of which move the opponent chooses in every move, the proponent can always find corresponding responses so that she eventually wins.

The next dialogue illustrates how negations—defined by absurdity—work. If anyone states $\bot$, then that can be attacked (this does not involve stating any formula). And there is no defense against this attack.

Proponent 1: I assert that $p \to \neg p \to q$ holds!

Opponent 1: I attack your claim by stating p.

Proponent 2: I defend myself by claiming $\neg p \to q$.

Opponent 2: I attack that claim by stating $\neg p$.

Proponent 3: I cannot state q as a defense yet. But I can attack your claim by stating p, and I can do that since you already stated p yourself.

Opponent 3: I defend by stating $\bot$.

Proponent 4: Ha! I attack that claim, and you have no response.

Any disjunction $\sigma \vee \tau$ can be attacked, and this does not involve stating any formula; it can be defended by stating σ or τ. Conjunctions can be attacked in two different ways (no formula is stated); each attack has a defense (by stating the attacked part of the conjunction). A propositional variable can neither be attacked nor defended.

We now proceed to formalize these notions. Each attack and defense consists of an indication of the attack or defense together with a formula. The symbol $-$ is used when there is no formula (we can think of it as some fixed propositional variable).

7.5.1. DEFINITION. A *dialogue* over φ is an empty, finite, or infinite sequence of *moves* $M_1, M_2, \ldots$, where the odd (resp. even) moves are called *proponent* (resp. *opponent*) moves, such that the following conditions hold.

- $M_1 = (\mathbf{D}, \varphi, 0)$. (Proponent begins.)
- Each opponent move has form $M_i = (X, \psi, i-1)$. For $i > 1$, each proponent move has form $M_i = (X, \psi, j)$, where $j < i$ and M_j is an opponent move. (Opponent refers to the immediately preceding move, proponent refers to any preceding opponent move.)
- For each proponent move $M_i = (X, p, j)$ stating some variable p, there is an opponent move $M_k = (Y, p, l)$ with $k < i$. (Proponent may assert a variable, if it is already asserted by opponent.)
- For all $i > 1$ with $M_i = (X, \psi, j)$ and $M_j = (Y, \rho, k)$, either $X : \psi$ is an attack on ρ (see left table[1]) or $M_k = (Z, \pi, l)$ and $Y : \rho$ is an attack

---
[1] A propositional variable cannot be attacked, so this case does not appear in the table. Similarly, there are no cases for $\bot$ and propositional variables in the right table.

7.5. Lorenzen dialogues

on π and $X : \psi$ is a defense of π against Y (see right table). We also say M_i is an attack on M_j, etc.

Formula	Attacks
$\sigma \to \tau$	$\mathbf{A} : \sigma$
$\sigma \wedge \tau$	$\mathbf{A_L} : -$, $\mathbf{A_R} : -$
$\sigma \vee \tau$	$\mathbf{A} : -$
$\bot$	$\mathbf{A} : -$

Formula	Attack	Defenses
$\sigma \to \tau$	$\mathbf{A}$	$\mathbf{D} : \tau$
$\sigma \wedge \tau$	$\mathbf{A_L}$	$\mathbf{D} : \sigma$
$\sigma \wedge \tau$	$\mathbf{A_R}$	$\mathbf{D} : \tau$
$\sigma \vee \tau$	$\mathbf{A}$	$\mathbf{D_L} : \sigma$, $\mathbf{D_R} : \tau$

(One can attack asserted formulas or defend against a matching attack.)

If $\mathbf{P}$ and $\mathbf{P}, M$ are dialogues over φ, then M is a *possible* move after $\mathbf{P}$.

Next we formalize the notion of a *proponent strategy* which, for technical reasons, is considered relative to a dialogue that has already begun.

7.5.2. DEFINITION. Let $\mathbf{P} = M_1, \ldots, M_n$ be a dialogue over φ.

(i) A *proponent strategy after* $\mathbf{P}$ is a tree labeled with moves such that

- The initial part of the tree from the root consists of a single path labeled $M_1, \ldots, M_n$.
- In every branch, each node after M_n labeled with a proponent move has one child for every possible opponent move (i.e. the strategy must account for all possible opponent moves).
- In each branch, each node after M_n labeled with an opponent move has one child, if there is a possible proponent move, and otherwise has no children.

(ii) A proponent strategy after $\mathbf{P}$ is *winning* if every path from the root is finite, and every leaf is a proponent move (i.e. all dialogues end with the opponent having no reply).

(iii) A *proponent strategy for* φ is a proponent strategy after the empty dialogue.

7.5.3. EXAMPLE. Collecting the two dialogues for $((p \to p) \to q) \to q$ into a tree yields a winning proponent strategy for $((p \to p) \to q) \to q$.

We will show that there is a winning proponent strategy for a formula φ iff $\vdash \varphi$ in LK. In fact, we translate strategies to derivations, and vice versa.

7.5.4. DEFINITION. Let $\mathbf{P}$ be a finite dialogue over ψ. The *position after* $\mathbf{P}$ is a pair of sets (written $\Gamma \vdash \Delta$) defined as follows. The position after the empty play is $\vdash \psi$. If $\Gamma \vdash \Delta$ is the position after $\mathbf{Q}$, the position after $\mathbf{Q}, M$ is

- $\Gamma \vdash \Delta$, if M is a proponent move;

- $\Gamma, \sigma \vdash \Delta$, if M is an opponent defense (X, σ, j); otherwise M is an opponent attack (X, σ, j) on $M_j = (Y, \varphi, k)$ and the position after **Q**,M is

$$\begin{array}{ll} \Gamma, \sigma \vdash \tau, \Delta & \text{if } \varphi = \sigma \to \tau; \\ \Gamma \vdash \tau_1, \Delta & \text{if } \varphi = \tau_1 \wedge \tau_2 \text{ and } X = \mathbf{A}_\mathrm{L}; \\ \Gamma \vdash \tau_2, \Delta & \text{if } \varphi = \tau_1 \wedge \tau_2 \text{ and } X = \mathbf{A}_\mathrm{R}; \\ \Gamma \vdash \tau_1, \tau_2, \Delta & \text{if } \varphi = \tau_1 \vee \tau_2; \\ \Gamma \vdash \Delta & \text{if } \varphi = \bot. \end{array}$$

7.5.5. LEMMA. *Let $\Gamma \vdash \Delta$ be the position after dialogue $M_1 \ldots M_n$ over ψ.*

(i) *$\sigma \in \Gamma$ iff there is an opponent move $M_i = (X, \sigma, j)$, for some X, j (where σ is a proper formula, i.e. not $-$).*

(ii) *$\tau \in \Delta$ iff $\tau = \psi$ or there is an opponent attack $M_i = (A, \sigma, j)$ on $M_j = (Y, \rho, k)$ such that $D : \tau$ is a defense of ρ against A, for some D.*

PROOF. Induction on n. □

The position $\Gamma \vdash \Delta$ after a dialogue **P** over ψ is an almost complete representation of the possibilities for the proponent to continue the dialogue: she may attack any formula in Γ or she may state any formula in Δ as a defense (except ψ, and for propositional variables only those that are in Γ).

The rules of LK can be seen as an axiomatization of existence of a winning proponent strategy. To make this even more explicit, consider the variant of LK in Figure 7.6. Every right rule and (Ax) expresses a condition for fusing winning proponent strategies together in a defense. For instance, the rule for conjunction states that if the proponent can win in the position $\Gamma \vdash \varphi, \Delta$ and $\Gamma \vdash \psi, \Delta$ then she can also win $\Gamma \vdash \varphi \wedge \psi, \Delta$ by stating $\varphi \wedge \psi$ in a defense, because the opponent must attack the conjunction, and this will lead to one of the considered positions. Similarly, each left rule and (L$\bot$) expresses a condition for fusing winning proponent strategies together in an attack. For instance, (L$\overset{\wedge}{\to}$) states that the proponent can win by attacking $(\varphi_1 \wedge \varphi_2) \to \psi$ provided she wins if the opponent defends, as well as if the opponent launches either of his two possible counter-attacks.

7.5.6. LEMMA. *The set of derivable sequents of LK is not affected by replacing* (Ax), (R$\vee$), *and* (L$\to$) *as indicated in Figure 7.6.*

PROOF. Show by induction on derivations that derivability in LK entails derivability in the modified system, and vice versa. For the former, show that (Ax) and the following variant of the left rule for implication

$$\frac{\Gamma_0 \vdash \Delta \quad \Gamma_1, \psi \vdash \Sigma}{\Gamma_0, \Gamma_1, \varphi \to \psi \vdash \Delta - \varphi, \Sigma}$$

are derived rules in the modified system. □

7.5. Lorenzen dialogues

Axioms:

$$\bot \vdash \ (L\bot) \qquad p \vdash p \ (Ax)$$

Structural Rules:

$$\frac{\Gamma \vdash \Sigma}{\Gamma, \varphi \vdash \Sigma} \ (LW) \qquad \frac{\Gamma \vdash \Pi}{\Gamma \vdash \varphi, \Pi} \ (RW)$$

$$\frac{\Gamma, \varphi, \psi, \Gamma' \vdash \Sigma}{\Gamma, \psi, \varphi, \Gamma' \vdash \Sigma} \ (LX) \qquad \frac{\Gamma \vdash \Delta, \varphi, \psi, \Delta'}{\Gamma \vdash \Delta, \psi, \varphi, \Delta'} \ (RX)$$

$$\frac{\Gamma, \varphi, \varphi \vdash \Sigma}{\Gamma, \varphi \vdash \Sigma} \ (LC) \qquad \frac{\Gamma \vdash \varphi, \varphi, \Delta}{\Gamma \vdash \varphi, \Delta} \ (RC)$$

Logical Rules:

$$\frac{\Gamma, \varphi \vdash \Sigma}{\Gamma, \varphi \wedge \psi \vdash \Sigma} \ (L\wedge) \frac{\Gamma, \psi \vdash \Sigma}{\Gamma, \varphi \wedge \psi \vdash \Sigma} \qquad \frac{\Gamma \vdash \varphi, \Delta \quad \Gamma \vdash \psi, \Delta}{\Gamma \vdash \varphi \wedge \psi, \Delta} \ (R\wedge)$$

$$\frac{\Gamma, \varphi \vdash \Sigma \quad \Gamma, \psi \vdash \Sigma}{\Gamma, \varphi \vee \psi \vdash \Sigma} \ (L\vee) \qquad \frac{\Gamma \vdash \varphi, \psi, \Delta}{\Gamma \vdash \varphi \vee \psi, \Delta} \ (R\vee)$$

$$\frac{\Gamma \vdash \varphi_1, \varphi_2, \Delta \quad \Gamma, \psi \vdash \Sigma}{\Gamma, (\varphi_1 \vee \varphi_2) \to \psi \vdash \Delta, \Sigma} (L_\to^\vee) \qquad \frac{\Gamma, \varphi \vdash \psi, \Delta}{\Gamma \vdash \varphi \to \psi, \Delta} \ (R\to)$$

$$\frac{\Gamma \vdash \varphi_1, \Delta \quad \Gamma \vdash \varphi_2, \Delta \quad \Gamma, \psi \vdash \Sigma}{\Gamma, (\varphi_1 \wedge \varphi_2) \to \psi \vdash \Delta, \Sigma} (L_\to^\wedge)$$

$$\frac{\Gamma, \varphi_1 \vdash \varphi_2, \Delta \quad \Gamma, \psi \vdash \Sigma}{\Gamma, (\varphi_1 \to \varphi_2) \to \psi \vdash \Delta, \Sigma} (L_\to^\to)$$

$$\frac{\Gamma, p, \psi \vdash \Sigma}{\Gamma, p, p \to \psi \vdash \Delta, \Sigma} (L_\to^P)$$

FIGURE 7.6: ALTERNATIVE RULES FOR LK.

7.5.7. THEOREM. *If ψ has a winning proponent strategy, then $\vdash_{LK} \psi$.*

PROOF. Let D be a winning proponent strategy for ψ. We show that if $\mathbf{P} = M_1, \ldots, M_n$ is a dialogue in D not ending in a proponent move, and $\Gamma \vdash \Delta$ is the position after $\mathbf{P}$, then $\Gamma \vdash_{LK} \Delta$ (using the rules of Figure 7.6), where Γ, Δ are regarded as sequences in the latter case. The proof is by induction on the subtree below $\mathbf{P}$ inside D.

Since $\mathbf{P}$ is winning, there is a proponent move $M_{n+1} = (X, \tau, j)$ just under $\mathbf{P}$. First assume the proponent move is an attack on $M_j = (Y, \varphi, k)$. By Lemma 7.5.5, we have $\varphi \in \Gamma$. We consider each shape of φ.

CASE 1: $\varphi = \bot$. Then $\Gamma \vdash_{LK} \Delta$ by $(L\bot)$ and structural rules.

CASE 2: $\varphi = \varphi_1 \vee \varphi_2$. Then the subtree below $\mathbf{P}$ is

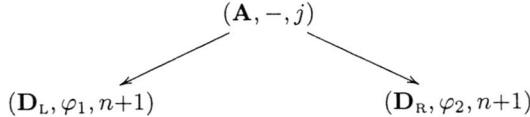

The positions after the leaves are $\Gamma, \varphi_1 \vdash \Delta$ and $\Gamma, \varphi_2 \vdash \Delta$, which are derivable by the induction hypothesis, so $\Gamma \vdash_{LK} \Delta$ by $(L\vee)$ and structural rules.

CASE 3: $\varphi = \varphi_1 \wedge \varphi_2$. Similar to the preceding case.

CASE 4: $\varphi = \tau \to \sigma$. Consider each shape of τ, e.g. $\varphi_1 \vee \varphi_2$. After the leaves in

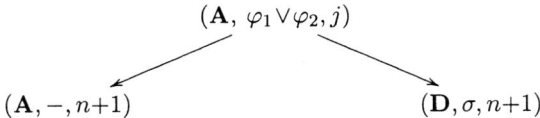

the positions are $\Gamma \vdash \varphi_1, \varphi_2, \Delta$ and $\Gamma, \sigma \vdash \Delta$, so $\Gamma \vdash_{LK} \Delta$ by the induction hypothesis, $(L\overset{\vee}{\to})$, and structural rules.

Next assume that the proponent move is a defense or the initial move. Then $\tau \in \Delta$ by Lemma 7.5.5. Also, $\tau \in \Gamma$ if τ is a propositional variable. We consider the cases where τ is a variable or a disjunction; absurdity, conjunction and implication are similar.

CASE 1: $\tau = p$. Then $\Gamma \vdash_{LK} \Delta$ by (Ax) and structural rules.

CASE 2: $\tau = \tau_1 \vee \tau_2$. Then the subtree below $\mathbf{P}$ is

$$(X, \tau_1 \vee \tau_2, j)$$
$$\downarrow$$
$$(\mathbf{A}, -, n+1)$$

The position after the leaf is $\Gamma \vdash \tau_1, \tau_2, \Delta$, so $\Gamma \vdash_{LK} \Delta$ by the induction hypothesis, $(R\vee)$, and structural rules. $\square$

7.5. LORENZEN DIALOGUES   187

Next we consider the opposite direction.

7.5.8. DEFINITION. A *possible position after* **P**, is a pair $\Gamma' \vdash \Delta'$, where $\Gamma' \subseteq \Gamma$ and $\Delta' \subseteq \Delta$, and where $\Gamma \vdash \Delta$ is the position after **P**.

7.5.9. THEOREM. *If* $\vdash_{\mathrm{LK}} \psi$, *then* ψ *has a winning proponent strategy.*

PROOF. We show that if $\Gamma \vdash_{\mathrm{LK}} \Delta$, and **P** is a dialogue over ψ, not ending with a proponent move, such that $\Gamma \vdash \Delta$ (regarded as a pair of sets) is a possible position after **P**, then there is a winning proponent strategy for ψ after **P**. We proceed by induction on the derivation of $\Gamma \vdash_{\mathrm{LK}} \Delta$ (using the rules of Figure 7.6). The structural rules are trivial. We consider the axioms and rules (R$\vee$) and (L$_\rightarrow^\vee$). The remaining cases are similar or easier.
CASE 1: The derivation consists of the axiom

$$\bot \vdash \quad (\mathrm{L}\bot)$$

Since $\bot \in \Gamma$, the formula $\bot$ must, by Lemma 7.5.5, have been asserted in an opponent move M_i, and

$$\mathbf{P}$$
$$\downarrow$$
$$(\mathbf{A}, -, i)$$

(where the top-most node should be construed as a branch labeled with the moves of **P**) is then a winning proponent strategy after **P**.
CASE 2: The derivation consists of the axiom

$$p \vdash p \quad (\mathrm{Ax})$$

By Lemma 7.5.5, the variable p must have been asserted in an opponent move. Moreover, ψ cannot be a variable, so by Lemma 7.5.5, the former can be stated in some defense D against some opponent attack M_i. Thus,

$$\mathbf{P}$$
$$\downarrow$$
$$(D, p, i)$$

is a winning proponent strategy after **P**.
CASE 3: The derivation ends in

$$\frac{\Gamma \vdash \varphi_1, \varphi_2, \Delta}{\Gamma \vdash \varphi_1 \vee \varphi_2, \Delta} \quad (\mathrm{R}\vee)$$

We consider three subcases.

3.1. $\varphi_1 \vee \varphi_2 \neq \psi$. Since $\Gamma \vdash \varphi_1 \vee \varphi_2, \Delta$ is a possible position after **P**, Lemma 7.5.5 implies that there is some defense D in which $\varphi_1 \vee \varphi_2$ can be stated against some opponent attack M_i. Thus,

$$M_{n+1} = (D, \varphi_1 \vee \varphi_2, i)$$

followed by
$$M_{n+2} = (\mathbf{A}, -, n+1)$$

(where n is the number of moves in **P**) are possible moves after **P**. By the induction hypothesis, there is a winning proponent strategy T after $\mathbf{P}, M_n, M_{n+1}$ Then

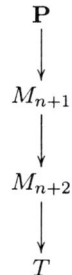

is a winning proponent strategy after **P**.

3.2. $\varphi_1 \vee \varphi_2 = \psi$ and **P** is the empty dialogue. Then

$$M_0 = (\mathbf{D}, \varphi_1 \vee \varphi_2, 0)$$

followed by
$$M_1 = (\mathbf{A}, -, 0)$$

are possible moves after **P**, and $\Gamma \vdash \varphi_1, \varphi_2, \Delta$ is a possible position after M_0, M_1. Now proceed as in Case 3.1.

3.3. $\varphi_1 \vee \varphi_2 = \psi$ and **P** is non-empty. Then $\mathbf{P} = M_0, (\mathbf{A}, -, 0), \mathbf{Q}$, and $\Gamma \vdash \varphi_1, \varphi_2, \Delta$ is a possible position after **P**, so by the induction hypothesis, there is a winning proponent strategy for ψ after **P**.

CASE 4: The derivation ends in

$$(\mathrm{L}^\vee_\rightarrow) \; \frac{\Gamma \vdash \varphi_1, \varphi_2, \Delta \quad \Gamma, \varphi_3 \vdash \Sigma}{\Gamma, (\varphi_1 \vee \varphi_2) \rightarrow \varphi_3 \vdash \Delta, \Sigma}$$

Since $\Gamma, (\varphi_1 \vee \varphi_2) \rightarrow \varphi_3 \vdash \Delta, \Sigma$ is a possible position after **P**, Lemma 7.5.5 implies that **P** must contain an opponent move

$$M_i = (X, (\varphi_1 \vee \varphi_2) \rightarrow \varphi_3, j).$$

Then
$$\mathbf{P}' = \mathbf{P}, (\mathbf{A}\to, \varphi_1 \vee \varphi_2, i), (\mathbf{A}\vee, -, n+1)$$
is a dialogue after which $\Gamma \vdash \varphi_1, \varphi_2, \Delta$ is a possible position, and
$$\mathbf{P}'' = \mathbf{P}, (\mathbf{A}\to, i, \varphi_1 \vee \varphi_2), (\mathbf{D}\to, \varphi_3, n+1)$$
is a dialogue after which $\Gamma, \varphi_3 \vdash \Sigma$ is a possible position. By the induction hypothesis, there are winning proponent strategies T' and T'' after $\mathbf{P}'$ and $\mathbf{P}''$, respectively. Then

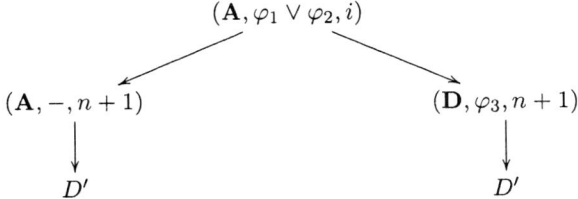

is the desired strategy. □

Thus, derivations (in our variant of LK) and strategies are isomorphic; when reading strategies as derivations we just have to add structural rules where needed, and when reading derivations as strategies we have to skip repititions of the initial formula and applications of the structural rules.

7.6. Notes

Sequent calculus systems for classical and intuitionstic logic were introduced by Gerhard Gentzen [171] in the same paper that introduced natural deduction systems. Gentzen arrived at natural deduction when trying to "set up a formalism that reflects as accurately as possible the actual logical reasoning involved in mathematical proofs." Instead of proving normalization for natural deduction, Gentzen found it convenient to introduce the corresponding sequent calculus and prove the cut elimination theorem. As corollaries he established a decision procedure for intuitionistic propositional logic, non-derivability of *tertium non datur* in the same system, and consistency of arithmetic without the induction axiom. The latter result was already known, but Gentzen was heading towards a stronger consistency result which would also cover the induction axiom [170].

Our proof of cut-elimination follows Kleene [269], except that Kleene introduces a new version of the system with (Cut) replaced by (Mix) (the so-called G2). Our proof of decidability also resembles Kleene's proof, except that he explicitly introduces a G3 system for this purpose. For more on the G3 variant that does not retain the main formula in the premises, see [489]. As mentioned, the intuitionistic version of this variant retains the main hypothesis in the left implication rule, but even this hidden contraction can be eliminated, as shown by Dyckhoff [141].

The correspondence between sequent calculus and natural deduction was established already by Gentzen [171], who related derivability in LK/LJ with NK/NJ.

Prawitz [403] extended the correspondence to relate not only derivability, but also proof reductions, by translating cut-free sequent calculus proofs to normal derivations in natural deduction, and vice versa. Our account of this result, which follows [29], is just the beginning of the story.

A main issue is that different cut-free sequent calculus derivations may correspond to the same normal derivation in natural deduction. Many authors investigate the problem of characterizing the conditions under which this can happen— see [402, 491, 511]. There are various attempts to explicitly define the equivalence relation between proofs representing the same deductions. Dyckhoff and Pinto [142], for instance, define a confluent and normalizing rewrite system for proofs, so that the induced equality holds exactly for proofs representing the same deduction. Mints [345] develops similar results.

Herbelin [224, 225] studies a modified sequent calculus LJT that limits the number of cut-free proofs so that the map to normal deductions becomes injective. It is probably the first sequent calculus in the literature where the corresponding term calculus is considered to have an explicit substitution operator. See also [422]. For a system based on de Bruijn notation relating sequent calculus to explicit substitution, see [499]. The note [498] discusses some subtleties related to term assignment and reductions for G3 systems. Some Curry-Howard interpretations of sequent calculus not focused on explicit substitution can be found in [54, 67, 161].

Explicit substitution calculi have been considered by several authors, see for instance [1, 50, 51, 94, 258, 309, 310, 335]. Failure of PSN for λx with permuting reductions [50] was first discovered by Melliès [335] in the context of the system $\lambda \sigma$.

Curien and Herbelin [96] introduce a very general calculus, which can be seen as a λ-calculus bearing the same relationship to LK, as $\lambda\mu$-calculus bears to NK. Their calculus also captures several symmetries between program and context, and between call-by-value and call-by-name. See also [369, 370, 503].

Strong cut elimination has been studied by a number of authors, beginning with Dragalin [134] and Zucker [511]. More recently, Dyckhoff and Urban [143] prove strong normalization of a variant of Herbelin's calculus. See also [492]. Our proof follows Dragalin's approach, in a form which we learned from Dyckhoff.

An alternative to explicitly showing how to eliminate cuts from a proof is to prove completeness of the cut-free fragment with respect to some semantics. This "semantic" approach was discovered by various authors in the 1950's, perhaps first published by Beth in 1955. Beth's work gave rise to the popular method of "semantic tableaux", a method for establishing the validity of a formula by constructing a tree of formulas, where all branches ultimately end in a certain way [152, 446]. The most well-known semantic cut-elimination proof is perhaps by Rasiowa and Sikorski [409] and it is a simplified version of [260]. The semantic approach was successfully used for higher-order logics, before Girard invented the method of candidates (see Chapter 11). More recent works are e.g. [226, 374]; see also [61].

Sequent calculus seems better suited for formalization of classical logic than natural deduction; part of the reason can be attributed to the presence of multiple succedents in sequent calculus sequents. It is natural to adapt the idea to natural deduction by permitting multiple succedents on the right hand side of natural deduction sequents; such systems are indeed studied by Ungar [491] and Parigot [380] (see also the notes to Chapter 6).

Sequent calculi may be called *resource-conscious;* i.e. one cannot use an assumption more than once—not without explicit duplication with the contraction rule. Similarly, one has to explicitly discard an assumption with the weakening rule. A more fine-grained analysis of the structural rules of sequent calculus and of the symmetries of sequent calculus is provided in linear logic (see Notes to Chapter 5). There is also an elaborate programme involving the study of games, strategies, etc. in the context of linear logic. The lecture notes by van Benthem [41] discuss many aspects of games relevant to logic, including the developments related to linear logic—see also [49, 5].

As mentioned in Chapter 4, the type of games considered here were introduced by Paul Lorenzen [315]. Completeness of dialogues with respect to formal systems was considered by Lorenz [314]; see also [147]. For more information see [34]. Other dialogues for classical logics and control operators are studied in [86, 225, 223, 375].

More information about Gentzen and his work can be found in [465]. Kleene [269] and Curry [106] are some classical texts which study cut elimination for sequent calculi along with applications; the latter proves normalization based on cut-elimination and contains a wealth of historical information. A newer reference is [489]. The book [162] emphasizes automated theorem proving and related notions of computation; in particular, the book contains a detailed account of SLD resolution and logic programming.

7.7. Exercises

7.1. Construct proofs in LJ for the formulas of Example 2.1.3.

7.2. Construct proofs in LK for the formulas of Example 2.1.4.

7.3. Show that $\Gamma \vdash \bot$ is derivable in LK (LJ) iff $\Gamma \vdash$ is derivable in LK (LJ).

7.4. Consider the system obtained from LJ by replacing (L$\bot$) and (RW) by:

$$\frac{\Gamma \vdash \bot}{\Gamma \vdash \varphi}$$

Prove that both systems prove the same sequents of the form $\Gamma \vdash \varphi$.

7.5. Consider the system LK¬ obtained from Figure 7.1 by replacing $\bot$ with ¬ in the propositional language, and by replacing (L$\bot$) by the two rules

$$(\text{L}\neg) \frac{\Gamma \vdash \varphi, \Delta}{\Gamma, \neg\varphi \vdash \Delta} \qquad (\text{R}\neg) \frac{\Gamma, \varphi \vdash \Delta}{\Gamma \vdash \neg\varphi, \Delta}$$

Define $\bot$ as $p \wedge \neg p$, for some p. Show that $\Gamma \vdash \Delta$ in LK¬ iff $\Gamma \vdash \Delta$ in LK (absurdity in LK¬ is defined in terms of negation, in LK the other way around).

7.6. Consider the system obtained from LK by replacing (L$\wedge$) and (R$\vee$) by:

$$\frac{\Gamma, \varphi, \psi \vdash \Sigma}{\Gamma, \varphi \wedge \psi \vdash \Sigma} \; (\text{L}'\wedge) \qquad \frac{\Gamma \vdash \varphi, \psi, \Delta}{\Gamma \vdash \varphi \vee \psi, \Delta} \; (\text{R}'\vee)$$

Show that $\Gamma \vdash \Delta$ in this system iff $\Gamma \vdash \Delta$ in LK.

7.7. Consider these variants of (L→):

$$\frac{\Gamma \vdash \varphi, \Sigma \quad \Gamma, \psi \vdash \Sigma}{\Gamma, \varphi \to \psi \vdash \Sigma} \; (\text{L}'\to) \qquad \frac{\Gamma \vdash \varphi, \Sigma \quad \Pi, \psi \vdash \Delta}{\Gamma, \Pi, \varphi \to \psi \vdash \Sigma, \Delta} \; (\text{L}''\to)$$

Show that replacing (L→) with any of these two rules does not affect the set of derivable sequents of LK.

7.8. We use the notation from Definition 7.3.2. Let $d(\varphi) = d+1$. Try to show that $\Gamma \vdash_L^d \varphi, \Delta$ and $\Gamma, \varphi \vdash_L^d \Sigma$ implies $\Gamma \vdash_L^d \Delta, \Sigma$. Where does the proof break down?

7.9. Consider the system obtained from LK by reading all left hand sides and right hand sides as sets and omitting the contraction and exchange rules. Show that $\Gamma \vdash \Delta$ in this system iff $\Gamma \vdash \Delta$ in LK (where Γ, Δ are understood as sets in the former sequent and sequences in the latter sequent).

7.10. Consider the system obtained from LK by omitting the weakening rules and replacing the axioms by:

$$\Gamma, \bot \vdash \Delta \ (L'\bot) \qquad\qquad \Gamma, \varphi \vdash \varphi, \Delta \ (Ax')$$

Show that $\Gamma \vdash \Delta$ in the resulting system iff $\Gamma \vdash \Delta$ in LK. Also show that in the revised system it suffices with the following rule instead of (Ax'):

$$\Gamma, p \vdash p, \Delta \ (Ax'')$$

7.11. Consider the system obtained from LK by omitting the contraction rules and (Cut), and modifying the logical rules to retain the principal formula:

$$\frac{\Gamma, \varphi \wedge \psi, \varphi \vdash \Sigma}{\Gamma, \varphi \wedge \psi \vdash \Sigma} \quad \frac{\Gamma, \varphi \wedge \psi, \psi \vdash \Sigma}{\Gamma, \varphi \wedge \psi \vdash \Sigma} \quad \frac{\Gamma \vdash \varphi, \varphi \wedge \psi, \Delta \quad \Gamma \vdash \psi, \varphi \wedge \psi, \Delta}{\Gamma \vdash \varphi \wedge \psi, \Delta}$$

$$\frac{\Gamma, \varphi \vee \psi, \varphi \vdash \Sigma \quad \Gamma, \varphi \vee \psi, \psi \vdash \Sigma}{\Gamma, \varphi \vee \psi \vdash \Sigma} \quad \frac{\Gamma \vdash \varphi, \varphi \vee \psi, \Delta}{\Gamma \vdash \varphi \vee \psi, \Delta} \quad \frac{\Gamma \vdash \psi, \varphi \vee \psi, \Delta}{\Gamma \vdash \varphi \vee \psi, \Delta}$$

$$\frac{\Gamma, \varphi \to \psi \vdash \varphi, \Sigma \quad \Gamma, \varphi \to \psi, \psi \vdash \Sigma}{\Gamma, \varphi \to \psi \vdash \Sigma} \quad \frac{\Gamma, \varphi \vdash \psi, \varphi \to \psi, \Delta}{\Gamma \vdash \varphi \to \psi, \Delta}$$

Show that $\Gamma \vdash \Delta$ in the resulting system iff $\Gamma \vdash \Delta$ in LK.

7.12. Consider a G3 variant obtained from LK by (see Exercises 7.6, 7.7, and 7.10)

- replacing (L∧), (R∨), (L→) by (L'∧), (R'∨), (L'→);
- replacing (L⊥) and (Ax) by (L'⊥) and (Ax'');
- omitting (Cut);
- reading antecedents and succedents as multisets;
- omitting the structural rules.

Show that $\Gamma \vdash \Delta$ in the resulting system iff $\Gamma \vdash \Delta$ in LK. For LJ make similar modifications, taking into account that we have at most one formula at the right; leave (R∨) in the usual form in LJ, and replace (L→) by (L'''→), where

$$\frac{\Gamma, \varphi \to \psi \vdash \varphi \quad \Gamma, \psi \vdash \sigma}{\Gamma, \varphi \to \psi \vdash \sigma} \ (L'''\to)$$

Show that $\Gamma \vdash \Delta$ in the resulting system iff $\Gamma \vdash \Delta$ in LJ.

7.7. Exercises

7.13. Show that provability in LJ can be decided in polynomial space. *Hint:* First do Exercises 7.4 and 7.9–7.11.

7.14. Show that every sequent that can be derived in LK, can be derived without the use of contraction. *Hint:* Study the proof of Proposition 7.2.3. Can you find a contraction-free proof of Peirce's law in LK that is also cut-free?

7.15. For a formula φ, let φ' arise from φ by replacing every conjunction by disjunction and vice versa. For a sequence Γ, let Γ' arise from Γ by replacing every element φ in the sequence by φ'. Show that if $\Gamma \vdash_{LK} \Delta$ then $\Delta' \vdash_{LK} \Gamma'$, provided no formula in Γ and Δ contain implications.

7.16. Show that $\vdash \psi$ in LP (resp. LM) implies $\vdash \psi[\bot := p]$ in LP (resp LM) for any fresh variable p.

7.17. Of course $\Gamma \vdash_{LK} \varphi, \Delta$ implies $\Gamma, \neg\Delta \vdash_C \varphi$, where $\neg\Delta$ arises by negating every formula in Δ and $\neg\varphi = \varphi \to \bot$ as usual.[2] But suppose we again define $\neg\varphi = \varphi \to \bot$ and take $\bot$ as some fixed propositional variable (see Notation 7.2.2). Can you show that $\Gamma \vdash_{LP} \varphi, \Delta$ implies $\Gamma, \neg\Delta \vdash_P \varphi$?

7.18. Suppose the conclusion of (Cut) in Figure 7.4 reads $\Gamma \vdash (\lambda x{:}\varphi.M)N : \sigma$ instead of $\Gamma \vdash M[x := N] : \sigma$. Can all simply typed λ-terms still be typed?

7.19. Consider these rules:

$$(L\wedge) \; \frac{\Gamma, x{:}\varphi_i \vdash M : \sigma}{\Gamma, y{:}\varphi_1 \wedge \varphi_2 \vdash M[x := \pi_i(y)] : \sigma} \qquad \frac{\Gamma \vdash M : \varphi \quad \Gamma \vdash N : \psi}{\Gamma \vdash \langle M, N \rangle : \varphi \wedge \psi} \; (R\wedge)$$

$$(L\vee) \; \frac{\Gamma, x{:}\varphi \vdash M : \sigma \quad \Gamma, y{:}\psi \vdash N : \sigma}{\Gamma + z{:}\varphi \vee \psi \vdash \textbf{case } z \textbf{ of } [x]M \textbf{ or } [y]N : \sigma} \qquad \frac{\Gamma \vdash M : \varphi_i}{\Gamma \vdash \text{in}_i(M) : \varphi_1 \vee \varphi_2} \; (R\vee)$$

The term assignment of Figure 7.4 can be extended to include conjunction and disjunction by adding the above rules.

(i) Generalize Propositions 7.4.2–7.4.3.

(ii) Find a β-normal form with no type in the cut-free fragment of this system (where β-reduction is defined as in Section 4.5).

(iii) Let $\to_c$ be the smallest compatible relation containing these rules (known as the *commuting conversions* or *permutative conversions*):

$$\pi_i(\textbf{case } M \textbf{ of } [x]P \textbf{ or } [y]Q) \;\to_c\; \textbf{case } M \textbf{ of } [x]\pi_i(P) \textbf{ or } [y]\pi_i(Q)$$

$$(\textbf{case } M \textbf{ of } [x]P \textbf{ or } [y]Q)N \;\to_c\; \textbf{case } M \textbf{ of } [x]PN \textbf{ or } [y]QN$$

$$\textbf{case } (\textbf{case } M \textbf{ of } [x]P \textbf{ or } [y]Q) \textbf{ of } [z]N \textbf{ or } [v]R \;\to_c\;$$
$$\textbf{case } M \textbf{ of } [x](\textbf{case } P \textbf{ of } [z]N \textbf{ or } [v]R) \textbf{ or } [y](\textbf{case } Q \textbf{ of } [z]N \textbf{ or } [v]R)$$

Show that Proposition 7.4.4 and Corollary 7.4.5 generalize when NF_β is replaced by $NF_{c\beta}$.

7.20. Modify the Lorenzen games thus:

---

[2] The equivalence between a formula on the right and its negation on the left can be used to formulate a version of LK that has only "one side" of $\vdash$ as opposed to two, known as *one-sided sequent calculus*—see for instance [489].

- There are no attacks on $\bot$.
- The proponent may only state $\bot$ if the opponent has stated $\bot$ earlier.

Show that there is a winning prover strategy for φ in this type of game iff $\vdash_{\mathrm{LP}} \varphi$.

7.21. Modify the Lorenzen games by stipulating that a proponent defense must refer to the most recent opponent attack. Show that there is a winning prover strategy for φ in this type of game iff $\vdash_{\mathrm{LJ}} \varphi$.

Chapter 8

First-order logic

The objects investigated by propositional logic are compound statements, built from some atomic statements (represented by propositional variables) by means of logical connectives. The objective is to understand relations between compound statements, depending on their structure, rather than on the actual "meaning" of the atoms occurring in these statements. For instance, one does not have to read [219] to assert the (constructive) correctness of the following reasoning by means of the propositional calculus.

If Yossarian is not insane then he will not be relieved of flying missions.
If Yossarian is insane then he will not be relieved of flying missions.
Therefore Yossarian will not be relieved of flying missions.

But mathematics always involves reasoning about *individual objects*, and statements about properties of objects sometimes cannot be adequately expressed in the propositional language. Consider another example:

Figaro shaves all men of Seville who do not shave themselves.
But he does not shave anyone who shaves himself.
Therefore Figaro is not a man of Seville.

To express this reasoning in the language of formal logic, we need to quantify over individual objects (humans), and of course we need predicates (relations) on individual objects. Logical systems involving these two features are known under the names *predicate logic*, *predicate calculus* or *first-order logic*.

8.1. Syntax of first-order logic

The first-order syntax is always defined with respect to a fixed *signature* Σ. Recall that a signature is a family of function and relation symbols (see Section A.2). Each of these symbols comes with a designated arity, possibly zero, i.e. the symbol can be nullary. Also recall that *algebraic terms* over Σ

are individual variables, and expressions of the form $(ft_1 \ldots t_n)$, where f is an n-ary function symbol in Σ, and $t_1, \ldots, t_n$ are algebraic terms over Σ. The set of all variables occurring in an algebraic term t is denoted by $\mathrm{FV}(t)$, and $t[a := t']$ is the result of substituting t' for a variable a in t.

8.1.1. DEFINITION.

(i) An *atomic formula* is either the logical constant $\bot$ or an expression of the form $(rt_1 \ldots t_n)$, where r is an n-ary relation symbol in Σ, and $t_1, \ldots, t_n$ are algebraic terms over Σ. In particular, each nullary relation symbol makes an atomic formula.

(ii) The set Φ_Σ of *first-order formulas* over Σ, is the least set such that:

- All atomic formulas are in Φ_Σ;
- If $\varphi, \psi \in \Phi_\Sigma$ then $(\varphi \to \psi), (\varphi \vee \psi), (\varphi \wedge \psi) \in \Phi_\Sigma$;
- If $\varphi \in \Phi_\Sigma$ and a is an individual variable, then $\forall a \varphi, \exists a \varphi \in \Phi_\Sigma$.

As usual, $\neg \varphi$ and $\varphi \leftrightarrow \psi$ abbreviate $\varphi \to \bot$ and $(\varphi \to \psi) \wedge (\psi \to \varphi)$.

(iii) The set $\mathrm{FV}(\varphi)$ of free variables of a formula φ is defined as follows:

- $\mathrm{FV}(rt_1 \ldots t_n) = \mathrm{FV}(t_1) \cup \ldots \cup \mathrm{FV}(t_n)$;
- $\mathrm{FV}(\varphi \to \psi) = \mathrm{FV}(\varphi \vee \psi) = \mathrm{FV}(\varphi \wedge \psi) = \mathrm{FV}(\varphi) \cup \mathrm{FV}(\psi)$;
- $\mathrm{FV}(\forall a \varphi) = \mathrm{FV}(\exists a \varphi) = \mathrm{FV}(\varphi) - \{a\}$.

That is, a quantifier $\forall a$ or $\exists a$ binds a within its scope. Formulas that differ only in their bound variables are identified. If Γ is a set of formulas then $\mathrm{FV}(\Gamma) = \bigcup \{\mathrm{FV}(\gamma) \mid \gamma \in \Gamma\}$.

(iv) A formula is *open* iff it contains no quantifiers. A *sentence*, also called a *closed formula*, is a formula without free variables. If $a_1, \ldots, a_k$ are the free variables of φ (listed in some fixed order) then the sentence $\forall a_1 \ldots \forall a_k \varphi$ is called the *universal closure* of φ.

It follows from the definition above that a formula is not merely an *expression* but rather an α-equivalence class of expressions, very much like a lambda-term, cf. Section 1.2. The way we handle α-conversion on formulas is compromising precision for simplicity. To give a more exact statement of this definition one needs a discussion similar to that in Section 1.2. A definition of a "pre-formula" has to be stated, etc. We believe the reader is able to reconstruct the missing parts of the formalism if needed.

In fact, it is not customary in logic to identify alpha-equivalent first-order formulas. Typically, such formulas are considered different, although they are equivalent with respect to all reasonable semantics, and one can be derived from another in all reasonable proof systems. But for uniformity of presentation we prefer to assume alpha-equivalence with respect to all kinds of variable binding in all the logics and lambda-calculi discussed in this book.

8.2. INFORMAL SEMANTICS

8.1.2. CONVENTION. The parentheses-avoiding conventions used for propositional formulas (Convention 2.1.2) apply as well to first-order formulas. In addition, we group together quantifiers, that is, we write $\forall ab\dots$ instead of $\forall a \forall b\dots$, and we sometimes use the dot notation, cf. Convention 1.2.2. That is, we can write $\forall a.\, \varphi \to \psi$ for $\forall a(\varphi \to \psi)$. But $\forall a \varphi \to \psi$, without the dot, should be understood as $(\forall a \varphi) \to \psi$.

8.1.3. DEFINITION. The *substitution* of a term for an individual variable, denoted $\varphi[a := t]$, is defined as follows:

- $\bot[a := t] = \bot$;
- $(rt_1 \dots t_n)[a := t] = rt_1[a := t]\dots t_n[a := t]$;
- $(\varphi \circ \psi)[a := t] = \varphi[a := t] \circ \psi[a := t]$, for $\circ \in \{\to, \vee, \wedge\}$;
- $(\nabla b\, \varphi)[a := t] = \nabla b\, \varphi[a := t]$ for $\nabla \in \{\forall, \exists\}$, if $b \neq a$, and $b \notin \mathrm{FV}(t)$.

Also the *simultaneous substitution*, written $\varphi[a_1, \dots, a_n := t_1, \dots, t_n]$, is defined in a similar way as for lambda-terms (see Definition 1.2.21).

8.1.4. CONVENTION. We sometimes write e.g. $\varphi(a, b, c)$ instead of φ, to stress that a, b, c may occur in φ. In this case, notation like $\varphi(s, t, u)$ is used as a shorthand for $\varphi[a, b, c := s, t, u]$.

Note that if we identify propositional variables with nullary relation symbols, we can interpret propositional formulas as a special case of open first-order formulas. That is, the propositional syntax can be seen as a fragment of the first-order syntax.

8.2. Informal semantics

The Brouwer-Heyting-Kolmogorov interpretation of propositional formulas (Chapter 2) can be extended to first-order logic. For this, one has to assume that individual variables and algebraic terms are interpreted in a certain domain of "objects." Since predicate logic expresses properties of such objects, it is natural to assume that the domain of interpretation should be non-empty, i.e. that there exists at least one object.[1]

As in Chapter 2, the interpretation of atoms is left unspecified. The informal understanding of propositional connectives remains of course the same. The BHK interpretation of quantified formulas is as follows:

- *A construction of $\forall a \varphi(a)$ is a method (function) transforming every object **a** into a construction of $\varphi(\mathbf{a})$.*

---
[1] In other logical systems involving quantification this assumption does not have to be *a priori* accepted. If we quantify over compound objects (like proofs) it may be reasonable to expect that some domains of quantification may be empty. See Remark 13.5.1.

- A construction of $\exists a\varphi(a)$ is a pair consisting of an object **a** and a construction of $\varphi(\mathbf{a})$.

8.2.1. EXAMPLE. Formulas of the following forms can be assigned constructions according to the BHK interpretation as above, if we agree that any term represents an "object".

(i) $\forall a\varphi \to \exists a\varphi$;

(ii) $\neg\exists a\varphi \leftrightarrow \forall a\neg\varphi$;

(iii) $\forall a(\varphi \to \psi) \to (\forall a\varphi \to \forall a\psi)$;

(iv) $\forall a(\varphi \to \psi) \to (\exists a\varphi \to \exists a\psi)$;

(v) $\forall a\varphi(a) \to \varphi(t)$;

(vi) $\varphi(t) \to \exists a\varphi(a)$;

(vii) $\psi \to \forall a\psi$, where $a \notin \mathrm{FV}(\psi)$;

(viii) $\forall a(\varphi \lor \neg\varphi) \land \neg\neg\forall a\varphi \to \forall a\varphi$;

(ix) $\exists a(\psi \land \varphi(a)) \leftrightarrow \psi \land \exists a\varphi(a)$, where $a \notin \mathrm{FV}(\psi)$;

(x) $\neg\neg\forall a\varphi \to \forall a\neg\neg\varphi$.

Formula (i) reflects the assumption about the non-empty domain. In order to obtain a construction of $\exists a\varphi$, one has to point out an arbitrary element of the domain and apply a construction of $\forall a\varphi$ to this element. Let us take formula (iii) as another example. Suppose we have a construction of $\forall a(\varphi \to \psi)$. We can describe a construction of $\forall a\varphi \to \forall a\psi$ as follows. Take any construction of $\forall a\varphi$. Our construction of $\forall a\psi$ begins with a request for an object **a**. When that is supplied, we insert it into our constructions of $\forall a\varphi$ and $\forall a(\varphi \to \psi)$. This will give us constructions of $\varphi(\mathbf{a})$ and of $\varphi(\mathbf{a}) \to \psi(\mathbf{a})$. It remains to apply the latter to the former. The reader can easily verify the other formulas of our example (Exercise 8.3).

REMARK. Universal quantification is often seen as a generalization of conjunction. The statement "*all cats have tails*" is quite like an infinite conjunction of statements concerning each individual cat separately. In the same spirit one can think of existential quantification as a generalized disjunction. This idea is reflected by the algebraic semantics, where we interpret quantifiers as (possibly infinite) joins and meets, see Definitions 8.4.5 and 8.5.1.

The BHK interpretation as presented above hints for another correspondence, between universal quantification and implication, because in both cases a construction is a function. This is consistent with our understanding of the cats' tails example. Indeed, a construction of an infinite conjunction must be an "infinite tuple" of constructions. And such an "infinite tuple" is like a function assigning a tail certificate to every cat.

8.2. INFORMAL SEMANTICS

8.2.2. EXAMPLE. Consider the following formulas:

(i) $\neg\neg\forall a(\varphi \vee \neg\varphi)$;

(ii) $\neg\forall a\varphi \leftrightarrow \exists a\neg\varphi$;

(iii) $\exists b(\varphi(b) \to \forall a\varphi(a))$;

(iv) $\exists a(\exists b\varphi(b) \to \varphi(a))$;

(v) $(\forall a\varphi(a) \to \psi) \to \exists a(\varphi(a) \to \psi)$, where $a \notin \mathrm{FV}(\psi)$;

(vi) $\exists a\exists b(\varphi(a) \to \psi(b)) \to \exists a(\varphi(a) \to \psi(a))$;

(vii) $\forall a(\varphi \vee \neg\varphi) \to \exists a\varphi \vee \forall a\neg\varphi$;

(viii) $\forall a(\varphi \vee \neg\varphi) \wedge \neg\neg\exists a\varphi \to \exists a\varphi$;

(ix) $\forall a(\psi \vee \varphi(a)) \leftrightarrow \psi \vee \forall a\varphi(a)$, where $a \notin \mathrm{FV}(\psi)$;

(x) $\forall a\neg\neg\varphi \to \neg\neg\forall a\varphi$.

All the above are schemes of classically valid formulas.[2] But one will have difficulties finding appropriate constructions, and we will soon see (Proposition 8.5.3) that none of them is valid in first-order intuitionistic logic. In some cases this should not be a big surprise. For instance, De Morgan's law (ii), is obviously a first-order version of the propositional formula (v) of Example 2.1.4, and we already know that the latter is not intuitionistically valid. There is a similar relationship between formula (iv) above and formula (ix) of Example 2.1.4.

However, not all of the formulas above can be explained this way. For instance, formula (i) demonstrates that Glivenko's Theorem 2.4.10 does not hold for first-order logic. Formulas (vii)–(x), including *Grzegorczyk's scheme* (ix) and *Markov's principle* (viii), also correspond to propositional formulas which are constructively valid (e.g. (ix) to a distributivity law.)

Formula (iii), contrary to a simple-minded intuition, states: *There is a certain b such that $\varphi(b)$ implies $\varphi(a)$ for all a*. Why? Because either all elements satisfy φ or there is one which does not, so the implication holds for free. But from the constructive point of view (iii) is doubtful. And so is even the seemingly innocent (iv) because one cannot know a before knowing b.

8.2.3. REMARK. Although Lorenzen games (see Chapter 7) can be generalized to first-order logic, we do not follow that path here. Nevertheless, it is illuminating to imagine the winning proponent strategies for the intuitionstic and classical tautologies of Examples 8.2.1 and 8.2.2. Consider, for instance, the implication from left to right in Example 8.2.2(ix). A winning proponent strategy could be as follows:

---

[2] We assume that the reader is generally familiar with classical first-order semantics. See Section 8.4 for a brief reminder, or e.g. [39, 113, 252, 337] for more.

Proponent 1: I assert that $\forall a(q \lor pa) \to q \lor \forall a\, pa$ holds!

Opponent 1: Really? If I state $\forall a(q \lor pa)$, can you infer the conclusion?

Proponent 2: No problem $q \lor \forall a\, pa$ holds!

Opponent 2: But which part?

Proponent 3: Hmm... $\forall a\, pa$.

Opponent 3: Oh yeah? Here is one term t, show me pt.

Proponent 4: Well, you claimed earlier that $\forall a(q \lor pa)$ holds. With this t, can you show $q \lor pt$?

At this point, the opponent will state pt or q. In the former case the proponent wins by stating pt. In the latter case she would say: "Oops, I changed my mind. When you asked me in your first step, which part of the disjunction holds, I meant q." Of course, this is again a *Catch-22* trick. Constructively, the proponent may lose.

8.3. Proof systems

The three main approaches, natural deduction, sequent calculus and Hilbert style, extend to first-order logic by adding suitable rules and axioms to the rules and axiom schemes for propositional logic. (If not clearly stated otherwise, we always mean intuitionistic logic.) The notation $\vdash_N, \vdash_L$, etc., is used as in the propositional case.

NATURAL DEDUCTION. We extend the system of propositional natural deduction (Definition 2.2.1) with the following rules to introduce and eliminate quantifiers. All other definitions and notations related to natural deduction apply without changes to the first-order system.

$$(\forall \text{I}) \; \frac{\Gamma \vdash \varphi}{\Gamma \vdash \forall a \varphi} \; (a \notin \text{FV}(\Gamma)) \qquad (\forall \text{E}) \; \frac{\Gamma \vdash \forall a \varphi}{\Gamma \vdash \varphi[a := t]}$$

$$(\exists \text{I}) \; \frac{\Gamma \vdash \varphi[a := t]}{\Gamma \vdash \exists a \varphi} \qquad (\exists \text{E}) \; \frac{\Gamma \vdash \exists a \varphi \quad \Gamma, \varphi \vdash \psi}{\Gamma \vdash \psi} \; (a \notin \text{FV}(\Gamma, \psi))$$

The quantifier rules are the same for classical and intuitionistic logic. The difference between the two natural deduction calculi lies in the choice of propositional rules for absurdity, either ($\neg$E) or ($\bot$E), cf. Section 6.1.

The reader should be warned that our rules rely on alpha-conversion of formulas. (Just try to derive $\forall b\, rb, pa \vdash \forall a\, ra$.) Otherwise, one has to modify rules ($\forall$I) and ($\exists$E) to work for any alpha-variant of the quantifier bindings (Exercise 8.7). Similar modifications would have to be applied to the other proof systems to follow.

8.3. Proof systems

TRANSLATION FROM CLASSICAL LOGIC. The double negation translations from classical to intuitionistic logic can be extended to the first-order case. Indeed, we can add the following clauses to the definition of the Kolmogorov translation of Chapter 6:

- $k(\alpha) := \neg\neg\alpha$, for atomic α;
- $k(\forall a \varphi) := \neg\neg \forall a\, k(\varphi)$;
- $k(\exists a \varphi) := \neg\neg \exists a\, k(\varphi)$.

8.3.1. PROPOSITION. *A formula φ is a classical theorem iff $k(\varphi)$ is an intuitionistic theorem.*

PROOF. The implication from right to left follows from the classical equivalence $k(\varphi) \leftrightarrow \varphi$. To prove the other implication, we show by induction with respect to derivations that if $\Gamma \vdash \varphi$ in classical logic then $k(\Gamma) \vdash k(\varphi)$ in intuitionistic logic. The propositional cases are as in the proofs of Theorems 6.4.3 and 6.7.4. Case ($\forall$I) follows easily from the induction hypothesis. For ($\exists$I) one first shows that $k(\varphi[x := t]) = k(\varphi)[x := t]$ holds for all φ and t.

Consider case ($\exists$E), i.e. suppose that we have derived $\Gamma \vdash \psi$ from $\Gamma \vdash \exists a \varphi$ and $\Gamma, \varphi \vdash \psi$. Let ψ' be such that $k(\psi) = \neg\psi'$, and let $\Gamma' = k(\Gamma), \exists a\, k(\varphi)$. From the induction hypothesis we obtain $\Gamma' \vdash k(\psi)$ by an application of rule ($\exists$E). Since $k(\Gamma) \vdash \neg\neg \exists a\, k(\varphi)$, we have $k(\Gamma) \vdash k(\psi)$, by Exercise 2.24(ii). Case ($\forall$E) is left to the reader. Use Exercise 8.21(i). □

The Kolmogorov translation k is obviously computable, and thus classical provability is reducible to intuitionistic provability. Since classical first-order logic is undecidable, we obtain undecidability of intuitionistic first-order logic as a consequence.

8.3.2. COROLLARY. *First-order intuitionistic logic is undecidable.*

In fact, the undecidability result holds already for a very restricted fragment of first-order intuitionistic logic, with $\forall$ and $\rightarrow$ as the only connectives and with no function symbols. In particular, there is no need for negation or falsity. See Corollary 8.8.3.

SEQUENT CALCULUS. The following are rules for quantifiers in classical sequent calculus. Note the symmetry between rules for the two quantifiers.

$$(\text{L}\forall)\frac{\Gamma, \varphi[a := t] \vdash \Sigma}{\Gamma, \forall a \varphi \vdash \Sigma} \qquad (\text{R}\forall)\frac{\Gamma \vdash \varphi, \Delta}{\Gamma \vdash \forall a \varphi, \Delta}\ (a \notin \text{FV}(\Gamma, \Delta))$$

$$(\text{L}\exists)\frac{\Gamma, \varphi \vdash \Sigma}{\Gamma, \exists a \varphi \vdash \Sigma}\ (a \notin \text{FV}(\Gamma, \Sigma)) \qquad (\text{R}\exists)\frac{\Gamma \vdash \varphi[a := t], \Delta}{\Gamma \vdash \exists a \varphi, \Delta}$$

To obtain the intuitionistic sequent calculus we restrict ourselves to at most one formula at the right-hand sides. (That is, Σ consists of at most one formula and Δ is always empty).

$$(\mathrm{L}\forall)\frac{\Gamma, \varphi[a := t] \vdash \sigma}{\Gamma, \forall a\varphi \vdash \sigma} \qquad (\mathrm{R}\forall)\frac{\Gamma \vdash \varphi}{\Gamma \vdash \forall a\varphi} \ (a \notin \mathrm{FV}(\Gamma))$$

$$(\mathrm{L}\exists)\frac{\Gamma, \varphi \vdash \sigma}{\Gamma, \exists a\varphi \vdash \sigma} \ (a \notin \mathrm{FV}(\Gamma, \sigma)) \qquad (\mathrm{R}\exists)\frac{\Gamma \vdash \varphi[a := t]}{\Gamma \vdash \exists a\varphi}$$

8.3.3. THEOREM (Cut elimination). *In both classical and intuitionistic sequent calculus, every derivable sequent has a cut-free proof.*

We omit the proof of first-order cut-elimination. It is similar to the proof given in Chapter 7. As a consequence of cut-elimination, first-order intuitionistic logic has the disjunction property, analogous to Corollary 7.3.5 (cf. Proposition 2.5.10). A similar result holds for the existential quantifier.

8.3.4. COROLLARY (Disjunction property). *If $\vdash \varphi \vee \psi$ then $\vdash \varphi$ or $\vdash \psi$.*

8.3.5. COROLLARY (Existence property). *If $\vdash \exists a\varphi$ then there exists a term t such that $\vdash \varphi[a := t]$.*

PROOF. The last rule in a cut-free proof of $\vdash \exists a\varphi$ must be (R$\exists$). □

We also have various conservativity results, in particular the following.

8.3.6. THEOREM. *The intuitionistic (resp. classical) predicate calculus is conservative over intuitionistic (resp. classical) propositional calculus.*

PROOF. If a propositional formula is provable in the first-order sequent calculus, then it is cut-free provable. It is not hard to see that a cut-free proof is done within the propositional fragment of the system. □

The proof of Theorem 8.3.6 uses a principle similar to the subformula property (Corollary 7.3.6). Note however that the notion of a "subformula" must be appropriately relaxed in first-order logic. For each algebraic term t, we must consider $\varphi(t)$ a subformula of both $\exists a\varphi(a)$ and $\forall a\varphi(a)$. It follows that there are infinitely many subformulas of any universal sentence, and here the decidability argument used in Corollary 7.3.9 breaks down (as it must, in view of Corollary 8.3.2).

8.3.7. REMARK. Note that this problem occurs even for function-free signatures. Although there are no complex terms, the number of variables occurring in a proof is not a priori bounded. This rules out naive attempts to

8.3. Proof systems

design a deciding algorithm based on estimating the proof size. For example suppose we try to reconstruct a proof of $\Gamma, \forall a \exists b(Sab) \vdash \varphi$. It may happen that the final part of the proof looks as follows. (Double lines indicate that structural rules are used in addition to the alternating (L∃) and (L∀).)

$$
\begin{array}{c}
\vdots \\
\Gamma, Sa_0a_1, Sa_1a_2, \ldots, Sa_{n-2}a_{n-1}, Sa_{n-1}a_n \vdash \varphi \\
\hline
\Gamma, Sa_0a_1, Sa_1a_2, \ldots, Sa_{n-2}a_{n-1}, \exists b(Sa_{n-1}b) \vdash \varphi \\
\hline
\Gamma, Sa_0a_1, Sa_1a_2, \ldots, Sa_{n-2}a_{n-1}, \forall a \exists b(Sab) \vdash \varphi \\
\vdots \\
\Gamma, Sa_0a_1, Sa_1a_2, \forall a \exists b(Sab) \vdash \varphi \\
\hline
\Gamma, Sa_0a_1, \exists b(Sa_1b), \forall a \exists b(Sab) \vdash \varphi \\
\hline
\Gamma, Sa_0a_1, \forall a \exists b(Sab) \vdash \varphi \\
\hline
\Gamma, \exists b(Sa_0b), \forall a \exists b(Sab) \vdash \varphi \\
\hline
\Gamma, \forall a \exists b(Sab) \vdash \varphi
\end{array}
$$

In general, the number n can be arbitrarily large, and it may happen that no shorter "chain" of variables is sufficient to derive φ. Thus we may be unable to set any upper bound for the size of the proof we are looking for.

HILBERT-STYLE PROOFS. There is a variety of Hilbert-style axioms discussed in the literature. But the following choice is fairly popular. We take as logical axioms all formulas matching the propositional axiom schemes of Chapter 5, and in addition all formulas of the form:

(Q1) $\forall a \varphi \to \varphi[a := t]$;

(Q2) $\varphi[a := t] \to \exists a \varphi$;

(Q3) $\forall a(\varphi \to \psi) \to \varphi \to \forall a \psi$, where $a \notin \mathrm{FV}(\varphi)$;

(Q4) $\forall a(\varphi \to \psi) \to \exists a \varphi \to \psi$, where $a \notin \mathrm{FV}(\psi)$.

As inference rules of our system we take *modus ponens* and the following *generalization rule*:

$$\frac{\varphi}{\forall a \varphi}$$

The use of generalization requires some caution (see Definition 8.3.8).

REMARK. Adding *tertium non datur* (or any other propositional axiom scheme of sufficient strength) to the system above results in a complete axiomatization of classical first-order logic. As with natural deduction, the difference between classical and intuitionistic logic occurs at the propositional level.

8.3.8. DEFINITION. A formal *proof* of a formula φ from a set Γ of assumptions (non-logical axioms) is a finite sequence of formulas $\psi_1, \psi_2, \ldots, \psi_n$, such that $\psi_n = \varphi$, and for all $i = 1, \ldots, n$, one of the following holds:

- ψ_i is a logical axiom;
- ψ_i is an element of Γ;
- there are $j, \ell < i$ such that $\psi_j = \psi_\ell \to \psi_i$ (that is, ψ_i is obtained from ψ_j and ψ_ℓ using *modus ponens*);
- there is $j < i$ such that $\psi_i = \forall a \psi_j$, for some $a \notin \mathrm{FV}(\Gamma)$ (that is, ψ_i is obtained from ψ_j by generalization).

A proof is always a finite sequence. Thus if $\Gamma \vdash \varphi$, for an infinite set Γ, then there is always a finite subset $\Gamma_0 \subseteq \Gamma$ such that in fact $\Gamma_0 \vdash \varphi$.

8.3.9. LEMMA (Deduction Theorem).
The conditions $\Gamma, \varphi \vdash_H \psi$ and $\Gamma \vdash_H \varphi \to \psi$ are equivalent.

PROOF. The proof is similar to that of Theorem 5.1.8. The part from right to left is again obvious, and for the converse we use induction with respect to proofs. There is one additional case to handle, namely when ψ was obtained by generalization. That is, $\psi = \forall a \vartheta$, with $a \notin \mathrm{FV}(\Gamma, \varphi)$, and we have $\Gamma, \varphi \vdash_H \vartheta$. From the induction hypothesis it follows that $\Gamma \vdash_H \varphi \to \vartheta$. This can be generalized to $\Gamma \vdash_H \forall a(\varphi \to \vartheta)$. Using axiom (Q3), one derives $\Gamma \vdash_H \varphi \to \forall a \vartheta$. □

WARNING. In some presentations, the definition of a Hilbert-style proof does not include the restriction on applicability of the generalization rule. For such systems the deduction theorem only holds under additional assumptions (typically one restricts attention to closed formulas).

8.3.10. THEOREM. *Natural deduction, sequent calculus, and the Hilbert-style proof system are all equivalert, i.e. $\Gamma \vdash_N \varphi$ and $\Gamma \vdash_L \varphi$ and $\Gamma \vdash_H \varphi$ are equivalent for all Γ and φ. The same holds for classical logic.*

PROOF. Similar to the proofs of Propositions 5.1.9, 6.1.8 and 7.2.5. □

From now on, we write $\Gamma \vdash \varphi$ to denote that an appropriate proof exists in either of the three systems.

8.3.11. DEFINITION. By a (first-order intuitionistic) *theory* we usually mean a set of sentences T such that T is closed under provable consequence, i.e. if $T \vdash \psi$ in first-order intuitionistic logic, for a sentence ψ, then $\psi \in T$. But it is also customary to talk about a theory as a set of axioms. In fact we often identify a set of axioms T with the theory consisting of all consequences of T.

Any sentence φ such that $T \vdash \varphi$ is called a *theorem of T*. And if $T, \Gamma \vdash \varphi$ then we also say that $\Gamma \vdash \varphi$ is a *provable judgement* of T. This terminology applies also to other logical systems, including first-order classical logic and second-order logics.

8.4. Classical semantics

In this section we make a detour into classical logic to prepare some background for Section 8.5. We begin with the standard definition of a classical first-order model. Assume that our signature Σ consists of the function symbols $f_1, \ldots, f_n$ and relation symbols $r_1, \ldots, r_m$.

8.4.1. DEFINITION. A *structure* or *model* for Σ is an algebraic system

$$\mathcal{A} = \langle A, f_1^{\mathcal{A}}, \ldots, f_n^{\mathcal{A}}, r_1^{\mathcal{A}}, \ldots, r_m^{\mathcal{A}} \rangle,$$

where $A \neq \varnothing$, the $f_i^{\mathcal{A}}$ and $r_i^{\mathcal{A}}$ are, respectively, operations and relations over A (of appropriate arities). If a function symbol f_i is nullary, the corresponding $f_i^{\mathcal{A}}$ is of course a constant, i.e. a distinguished element of the domain. A nullary relation symbol r_i is interpreted as a nullary relation, which is is simply a truth value, cf. Section A.1.

A special case of a structure is a *term model*. The domain A of a term model $\mathcal{A}$ is the set of all terms over Σ. The operations are as simple as possible: if f is a k-ary function symbol then $f^{\mathcal{A}}(t_1, \ldots, t_k)$ is just the term $(ft_1 \ldots t_k)$. Relations in a term model can be arbitrary.

We may think of relations in a structure $\mathcal{A}$ as functions ranging over the set $\{0, 1\}$ rather than as sets of tuples. This may be generalized so that relation symbols are interpreted by functions ranging over any fixed set of values, leading to the following definition.

8.4.2. DEFINITION. Let D be an arbitrary non-empty set. A *D-structure* (or *D-model*) for Σ is defined as in Definition 8.4.1, with the following modification. Any k-ary relation symbol r_i is interpreted as a k-ary function

$$r_i^{\mathcal{A}} : A^k \to D.$$

In an analogous way we define a *term D-model*. Structures in the sense of Definition 8.4.1 are thus the same as $\{0, 1\}$-structures, and will sometimes be called *ordinary* or *classical*. Typical notational conventions are to write f rather than $f^{\mathcal{A}}$, and to identify $\mathcal{A}$ and A.

An arbitrary term can be interpreted in a D-structure as an element of the domain, provided we first assign values to individual variables.

8.4.3. DEFINITION.

(i) Let $\mathcal{A}$ be an arbitrary D-structure (in particular an ordinary model). An *individual valuation* (or simply a *valuation*) in $\mathcal{A}$ is any function $\varrho : V \to \mathcal{A}$, where V is the set of all individual variables.

(ii) For any term t we define the *value* $[\![t]\!]_\varrho$ of t with respect to a valuation ϱ:

- $[\![a]\!]_\varrho = \varrho(a)$, if a is a variable;
- $[\![ft_1 \ldots t_n]\!]_\varrho = f^{\mathcal{A}}([\![t_1]\!]_\varrho, \ldots, [\![t_n]\!]_\varrho)$.

8.4.4. LEMMA.

(i) *The value of a term depends only on the values of its free variables. That is, if $\varrho_1(a) = \varrho_2(a)$, for all $a \in \mathrm{FV}(t)$, then $[\![t]\!]_{\varrho_1} = [\![t]\!]_{\varrho_2}$.*

(ii) *Substitution "commutes" with semantics:* $[\![t[a := s]]\!]_\varrho = [\![t]\!]_{\varrho(a \mapsto [\![s]\!]_\varrho)}$.

PROOF. Routine induction with respect to t. □

Formulas are assigned the truth values 0 and 1, relative to a given valuation. In order to make the following definition easier to understand, recall from Chapter 2 that $a \Rightarrow b$ is defined in a Boolean algebra as $-a \sqcup b$, which is 1 when $a = 0$ or $b = 1$. Observe also that for $X \subseteq \{0, 1\}$:

$$\inf X = 0 \quad \text{iff} \quad 0 \in X;$$
$$\sup X = 1 \quad \text{iff} \quad 1 \in X.$$

8.4.5. DEFINITION. Let $\mathcal{A}$ be an ordinary model and let ϱ be a valuation in $\mathcal{A}$. For each formula φ, the *value* $[\![\varphi]\!]_\varrho$ of φ with respect to ϱ is given by:

$$\begin{aligned}
[\![rt_1 \ldots t_n]\!]_\varrho &= r^{\mathcal{A}}([\![t_1]\!]_\varrho, \ldots, [\![t_n]\!]_\varrho); \\
[\![\bot]\!]_\varrho &= 0; \\
[\![\varphi \vee \psi]\!]_\varrho &= [\![\varphi]\!]_\varrho \sqcup [\![\psi]\!]_\varrho; \\
[\![\varphi \wedge \psi]\!]_\varrho &= [\![\varphi]\!]_\varrho \sqcap [\![\psi]\!]_\varrho; \\
[\![\varphi \to \psi]\!]_\varrho &= [\![\varphi]\!]_\varrho \Rightarrow [\![\psi]\!]_\varrho; \\
[\![\forall b \varphi]\!]_\varrho &= \inf\{[\![\varphi]\!]_{\varrho(b \mapsto \mathbf{a})} \mid \mathbf{a} \in \mathcal{A}\}; \\
[\![\exists b \varphi]\!]_\varrho &= \sup\{[\![\varphi]\!]_{\varrho(b \mapsto \mathbf{a})} \mid \mathbf{a} \in \mathcal{A}\},
\end{aligned}$$

where the operations $\sqcup$, $\sqcap$, $\Rightarrow$, inf and sup, and the constant 0 are in the two-element Boolean algebra of truth values.

8.4.6. NOTATION. We write:

- $\mathcal{A}, \varrho \models \varphi$, when $[\![\varphi]\!]_\varrho = 1$;
- $\mathcal{A} \models \varphi$, when $\mathcal{A}, \varrho \models \varphi$, for all ϱ (that is, when φ is *true in* $\mathcal{A}$);

- $\mathcal{A}, \varrho \models \Gamma$, when $\mathcal{A}, \varrho \models \varphi$, for all $\varphi \in \Gamma$;
- $\mathcal{A} \models \Gamma$, when $\mathcal{A} \models \varphi$, for all $\varphi \in \Gamma$;
- $\models \varphi$, when $\mathcal{A} \models \varphi$, for all $\mathcal{A}$ (that is, when φ is a classical *tautology*);
- $\Gamma \models \varphi$, when $\mathcal{A}, \varrho \models \varphi$, for all $\mathcal{A}$ and ϱ with $\mathcal{A}, \varrho \models \Gamma$.

For a given model $\mathcal{A}$, the symbol $\text{Th}(\mathcal{A})$ denotes the *theory of* $\mathcal{A}$, i.e. the set of all sentences true in $\mathcal{A}$. That is, $\text{Th}(\mathcal{A}) = \{\varphi \mid \text{FV}(\varphi) = \varnothing \text{ and } \mathcal{A} \models \varphi\}$.

8.4.7. THEOREM (Completeness). *In classical predicate calculus the conditions $\Gamma \vdash \varphi$ and $\Gamma \models \varphi$ are equivalent.*

PROOF. Exercise 8.16. □

The above applies to $\{0,1\}$-structures, but can be easily generalized to $\mathcal{P}(X)$-structures, for an arbitrary set X. The only difference is that we must use the Boolean algebra operations in $\mathcal{P}(X)$ to define values of formulas, rather than operations in $\{0,1\}$. What concerns sup and inf we take the generalized union $\bigcup$ and intersection $\bigcap$.

One can go further and postulate values of formulas in an arbitrary Boolean algebra $\mathcal{B}$. This will work as well, provided $\mathcal{B}$ is a *complete* algebra, i.e. all infinite sup's and inf's exist in $\mathcal{B}$. (Otherwise, values of some quantified formulas may be undefined.)

The reader is invited to verify (Exercises 8.9 and 8.10) that these generalizations do not change the set of classically correct entailments $\Gamma \models \varphi$, i.e. the completeness theorem remains true.

8.5. Algebraic semantics of intuitionistic logic

An obvious idea for adapting the algebraic approach to intuitionistic logic is to replace complete Boolean algebras by complete Heyting algebras.

8.5.1. DEFINITION. Let $\mathcal{H} = \langle H, \sqcup, \sqcap, \Rightarrow, -, 0, 1 \rangle$ be a complete Heyting algebra, and let $\mathcal{A}$ be an H-structure.

(i) The values of terms and formulas with respect to any valuation ϱ are defined as in Definition 8.4.5, but the operations $\sqcup$, $\sqcap$, inf, sup and $\Rightarrow$ and constants $0, 1$ are in $\mathcal{H}$.

(ii) The terminology and notation of 2.4.6 and 8.4.6 is used accordingly.

An example of a complete Heyting algebra is the algebra

$$\mathcal{O}(\mathcal{T}) = \langle \mathcal{O}(\mathcal{T}), \cap, \cup, \Rightarrow, \sim, \varnothing, \mathcal{T} \rangle$$

of open sets of a topological space T (in particular a metric space). Recall from Chapter 2 that the lattice operations ($\cap$ and $\cup$) in $\mathcal{O}(T)$ are set-theoretic, and that

$$A \Rightarrow B = \text{Int}(-A \cup B) \quad \text{and} \quad \sim A = \text{Int}(-A).$$

The supremum in $\mathcal{O}(T)$ is the set-theoretic $\bigcup$, and

$$\inf\{A_i \mid i \in I\} = \text{Int}(\bigcap\{A_i \mid i \in I\}).$$

8.5.2. NOTATION. The symbol $\models_T$ is understood as $\models$ restricted to Heyting algebras of the form $\mathcal{O}(T)$, where T is a topological space. For instance, we write $\models_T \varphi$ when $\mathcal{O}(T) \models \varphi$ holds for all topological spaces T.

8.5.3. PROPOSITION. *Let p be a nullary and let q, r be unary relation symbols. The following formulas (cf. Example 8.2.2) are not valid.*

(i) $\neg\neg\forall a(ra \vee \neg ra)$;

(ii) $\neg\forall a r a \to \exists a \neg r a$;

(iii) $\exists b(rb \to \forall a r a)$;

(iv) $\exists a(\exists b r b \to r a)$;

(v) $(\forall a r a \to p) \to \exists a(ra \to p)$;

(vi) $\exists a \exists b(ra \to qb) \to \exists a(ra \to qa)$.

(vii) $\forall a(ra \vee \neg ra) \to \exists a r a \vee \forall a \neg r a$;

(viii) $\forall a(ra \vee \neg ra) \wedge \neg\neg\exists a r a \to \exists a r a$;

(ix) $\forall a(p \vee ra) \to p \vee \forall a r a$;

(x) $\forall a \neg\neg r a \to \neg\neg \forall a r a$;

PROOF. We show counterexamples to (i), (vii) and (ix), leaving the rest to the reader (Exercise 8.20). In each case we take $A = \mathbb{N} - \{0\}$ as the domain of our $\mathcal{H}$-structure $\mathcal{A}$.

(i) Let $\mathcal{H} = \mathcal{O}(\mathbb{R})$ and let $\{w_i \mid i \in \mathbb{N} - \{0\}\}$ be some enumeration of all rational numbers. Define $r^{\mathcal{A}}$ so that $r^{\mathcal{A}}(n) = \mathbb{R} - \{w_n\}$, a full line without one rational number. Then $\sim r^{\mathcal{A}}(n) = \emptyset$, and thus $\sim r^{\mathcal{A}}(n) \cup r^{\mathcal{A}}(n) = r^{\mathcal{A}}(n)$. The intersection $\bigcap\{r^{\mathcal{A}}(n) \mid n \in \mathbb{N}\}$ has an empty interior, so the value of the formula $\forall a(ra \vee \neg ra)$ is the empty set. It remains to observe that $\sim\sim\emptyset = \emptyset$.

(vii) This time choose $\mathcal{H} = \mathcal{O}(\mathbb{Q})$. Take an increasing sequence ζ_n of negative irrational numbers, converging to zero. Let $r^{\mathcal{A}}(n) = (-\infty, \zeta_n) \cap \mathbb{Q}$. Then $\sim r^{\mathcal{A}}(n) = (\zeta_n, \infty) \cap \mathbb{Q}$ and $r^{\mathcal{A}}(n) \cup \sim r^{\mathcal{A}}(n) = \mathbb{Q}$. The intersection of all these sets is thus also equal to $\mathbb{Q}$. The union of all $r^{\mathcal{A}}(n)$ (the value of $\exists a(ra)$)

is the set $(-\infty, 0) \cap \mathbb{Q}$. The intersection of all $\sim r^A(n)$ is $[0, \infty) \cap \mathbb{Q}$, but this set is not open, so the value of $\forall a(\neg ra)$ is only $(0, \infty) \cap \mathbb{Q}$. We conclude that zero is missing at the right hand side.

(ix) Take $\mathcal{H} = \mathcal{O}(\mathbb{R})$ again. Let $p^A = \mathbb{R} - \{0\}$ and let $r^A(n) = (-\frac{1}{n}, \frac{1}{n})$. The value of the left-hand side is $\mathbb{R}$ because each set $p^A \cup r^A(n)$ covers the whole line. The value of the right-hand side is p^A, because the infimum of all $r^A(n)$ is empty. (Their intersection, the one-point set $\{0\}$, is not open.) □

Some preparation is needed to prove completeness. The following is an extension of Lemma 8.4.4.

8.5.4. LEMMA.

(i) If $\varrho_1(a) = \varrho_2(a)$ for all $a \in \mathrm{FV}(\varphi)$, then $[\![\varphi]\!]_{\varrho_1} = [\![\varphi]\!]_{\varrho_2}$.

(ii) For all φ and t we have $[\![\varphi[a := t]]\!]_{\varrho} = [\![\varphi]\!]_{\varrho(a \mapsto [\![t]\!]_\varrho)}$.

PROOF. Induction with respect to φ. As an example we consider the case of $\varphi = \forall b \psi$ in part (ii), where it is of course assumed that $b \notin \mathrm{FV}(t)$. Thus,

$$[\![\varphi[a := t]]\!]_\varrho = [\![\forall b \psi[a := t]]\!]_\varrho = \inf\{[\![\psi[a := t]]\!]_{\varrho(b \mapsto \mathbf{b})} \mid \mathbf{b} \in \mathcal{A}\}.$$

By the induction hypothesis $[\![\psi[a := t]]\!]_{\varrho(b \mapsto \mathbf{b})}$ is the same as $[\![\psi]\!]_{\varrho'}$ where $\varrho' = \varrho(b \mapsto \mathbf{b})(a \mapsto [\![t]\!]_{\varrho(b \mapsto \mathbf{b})})$. But $b \notin \mathrm{FV}(t)$, so $[\![t]\!]_{\varrho(b \mapsto \mathbf{b})} = [\![t]\!]_\varrho$, and thus $\varrho' = \varrho(b \mapsto \mathbf{b})(a \mapsto [\![t]\!]_\varrho) = \varrho(a \mapsto [\![t]\!]_\varrho)(b \mapsto \mathbf{b})$.

On the other hand, we have $[\![\varphi]\!]_{\varrho(a \mapsto [\![t]\!]_\varrho)} = \inf\{[\![\psi]\!]_{\varrho(a \mapsto [\![t]\!]_\varrho)(b \mapsto \mathbf{b})} \mid \mathbf{b} \in \mathcal{A}\}$, and we happily discover that we deal with the same set again. □

The next lemma is a crucial step in the completeness proof to follow. It guarantees that whenever we deal with an arbitrary Heyting algebra $\mathcal{H}$, we can always "add" all the missing sup's and inf's, without affecting those that are already available in $\mathcal{H}$. Note the subtlety: an embedding of a Heyting algebra into a complete Heyting algebra does not in general have to preserve all infinite sup's and inf's (Exercise 8.11).

8.5.5. LEMMA. *Let $\mathcal{H}$ be a Heyting algebra. There exists a complete Heyting algebra $\mathcal{G}$, such that $\mathcal{H}$ is a subalgebra of $\mathcal{G}$, and, in addition, $\inf_\mathcal{G} A = \inf_\mathcal{H} A$ (resp. $\sup_\mathcal{G} A = \sup_\mathcal{H} A$), whenever $\inf_\mathcal{H} A$ (resp. $\sup_\mathcal{H} A$) exists.* □

PROOF. As this result is quite technical and requires a bit of effort, we omit the proof here. See Exercises 8.12–8.13. □

8.5.6. THEOREM (Completeness).
The conditions $\Gamma \vdash \varphi$ and $\Gamma \models \varphi$ are equivalent.

210   CHAPTER 8. FIRST-ORDER LOGIC

PROOF. The proof of soundness (the left-to-right part) is similar to the proof of the corresponding part of Theorem 2.4.7. By induction with respect to proofs, we show that $\Gamma \vdash \varphi$ implies $[\![\Gamma]\!]_\varrho \leq [\![\varphi]\!]_\varrho$, for any valuation ϱ in any $\mathcal{H}$-structure $\mathcal{A}$. We consider the case when the proof ends with an application of rule ($\exists$E), leaving other cases to the reader. (Use Lemma 8.5.4.)

Suppose $\Gamma \vdash \varphi$ was obtained from $\Gamma \vdash \exists a\psi$ and $\Gamma, \psi \vdash \varphi$. Let $\mathbf{a} \in \mathcal{A}$. Since $a \notin \mathrm{FV}(\Gamma, \varphi)$, the induction hypothesis applied to $\Gamma, \psi \vdash \varphi$ with $\varrho(a \mapsto \mathbf{a})$ yields $[\![\Gamma]\!]_\varrho \sqcap [\![\psi]\!]_{\varrho(a \mapsto \mathbf{a})} \leq [\![\varphi]\!]_\varrho$, whence $[\![\psi]\!]_{\varrho(a \mapsto \mathbf{a})} \leq [\![\Gamma]\!]_\varrho \Rightarrow [\![\varphi]\!]_\varrho$. As this holds for all $\mathbf{a}$, we have $\sup\{[\![\psi]\!]_{\varrho(a \mapsto \mathbf{a})} \mid \mathbf{a} \in \mathcal{A}\} \leq [\![\Gamma]\!]_\varrho \Rightarrow [\![\varphi]\!]_\varrho$.

On the other hand, $[\![\Gamma]\!]_\varrho \leq \sup\{[\![\psi]\!]_{\varrho(a \mapsto \mathbf{a})} \mid \mathbf{a} \in \mathcal{A}\}$, by the induction hypothesis applied to $\Gamma \vdash \exists a\psi$. We can conclude that $[\![\Gamma]\!]_\varrho \leq [\![\Gamma]\!]_\varrho \Rightarrow [\![\varphi]\!]_\varrho$, from which $[\![\Gamma]\!]_\varrho \leq [\![\varphi]\!]_\varrho$ follows easily.

To prove the "hard" part, we investigate the properties of the Lindenbaum algebra $\mathcal{L}_\Gamma$, defined in exactly the same way as for the propositional case (Section 2.4). We consider the equivalence relation

$$\varphi \sim_\Gamma \psi \quad \text{iff} \quad \Gamma \vdash \varphi \to \psi \text{ and } \Gamma \vdash \psi \to \varphi,$$

in the set Φ_Σ of all formulas. The elements of $\mathcal{L}_\Gamma$ are equivalence classes of $\sim_\Gamma$ (abbreviated as $\sim$), and the relation $\leq_\Gamma$ (abbreviated $\leq$), defined by

$$[\varphi]_\sim \leq_\Gamma [\psi]_\sim \quad \text{iff} \quad \Gamma \vdash \varphi \to \psi$$

is a partial order with the least element $[\bot]_\sim$, and the top element $[\top]_\sim$, where $\top = \bot \to \bot$. Algebraic operations are defined as expected:

$$\begin{aligned}
{[\varphi]_\sim \sqcup [\psi]_\sim} &= [\varphi \vee \psi]_\sim; \\
{[\varphi]_\sim \sqcap [\psi]_\sim} &= [\varphi \wedge \psi]_\sim; \\
{[\varphi]_\sim \Rightarrow [\psi]_\sim} &= [\varphi \to \psi]_\sim; \\
-[\varphi]_\sim &= [\neg \varphi]_\sim; \\
0 &= [\bot]_\sim; \\
1 &= [\top]_\sim.
\end{aligned}$$

Quite like in Chapter 2, the reader may now easily check that the structure $\mathcal{L}_\Gamma = \langle \Phi_\Sigma/\sim, \sqcup, \sqcap, \Rightarrow, -, 0, 1 \rangle$ is a Heyting algebra. In addition, we have:

$$[\forall a \varphi]_\sim = \inf\{[\varphi[a := t]]_\sim \mid t \text{ is a term}\}; \tag{i}$$
$$[\exists a \varphi]_\sim = \sup\{[\varphi[a := t]]_\sim \mid t \text{ is a term}\}. \tag{ii}$$

We check property (i). Of course $[\forall a \varphi]_\sim$ is a lower bound (axiom (Q1)). To show that it is the greatest one, assume that $[\psi]_\sim \leq [\varphi[a := t]]_\sim$ holds for all t. In particular, $[\psi]_\sim \leq [\varphi[a := b]]_\sim$, where b is a fresh variable.[3]

---

[3] One can assume without loss of generality that there exist variables not occurring free in Γ nor φ (Exercise 8.14).

8.5. Algebraic semantics of intuitionistic logic

This means that $\Gamma \vdash \psi \to \varphi[a := b]$, and since $b \notin \mathrm{FV}(\psi)$, we also have $\Gamma \vdash \psi \to \forall b \varphi[a := b]$, which is equivalent to $[\psi]_\sim \leq [\forall a \varphi]_\sim$. The proof of (ii) is similar, and uses axioms (Q2) and (Q4).

Unfortunately, the algebra $\mathcal{L}_\Gamma$ does not have to be complete. But by Lemma 8.5.5 we can assume that it is a subalgebra of a complete algebra $\mathcal{H}$ and that all the sup's and inf's are the same.

Once we have arrived at this point, we can begin the proper proof. Suppose that $\Gamma \nvdash \varphi$. Then $[\varphi]_\sim \neq 1$. In order to show that $\Gamma \nvDash \varphi$ we construct a term $\mathcal{H}$-model $\mathcal{A}$. For any k-ary relation symbol, we define

$$r^{\mathcal{A}}(t_1, \ldots, t_k) = [rt_1, \ldots, t_k]_\sim.$$

Then we consider the simplest possible valuation in $\mathcal{A}$, namely $\varrho(a) = a$, for all variables a. Clearly, $[\![t]\!]_\varrho = t$ for any term t. For all formulas ψ we prove

$$[\![\psi]\!]_\varrho = [\psi]_\sim \qquad (*)$$

by induction. If ψ is an atomic formula $rt_1 \ldots t_k$ then straight from the definitions we have $[\![\psi]\!]_\varrho = r^{\mathcal{A}}(t_1, \ldots, t_k) = [rt_1, \ldots, t_k]_\sim$. Other cases are equally routine. For instance, if $\psi = \exists a \vartheta$ then

$$[\![\psi]\!]_\varrho = \sup\{[\![\vartheta]\!]_{\varrho(a \mapsto t)} \mid t \text{ a term}\} = \sup\{[\![\vartheta[a := t]]\!]_\varrho \mid t \text{ a term}\},$$

by Lemma 8.5.4. From the induction hypothesis, the latter is the same as $\sup\{[\vartheta[a := t]]_\sim \mid t \text{ a term}\}$, and thus $[\![\psi]\!]_\varrho = [\exists a \vartheta]_\sim$ by property (ii).

In particular, it follows from $(*)$ that $[\![\varphi]\!]_\varrho = [\varphi]_\sim \neq 1$. Thus $\mathcal{A}, \varrho \nvDash \varphi$. But $\mathcal{A}, \varrho \vDash \Gamma$, so we conclude that $\Gamma \nvDash \varphi$. □

All counterexamples for Proposition 8.5.3 were found in algebras $\mathcal{O}(\mathcal{T})$ of open subsets of various topological spaces $\mathcal{T}$. In fact, if there is a counterexample at all then there is one in a topological space. The completeness theorem 8.5.6 can be strengthened as follows.

8.5.7. THEOREM. *The conditions $\Gamma \vdash \varphi$ and $\Gamma \vDash_T \varphi$ are equivalent.* □

The proof of Theorem 8.5.7 is based on the existence of embeddings of countable Heyting algebras into certain metric spaces. We refer the interested reader to Chapter X of [410].

In many cases it is convenient to work with open subsets of $\mathbb{R}$. For the propositional case this is perfectly sufficient, cf. Theorem 2.4.11. However, in the first-order case the algebra $\mathcal{O}(\mathbb{R})$ is not enough, see Exercise 8.22.

8.5.8. REMARK. An easy variation of the proof of Theorem 8.5.6 shows the completeness of classical predicate calculus with respect to Boolean algebra semantics. This can further be refined to prove completeness with respect to the two-valued semantics (see Exercises 8.15 and 8.16).

8.6. Kripke semantics

The algebraic semantics for intuitionistic logic discussed in Section 8.5 required a generalization of classical first-order models to $\mathcal{H}$-structures. An alternative way of relaxing the definition of classical semantics is to keep the ordinary notion of a model, but assume that a model (the "world") may change as our knowledge progresses. We may add new elements to the domain, as well as accept more atomic propositions. First-order Kripke models are built from such "possible worlds".

8.6.1. DEFINITION. A structure $\mathcal{B} = \langle B, f_1^\mathcal{B}, \ldots, f_n^\mathcal{B}, r_1^\mathcal{B}, \ldots, r_m^\mathcal{B} \rangle$ is an *extension* of $\mathcal{A} = \langle A, f_1^\mathcal{A}, \ldots, f_n^\mathcal{A}, r_1^\mathcal{A}, \ldots, r_m^\mathcal{A} \rangle$ iff the following hold:

- $A \subseteq B$;
- $r_i^\mathcal{A} \subseteq r_i^\mathcal{B}$, for all i;
- $f_i^\mathcal{A} \subseteq f_i^\mathcal{B}$, for all i.

Thus, $\mathcal{B}$ extends the "world" $\mathcal{A}$ by enriching the domain of available objects and by adding more information about known objects. We write $\mathcal{A} \subseteq \mathcal{B}$ to express that $\mathcal{B}$ is an extension of $\mathcal{A}$.

8.6.2. DEFINITION. A *Kripke model* for first-order logic is a triple of the form $\mathcal{C} = \langle C, \leq, \{\mathcal{A}_c \mid c \in C\}\rangle$, where C is a non-empty set of *states*, $\leq$ is a partial order in C, and the $\mathcal{A}_c$ are classical structures such that

$$\text{if } c \leq c' \text{ then } \mathcal{A}_c \subseteq \mathcal{A}_{c'}.$$

Now let φ be a formula such that all free variables of φ are among $\vec{a}$, and let ϱ be a valuation in $\mathcal{A}_c$. The relation $c, \varrho \Vdash \varphi$ is defined by induction.

- $c, \varrho \Vdash rt_1 \ldots t_n$ iff $\mathcal{A}_c, \varrho \models rt_1 \ldots t_n$ (classically);
- $c, \varrho \not\Vdash \bot$;
- $c, \varrho \Vdash \varphi \vee \psi$ iff $c, \varrho \Vdash \varphi$ or $c, \varrho \Vdash \psi$;
- $c, \varrho \Vdash \varphi \wedge \psi$ iff $c, \varrho \Vdash \varphi$ and $c, \varrho \Vdash \psi$;
- $c, \varrho \Vdash \varphi \to \psi$ iff $c', \varrho \Vdash \psi$, for all $c' \geq c$ such that $c', \varrho \Vdash \varphi$;
- $c, \varrho \Vdash \exists a \varphi$ iff $c, \varrho(a \mapsto \mathbf{a}) \Vdash \varphi$, for some $\mathbf{a} \in \mathcal{A}_c$;
- $c, \varrho \Vdash \forall a \varphi$ iff $c', \varrho(a \mapsto \mathbf{a}) \Vdash \varphi$, for all $c' \geq c$ and all $\mathbf{a} \in \mathcal{A}_{c'}$.

The symbol $\Vdash$ is now used as usual, in particular $\Gamma \Vdash \varphi$ means that for every Kripke model $\mathcal{C}$ and every $c \in C$, the condition $c, \varrho \Vdash \Gamma$ implies $c, \varrho \Vdash \varphi$.

8.6. KRIPKE SEMANTICS

8.6.3. EXAMPLE. Let $\mathcal{C} = \langle \mathbb{N}, \leq, \{\mathcal{A}_k \mid k \in \mathbb{N}\}\rangle$, where $\leq$ is the standard order on $\mathbb{N}$. For $k \in \mathbb{N}$, let $\mathcal{A}_k = \langle A_k, r_k, p_k \rangle$, where $A_k = \{0, \ldots, k^2\}$, $r_k = \{n \in A_k \mid n < k\}$ and $p_k = \{(n, m) \in A_k \times A_k \mid 2n = m\}$. Then $\mathcal{C} \Vdash \forall a(ra \to \exists b p a b)$, because $2n < k^2$ for all $n < k$.

As expected, substitution again "commutes" with semantics.

8.6.4. LEMMA. *The conditions $c, \varrho \Vdash \varphi[a := t]$ and $c, \varrho(a \mapsto \llbracket t \rrbracket_\varrho) \Vdash \varphi$ are equivalent.*

PROOF. Induction with respect to φ. □

As mentioned above, the idea behind Kripke models is to represent possible increase in knowledge about facts by considering many possible worlds. In the first-order case, expanding the knowledge includes also expanding the set of known objects (the domain of $\mathcal{A}_c$ may grow as c grows). This generalization is necessary for completeness, see Exercise 8.26.

Observe that classical first-order semantics is a special case of Kripke semantics, obtained by restricting attention to one-state Kripke models only. Also, the Kripke models for propositional logic can be seen as a special case of the above, when the (infinite!) signature consists of all propositional variables understood as nullary relation symbols.

Unfortunately, the translation from Heyting semantics to Kripke semantics used in the proof of Theorem 2.5.9 does not generalize to first-order logic. In a sense, this translation is too "rough" to properly handle the upper and lower bounds used to interpret quantifiers. The first-order case is more delicate, and we need a separate proof.

8.6.5. DEFINITION. Let Γ be a set of formulas over a signature Σ. We say that Γ is *prime with respect to* Σ when the following conditions hold for all formulas over Σ:

(i) If $\Gamma \vdash \varphi \vee \psi$ then either $\Gamma \vdash \varphi$ or $\Gamma \vdash \psi$;

(ii) If $\Gamma \vdash \exists a \varphi(a)$ then $\Gamma \vdash \varphi(o)$, for some constant $o \in \Sigma$.

The following is the crucial step in our completeness proof.

8.6.6. LEMMA (Henkin). *Assume that $\Gamma \nvdash \varphi$, for some Γ and φ over a signature Σ, and let C be a countably infinite set of constants not in Σ. Then there exists a set Γ' of formulas (over the signature $\Sigma \cup C$) which is prime with respect to $\Sigma \cup C$ and such that $\Gamma \subseteq \Gamma'$ and $\Gamma' \nvdash \varphi$.*

PROOF. Let $\{\zeta_m \mid m \in \mathbb{N}\}$ be an enumeration of all formulas over the signature $\Sigma \cup C$. We define Γ' as the union of an increasing sequence of sets Γ_n, beginning with $\Gamma_0 = \Gamma$. The definition of Γ_n is by induction and the induction hypothesis consists of two conditions. The first condition is $\Gamma_n \nvdash \varphi$, and the second is that only a finite number of constants from C occurs in Γ_n. Suppose Γ_n has already been defined so that these conditions hold. Let $m = \mathbf{m}(n)$ be the least number such that $\Gamma_n \vdash \zeta_m$, and

a) Either $\zeta_m = \exists a\psi(a)$ and $\Gamma_n \nvdash \psi(o)$ for all $o \in C$;

b) Or $\zeta_m = \psi \vee \vartheta$ and $\Gamma_n \nvdash \psi$ and $\Gamma_n \nvdash \vartheta$.

That is, ζ_m is the first formula that violates the primeness of Γ_n (cf. Definition 8.6.5). If such an m does not exist, we can simply take $\Gamma_{n+1} = \Gamma_n$. Otherwise, in case (a) we define $\Gamma_{n+1} = \Gamma_n \cup \{\psi(o)\}$, where o is any constant from C, not occurring in Γ_n and ζ_m. There must be such a constant because only finitely many are in Γ_n. In case (b) we define $\Gamma_{n+1} = \Gamma_n \cup \{\psi\}$, unless it happens that $\Gamma_n, \psi \vdash \varphi$. Then we choose $\Gamma_{n+1} = \Gamma_n \cup \{\vartheta\}$.

Clearly, Γ_{n+1} has at most one more constant than Γ_n, thus still finitely many. Suppose $\Gamma_{n+1} \vdash \varphi$. In case (a) this means that $\Gamma_n, \psi(o) \vdash \varphi$. Since o is a fresh constant, we can equally well state that $\Gamma_n, \psi(a) \vdash \varphi$, where a is a fresh variable. But then we would have $\Gamma_n, \exists a\psi(a) \vdash \varphi$. Since $\Gamma_n \vdash \exists a\psi(a)$, we get $\Gamma_n \vdash \varphi$, contradicting the induction hypothesis.

In case (b) we would have $\Gamma_n, \psi \vdash \varphi$ and $\Gamma_n, \vartheta \vdash \varphi$. It follows that $\Gamma_n, \psi \vee \vartheta \vdash \varphi$, so that $\Gamma_n \vdash \varphi$ and we again have a contradiction.

Finally, define $\Gamma' = \bigcup \{\Gamma_n \mid n \in \mathbb{N}\}$. Suppose $\Gamma' \vdash \varphi$. A proof must be finite, so $\Gamma_n \vdash \varphi$ would have to hold for some n.

Assume now that $\Gamma' \vdash \psi \vee \vartheta$, but $\Gamma' \nvdash \psi$ and $\Gamma' \nvdash \vartheta$. Again, we must have $\Gamma_n \vdash \psi \vee \vartheta$, for some n. Let m be such that $\zeta_m = \psi \vee \vartheta$. We claim that $\Gamma_{n+i} \vdash \psi$ or $\Gamma_{n+i} \vdash \vartheta$, for some $i \leq m$. Indeed, consider the sequence of numbers $\mathbf{m}(n), \mathbf{m}(n+1), \mathbf{m}(n+2), \ldots$. These numbers are all distinct, so eventually $\mathbf{m}(n+i) > m$, for some i. But this means that ζ_m is no longer a witness of non-primality for Γ_{n+i}. In a similar way we show that $\Gamma' \vdash \exists a\varphi(a)$ implies $\Gamma' \vdash \varphi(o)$, for some o. □

8.6.7. THEOREM. *The conditions $\Gamma \vdash \varphi$ and $\Gamma \Vdash \varphi$ are equivalent.*

PROOF. For the implication from left to right we proceed in a similar way as in the proof of Theorem 2.5.9. For the other direction, we assume that $\Gamma \nvdash \varphi$ and we show $\Gamma \nVdash \varphi$ by constructing an appropriate Kripke model.

Let $C_1 \subseteq C_2 \subseteq C_3 \subseteq \cdots$ be an infinite sequence of countably infinite sets of constants, not in Σ, such that each $C_{n+1} - C_n$ is infinite. The states of our Kripke model $\mathcal{C}$ are all pairs $\langle \Delta, C_n \rangle$, where Δ is a set of formulas in the signature $\Sigma \cup C_n$, prime with respect to $\Sigma \cup C_n$. The ordering relation is

$$\langle \Delta, C_n \rangle \leq \langle \Delta', C_m \rangle \quad \text{iff} \quad \Delta \subseteq \Delta' \text{ and } n \leq m.$$

The structure associated to a state $\langle \Delta, C_n \rangle$ is the classical term model over the signature $\Sigma \cup C_n$, where

$$r^{\mathcal{A}}(t_1, \ldots, t_k) \text{ holds if and only if } \Delta \vdash rt_1, \ldots, t_k,$$

for all k and all k-ary relation symbols r. Let $\varrho(a) = a$, for all variables a. By induction with respect to the number of logical symbols in ψ we now prove the following claim:

$$\langle \Delta, C_n \rangle, \varrho \Vdash \psi \text{ if and only if } \Delta \vdash \psi, \qquad (*)$$

for all formulas ψ over the signature $\Sigma \cup C_n$.

For atomic ψ the claim follows straight from the definition. The case when ψ is a conjunction is a routine application of the induction hypothesis, and similarly for disjunction, but now we use the fact that Δ is prime.

Suppose $\psi = \psi_1 \to \psi_2$, and assume $\Delta \nvdash \psi$. Then $\Delta, \psi_1 \nvdash \psi_2$. From Lemma 8.6.6 we know that there is a Δ', prime with respect to $\Sigma \cup C_{n+1}$ and such that $\Delta \cup \{\psi_1\} \subseteq \Delta'$ and $\Delta' \nvdash \psi_2$. By the induction hypothesis we have $\langle \Delta', C_{n+1} \rangle, \varrho \Vdash \psi_1$ and $\langle \Delta', C_{n+1} \rangle, \varrho \nVdash \psi_2$. Thus $\langle \Delta, C_n \rangle, \varrho \nVdash \psi$. The converse implication for $\psi = \psi_1 \to \psi_2$ is easy.

Now let $\psi = \forall a \psi_1(a)$, and assume that $\Delta \nvdash \psi$. We claim that $\Delta \nvdash \psi_1(o)$, for any constant $o \in C_{n+1} - C_n$. Indeed, otherwise there is a finite subset $\Delta' \subseteq \Delta$ with $\Delta' \vdash \psi_1(o)$. We can replace o by a variable $a \notin \mathrm{FV}(\Delta', \psi_1)$ and generalize to $\Delta \vdash \forall a \psi_1(a)$. Using Lemma 8.6.6, we now extend the set Δ to a set Δ', prime with respect to $\Sigma \cup C_{n+2}$, and such that $\Delta' \nvdash \psi_1(o)$. From the induction hypothesis we infer $\langle \Delta', C_{n+2} \rangle, \varrho \nVdash \psi_1(o)$, so that $\langle \Delta, C_n \rangle, \varrho \nVdash \psi$. The converse implication is easy.

For existential formulas the implication from right to left follows from primality. We prove the other implication. Assume that $\langle \Delta, C_n \rangle, \varrho \Vdash \exists a \psi_1$. We have $\langle \Delta, C_n \rangle, \varrho(a \mapsto t) \Vdash \psi_1$, where t is a term over $\Sigma \cup C_n$. By Lemma 8.6.4, we have $\langle \Delta, C_n \rangle, \varrho \Vdash \psi_1[a := t]$ and thus $\Delta \vdash \psi_1[a := t]$ by the induction hypothesis. We conclude that $\Delta \vdash \exists a \psi_1$. This completes the proof of $(*)$.

Now if $\Gamma \nvdash \varphi$ then $\Gamma \subseteq \Delta$ for some Δ, prime with respect to $\Sigma \cup C_1$, and such that $\Delta \nvdash \varphi$. From $(*)$ it follows that $\langle \Delta, C_1 \rangle \nVdash \varphi$. Since $\langle \Delta, C_1 \rangle, \varrho \Vdash \Gamma$, we conclude that $\Gamma \nVdash \varphi$. □

8.7. Lambda-calculus

In this section we assign a Curry-Howard interpretation to first-order intuitionistic logic. That is, we define a typed lambda-calculus λP_1, such that the inhabited types of λP_1 correspond to first-order intuitionistic theorems.

In order to understand the computational contents of a constructive proof in predicate calculus, we must be able to interpret first-order formulas as types. Let us begin with atomic formulas. Suppose r is a unary relation symbol. Then, for any algebraic term t, the expression rt should be a type. This type *depends* on the choice of t, so we call it a *dependent type*.

The relation r itself is not a type, but a *type constructor*, i.e. a function yielding a type for a given individual argument. (More generally, we use the expression *type constructor* to denote any operation returning types as results, no matter what its arguments are. We will see constructors of other kinds in the chapters to follow.)

For instance, one may consider a type **array**(n) of integer arrays of length n. This type depends on a choice of n. The operator **array** makes a type from an integer, and corresponds, under the Curry-Howard isomorphism, to a unary predicate.

A definition of an object of type **array**(n) may happen to be uniform in n, i.e. we may have a generic procedure *Zeros* that turns any n : **int** into an array of zeros of length n. The type of such a procedure could be written as $\forall a$:**int**. **array**(a). Assuming for simplicity that all individual objects are of type **int**, we could as well write it as $\forall a.$ **array**(a).

In general, a type of the form $\forall a \varphi(a)$ is a type of a function applicable to individuals[4] and returning an object of type $\varphi(t)$, for each argument t. That is, a universal type is similar to a set-theoretic *product* of the form $\Pi_t \varphi(t)$, and can be justifiably called a *product type*. Expressions of a product type can be constructed with help of λ-abstraction over individual variables, and may be applied to individual arguments quite like proof terms of type $\varphi \to \psi$ are applied to proof terms of type ψ. Thus, product behaves very much like a function type. Recalling again the analogy with set theory, the set of functions from I to B is a special case of a product $\Pi_{i \in I} B_i$, where all B_i are equal to B. We will see later (in Chapters 13 and 14) that in certain type systems, implication is also definable from the universal quantifier.

The interpretation of the existential quantifier is perhaps less obvious. The BHK interpretation suggests that $\exists a \varphi(a)$ should be understood as a *disjoint union* or *co-product* of types $\varphi(t)$, for all individual objects t. That is, objects of type $\exists a \varphi(a)$ are pairs consisting of an individual t and a proof M of $\varphi(t)$. This may be syntactically represented as $[t, M]_{\exists a \varphi}$. (In order to distinguish such pairs from members of ordinary product types, we use square brackets instead of angles.) However, the interface $\exists a \varphi$ does not reveal the actual object t used to construct M. The value of t is locally *encapsulated*, and not visible to the outside world. We can thus say that an object of an existential type is like an abstract type or a module with a "private" implementation of a "public" specification. An access to such an object is only possible via this public specification. We model it by an elimination operator of the form **let** $[a, y{:}\varphi] = M$ **in** N, where N is the external "context" communicating with a module $M : \exists a \varphi$ via the identifiers a and y. This operator is executed by "opening" the package M as follows. If $M = [t, M']_{\exists a \varphi}$ then a in N becomes bound to the value t and y to the implementation M' of the concrete type $\varphi(t)$. The body of N is then evaluated under these bindings.

8.7.1. EXAMPLE. Suppose that **int** is the only individual type available in a language and that **array** is a unary dependent type constructor. That is,

---

[4] All individuals are assumed here to be of the same base type, like in the one-sorted first-order logic.

8.7. LAMBDA-CALCULUS

array(n) is a type for each $n : $ **int**.

Then a function that adds all values in a given array could be given the following type:
$$sum_1 : \forall n \, (\mathbf{array}(n) \to \mathbf{int}).$$

A correct application of sum_1 takes the form $sum_1 \, n \, A$: **int**, where A is of type **array**(n). Alternatively, one can define **Array** $= \exists n \, \mathbf{array}(n)$ and design a summator of a different type:
$$sum_2 : \mathbf{Array} \to \mathbf{int}.$$

But now to sum up the entries of an array $A : \mathbf{array}(n)$ we should evaluate the expression $sum_2[n, A]$.

Similarly, the operation *append* that concatenates two arrays could be implemented as either of the following:

$$append_1 : \forall n \forall m \, (\mathbf{array}(n) \to \mathbf{array}(m) \to \mathbf{array}(n+m)),$$
$$append_2 : \mathbf{Array} \to \mathbf{Array} \to \mathbf{Array}.$$

See also Exercises 8.30 and 8.31.

GLOSSARY. In the following table we put together the corresponding notions from first-order logic and the system of lambda-calculus to be defined below. Compare this to the similar table in Chapter 4.

first-order logic	lambda-calculus
formulas	types
proofs	terms
relations	type constructors
atomic formula	dependent type
universal formula	product type
proof by generalization	abstraction $\lambda a.M$
proof by specialization	application Mt
existential type	abstract type or module
proof by $\exists$-introduction	encapsulation
proof by $\exists$-elimination	opening a package

We now define our (Church style) calculus λP_1.

8.7.2. DEFINITION.

(i) In system λP_1, we have two sorts of variables. Apart from individual variables used in types (formulas), we also have *proof variables*, or just "variables", occurring in terms (proofs).

(ii) By an *environment* we mean a set of declarations $(x : \varphi)$, where x is a proof variable and φ is a formula.

(iii) Types of system λP_1 are first-order formulas over a signature Σ containing infinitely many relation and function symbols of all arities.

Typing rules for the propositional connectives are as in the simply typed calculus (Figure 4.1). For quantifiers we choose the following syntax:

$$(\forall I) \; \frac{\Gamma \vdash M : \varphi}{\Gamma \vdash (\lambda a\, M) : \forall a \varphi} \qquad (\forall E) \; \frac{\Gamma \vdash M : \forall a \varphi}{\Gamma \vdash (Mt) : \varphi[a := t]}$$

$$(\exists I) \; \frac{\Gamma \vdash M : \varphi[a := t]}{\Gamma \vdash [t, M]_{\exists a \varphi} : \exists a \varphi} \qquad (\exists E) \; \frac{\Gamma \vdash M : \exists a \varphi \quad \Gamma, y : \varphi \vdash N : \psi}{\Gamma \vdash (\mathbf{let}\; [a, y{:}\varphi] = M \;\mathbf{in}\; N) : \psi}$$

where, as usual, $a \notin \mathrm{FV}(\Gamma)$ in rule $(\forall I)$, and $a \notin \mathrm{FV}(\Gamma) \cup \mathrm{FV}(\psi)$ in rule $(\exists E)$. That is, we have a new form of application, abstraction and pair. If there is no risk of a confusion, we abbreviate $[t, M]_{\exists a \varphi}$ by $[t, M]$, as we did already with $\mathbf{in}_1$ and $\mathbf{in}_2$.

8.7.3. EXAMPLE. The following terms are inhabitants of the Hilbert-style axioms (Q1)–(Q4):

(Q1) $\vdash \lambda x^{\forall a \varphi(a)}.xt \;:\; \forall a \varphi(a) \to \varphi(t)$;

(Q2) $\vdash \lambda x^{\varphi(t)} [t, x]_{\exists a \varphi} \;:\; \varphi(t) \to \exists a \varphi(a)$;

(Q3) $\vdash \lambda x^{\forall a(\varphi \to \psi)} \lambda y^\varphi \lambda a.\, xay \;:\; \forall a (\varphi \to \psi) \to \varphi \to \forall a \psi$;

(Q4) $\vdash \lambda x^{\forall a(\varphi \to \psi)} \lambda y^{\exists a \varphi}.\mathbf{let}\; [a, z{:}\varphi] = y \;\mathbf{in}\; xaz \;:\; \forall a (\varphi \to \psi) \to \exists a \varphi \to \psi$.

We have both individual and proof variables occurring free in proof terms. As these belong to two disjoint categories, the following definition of $\mathrm{FV}(M)$ does not cause any real loss of information.

$$\begin{aligned}
\mathrm{FV}(x) &= \{x\}; \\
\mathrm{FV}(\lambda x{:}\varphi\, M) &= \mathrm{FV}(\varphi) \cup (\mathrm{FV}(M) - \{x\}); \\
\mathrm{FV}(MN) &= \mathrm{FV}(M) \cup \mathrm{FV}(N); \\
\mathrm{FV}(\langle M, N \rangle) &= \mathrm{FV}(M) \cup \mathrm{FV}(N); \\
\mathrm{FV}(\pi_i(M)) &= \mathrm{FV}(M); \\
\mathrm{FV}(\mathbf{in}_1^{\varphi \vee \psi} M) &= \mathrm{FV}(M) \cup \mathrm{FV}(\varphi \vee \psi); \\
\mathrm{FV}(\mathbf{in}_2^{\varphi \vee \psi} M) &= \mathrm{FV}(M) \cup \mathrm{FV}(\varphi \vee \psi); \\
\mathrm{FV}(\mathbf{case}\; M \;\mathbf{of}\; [x]P \;\mathbf{or}\; [y]Q) &= \mathrm{FV}(M) \cup (\mathrm{FV}(P) - \{x\}) \cup (\mathrm{FV}(Q) - \{y\}); \\
\mathrm{FV}(\lambda a\, M) &= \mathrm{FV}(M) - \{a\}; \\
\mathrm{FV}(Mt) &= \mathrm{FV}(M) \cup \mathrm{FV}(t); \\
\mathrm{FV}([t, M]_{\exists a \varphi}) &= \mathrm{FV}(M) \cup \mathrm{FV}(t) \cup \mathrm{FV}(\exists a \varphi); \\
\mathrm{FV}(\mathbf{let}\; [a, y{:}\varphi] = M \;\mathbf{in}\; N) &= \mathrm{FV}(M) \cup (\mathrm{FV}(N) - \{a, y\}) \cup (\mathrm{FV}(\varphi) - \{a\}).
\end{aligned}$$

8.7. LAMBDA-CALCULUS

In addition to lambda-abstractions and **case**, variables may be bound also by **let**. The scope of y in **let** $[a, y:\varphi] = M$ **in** N is the term N, while the scope of a is N and φ. For instance, variables a, y are bound and b, x, z are free in the expression **let** $[a, y: P(a)] = [b, x]_{\exists a\, Pa}$ **in** zay. Alpha-conversion is defined so that the scope rules above are respected. And, as usual, we identify alpha-convertible terms. Details are left to the reader.

Beta-reduction is extended by the following rules. The first one is an obvious analogue of the ordinary beta rule. We call it *first-order beta-reduction*. The other represents the execution of a program N using an externally defined "module" $[t, M]$ accessible in N via an interface y of type $\varphi(a)$.

$$(\lambda a M)t \;\to\; M[a := t]$$
$$\textbf{let } [a, y:\varphi] = [t, M]_{\exists a\varphi} \textbf{ in } N \;\to\; N[a := t][y := M]$$

The following states the Curry-Howard correspondence between $\lambda \mathrm{P}_1$ and first-order logic.

8.7.4. PROPOSITION. *A first-order formula φ is intuitionistically valid if and only if there exists M such that $\vdash M : \varphi$ is derivable in $\lambda \mathrm{P}_1$.* □

PROOF. The proof is quite routine. Note however that one cannot in general require that the term M is closed, as in Propositions 4.1.1 and 6.2.6. Indeed, a normal inhabitant of $\forall a\varphi \to \exists a\varphi$ must be as follows:

$$\lambda x^{\forall a\varphi}[t, xt],$$

where t is an algebraic term. If there are no constants in the signature (and thus no closed algebraic terms) one must use a free variable. □

CONTRACTING MAP. First-order formulas are meant to express properties of individual objects. But the information contained in a first-order formula, and in its proof, can be split into two layers: the essentially "first-order" aspect (quantifiers and atoms) and the "propositional" aspect. An important idea, which we are now about to discuss for the first time, is that certain properties of predicate logic can be studied at the propositional level. To implement this idea we define a translation from first-order to propositional formulas, and a matching translation on proofs.

8.7.5. NOTATION. Assume that a fixed propositional variable p_r is associated with every relation constant r in the signature, and that this association is injective. In what follows, we simply identify r and p_r.

8.7.6. DEFINITION. The *contracting map* $\flat$ applies to types and proof terms, and is extended to contexts in the obvious way.

$$\begin{aligned}
\flat(rt_1\ldots t_n) &= r, \text{ a propositional variable;} \\
\flat(\bot) &= \bot; \\
\flat(\varphi \circ \psi) &= \flat(\varphi) \circ \flat(\psi), \text{ for } \circ \in \{\to, \vee, \wedge\}; \\
\flat(\nabla a \varphi) &= \flat(\varphi), \text{ for } \nabla \in \{\exists, \forall\}; \\
\flat(x) &= x; \\
\flat(\lambda x{:}\varphi\, M) &= \lambda x{:}\flat(\varphi).\flat(M); \\
\flat(MN) &= \flat(M)\flat(N); \\
\flat(\langle M, N \rangle) &= \langle \flat(M), \flat(N) \rangle; \\
\flat(\pi_i(M)) &= \pi_i(\flat(M)); \\
\flat(\text{in}_i\, M) &= \text{in}_i\, \flat(M); \\
\flat(\text{case } M \text{ of } [x]P \text{ or } [y]Q) &= \text{case } \flat(M) \text{ of } [x]\flat(P) \text{ or } [y]\flat(Q); \\
\flat(\varepsilon_\psi(M)) &= \varepsilon_{\flat(\psi)}(\flat(M)); \\
\flat(\lambda a\, M) &= \flat(M); \\
\flat(Mt) &= \flat(M); \\
\flat([t, M]) &= \flat(M); \\
\flat(\text{let } [a, y{:}\varphi] = M \text{ in } N) &= (\lambda y{:}\flat(\varphi).\flat(N))\flat(M).
\end{aligned}$$

8.7.7. LEMMA. *The operation* $\flat$ *commutes with substitution:*

(i) $\flat(\varphi[a := t]) = \flat(\varphi)$;

(ii) $\flat(M[a := t]) = \flat(M)$;

(iii) $\flat(M[y := N]) = \flat(M)[y := \flat(N)]$.

PROOF. A routine but lengthy induction argument. □

The following lemmas are easy to prove using Lemma 8.7.7.

8.7.8. LEMMA. *If* $\Gamma \vdash M : \varphi$ *in* λP_1 *then* $\flat(\Gamma) \vdash \flat(M) : \flat(\varphi)$ *in simple types.*

8.7.9. LEMMA. *If* $M \to_\beta N$, *then* $\flat(M) \twoheadrightarrow_\beta \flat(N)$. *In addition, the reduction* $\flat(M) \twoheadrightarrow \flat(N)$ *takes at least one step in all cases except when* $M \to_\beta N$ *is an application of first-order beta-reduction.*

8.7.10. THEOREM. *System* λP_1 *has the strong normalization property.*

PROOF. If $M_1 \to M_2 \to \cdots$ then $\flat(M_1) \twoheadrightarrow \flat(M_2) \twoheadrightarrow \cdots$ This reduction sequence must terminate (Theorem 4.5.4), unless, for some m, all the steps $M_m \to M_{m+1} \to \cdots$ are first-order beta-reductions. This cannot happen, because the number of lambdas is decreasing at every such step. □

8.7.11. THEOREM. *System* λP_1 *has the Church-Rosser property.*

PROOF. One can easily check that in λP_1 there are no critical pairs in the sense of Remark 6.3.10. The hypothesis thus follows from strong normalization and Newman's Lemma 3.6.2. □

8.8. Undecidability

From the strong normalization of system $\lambda \mathrm{P}_1$ we can infer a syntactic proof of the undecidability of first-order intuitionistic logic. Indeed, by the Curry-Howard isomorphism, it is enough to show undecidability of the existence of normal inhabitants.

The undecidability proof below works for the $(\forall, \to)$-fragment of first-order logic, and uses only a restricted class of formulas. For the later consideration in Chapter 11 it is useful to characterize these formulas.

8.8.1. DEFINITION. Let $\mathbf{ff}$ be a distinguished nullary relation symbol. Formulas of the following forms, where all relation symbols are at most binary, are called *special*:

(i) Atomic formulas, different than $\mathbf{ff}$;

(ii) Closed formulas of the form $\forall \vec{a}(\alpha_1 \to \cdots \to \alpha_n \to \beta)$, where α_i and β are atomic formulas, all α_i are different from $\mathbf{ff}$, and each variable free in β occurs in some α_i.

(iii) Closed formulas of the form $\forall a(\forall b(pab \to \mathbf{ff}) \to \mathbf{ff})$, where p is a binary relation symbol.

The intuition to be associated with formulas of type (iii) comes from the classical equivalence between $\forall a(\forall b(pab \to \bot) \to \bot)$ and $\forall a \exists b\, pab$.

8.8.2. THEOREM. *Given a set Γ of special formulas, it is undecidable whether $\Gamma \vdash \mathbf{ff}$ holds in the $(\forall, \to)$ fragment of first-order intuitionistic logic.*

PROOF. We reduce the halting problem of two-counter automata, as defined in Section A.4, to first-order provability. Given a two-counter automaton $\mathcal{A} = \langle Q, q_0, q_f, I \rangle$, we choose a signature Σ consisting of the symbol $\mathbf{ff}$ and:

- A unary relation symbol Z;

- A binary relation symbol S;

- Binary relation symbols R_q, for all $q \in Q$.

We now define a set of formulas Γ_I representing machine instructions. For each instruction in I there are one or two formulas in Γ_I:

- $\forall abc\,(R_q ab \to Sac \to R_p cb)$, whenever $(q : c_1 := c_1 + 1;\ \text{goto}\ p) \in I$;

- $\forall abc\,(R_q ab \to Sca \to R_p cb)$, whenever $(q : c_1 := c_1 - 1;\ \text{goto}\ p) \in I$;

- $\forall ab(R_q ab \to Za \to R_p ab)$ and $\forall abc\,(R_q ab \to Sca \to R_r ab)$, whenever $(q : \text{if}\ c_1 = 0\ \text{then goto}\ p\ \text{else goto}\ r) \in I$;

- $\forall abc\,(R_qab \to Sbc \to R_pac)$, whenever $(q : c_2 := c_2 + 1; \text{ goto } p) \in I$;

- $\forall abc\,(R_qab \to Scb \to R_pac)$, whenever $(q : c_2 := c_2 - 1; \text{ goto } p) \in I$;

- $\forall ab(R_qab \to Zb \to R_pab)$ and $\forall abc\,(R_qab \to Scb \to R_rab)$, whenever $(q : \text{if } c_2 = 0 \text{ then goto } p \text{ else goto } r) \in I$.

We will also use the following extension of Γ_I:

$$\Gamma_{\mathcal{A}} = \Gamma_I \cup \{Za_0, \forall a(\forall b(Sab \to \mathbf{ff}) \to \mathbf{ff}), R_{q_0}a_0a_0, \forall ab(R_{q_f}ab \to \mathbf{ff})\},$$

where a_0 is an individual variable. (The second formula above is used as a substitute for $\forall a \exists b\, Sab$.) Our goal is to prove the following:

$$\text{The automaton halts if and only if } \Gamma_{\mathcal{A}} \vdash \mathbf{ff}. \qquad (*)$$

Assume first that the automaton halts, and let $a_1, a_2, \ldots$ be an infinite sequence of distinct variables. Take

$$\Delta_k = \{Za_0, Sa_0a_1, Sa_1a_2, \ldots, Sa_{k-1}a_k\}.$$

The intended meaning of a_i is to represent the number i, and the assumptions in Δ_k state that there are sufficiently large numbers available.[5]

A configuration $\mathcal{C} = \langle q, m, n \rangle$ is represented by the formula $\hat{\mathcal{C}} = R_q a_m a_n$. We claim that

$$\mathcal{C}_1 \twoheadrightarrow_{\mathcal{A}} \mathcal{C}_2 \quad \text{implies} \quad \Gamma_I \cup \Delta_k, \hat{\mathcal{C}}_1 \vdash \hat{\mathcal{C}}_2, \text{ for some } k. \qquad (**)$$

The proof of the implication above is by induction with respect to the number of machine steps needed to reach $\mathcal{C}_2$ from $\mathcal{C}_1$. In case of zero steps the claim is obvious. Otherwise, we have $\mathcal{C}_1 \twoheadrightarrow_{\mathcal{A}} \mathcal{C}_3$ and $\mathcal{C}_3 \to_{\mathcal{A}} \mathcal{C}_2$ and the induction hypothesis yields $\Gamma_I \cup \Delta_{k_1}, \hat{\mathcal{C}}_1 \vdash \hat{\mathcal{C}}_3$. By inspection of the eight possible cases we also obtain $\Gamma_I \cup \Delta_{k_2}, \hat{\mathcal{C}}_3 \vdash \hat{\mathcal{C}}_2$, for some k_2, and thus $\Gamma_I \cup \Delta_k, \hat{\mathcal{C}}_1 \vdash \hat{\mathcal{C}}_3$, where $k = \max\{k_1, k_2\}$.

For example, assume that $\mathcal{C}_3 = \langle q, m+1, n \rangle$ and $\mathcal{C}_2 = \langle p, m, n \rangle$ is obtained from $\mathcal{C}_3$ by applying $q : c_1 := c_1 - 1; \text{ goto } p$. Then $\hat{\mathcal{C}}_3 = R_q a_{m+1} a_n$ and if $k_2 > m$ then Δ_{k_2} contains the assumption $Sa_m a_{m+1}$. Using the formula $\forall abc\,(R_qab \to Sca \to R_pcb)$, one derives $\hat{\mathcal{C}}_2 = R_p a_m a_n$.

Since the automaton halts, the final configuration $\mathcal{C} = \langle q_f, m, n \rangle$ is reachable from $\langle q_0, 0, 0 \rangle$. By $(**)$ we have

$$\Gamma_I \cup \Delta_k, R_{q_0}a_0a_0 \vdash R_{q_f}a_m a_n,$$

for some k, and thus $\Gamma_{\mathcal{A}} \cup \{Sa_0a_1, Sa_1a_2, \ldots, Sa_{k-1}a_k\} \vdash \mathbf{ff}$. It follows that $\Gamma_{\mathcal{A}} \cup \{Sa_0a_1, Sa_1a_2, \ldots, Sa_{k-2}a_{k-1}\} \vdash Sa_{k-1}a_k \to \mathbf{ff}$, which can be generalized to $\Gamma_{\mathcal{A}} \cup \{Sa_0a_1, Sa_1a_2, \ldots, Sa_{k-2}a_{k-1}\} \vdash \forall b(Sa_{k-1}b \to \mathbf{ff})$. We

---

[5] Observe that there is a certain similarity between our present construction and Remark 8.3.7. We cannot set an a priori bound for the values of counters.

8.8. UNDECIDABILITY

can now obtain $\Gamma_A \cup \{Sa_0a_1, Sa_1a_2, \ldots, Sa_{k-2}a_{k-1}\} \vdash \mathbf{ff}$ using the formula $\forall a(\forall b(Sab \to \mathbf{ff}) \to \mathbf{ff})$. By repeating this trick k times we get rid of all the assumptions Sa_ia_{i+1} and finally we obtain $\Gamma_A \vdash \mathbf{ff}$. This completes the proof of the left-to-right direction in $(*)$.

The proof of the other implication begins with a definition. A set Δ of assumptions of the form Sab is *good* iff $Sab, Sa'b \in \Delta$ implies $a = a'$. We then have a partial function p_Δ such that $p_\Delta(b) = a$ iff $Sab \in \Delta$. We write $|b| = n$ if $p_\Delta^n(b) = a_0$, where a_0 is the variable occurring free in Γ_A. (For instance, the sets Δ_k above are good with $|a_n| = n$.)

Assume that $\Gamma_A \vdash \mathbf{ff}$. We first show that

$$\Gamma_A, \Delta \vdash R_{q_f}ab, \quad \text{for some good } \Delta \text{ and some } a, b. \qquad (***)$$

For this we prove a more general claim: If $\Gamma_A, \Delta \vdash \mathbf{ff}$, for some good Δ then $\Gamma_A, \Delta' \vdash R_{q_f}ab$, for some a, b and some good set Δ' such that $\Delta \subseteq \Delta'$. It is now convenient to identify sets of formulas with λP_1 environments. In particular, assume that Γ_A contains the declarations

$$x : \forall a(\forall b(Sab \to \mathbf{ff}) \to \mathbf{ff}) \text{ and } y : \forall ab(R_{q_f}ab \to \mathbf{ff}).$$

Suppose that $\Gamma_A, \Delta \vdash M : \mathbf{ff}$, where, by Theorem 8.7.10, the term M can be assumed normal. We proceed by induction with respect to the size of M. The formula $\mathbf{ff}$ is atomic and there is no declaration of type $\mathbf{ff}$, so M must be of the form $ya'b'N$ or $xa'N$. In the first case we have $\Gamma_A, \Delta \vdash N : R_{q_f}a'b'$, so let us consider the second case. Then $\Gamma_A, \Delta \vdash N : \forall b(Sa'b \to \mathbf{ff})$ and we can see that N must be an abstraction $\lambda b \lambda z^{Sa'b}.M'$, where b may be assumed fresh. It follows that $\Gamma_A, \Delta, z : Sa'b \vdash M' : \mathbf{ff}$. The set $\Delta, Sa'b$ is still good, because b was new, and we can apply induction to M'.

Now we know $(***)$ and we want the automaton to halt. We need again a more general statement. Let Δ be good. Then

$$\Gamma_A, \Delta \vdash R_pab \quad \text{implies} \quad \langle q_0, 0, 0 \rangle \twoheadrightarrow_A \langle p, m, n \rangle,$$

for some m and n with $|a| = m$ and $|b| = n$. (In particular, $|a|$ and $|b|$ are defined.) Let $\Gamma_A, \Delta \vdash M : R_pab$ and let M be normal. If M is a variable then its type must be $R_{q_0}a_0a_0$. In other words, $\langle p, m, n \rangle$ is the initial configuration. Otherwise, M is an application, beginning with a variable z declared in Γ_I.

Depending on the type of z we now have eight cases of which we consider one for example. Let $M = za'b'c'NP$ and $z : \forall abc(R_qab \to Scb \to R_pac)$. Then $a' = a$, $c' = b$ and N has type R_qab'. By the induction hypothesis for N we have $\langle q_0, 0, 0 \rangle \twoheadrightarrow_A \langle q, m, n \rangle$, where $|a| = m$ and $|b'| = n$. Since P has type Sbb', it must simply be a variable declared in Δ. Thus $|b| = n - 1$, and we conclude that $\langle q_0, 0, 0 \rangle \twoheadrightarrow_A \langle q, m, n \rangle \to_A \langle p, m, n - 1 \rangle$. □

8.8.3. COROLLARY. *The $(\forall, \to)$ fragment of the first-order intuitionistic logic (i.e. the inhabitation problem for system λP_1) is undecidable.*

In our proof of Theorem 8.8.2, the signature depends on the automaton. One can improve that to obtain undecidability of first-order logic for a fixed signature (Exercise 8.34). The proof can also be significantly simplified if function symbols can be used (Exercise 8.33).

8.9. Notes

The use of quantifiers in logic dates back to *Begriffsschrift*, the fundamental of Frege of 1879 [218], but the idea can also be found in the work of others, especially Peirce, who was perhaps the author of the term "first-order logic." In the book [231] by Hilbert and Ackermann, first published in 1928, first-order logic (called the "engere Funktionenkalkül," the *narrower* function calculus) is recognized as a fundamental logical system. In that book, two essential questions concerning first-order classical logic are posed: The question of completeness and the *Entscheidungsproblem* ("decision problem").

Before the next edition appeared in 1938, these two milestone questions were answered. First, Gödel announced his completeness (and also compactness) proof in 1929, and then the undecidability of the Entscheidungsproblem was shown by Church in 1936.

Other completeness proofs for first-order classical logic were later given by various authors, e.g. by Henkin [220] and Rasiowa and Sikorski [408]. The latter proof is based on the algebraic approach and uses what we now call "the Rasiowa-Sikorski lemma" (Exercise 8.15). Note that the very statement of the completeness theorem requires a notion of validity of a formula. This notion, although intuitively understood much earlier, was precisely defined only in 1933 in Tarski's famous booklet [472] on the notion of truth, after Gödel's completeness paper was already published.

First-order intuitionistic logic is as old as intuitionistic logic as such, and an axiom system was postulated by Heyting as early as 1929 in the same issue of the *Sitzungsberichte der Preussischen Akademie der Wissenschaften* in which his propositional axioms were published. Topological semantics of predicate calculus was probably first used by Mostowski to exhibit counterexamples. Algebraic completeness proofs were given independently by Henkin [221] and Rasiowa [407]. Rasiowa's PhD thesis contains the first explicit statement of Lemma 8.5.5, although Rasiowa points out that it was implicit in earlier work of McKinsey and Tarski. The reader might have noticed that our completeness proof is not constructive. In a sense this has to be the case, see [279, 333]. More about semantics can be found in [134, 410, 488].

Henkin's completeness proof can be applied to classical logic as well. Since a classical prime theory must be complete (prove either φ or $\neg\varphi$ for all sentences φ) and admits only conservative extensions, the Kripke model constructed in the proof of Theorem 8.6.7 collapses to one state, and we obtain a classical structure. In case of countable signatures, this construction can be further simplified and made more abstract (Exercise 8.16). Instead of a prime theory, one can consider a prime filter in the Lindenbaum algebra, and then a quotient construction yields a $\{0,1\}$-structure.

A similar "algebraization" is also possible for propositional intuitionistic logic. Recall that in the proof of Lemma 2.5.8 states of the Kripke models were determined by prime filters. One may wonder if the same is possible at the first-order level. This however would require an adequate reformulation of the Rasiowa-Sikorski lemma (Exercise 8.15), to guarantee that there exist prime filters of the desired properties. But as shown in [411], the Rasiowa-Sikorski lemma does not hold in general for Heyting algebras, unless they satisfy Grzegorczyk's scheme. We do not know how to extend the algebraic approach beyond the logic of constant domains [198, 209].

The question whether intuitionistic predicate logic is complete for the real topology (Exercise 8.22) was mentioned as an open problem in [410] and, surprisingly, our solution seems to be new. In case the reader feels upset about this incompleteness, here is some good news: There is a result of Moerdijk [351] stating that reals *are* complete for intuitionistic predicate logic *with equality* under a different notion of model, called *sheaf model*.

The computational contents of first-order proofs was investigated by many logicians, beginning with Kleene's realizability idea, and followed by Kreisel, Läuchli, Goodman and others. In Howard's seminal paper [247], first-order features are first-class citizens. Howard discusses two ways of handling existential quantifier. One of these is the so-called "strong sum", with which it is possible to explicitly extract t and M from $[t, M]_{\exists a \varphi}$. Our term syntax for existential quantification represents information hiding, and does not permit such operations. However, as shown by Tait [469], such an extension is conservative for first-order logic. It is not the case though with more powerful systems, like λC (cf. Section 14.7).

In the language Automath of de Bruijn [58, 60] as well as in Edinburgh LCF, Constable's PRL and Martin-Löf's type theory, the correspondence between first-order predicates and dependent types was fully employed. A type theory called Edinburgh Logical Framework was developed in [214] as a formal basis for LCF and similar calculi. First-order and higher-order logic can be represented in this type theory.

Strong normalization for systems of first-order proof notation, like our λP_1, is usually shown using contracting maps [432] or the reducibility method [161]. The idea of the contracting map occurs in works by various authors, cf. [189, 191, 214, 304, 308, 386]. Martin-Löf [323] probably first used such a map explicitly. Our presentation was inspired by Leivant's papers [304, 308]. The proof of Theorem 8.8.2 is based on [46].

For a general exposition of classical predicate logic, the reader may want to consult for instance [6, 39, 113, 252, 337]. See [53, 252, 337] for proofs of undecidability, and [161, 471, 489] for cut-elimination.

8.10. Exercises

8.1. Consider the two arguments from the beginning of this chapter. Are they correct classically? Constructively? What will change if we replace everywhere *"will not be relieved of flying missions"* by *"will continue flying missions"* in the first of these arguments?

8.2. Is the following a constructively correct argument? Is it correct classically?

> *Figaro shaves all men of Seville who do not shave themselves.*
> *Figaro is a man of Seville.*
> *Therefore Figaro shaves himself.*

8.3. Find constructions for the formulas in Example 8.2.1, and do not find constructions for the formulas in Example 8.2.2.

8.4. A first-order formula is in *prenex normal form* iff it consists of a sequence of quantifiers followed by an open formula. Consider a signature with only relation symbols, and let φ be an intuitionistic theorem in prenex normal form. Show that there is an open formula φ' obtained from φ by removing quantifiers and by replacing some existentially quantified variables by other variables, and such that $\vdash \varphi'$. *Hint:* Use the existence property (Proposition 8.3.5).

8.5. (V.P. Orevkov). Apply Exercise 8.4 to show that the prenex fragment of intuitionistic first-order logic over function-free signatures is decidable.[6]

8.6. Prove that every first-order formula is classically equivalent to a formula in prenex normal form. Then prove that intuitionistic first-order logic does not have this property.

8.7. What kind of adjustment should be made in the natural deduction proof system for first-order logic (intuitionistic or classical) if alpha-equivalent formulas are not identified?

8.8. Prove the equivalence between natural deduction, sequent calculus, and Hilbert style (Theorem 8.3.10) for intuitionistic first-order logic. Then show the same for first-order classical logic.

8.9. Let X be an arbitrary non-empty set. Show that the semantics of classical logic where values of formulas are taken in the field of sets $\mathcal{P}(X)$ is equivalent to the ordinary $\{0, 1\}$-semantics. That is, the sets of tautologies are the same.

8.10. Define semantics of classical logic (a notion of $\Gamma \models \varphi$) in $\mathcal{B}$-structures, where $\mathcal{B}$ is an arbitrary complete Boolean algebra. Show that this semantics is adequate, i.e. $\Gamma \vdash \varphi$ implies $\Gamma \models \varphi$. From the completeness theorem for the ordinary $\{0, 1\}$-semantics (Theorem 8.4.7) conclude that the two notions of $\models$ are equivalent.

8.11. Give an example of a complete Heyting algebra $\mathcal{G}$ and a subalgebra $\mathcal{H}$ of $\mathcal{G}$ such that $\inf_\mathcal{H} A$ exists for some $A \subseteq \mathcal{H}$, but $\inf_\mathcal{G} A \neq \inf_\mathcal{H} A$. *Hint:* Find a subset $A \subseteq \mathcal{P}(\mathbb{R})$, such that $\bigcap A = \varnothing$, but $A \subseteq F$ for a certain prime filter F. Then use Exercise 2.15.

8.12. Let $\mathcal{H}$ be a Heyting algebra. For $X \subseteq \mathcal{H}$, let $X{\Downarrow}$ and $X{\Uparrow}$ stand, respectively, for the sets of all lower and upper bounds of X. For all $X, Y \subseteq \mathcal{H}$, we define $X \cdot Y = \{x \sqcap y \mid x \in X \text{ and } y \in Y\}$, and $con(X) = X{\Uparrow}{\Downarrow}$. A non-empty set X is a *cone* iff $con(X) = X$. (If $p \in \mathcal{H}$ then $\{p\}{\Downarrow}$ (abbr. $p{\Downarrow}$) is a cone.) Prove that:

(i) The set $con(X)$ is the least cone containing X.

(ii) The equation $con(X) \cap con(Y) = con(X \cdot Y)$ holds for all $X, Y \subseteq \mathcal{H}$.

(iii) The set Con of all cones is a complete Heyting algebra with the pseudo-complement $\Rightarrow$ defined by $X \Rightarrow Y = \{p \in \mathcal{H} \mid X \cap p{\Downarrow} \subseteq Y\}$.

8.13. Prove Lemma 8.5.5: Every Heyting algebra can be embedded into a complete Heyting algebra, so that all existing joins and meets are preserved. *Hint:* Use the algebra Con from Exercise 8.12.

8.14. Explain footnote 3. What if *all* variables occur in Γ or φ?

8.15. Prove the Rasiowa-Sikorski lemma: Let $\mathcal{B}$ be a Boolean algebra and assume that $a_n = \inf_\mathcal{B} A_n$, for $n \in \mathbb{N}$. For $a \neq 0$ there exists a prime filter F in $\mathcal{B}$ with $a \in F$ and such that $A_n \subseteq F$ implies $a_n \in F$, for all $n \in \mathbb{N}$.

---
[6]This remains true even with function symbols in the signature, see [125].

8.10. EXERCISES

8.16. Prove the completeness of first-order classical logic (over a countable signature) with respect to the two-valued semantics. *Hint:* Apply Exercise 8.15 to the Lindenbaum algebra.

8.17. A set Γ of formulas is (classically) *satisfiable* when $\mathcal{A}, \varrho \models \Gamma$, for some $\mathcal{A}$ and ϱ. Prove the following version of the compactness theorem for classical logic: *If every finite subset of a set Γ of formulas is satisfiable then Γ is satisfiable.* Then as a corollary show that there is no set of first-order sentences, that would be satisfied exactly in all finite models.

8.18. Verify that all the formulas of Example 8.2.1 are intuitionistically valid.

8.19. Verify that all the formulas of Example 8.2.2 are classical tautologies.

8.20. Complete the proof of Proposition 8.5.3, using topological counterexamples.

8.21. Show that the following are intuitionistically provable:

(i) $\neg\neg\forall a \neg\neg r a \leftrightarrow \forall a \neg\neg r a$;

(ii) $\neg\neg \exists a \neg\neg r a \leftrightarrow \neg\neg \exists a r a$.

8.22. Show that the formula $\forall a(\varphi \vee \neg\varphi) \to \exists a\varphi \vee \forall a(\neg\varphi)$ from Example 8.2.2(vii) is valid in all $\mathcal{O}(\mathbb{R})$-structures.

8.23. Show that Markov's principle:

$$\forall a(\varphi \vee \neg\varphi) \wedge \neg\neg\exists a\varphi \to \exists a\varphi$$

is valid in all $\mathcal{O}(\mathbb{R})$-structures. *Hint:* Use the previous exercise.

8.24. Prove Proposition 8.5.3, using Kripke models.

8.25. In classical logic, the existential and universal quantifiers are interdefinable. Is this the case in first-order intuitionistic logic?

8.26. (Grzegorczyk, Görnemann) A Kripke model $\langle C, \leq, \{\mathcal{A}_c : c \in C\}\rangle$ has *constant domains* iff all the $\mathcal{A}_c$ are the same. Prove that Grzegorczyk's scheme

$$\forall a(\psi \vee \varphi(a)) \leftrightarrow \psi \vee \forall a \varphi(a),$$

(where $a \notin \mathrm{FV}(\psi)$) is valid in all models with constant domains.

8.27. Show that if $\mathcal{H}$ is a finite Heyting algebra then Grzegorczyk's scheme is valid in all $\mathcal{H}$-structures.

8.28. Prove that the formula $\neg\neg\forall a(\varphi \vee \neg\varphi)$ (formula (i) of Example 8.2.2) is valid in all Kripke models with finite sets of states.

8.29. Consider a signature with a distinguished relation symbol $=$, which is traditionally written as an infix operator (i.e. we write $t = s$ instead of $=ts$). Add the following formulas to the axioms of first-order logic:

(E1) $\forall a(a = a)$

(E2) $\forall ab(a = b \to \varphi[c := b] \to \varphi[c := a])$.

Derive transitivity and symmetry laws for equality:

(E3) $\forall cab(a = b \to b = c \to a = c)$;

(E4) $\forall ab(a = b \to b = a)$.

Also show that it suffices to assume (E2) for atomic formulas (including equations).

8.30. Under the assumptions of Example 8.7.1, what could be a type of an operator *push* which appends an additional integer value at the end of an array?

8.31. Assume that a procedure sum_1 as in Example 8.7.1 has been defined. Using this, write a definition of sum_2. Then assume sum_2 and define sum_1. Compare this to the equivalence $\forall a(pa \to q) \leftrightarrow (\exists a\, pa \to q)$.

8.32. Write normal inhabitants in λP_1 for the formulas in Example 8.2.1.

8.33. Simplify the proof of Corollary 8.8.3 using function symbols. *Hint:* Encode a configuration $\langle q, m, n \rangle$ as $R_q(s^m(0), s^n(0))$, where s is a unary function symbol and 0 is a constant.

8.34. Show that first-order logic is undecidable for some fixed signature Σ.

Chapter 9

First-order arithmetic

Arithmetic is an indispensable part of all branches of mathematics. The language of arithmetic is capable of expressing properties of finite objects of diverse nature. In particular, many issues in the theory of computation can be studied in the framework of arithmetic. Since the fundamental work of Dedekind and Peano in the end of the 19th century, arithmetic has been a primary subject of research in logic and foundations of mathematics. In this chapter we discuss the fundamental properties of first-order classical (Peano) and intuitionistic (Heyting) arithmetic. We consider various ways in which algorithms (recursive functions) can be expressed in arithmetic, and we also show how formulas of arithmetic can be *realized* by recursive functions.

9.1. The language of arithmetic

When we say "arithmetic", we usually assume that the underlying signature consists of two binary function symbols $+$ and $\cdot$ for addition and multiplication, one unary function symbol s for the successor, and the constant 0. Then we have the symbol $=$ for equality. We follow this choice for the beginning of our exposition, but later it will be convenient to (conservatively) extend the signature with additional function symbols.

9.1.1. DEFINITION. The *standard model* of arithmetic is the set $\mathbb{N}$ of natural numbers with the ordinary understanding of these symbols, i.e. the structure:

$$\mathcal{N} = \langle \mathbb{N}, +, \cdot, s, 0, = \rangle.$$

Every number $n \in \mathbb{N}$ has a name in the language of arithmetic. The *numeral* $\underline{n}$ is the term $s(s(\cdots(s0)\cdots))$, with exactly n occurrences of s. In particular $\underline{0}$ is simply 0. Commonly used abbreviations are $t \neq s$ for $\neg(t = s)$ and $t \leq s$ for $\exists a(t + a = s)$. Of course $t < s$ means as usual $(t \leq s) \wedge (t \neq s)$.

The classical theory $\text{Th}(\mathcal{N})$ of the standard model (see Notation 8.4.6) is called the *complete arithmetic*.

The following result demonstrates an important limitation of first-order logic. The standard model of arithmetic is not definable by any set of first-order formulas. Even the whole set $\text{Th}(\mathcal{N})$, containing all possible first-order properties of $\mathbb{N}$ does not uniquely determine the model.

9.1.2. THEOREM (Skolem). *There exists a non-standard model of arithmetic, i.e. a countable structure* $\mathcal{M} = \langle \mathbb{M}, \oplus, \otimes, \mathbf{s}, \mathbf{0}, = \rangle$ *such that* $\text{Th}(\mathcal{M}) = \text{Th}(\mathcal{N})$, *but* $\mathcal{M}$ *and* $\mathcal{N}$ *are not isomorphic.*

PROOF. Exercise 9.1. □

However, the definitional strength of first-order logic over the standard model is non-trivial. The class of definable sets includes all recursively enumerable sets. In order to prove this fact, we employ a certain number-theoretic trick, a variant of a classical result known as the *Chinese remainder theorem*. Consider Gödel's *beta function*.

$$\beta(m, n, j) = m \bmod (n(j+1) + 1).$$

The crucial property of the beta function is as follows.

9.1.3. LEMMA. *For every finite sequence* $k_0, k_1, \ldots, k_r$ *there exist two numbers* m, n, *such that* $\beta(m, n, j) = k_j$, *for* $j = 0, \ldots, r$.

PROOF. Take $n_0 = \max\{r, k_0, k_1, \ldots, k_r\}$ and define n as the factorial $n_0!$ of n_0. Let $a_j = n(j+1) + 1$ for $j = 0, \ldots, r$, so that $\beta(m, n, j) = m \bmod a_j$.

The numbers a_j are pairwise *relatively prime*, i.e. the only common divisor of each pair of them is 1. Indeed, suppose p is a prime divisor of some a_j and a_{j+q}. Then p divides the number $a_{j+q} - a_j = q \cdot n_0!$ and thus $p \leq n_0$. But 1 is the only divisor of a_j that is less than or equal to n_0.

For $\ell = 0, \ldots, r$ we define (by induction with respect to ℓ) a number m_ℓ such that $\beta(m_\ell, n, j) = k_j$ holds for all $j \leq \ell$. We begin with $m_0 = k_0$. Then assume that m_ℓ has been already defined for some $\ell < r$. Let $a = a_0 \cdots a_\ell$. The following are equivalent for any given numbers m_1 and m_2:

- $m_1 \bmod a = m_2 \bmod a$;
- $m_1 \bmod a_j = m_2 \bmod a_j$, for all $j = 0, \ldots, \ell$.

This claim follows from the simple fact that the difference $m_1 - m_2$ is divisible by a if and only if it is divisible by all a_j.

Now observe that a and $a_{\ell+1}$ are relatively prime (because the divisors of a are relatively prime to $a_{\ell+1}$). Thus there are numbers u and v such that $u \cdot a_{\ell+1} - v \cdot a = 1$ (see Exercise 9.6). A simple multiplication gives

$$u' \cdot a_{\ell+1} - v' \cdot a = m_\ell - k_{\ell+1},$$

for some u', v'. Define $m_{\ell+1} = u' \cdot a_{\ell+1} + k_{\ell+1} = m_\ell + v' \cdot a$. It is now easy to see that $m_{\ell+1} \bmod a = m_\ell \bmod a$, and $m_{\ell+1} \bmod a_{\ell+1} = k_{\ell+1}$. □

9.1. THE LANGUAGE OF ARITHMETIC 231

9.1.4. DEFINITION. We say that a k-ary relation r over $\mathbb{N}$ is *arithmetical* if there is a formula $\varphi(\vec{a})$ with k free variables $\vec{a}$, such that, for every $\vec{n} \in \mathbb{N}$,

$$r(\vec{n}) \text{ holds } \quad \text{iff} \quad \mathcal{N} \models \varphi(\underline{\vec{n}}).$$

A function is *arithmetical* iff it is arithmetical as a relation.

9.1.5. THEOREM (Gödel). *All recursive functions are arithmetical.*

PROOF. It is easy to see that the initial functions are arithmetical. For instance, the following formula $\varphi(a_1, a_2, a_3, b)$ defines a ternary projection on the second coordinate:

$$a_1 = a_1 \wedge a_3 = a_3 \wedge b = a_2.$$

(The first two equations are here because Definition 9.1.4 requires that all the four variables are free in φ.)

The reader will easily check that a composition of arithmetical functions is arithmetical. For a function f defined by minimization

$$f(\vec{n}) = \mu y [g(\vec{n}, y) = 0],$$

with g defined by a formula $\psi(\vec{a}, b, c)$, we can write a formula $\varphi(\vec{a}, b)$:

$$\psi(\vec{a}, b, 0) \wedge \forall c(c < b \rightarrow \neg \psi(\vec{a}, b, 0)).$$

It remains to consider the case of a function defined by primitive recursion:

$$\begin{aligned} f(0, \vec{n}) &= g(\vec{n}); \\ f(m+1, \vec{n}) &= h(\vec{n}, m, f(m, \vec{n})). \end{aligned}$$

Here we apply Lemma 9.1.3. It should be easy to see that the beta function is arithmetical. Let $B(a, b, i, c)$ be an appropriate formula and let f be as above. Assume that the formulas $\varphi(\vec{a}, b)$ and $\psi(\vec{a}, b, c, d)$ define respectively g and h. Then the following formula (with free variables $a, \vec{b}, c$) defines f:

$$\exists mn (\exists w(B(m, n, 0, w) \wedge \varphi(\vec{b}, w)) \wedge B(m, n, a, c)$$
$$\wedge \forall i(i < a \rightarrow \exists uv(B(m, n, i, u) \wedge B(m, n, s(i), v) \wedge \psi(\vec{b}, i, u, v)))).$$

This formula expresses the following equivalence: $f(a, \vec{b}) = c$ iff there are numbers $k_0, k_1, \ldots, k_a$, such that $k_0 = g(\vec{b})$ and $k_a = c$, and $k_{i+1} = h(\vec{b}, i, k_i)$ holds for all $i = 0, \ldots, a - 1$. □

9.1.6. COROLLARY. *All recursively enumerable sets and all partial recursive functions are arithmetical.*

PROOF. An immediate consequence of Corollary A.3.9. □

9.1.7. COROLLARY. Th($\mathcal{N}$) *is undecidable.*

PROOF. Otherwise, membership in any r.e. set would also be decidable. □

Gödel's theorem states that every recursive function can be given an *explicit* definition by a first-order formula. For instance, we have an explicit definition of exponentiation, and thus no specific symbol for exponentiation is present in the signature of arithmetic. In a sense there is no need for it, as every use of such a symbol in a formula can be replaced by its definition.

Observe that an explicit definition is something else than an *inductive* definition. For instance, the following definition of multiplication

$$n \cdot 0 = 0;$$
$$n \cdot sm = n \cdot m + n,$$

does not provide the possibility to eliminate the multiplication symbol from the language. It turns out that an explicit definition of multiplication in terms of addition is not possible. Indeed, the theory of $\langle \mathbb{N}, +, s, 0, = \rangle$, called *Presburger arithmetic*, is decidable [53, Chapter 24], while Th($\mathcal{N}$) is not, by Corollary 9.1.7.

This difference may seem paradoxical, as multiplication is an iterated addition in very much the same way as exponentiation is an iterated multiplication. But once we have multiplication we have reached the "critical mass" of expressive power, for which successor and addition are not enough.

Corollary 9.1.6 can be read as follows: First-order arithmetic is capable of expressing the fundamental notions related to computations, including termination of programs. The reader familiar with Hoare logic should now notice that we have just shown the following result: The standard model of arithmetic is "expressive" and thus by Cook's Theorem, Hoare logic [12] is relatively complete with respect to $\mathcal{N}$.

9.2. Peano Arithmetic

Before Gödel, people thought that it might be possible to axiomatize Th($\mathcal{N}$), i.e. to effectively describe a set of axioms A such that all sentences of Th($\mathcal{N}$) are consequences of A. Peano Arithmetic, abbreviated PA, was a plausible candidate. Presently, most authors define PA as a first-order classical theory (cf. Definition 8.3.11) of the following axioms (where φ is arbitrary).

- $\forall a\, (a = a)$;
- $\forall a \forall b\, (a = b \to b = a)$;
- $\forall a \forall b \forall c\, (a = b \to b = c \to a = c)$;
- $\forall a \forall b\, (a = b \to sa = sb)$;

9.2. PEANO ARITHMETIC

- $\forall a \forall b (sa = sb \rightarrow a = b)$;
- $\forall a (sa = 0 \rightarrow \bot)$;
- $\forall a (a + 0 = a)$;
- $\forall a \forall b (a + sb = s(a+b))$;
- $\forall a (a \cdot 0 = 0)$;
- $\forall a \forall b (a \cdot sb = (a \cdot b) + a)$;
- $\forall a (\varphi(a) \rightarrow \varphi(sa)) \rightarrow \varphi(0) \rightarrow \forall a \varphi(a)$.

The first four on the list are axioms for equality. These axioms ensure that equality is an equivalence relation preserved by successor, and imply the following properties (Exercise 9.3).

$$\forall a \forall b \forall a' \forall b' (a = a' \wedge b = b' \rightarrow a + b = a' + b'), \tag{9.1}$$
$$\forall a \forall b \forall a' \forall b' (a = a' \wedge b = b' \rightarrow a \cdot b = a' \cdot b'), \tag{9.2}$$
$$\forall a \forall b (a = b \rightarrow \varphi[c := b] \rightarrow \varphi[c := a]). \tag{9.3}$$

The next two axioms state that the successor is a one to one function and that zero is not in the range of successor. Then we have inductive definitions of addition and multiplication. The last item (called the *induction scheme*) is not a single axiom. Although the set of axioms is thus infinite, it is still recursive, and the set of theorems (derivable sentences) of PA is r.e.

Despite their simplicity, the axioms of PA are very strong. Most of the principal results in number theory can be derived from these axioms, usually by a formalization of ordinary metamathematical proofs. The tools used in such proofs are with more or less effort expressible in the language of PA. In particular, a very universal tool—induction—is readily available. Here is a simple example of how it works. More to be found in Exercises 9.2–9.6.

9.2.1. EXAMPLE. The formula

$$\forall a (\neg(a = 0) \rightarrow \exists b (a = sb))$$

is provable in PA with the help of the induction scheme, where $\varphi(a)$ is instantiated to $\neg(a = 0) \rightarrow \exists b(a = sb)$. Indeed, it is enough to show that the following formulas are provable:

$$\neg(0 = 0) \rightarrow \exists b(0 = sb);$$

$$\forall a [(\neg(a = 0) \rightarrow \exists b(a = sb)) \rightarrow \neg(sa = 0) \rightarrow \exists b(sa = sb)].$$

The first formula is proved using the axiom $\forall a(a = a)$ instantiated to $0 = 0$. The second formula is easily derived from $\exists b(sa = sb)$, and the latter is a consequence of $sa = sa$, another instance of $\forall a(a = a)$.

9.3. Gödel's theorems

Of course the standard model of arithmetic is a model of PA, in other words we have the inclusion PA $\subseteq$ Th($\mathcal{N}$). Gödel's famous incompleteness theorem asserts that the converse inclusion does not hold. In other words, PA is not a *complete* theory. (A theory T is *complete* iff for all sentences ψ in the signature of T, either $T \vdash \psi$ or $T \vdash \neg\psi$. Clearly, every classical theory of the form Th($\mathcal{A}$) is complete.)

In fact, a careful reader should be able to give a proof of Gödel's theorem on the basis of what has already been said in this chapter (Exercise 9.7). But we find it worthwhile to sketch the original approach, if only for its beauty.

The main idea of Gödel's proof is *arithmetization*. All the symbols in the language of arithmetic are numbered, for instance this way:

Symbol:	0	s	+	$\cdot$	$\perp$	$\rightarrow$	=	$\forall$	(	)	a_0	a_1	...
Number:	1	2	3	4	5	6	7	8	9	10	11	12	...

where $a_0, a_1, \ldots$ is a countably infinite supply of individual variables. Now we assign a numeric code (*a Gödel number*) to every string of symbols (in particular every formula, proof, etc.) If we denote by $\#s$ the number of a symbol s, then the Gödel number of the string $s_1 s_2 \ldots s_n$ is

$$Code(s_1 s_2 \ldots s_n) = 2^{\#s_1} 3^{\#s_2} 5^{\#s_3} 7^{\#s_4} \cdots p_n^{\#s_n},$$

where p_n denotes the n-th prime number. The discovery of Gödel is that properties of arithmetical formulas can be expressed in the very language of arithmetic as number-theoretic properties of codes. Instead of talking about properties of the formula $\forall a_0 ((a_1 + a_0 = 0) \rightarrow \perp)$ one can discuss properties of its Gödel number, i.e. of the integer

$$Code(\forall a_0((a_1+a_0=0) \rightarrow \perp)) = 2^8 3^{11} 5^9 7^9 11^{12} 13^3 17^{11} 19^7 23^1 29^{10} 31^6 37^5 41^{10}$$

As long as these properties are recursively enumerable, one can now write formulas expressing them in the language of arithmetic (Corollary 9.1.6). Starting with a very simple example, the formula $\exists a\, (\underline{n} = a \cdot \underline{13})$ states that "*any string coded by the number n consists of at least 6 symbols*". To express "*the number n codes a string*" one writes $\neg(n = 0) \wedge \forall abc\, (prime(a) \rightarrow prime(b) \rightarrow a < b \rightarrow n = b \cdot c \rightarrow \exists d\, (n = a \cdot d))$, where *prime* is a formula expressing primality. With some additional effort (and using the beta function many times) one can then write a formula $F(a)$ with one free variable a, such that for every $n \in \mathbb{N}$ the following holds:

$\mathcal{N} \models F(\underline{n})$ iff n is the Gödel number of a formula with one free variable.

Then various properties of formulas can be expressed. In particular, there is a formula $S(a, b, c)$ such that $\mathcal{N} \models S(\underline{n}, \underline{m}, \underline{k})$, iff

9.3. GÖDEL'S THEOREMS

- m is the Gödel number of a formula $\zeta(a)$ with one free variable;
- n is the Gödel number of the sentence $\zeta(\underline{k})$.

We can abbreviate the above as follows:

$$\mathcal{N} \models S(\underline{n}, \underline{m}, \underline{k}) \quad \text{iff} \quad n \text{ is the Gödel number of the sentence } \varphi_m(\underline{k}),$$

if we agree to write $\varphi = \varphi_m$ when the Gödel number of φ is m. But not everything can be expressed in the language of arithmetic.

9.3.1. THEOREM (Tarski [472]). *Validity of formulas (in the standard model) is not definable in arithmetic. That is, there is no formula $P(a)$ satisfying:*

$$\mathcal{N} \models P(\underline{n}) \quad \text{iff} \quad n \text{ is the Gödel number of a sentence } \zeta \in \text{Th}(\mathcal{N}).$$

PROOF. Suppose otherwise. We obtain a contradiction by expressing *liar's paradox*[1] in the language of arithmetic. Consider the formula

$$T(a) = \exists b(S(b, a, a) \wedge \neg P(b)).$$

Then $\mathcal{N} \models T(\underline{n})$ iff

- n is the Gödel number of a formula $\zeta(a)$ with one free variable;
- the sentence $\zeta(\underline{n})$ is false in $\mathcal{N}$.

This can be abbreviated as:

$$\mathcal{N} \models T(\underline{n}) \quad \text{iff} \quad \mathcal{N} \not\models \varphi_n(\underline{n}).$$

The formula $T(x)$ has a Gödel number too, say $T(a) = \varphi_k(a)$, and we have

$$\mathcal{N} \models T(\underline{k}) \quad \text{iff} \quad \mathcal{N} \not\models \varphi_k(\underline{k}).$$

But observe that $\varphi_k(\underline{k})$ is just the formula $T(\underline{k})$. Thus,

$$\mathcal{N} \models T(\underline{k}) \quad \text{iff} \quad \mathcal{N} \not\models T(\underline{k}).$$

Oops! The sentence $T(\underline{k})$ says *"I am not true!"*. That is a contradiction. □

We obtain Gödel's incompleteness theorem after a slight modification of the argument above. Instead of the undefinable notion of truth, we will use the definable notion of *provability*. The same "diagonal" construction will result in a sentence Z, saying *"I have no proof in* PA!*"*.

---

[1] The sentence: *"This sentence is false"* cannot be true and cannot be false. Theorem 9.3.1 helps to explain this paradox. The "liar" sentence is ambiguous. In order to classify it as "true" or "false" we must understand (be able to explain in English) what it means for an English sentence to be true or false. But the notion of "truth" is external to the language, and cannot be precisely defined within it. Note that the problem here is not just the self-application but a reference to the notion of truth. For instance, we can say *"This sentence consists of six words"* without creating any ambiguity.

9.3.2. THEOREM (Gödel incompleteness). *There is a sentence Z such that neither* $\text{PA} \vdash Z$ *nor* $\text{PA} \vdash \neg Z$.

PROOF. The proof goes along the same lines as the proof of Theorem 9.3.1, but instead of the formula $P(a)$ we take a formula $Q(a)$, such that

$$\mathcal{N} \models Q(\underline{n}) \quad \text{iff} \quad n \text{ is the Gödel number of a sentence provable in PA.}$$

The existence of $Q(a)$ follows from Corollary 9.1.6, because the property in question is recursively enumerable. Repeating the argument in Theorem 9.3.1, we arrive at this conclusion

$$\mathcal{N} \models T(\underline{k}) \quad \text{iff} \quad \text{PA} \nvdash T(\underline{k}).$$

Take $Z = T(\underline{k})$. Then

$$\mathcal{N} \models Z \quad \text{iff} \quad \text{PA} \nvdash Z.$$

Now if $\text{PA} \vdash Z$ then $\mathcal{N} \models Z$, because $\mathcal{N}$ is a model of PA, and thus $\text{PA} \nvdash Z$, by the property of Z. So it must be the case that $\text{PA} \nvdash Z$ and $\mathcal{N} \models Z$. Since $\text{PA} \vdash \neg Z$ would imply $\mathcal{N} \models \neg Z$, we conclude that Z can be neither proved nor disproved within PA. □

The importance of Gödel's theorem is not only in that PA is incomplete. If we extend the set of axioms then, as long as the new axioms are effectively given (the set of axioms is r.e.), we can repeat the whole argument, again discovering a sentence independent of our axioms. Paradoxically, incompleteness occurs not because the theory PA is too weak, but because it is too strong. This is the real surprise: Every recursively enumerable consistent theory strong enough to express elementary arithmetic must be incomplete.

With the Gödel numbers technique, one can express the consistency of PA. Indeed, let $R(x)$ be a formula such that:

$$\mathcal{N} \models R(\underline{n}) \quad \text{iff} \quad \text{PA} \vdash \varphi_n,$$

for every sentence φ_n. Let k be the number of the equation $0 = 1$ and let **Con** be the formula $\neg R(\underline{k})$. Then **Con** expresses consistency of PA:

$$\mathcal{N} \models \textbf{Con} \quad \text{iff} \quad \text{PA is consistent.}$$

The proof of Theorem 9.3.2 can be refined to yield a sentence Z, provably equivalent to its own unprovability, and such that

$$\text{PA} \vdash \textbf{Con} \to Z.$$

It follows that one cannot prove consistency of PA within PA, unless PA is actually inconsistent. This is Gödel's *second incompleteness theorem*. Proofs can be found e.g. in [6, 337, 355, 443].

9.3.3. THEOREM (Gödel). *If* PA *is consistent then* $\text{PA} \nvdash \textbf{Con}$.

Gödel's second theorem has an important consequence: To prove consistency of arithmetic, one must necessarily use tools from outside the arithmetic. That is, no finitary proof of consistency is possible (cf. Exercise 9.11.)

9.4. Representable and provably recursive functions

We now consider two properties of functions that are stronger than being arithmetical. We not only want our functions to be definable over the standard model, but we want to prove in PA (or some other theory) that the appropriate formula actually defines a function.

In the definition below, the symbol $\exists!$ should be read as "there exists exactly one". Formally, $\exists! a \varphi(a)$ abbreviates $\exists a(\varphi(a) \land \forall b(\varphi(b) \to a = b))$.

9.4.1. DEFINITION. A k-ary total function f over $\mathbb{N}$ is *representable* in PA iff there exists a formula $\varphi(\vec{a}, b)$, with $k+1$ free variables $\vec{a}, b$, such that:

1) $f(\vec{n}) = m$ implies PA $\vdash \varphi(\underline{\vec{n}}, \underline{m})$, for all $\vec{n}, m \in \mathbb{N}$;

2) PA $\vdash \exists! b \varphi(\underline{\vec{n}}, b)$, for all $\vec{n} \in \mathbb{N}$.

A function is *strongly representable* in PA, if (1) holds and

3) PA $\vdash \forall \vec{a} \exists! b \varphi(\vec{a}, b)$.

Each representable function is in fact strongly representable, but proving that (2) implies (3) is a brutal application of *tertium non datur*.

9.4.2. PROPOSITION (V. Huber-Dyson [337]). *Every representable function is strongly representable.*

PROOF. Suppose a function f is representable by a formula φ. Then f is strongly representable by the formula

$$\psi(\vec{a}, b) = (\exists! c \varphi(\vec{a}, c) \land \varphi(\vec{a}, b)) \lor (\neg \exists! c \varphi(\vec{a}, c) \land b = 0).$$

Indeed, if $f(\vec{n}) = m$ then both $\exists! b \varphi(\underline{\vec{n}}, b)$ and $\varphi(\underline{\vec{n}}, \underline{m})$ are provable, whence PA $\vdash \psi(\underline{\vec{n}}, \underline{m})$. To show that PA $\vdash \forall \vec{a} \exists! b \psi(\vec{a}, b)$ we begin with PA $\vdash \vartheta \lor \neg \vartheta$, where $\vartheta = \exists! c \varphi(\vec{a}, c)$. In addition, we have PA $\vdash \vartheta \to \exists! b(\vartheta \land \varphi(\vec{a}, b))$ and PA $\vdash \neg \vartheta \to \exists! b(\neg \vartheta \land b = 0)$. Thus PA $\vdash \exists! b(\vartheta \land \varphi(\vec{a}, b)) \lor \exists! b(\neg \vartheta \land b = 0)$, whence PA $\vdash \exists! b \psi$, because the two disjuncts contradict each other. □

The following is a stronger version of Theorem 9.1.5:

9.4.3. THEOREM (Gödel). *A function is representable in PA if and only if it is recursive.*

PROOF. The implication from left to right is easy: If a function f is representable, then we can find $f(n)$ for any n, by an exhaustive proof search. For the converse one shows that the construction in the proof of Theorem 9.1.5 can be formalized in arithmetic. Details can be found in [337, 355]. □

PRIMITIVE RECURSIVE FUNCTIONS. A careful analysis of a detailed proof of Theorem 9.4.3 reveals that a stronger property holds for primitive recursive functions, or more precisely, primitive recursive *algorithms*.[2]

9.4.4. THEOREM. *There is a uniform way to assign a formula φ_f to every primitive recursive function f, in such a way that φ_f strongly represents f in* PA *and if f is defined by*

$$f(0, \vec{n}) = g(\vec{n});$$
$$f(m+1, \vec{n}) = h(\vec{n}, m, f(m, \vec{n})),$$

then the following are provable in PA:

$$\varphi_f(0, \vec{b}, c) \leftrightarrow \varphi_g(\vec{b}, c);$$
$$\varphi_f(sa, \vec{b}, c) \leftrightarrow \exists u(\varphi_f(a, \vec{b}, u) \wedge \varphi_h(\vec{b}, a, u, c)).$$

As we said before, the signature of arithmetic does not contain a symbol for exponentiation, because addition and multiplication are sufficient. However, an additional symbol for exponentiation can sometimes be useful. More generally, one can extend the language of arithmetic by new function symbols representing arbitrary primitive recursive functions. The set of axioms is then extended by the equations defining the new identifiers. For instance, if f is defined from g and h by primitive recursion, then the new axioms are:

$$\forall \vec{b}(f 0 \vec{b} = g\vec{b}); \tag{9.4}$$

$$\forall a\vec{b}(f(sa)\vec{b} = h\vec{b}a(fa\vec{b})). \tag{9.5}$$

For initial functions and functions defined by compositions these axioms are even simpler. This extension does not affect any significant property of Peano Arithmetic, and may be considered "syntactic sugar".

9.4.5. THEOREM. *The extension of* PA *by names and defining equations for all primitive recursive functions is conservative over* PA, *that is, if a formula without the additional symbols is provable in the extended system then it is a theorem of* PA.

PROOF. Let φ_f be as in Theorem 9.4.4, and consider an extension of PA obtained by adding the new symbols together with the explicit definitions of the form $f(\vec{a}) = b \leftrightarrow \varphi_f(\vec{a}, b)$ as axioms. This theory is conservative over PA (cf. e.g. [443, 489]) and by Theorem 9.4.4 it has the same provable consequences as our extension. □

---

[2]Primitive recursive functions are defined from the initial functions by composition and primitive recursion. But each of them has many such definitions. When we refer to a primitive recursive "function" we usually mean a particular definition.

9.4. Representable and provably recursive functions

By Theorem 9.4.5 we can now assume that the additional symbols are actually present in the language of PA, i.e. that the signature is infinite, and that all the equational axioms are part of the theory. It is convenient to interpret definitions such as (9.4) and (9.5) as *term rewriting rules*:

$$f\underline{0}\vec{b} \Rightarrow g\vec{b}; \tag{9.6}$$
$$f(sa)\vec{b} \Rightarrow h\vec{b}a(fa\vec{b}). \tag{9.7}$$

An algebraic term t rewrites to another term u in one step (notation $t \to u$) if u is obtained from t by replacing a subterm matching a left-hand side of a rule with the corresponding instance of the right hand side. It is not difficult to see that we have the following fact.

9.4.6. PROPOSITION. *The rewrite system given by defining equations for primitive recursive functions has the CR and SN properties. Thus every term has exactly one normal form, and every closed term rewrites to a numeral.*

9.4.7. DEFINITION. If t and s are closed algebraic terms in the extended signature of arithmetic then we write $t \approx s$ if t and s rewrite to the same numeral. (Clearly, if $t \approx s$ then $\text{PA} \vdash t = s$.)

PROVABLE TOTALITY. Theorem 9.4.3, together with Proposition 9.4.2, implies that the totality of every recursive function can be proven in PA in a uniform way. However, the excluded middle trick used in the proof of Proposition 9.4.2 shows that such proofs are not necessarily informative. This calls for a finer notion of provable totality.

Recall that, by Kleene's Theorem A.3.8, every partial recursive function f can be written as

$$f(\vec{n}) = \ell(\mu y[t_f(\vec{n}, y) = 0]),$$

where ℓ and t_f are primitive recursive. The function t_f (see Notation A.3.7) describes a particular algorithm computing f. Termination of this algorithm can be expressed by a formula of the form

$$\forall \vec{x} \, \exists y \, (t_f(\vec{x}, y) = 0).$$

If this formula has a proof, one can say that the function is provably total.

9.4.8. DEFINITION. A recursive function f is said to be *provably total* (or *provably recursive*) in a theory T iff

$$T \vdash \forall \vec{x} \, \exists y \, (t_f(\vec{x}, y) = 0).$$

It is customary to talk about provably recursive *functions*, but what we actually deal with is the notion of a provably recursive *algorithm*. A *function* should be regarded provably total if one of its algorithms is provably total.

The class of functions provably total in PA is very large and includes most of the commonly considered functions, and much more, up to unthinkable complexity. But not all recursive functions are provably total.

9.4.9. PROPOSITION. *Not all recursive functions are provably total in* PA.

PROOF. A systematic listing of proofs of formulas $\forall \vec{x} \exists y \, (t_f(\vec{x}, y) = 0)$ yields an effective enumeration $f_0, f_1, f_2, \ldots$ of all provably recursive functions. The function $f(n) = f_n(n) + 1$ is recursive but different from all the f_n's. □

REMARK. Our definition of a provably total function does not depend on the availability of a special symbol for t_f. One can replace the equation $t_f(\vec{x}, y) = 0$ by the equivalent formula written in the "core" language of arithmetic, without any loss of generality.

9.5. Heyting Arithmetic

By *Heyting Arithmetic* (HA), we mean an intuitionistic first-order theory (see Definition 8.3.11) based on the following axioms:

- All axioms of Peano Arithmetic;
- Defining equations for all primitive recursive functions.

That is, HA is a theory in the language of arithmetic, extended by new function symbols for all primitive recursive functions. This extension is not essential. Indeed, Theorem 9.4.5 works in the intuitionistic case (see [269, §49] and [232, §8.3]) because the whole argument can be formalized constructively. But the trick is technically very useful. In particular, note that any primitive recursive relation $P(\vec{n})$ over $\mathbb{N}$ can be expressed by an equation $\chi_P(\vec{n}) = 0$, where χ_P is the appropriate primitive recursive characteristic function. This is as good as assuming that a dedicated relation symbol is available for every primitive recursive relation.

We begin our study of HA with some basic properties.

9.5.1. PROPOSITION.

(i) $\text{HA} \vdash \forall a \forall b \, (a = b \vee \neg(a = b))$.

(ii) $\text{HA} \vdash \forall a \forall b \, (\neg \neg a = b \to a = b)$.

(iii) $\text{HA} \vdash (\varphi \vee \psi) \leftrightarrow \exists a \, ((a = 0 \to \varphi) \wedge (\neg(a = 0) \to \psi))$, for all φ, ψ.

PROOF. Part (i) is shown by (formalized) induction (see Exercise 9.16). Part (ii) is an immediate consequence of part (i), because the propositional scheme $(p \vee \neg p) \to \neg \neg p \to p$ is intuitionistically valid (Example 2.1.3(x)).

For part (iii), let α denote the formula $(a = 0 \to \varphi) \wedge (\neg(a = 0) \to \psi)$. Since $a = 0, \alpha \vdash \varphi$ and $a \neq 0, \alpha \vdash \psi$, also $a = 0 \vee a \neq 0, \alpha \vdash \varphi \vee \psi$. From part (i) we have $\alpha \vdash \varphi \vee \psi$, so $\exists a \alpha \vdash \varphi \vee \psi$ holds too.

For the converse implication we have $\varphi \vdash \alpha[a := \underline{0}]$ and $\psi \vdash \alpha[a := \underline{1}]$. Thus $\varphi \vdash \exists a \alpha$ and $\psi \vdash \exists a \alpha$. It follows that $\varphi \vee \psi \vdash \exists a \alpha$. □

9.5. HEYTING ARITHMETIC

9.5.2. PROPOSITION. *The Kolmogorov translation (see Proposition 8.3.1) interprets* PA *in* HA, *i.e.* PA $\vdash \varphi$ *if and only if* HA $\vdash k(\varphi)$.

PROOF. The proof is similar to that of Proposition 8.3.1 but one has to show in addition that the translations of all axioms of PA are provable in HA. The most complicated case is that of induction. The translation of

$$\forall a(\phi(a) \to \phi(sa)) \to \phi(0) \to \forall a \phi(a)$$

is the following formula:

$$\neg\neg(\neg\neg\forall a \neg\neg (k(\phi(a)) \to k(\phi(sa))) \to \neg\neg(k(\phi(0)) \to \neg\neg\forall a k(\phi(a)))).$$

This formula can be quite simplified if we remember that $k(\varphi(a))$ begins with a double negation and that the following are intuitionistically provable:

- $\neg\neg(\varphi \to \psi) \leftrightarrow (\neg\neg\varphi \to \neg\neg\psi)$ (Exercise 2.23);
- $\neg\neg\neg\neg\varphi \leftrightarrow \neg\neg\varphi$ (Example 2.1.3(v));
- $\neg\neg\forall a \neg\neg\varphi \leftrightarrow \forall a \neg\neg\varphi$ (Exercise 8.21).

Using the above as rewrite rules oriented from left to right, we obtain another instance of the induction scheme:

$$\forall a(k(\phi(a)) \to k(\phi(sa))) \to k(\phi(0)) \to \forall a k(\phi(a)). \qquad \Box$$

9.5.3. COROLLARY. *Consistency of* HA *implies consistency of* PA.

PROOF. If PA $\vdash \bot$ then HA $\vdash k(\bot)$. But $k(\bot) = \neg\neg\bot$ implies $\bot$. $\qquad \Box$

Our goal is now to show that certain proofs in classical arithmetic do in fact have a constructive contents.

9.5.4. LEMMA. *If* PA $\vdash \exists b P(\vec{a}, b)$, *for a primitive recursive relation* P, *then* HA $\vdash \neg\neg\exists b P(\vec{a}, b)$.

PROOF. From Proposition 9.5.2 we have HA $\vdash \neg\neg\exists b k(P(\vec{a}, b))$. Primitive recursive relations are expressed as single equations, and thus treated as atomic formulas. Thus $k(P(\vec{a}, b)) = \neg\neg P(\vec{a}, b)$ is equivalent to $P(\vec{a}, b)$. $\qquad \Box$

We now describe a simple, but surprisingly efficient trick, called *Friedman's translation*. Let ϱ be a fixed formula. For any formula φ, we denote by φ^ϱ the formula obtained from φ by replacing every atomic subformula ν (including $\bot$) by the disjunction $\nu \vee \varrho$. The replacement must be done without variable capture, so whenever we consider e.g. $(\forall a \varphi)^\varrho$, we must assume that a is not free in ϱ.

9.5.5. LEMMA (Friedman's translation). *If* HA ⊢ φ *then also* HA ⊢ φ^ϱ.

PROOF. First we check that HA ⊢ φ^ϱ for all axioms φ of HA. Then we prove that $\Gamma \vdash \varphi$ implies $\Gamma^\varrho \vdash \varphi^\varrho$, where $\Gamma^\varrho = \{\gamma^\varrho \mid \gamma \in \Gamma\}$. To handle ⊥-elimination observe that $\varrho \vdash \varphi^\varrho$ for all φ. □

9.5.6. THEOREM (Kreisel, 1958). *If* PA ⊢ $\forall \vec{a} \exists b P(\vec{a}, b)$, *where P is primitive recursive, then* HA ⊢ $\forall \vec{a} \exists b P(\vec{a}, b)$.

PROOF. Assume PA ⊢ $\forall \vec{a} \exists b P(\vec{a}, b)$, and let $\varrho = \exists b P(\vec{a}, b)$. Of course PA ⊢ ϱ, so HA ⊢ $\neg\neg\varrho$, by Lemma 9.5.4. Recall that $\neg\neg\varrho = (\varrho \to \bot) \to \bot$.

From Friedman's Lemma 9.5.5, we have HA ⊢ $(\varrho^\varrho \to \varrho) \to \varrho$. The formula $P(\vec{a}, b)$ is atomic, and thus ϱ^ϱ is of the form

$$\exists b (P(\vec{a}, b) \vee \exists b P(\vec{a}, b)),$$

and is equivalent to ϱ. Thus HA ⊢ $(\varrho \to \varrho) \to \varrho$. Since ⊢ $\varrho \to \varrho$, we finally get HA ⊢ ϱ and it remains to generalize over $\vec{a}$. □

9.5.7. COROLLARY. *A recursive function is provably total in Peano Arithmetic iff it is provably total in Heyting Arithmetic.*

PROOF. The right-to-left part is immediate. The left-to-right part follows from Theorem 9.5.6. □

Kreisel's theorem has the following consequence: classical termination proofs can be made constructive.

9.5.8. EXAMPLE. Consider a classically provable formula $\forall a \exists b P(a, b)$ with primitive recursive P. It can be seen as a specification for an input-output relation of a program. By Theorem 9.5.6, this specification can be given a constructive proof. A program (a recursive function) satisfying the specification can be extracted from this proof. To see this, let us assume that the signature of arithmetic (and perhaps a bit more) has been added to the system λP_1 of Chapter 8. Then, for example, the formula $\forall n \exists m (n = \underline{2} \cdot m \vee n = \underline{2} \cdot m + 1)$ is inhabited by a proof, i.e. a lambda term, say M. Applied to any specific $\underline{n}$, it will evaluate to a normal form $[\underline{m}, N]$, for some m. Thus M is actually a program for dividing numbers by 2.

The little missing point in the example above is the "bit more" to be added to the lambda calculus λP_1. There are specific axioms of arithmetic, in particular the induction scheme. There are also axioms for equality. All these axioms should be represented as constants of appropriate types, and equipped with meaningful reduction rules. The resulting system becomes thus quite complicated, see [308, 493]. One way to avoid this is to consider a propositional version of such calculus obtained via an appropriate extension of the contracting map $\mathfrak{b}$ of Chapter 8. We will discuss this in Chapter 10. But first let us mention another method to exhibit the computational meaning of intuitionistic arithmetic.

9.6. Kleene's realizability interpretation

The realizability interpretation of Kleene [268] can be seen as a translation from intuitionistic arithmetic to recursive function theory, in other words, as a model of Heyting Arithmetic built with help of recursive functions.

The idea is that a formula $\varphi \to \psi$ is justified, or *realized*, by a recursive function mapping realizers of φ (encoded as numbers) to realizers of ψ. The function itself is then represented by its Gödel number. Similarly, a conjunction is realized by a pair and so on.

9.6.1. DEFINITION. Let $p \in \mathbb{N}$ and φ be a closed formula of arithmetic. Then the circumstances under which p *realizes* φ are defined as follows.[3]

- p does not realize $\bot$;
- p realizes an equation $t = s$ iff t and s rewrite to the same numeral;
- p realizes $\varphi_1 \wedge \varphi_2$ if $p = 2^q \cdot 3^r$ where q realizes φ_1 and r realizes φ_2;
- p realizes $\varphi_1 \vee \varphi_2$ if $p = 3^q$ and q realizes φ_1, or $p = 2 \cdot 3^q$ and q realizes φ_2;
- p realizes $\varphi_1 \to \varphi_2$ if p is the Gödel number of a partial recursive function f of one argument such that, whenever q realizes φ_1, then $f(q)$ realizes φ_2;
- p realizes $\exists a \varphi(a)$, where $\varphi(a)$ is a formula containing only a free, if $p = 2^n \cdot 3^q$ where q realizes $\varphi(\underline{n})$;
- p realizes $\forall a \varphi(a)$, where $\varphi(a)$ is a formula containing only a free, if p is the Gödel number of a total recursive function f of one argument such that $f(n)$ realizes $\varphi(\underline{n})$, for every n.

Finally we say that a number p *realizes* an arbitrary formula φ when p realizes the universal closure of φ. A formula φ is *realizable* if there exists a number p which realizes φ.

9.6.2. THEOREM. *If* $\mathsf{HA} \vdash \varphi$ *then* φ *is realizable.*

PROOF. By induction with respect to proofs in HA we show a slightly more general statement, namely that $\mathsf{HA}, \gamma_1, \ldots, \gamma_r \vdash \varphi$ implies the realizability of $\gamma_1 \to \cdots \to \gamma_r \to \varphi$. The proof is quite routine (see §82 of Kleene's classic [269] for details) and we consider only a few example cases. Below, the notation $\{p\}$ stands for the (unary) partial recursive function whose Gödel number is p. We write $\{p\}(k_1, k_2, \ldots, k_r)$ for $\{\cdots \{\{p\}(k_1)\}(k_2) \cdots \}(k_r)$.

---

[3]Note that $2^p 3^q$ is an injective pairing function.

We begin with the natural deduction axiom $\Gamma, \varphi \vdash \varphi$, which is clearly realizable by the Gödel number of an appropriate (curried) projection. The next step is to see that all the specific axioms of arithmetic are realizable. The only non-immediate case is the induction scheme. Indeed, each of the other axioms is realized by the Gödel number of a constant function zero (with an appropriate number of arguments). To show that induction is realizable, assume that all free variables of $\varphi(a)$ are among $a, b_1, \ldots, b_d$. Let $n, n_1, \ldots, n_d$ be any numbers and let $\varphi' = \varphi[b_1 := \underline{n}_1, \ldots, b_d := \underline{n}_d]$. Assume that m and k realize, respectively, the formulas $\forall a (\varphi'(a) \to \varphi'(sa))$ and $\varphi'(0)$. Define $k_0 = k$ and $k_{i+1} = \{\{m\}(i)\}(k_i)$. Then k_n realizes the formula $\varphi'(\underline{n})$, and the process of computing k_n from $n_1, \ldots, n_d, m, k$ and n is effective. The induction axiom is realized by the Gödel number of the Turing Machine implementing this process.

The induction step in our proof splits into cases depending on the last rule used in the proof. For instance, assume that the last step of the proof was ∃-elimination, where $\Gamma = \{\gamma_1, \ldots, \gamma_r\}$:

$$\frac{\Gamma \vdash \exists a \psi, \quad \Gamma, \psi \vdash \vartheta}{\Gamma \vdash \vartheta}$$

Let $b_1, \ldots, b_d$ be all the variables, other than a, which are free in Γ, ψ and ϑ. By the induction hypothesis there are numbers m and k realizing respectively the judgements $\Gamma \vdash \exists a \psi$ and $\Gamma, \psi \vdash \vartheta$. The conclusion $\Gamma \vdash \vartheta$ is realized by the Gödel number of the following algorithm. Given $n_1, \ldots, n_d$ and $k_1, \ldots, k_r$, compute the realizer $\{m\}(n_1, \ldots, n_d, k_1, \ldots, k_r)$ of the formula $\exists a \psi$. This number should be of the form $2^n 3^q$ for some n and q. Then compute the number $\{k\}(n_1, \ldots, n_d, n, k_1, \ldots, k_r, q)$ realizing ϑ.

We invite the reader to try a few more examples (Exercise 9.19). □

9.6.3. COROLLARY. HA *is consistent.*

PROOF. There is no number realizing ⊥. □

9.6.4. COROLLARY. PA *is consistent.*

PROOF. Immediate from Corollaries 9.5.3 and 9.6.3. □

There is an obvious similarity between realizers and proofs. But there are important differences, the following in the first place.

9.6.5. PROPOSITION.

(i) *For every closed φ, either φ or $\neg \varphi$ is realizable.*

(ii) *It is undecidable whether a given number p realizes a given formula φ.*

PROOF. For part (i) suppose that there is no realizer of φ. Then an arbitrary number realizes $\neg\varphi$. For part (ii) observe that a number p realizes $\forall a(a = a)$ if and only if it is the Gödel number of a total recursive function. □

A slight modification of the realizability relation makes it possible to derive the disjunction and existence properties for HA.

9.6.6. PROPOSITION (Harrop, 1956).

(i) *If* $\text{HA} \vdash \exists a \varphi(a)$, *where* $\exists a \varphi(a)$ *is a closed formula, then* $\text{HA} \vdash \varphi(\underline{k})$, *for some* $k \in \mathbb{N}$.

(ii) *If* $\text{HA} \vdash \varphi \vee \psi$, *for closed* φ *and* ψ, *then either* $\text{HA} \vdash \varphi$ *or* $\text{HA} \vdash \psi$.

PROOF. Part (ii) is a direct consequence of part (i) and Proposition 9.5.1(iii). To prove part (i) we define a notion of $\vdash$-*realizability* by modifying Definition 9.6.1 as follows:

- Replace "realizes" by "$\vdash$-realizes" throughout;
- Add "and $\text{HA} \vdash \varphi_1 \to \varphi_2$" at the end of the fifth clause;
- Add "and $\text{HA} \vdash \forall a \varphi(a)$" at the end of the last clause.

Now, if φ is a closed formula then

$$\text{HA} \vdash \varphi \quad \text{iff} \quad \varphi \text{ is } \vdash\text{-realizable.}$$

The left to right implication is a simple adaptation of Theorem 9.6.2, and the other implication is shown by induction (for the second clause observe that $t \approx s$ implies $\text{HA} \vdash t = s$). To conclude the proof, let us suppose that $\text{HA} \vdash \exists a \varphi(a)$. Then some $\varphi(\underline{n})$ is $\vdash$-realizable, and thus provable. □

9.7. Notes

Giuseppe Peano is usually considered the inventor of modern arithmetic. But similar ideas were independently formulated by other authors, like Dedekind or Grassmann. We see it as another example of a recurrent pattern: Great discoveries are often made independently by many people at the same time. However, Peano was the first to explicitly formulate the famous five axioms.

"Peano axioms" can be found today in numerous textbooks in a form similar to our list in Section 9.2. But the original Peano axioms were quite different. First of all, the notion of a natural number (a member of the set $\mathbb{N}$) was taken as a primitive, and there were explicit axioms stating that zero and its successors are natural numbers. Also Peano used a language richer than first-order logic. There was a single induction axiom, involving quantification over sets, and there were no axioms for addition and multiplication (Example 12.2.2(iv) explains why). Peano's system can thus be more adequately represented in second-order logic (see Chapters 11 and 12). A minor difference is that his numbers began with 1, not 0.

In [388], a booklet of 1889, Peano stated nine axioms of arithmetic, including four axioms of equality and five "proper" axioms of arithmetic. Two years later, in the first part of the article [389], the five axioms are listed separately.[4]

1. $1 \in N$.
2. $s \in N^N$.
3. $\forall ab(a, b \in N \to s(a) = s(b) \to a = b)$.
4. $1 \notin s(N)$.
5. $\forall s \in K(1 \in s \to s(s) \subseteq s \to N \subseteq s)$.

Interestingly, the induction axiom changed slightly between 1889 and 1891. Indeed, the 1889 version was as as follows:

9. $\forall k \in K(1 \in k \to \forall a(a \in N \to a \in k \to s(a) \in k) \to N \subseteq k)$.

This axiom corresponds to a first-order induction scheme of the form

$$\varphi(1) \to \forall a(\text{int}\, a \to \varphi(a) \to \varphi(s\, a)) \to \forall a(\text{int}\, a \to \varphi(a)),$$

where (to fit everything into the first-order syntax) we replaced $a \in N$ by a unary predicate $\text{int}\, a$. The 1891 axiom translates to a different scheme

$$\varphi(1) \to \forall a(\varphi(a) \to \varphi(s\, a)) \to \forall a(\text{int}\, a \to \varphi(a)),$$

which is slightly weaker, because the first assumption is stronger. The advantage of the new axiom is that it can be turned into a second-order definition (cf. Example 12.2.2(iii)) stating that *a natural number is an object which belongs to all sets containing zero and closed under successor.* Another aspect of this difference will be considered in Section 10.2.

It is commonly believed that the first "natural" mathematical problem independent from PA was a theorem in finite combinatorics discussed by Paris and Harrington in [385]. But this belief is not quite true. We will see in the next chapter an example of a mathematically meaningful result (Theorem 10.3.8) which cannot be derived in PA. This result (anticipated by Gödel himself in 1941) is due to Tait [467], who also made an explicit statement about its unprovability. Being a tool used in a consistency proof, Tait's theorem was not recognized as truly "mathematical", for a long time. But one should also read it as a termination property for a reduction system representing a certain functional programming language—undoubtedly a "mathematical" result on its own.

Gödel's incompleteness theorems of 1930–31 (especially the second theorem) caused the failure of Hilbert's programme. The idea of mathematics built on finitary methods and guarded by finitary consistency proofs turned out to be another form of utopia, no less naive than naive set theory. Logicians of the 30's and 40's were left with a less spectacular but more practical goal: To establish consistency of mathematics, in particular arithmetic, on as restricted means as possible. Since finitary methods are not sufficient, one should still aim at choosing the weakest and most intuitively basic tools from what is available.

---

[4]It is quite common in the literature to mention e.g. the "third" or "fourth" Peano axiom, and this is usually a reference to the list above.

The first consistency proof of PA was given by Gentzen in 1935. Proofs were assigned certain ordinal numbers and it was shown that any proof of $\bot$ could always be reduced to another proof assigned a smaller ordinal. This transfinite induction is clearly not arithmetical. Other approaches included Gödel's System **T** (see the next chapter) and Kleene's realizability.

The realizability approach was introduced by Kleene [268] in 1945 in a form quite similar to that discussed in Section 9.6. In the same paper he proposed a formalized or "internalized" notion of realizability, in the form of a definable predicate $n\,\mathbf{r}\,m$ read as "n realizes the formula φ_m." Since then many different notions of realizability, most of them in the formalized style, have been invented and proven very useful. In Chapter 10, we will discuss one of them, the *modified realizability* of Kreisel, but for simplicity we will not make it formalized.

There is an evident similarity between realizability and the BHK interpretation as described in Sections 2.1 and 8.2. The BHK interpretation is an informal but general explanation of intuitionistic logic and admits many different ways of reading. Realizability can be seen as one of these, and here we read "construction" as "realizer". This is a specific reading, so it defines a particular "model" of HA rather than a complete semantics. The properties of this model are quite unusual. By Proposition 9.6.5(i), it is to some extent "classical". On the other hand, it is not a model of classical arithmetic: Exercise 9.20 gives an example of a sentence φ, true in the standard model (and even provable in PA), and such that $\neg\varphi$ is realizable.

The so-called Friedman's translation was invented independently by Friedman [154] and Dragalin [133]. The Friedman-Dragalin translation is one of a variety of *syntactic translations* occurring in the literature. See for instance [89, 303].

More information about the *arithmetical hierarchy* of all the arithmetical sets and functions can be found in [276, 419]. We recommend [6, 337, 355, 443] for further reading on Gödel's theorems. The monumental edition [195] of Gödel's collected works contains nearly everything Gödel has ever written, accompanied by excellent comments. Our refeerences to the work of Peano are [262, 390]. The crucial articles by Peano and Gödel can also be found in [218]. An interested reader will find a lot about realizability in the articles [377, 486].

9.8. Exercises

9.1. Prove the existence of a non-standard model of arithmetic (Theorem 9.1.2). *Hint:* Use the compactness theorem for classical logic (Exercise 8.17).[5]

9.2. Assuming derivability of (9.1) and (9.2), show that the following theorems are derivable in PA and HA.

(i) $\forall a \forall b \forall c ((a + b) + c = a + (b + c))$;

(ii) $\forall a \forall b (a + b = b + a)$;

(iii) $\underline{2} + \underline{2} = \underline{4}$;

(iv) $\forall a \exists b (a = \underline{2} \cdot b \vee a = s(\underline{2} \cdot b))$.

---

[5]The proof is of course not constructive, and the reader may wonder how an actual non-standard model may look like. Here is the bad news: Although there exist countable non-standard models, none of them is recursive (see Chapter 25 in [53]), i.e. there is no effective definition of the non-standard addition and multiplication.

Try other common arithmetical properties.

9.3. Show that axioms (9.1)–(9.3) are derivable in PA and HA.

9.4. Define $a \leq b$ as $\exists c(a + c = b)$ and $a < b$ as $a \leq b \wedge \neg a = b$. Show that the statement "$\leq$ *is a total order*" can be expressed in the language of arithmetic and proven in PA and HA. Then derive the formula $\forall ab(a \leq b \leftrightarrow a < sb)$.

9.5. Show that this variant of the induction scheme is provable in PA and HA:

$$\forall a(\forall b(b < a \to \varphi(b)) \to \varphi(a)) \to \forall a \varphi(a).$$

9.6. Prove that if $n, m \geq 1$ are relatively prime numbers then there exist u and v such that $u \cdot n - v \cdot m = 1$. Then formalize your proof in PA and HA.

9.7. Prove Theorem 9.3.2 (PA is incomplete) using results of Sections 9.1 and 9.2.

9.8. Theorem 9.3.1 applies to any mathematical theory strong enough to express arithmetic. So does it make any sense at all to study mathematics (e.g. arithmetic) if we cannot tell which mathematical statements are true and which are not?

9.9. A set $A \subseteq \mathbb{N}$ is *weakly representable* in PA iff there exists a formula $\varphi(a)$ such that for all $n \in \mathbb{N}$, the following three conditions are equivalent:

- $n \in A$;
- $\mathcal{N} \models \varphi(\underline{n})$;
- $\text{PA} \vdash \varphi(\underline{n})$.

Show that the weakly representable subsets of $\mathbb{N}$ are exactly the r.e. sets.

9.10. Prove that PA is undecidable. *Hint:* Use Exercise 9.9.

9.11. Is consistency of first-order logic provable in a finitary way?

9.12. (Schwichtenberg) Does there exist a primitive recursive predicate $P(a)$, such that the formula $\forall a P(a)$ holds in the standard model of arithmetic, but is not provable in PA?

9.13. Show that a function f is provably recursive in PA if and only if it is representable in PA by a formula of the form $\exists c P(\vec{a}, b, c)$, where P is a primitive recursive relation such that $\text{PA} \vdash \forall \vec{a} \exists b \exists c P(\vec{a}, b, c)$.

9.14. Assume that a formula φ does not contain $\vee$ nor $\exists$. Prove that $\text{PA} \vdash \varphi$ iff $\text{HA} \vdash \varphi$. *Hint:* Use Exercises 2.23 and 8.21(i) together with the Kolmogorov translation.

9.15. Is PA a proper extension of HA, i.e. is there any formula φ such that $\text{PA} \vdash \varphi$ but $\text{HA} \nvdash \varphi$?

9.16. Prove that $\text{HA} \vdash \forall a \forall b(a = b \vee \neg(a = b))$.

9.17. Prove that HA is closed under *Markov's rule* for formulas with one free variable: If $\text{HA} \vdash \forall a(\varphi(a) \vee \neg \varphi(a))$ and $\text{HA} \vdash \neg\neg \exists a \varphi(a)$ then $\text{HA} \vdash \exists a \varphi(a)$.

9.18. Prove that HA is undecidable. *Hint:* First do Exercise 9.10.

9.19. Recall that $\{p\}$ denotes the partial recursive function which Gödel number is p. Write $\Lambda x.expression(x)$ for the Gödel number of the unary function mapping any x to $expression(x)$. Using this notation describe e_1 and e_2 such that

- e_1 realizes $\varphi \to \psi \to \varphi$;
- e_2 realizes $(\varphi \to \psi \to \vartheta) \to (\varphi \to \psi) \to \varphi \to \vartheta$.

9.8. EXERCISES

9.20. Consider the formula

$$\varphi = \forall a[\exists c T(a,a,c) \vee \forall c \neg T(a,a,c)],$$

where T is Kleene's predicate (Lemma A.3.5). Show that φ is provable in **PA**, and that $\neg \varphi$ is realizable.

9.21. Can Theorem 9.5.6 be generalized for arbitrary formulas of the form $\forall \vec{a} \exists b \varphi$?

9.22. We know from Theorem 9.3.3 that consistency of PA is not provable in PA. Which part of the proof of Corollary 9.6.4 cannot be formalized in PA?

9.23. (Schwichtenberg's paradox) It is quite common to identify BHK constructions with proofs. It is also common to assume that a construction of an equality or non-equality between two given closed arithmetical expressions is simply to calculate and to compare the values of these expressions. So here is a proof of Fermat's Last Theorem. Given numbers $m, n, k, \ell \geq 3$, calculate the values $m^n + m^k$ and m^ℓ and check that these values are different numbers. What's wrong with this "proof"?

Chapter 10

Gödel's system T

We have seen in Chapter 8 that first order logic and the related calculus λP_1 correspond under a contracting map to the simply typed lambda calculus. A similar correspondence occurs between first-order arithmetic and an extension of $\lambda_\to$ obtained by adding an appropriate recursion scheme. This system, known as Gödel's system **T**, was conceived and used by Gödel to prove the consistency of arithmetic.

System **T** is far more expressible than $\lambda_\to$, yet satisfies strong normalization. It can be seen as a powerful programming language in which all well-typed programs terminate. Terms of system **T** represent the computational contents of arithmetical reasoning, and the contracting map can be used to exhibit this contents.

10.1. From Heyting Arithmetic to system T

We would like to develop a technique for extracting programs from proofs in Heyting Arithmetic. For this, one can define a system of lambda-calculus such that terms in this system exactly represent proofs in HA, just like terms of λP_1 represent proofs in predicate logic (Example 9.5.8).

In such a calculus, the specific axioms of arithmetic correspond to constants of appropriate types. The proof reduction rules, naturally associated with the induction axiom, represent a general form of recursive computation. After proving strong normalization, one applies a contracting map to a normal proof to obtain a lambda-term [247, 308, 493].

We prefer to avoid the technical complexity of this construction, so instead of doing it in all detail, we use a simpler method, called *modified realizability*, for program extraction.

However, we find it useful to discuss the idea of "Heyting Arithmetic as a lambda-calculus" on an informal level. This will help understanding Gödel's system **T**, as the image of such a calculus under the contracting map.

As noted above, one may think of the axioms of HA as of constants of appropriate types. Reduction rules for such constants should be inspired by proof reduction rules corresponding to the use of these axioms in proofs. Of course, the most important axiom scheme is induction:

$$\forall a(\varphi(a) \to \varphi(sa)) \to \varphi(0) \to \forall a \varphi(a).$$

One can think of induction as an additional introduction rule for the universal quantifier, that might be written as

$$\frac{\forall a(\varphi(a) \to \varphi(sa)) \quad \varphi(0)}{\forall a \varphi(a)}$$

It may happen that the quantifier introduced by an inductive proof is then eliminated by applying the universal statement to a particular argument. This resembles very much an ordinary first-order beta-redex, which is considered a redundant part of the proof and may be normalized. In case of the universal quantifier introduced by induction it is also possible to eliminate the redundancy.

For instance, there is no point in proving $\forall a \varphi(a)$ if all we need is $\varphi(0)$. This calls for a proof reduction rule that replaces induction by its base step. To prove $\varphi(\underline{3})$ it is also not necessary to know that $\forall a \varphi(a)$. One can prove the base step $\varphi(0)$ and then apply the induction step three times.

There are two premises of the induction scheme, and a normalization step is possible when it is known which of the two is to be used. In other words, it is applicable whenever the argument is either zero or it is of the form st. In the latter case we do not eliminate induction altogether, but we can reduce its use to proving $\varphi(t)$, making the last induction step explicit.

Thus, if we assume an induction constant[1]

$$\mathsf{Ind}^\varphi : \forall a(\varphi(a) \to \varphi(sa)) \to \varphi(0) \to \forall a \varphi(a),$$

we could postulate the following reduction rules:

$$\mathsf{Ind}^\varphi MN0 \Rightarrow N;$$
$$\mathsf{Ind}^\varphi MN(st) \Rightarrow Mt(\mathsf{Ind}^\varphi MNt).$$

Hovewer, too much information will be lost if we apply a contracting map to the rules above (Exercise 10.1). Therefore it is more convenient to use a form of induction axiom closer to the original Peano style (cf. Section 9.7):

$$\mathsf{Ind}^\varphi : \forall a(\mathbf{int}\, a \to \varphi(a) \to \varphi(sa)) \to \varphi(0) \to \forall a(\mathbf{int}\, a \to \varphi(a)),$$

---

[1] Of course, in fact we have an infinite family of constants, one for each formula φ.

10.2. Syntax

where **int** is a unary relation symbol. Recall that a proof of $\textbf{int}\,\underline{n}$ can be obtained with the help of Peano's first and second axioms. Of course we see these as two typed constants:

$$P_1 : \textbf{int}\, 0;$$
$$P_2 : \forall a(\textbf{int}\, a \to \textbf{int}(sa)).$$

The reduction rules for the induction constant should now take such proofs into account. One rule can be as follows:

$$\textsf{Ind}^\varphi MN0P_1 \Rightarrow N. \tag{10.1}$$

The term M is of type $\forall a(\textbf{int}\, a \to \varphi(a) \to \varphi(sa))$ and represents the induction step. The term N, of type $\varphi(0)$, is the base step and P_1 is a proof that zero is a natural number.

The other rule represents the use of induction to prove statements about non-zero terms:

$$\textsf{Ind}^\varphi MN(st)(P_2 tQ^{\textbf{int}\, t})^{\textbf{int}(st)} \Rightarrow MtQ(\textsf{Ind}^\varphi MNtQ) : \varphi(st). \tag{10.2}$$

At the left hand side, M and N are again the induction step and the base step. The expression $P_2 tQ^{\textbf{int}\, t}$ is a proof that the individual argument st is a natural number. Not just any proof of $\textbf{int}(st)$, but one that contains a proof $Q : \textbf{int}\, t$. This proof is used at the right hand side.

We are not going to further investigate these reduction rules, nor even to formally define an appropriate lambda-calculus. Instead, we only consider the propositional (i.e. simply typed) variant.

10.2. Syntax

Our plan is to extend the simply typed lambda-calculus by a new constant, the *recursor*, the type of which is the image of the induction axiom under a contracting map. More precisely, as there are infinitely many instances of the induction scheme, we should also have a separate recursor $\mathbf{R}_\sigma$, for every type σ. The first and second Peano axioms are types of two other constants.

10.2.1. DEFINITION. Types of system **T** are defined by the grammar

$$\tau ::= \textbf{int} \mid \tau \to \tau \mid \tau \wedge \tau,$$

where **int** is a type constant. There are no type variables. Terms are like in the Church-style simply typed lambda-calculus with pairs and projections, and with the addition of new constants $\mathbf{0}$, $\mathbf{s}$ and $\mathbf{R}_\sigma$, for all types σ, together with the following axioms:

⊢ **0** : **int**;

⊢ **s** : **int** → **int**;

⊢ **R**$_\sigma$: (**int** → σ → σ) → σ → **int** → σ.

The reduction rules of system **T** are the ordinary beta-rules of the simply typed lambda-calculus with products, and the following two in addition:

$$\mathbf{R}_\sigma MN\mathbf{0} \Rightarrow N;$$
$$\mathbf{R}_\sigma MN(\mathbf{s}P) \Rightarrow MP(\mathbf{R}_\sigma MNP).$$

The symbols →$_T$, ↠$_T$ and =$_T$ refer to reductions and equality of system **T**.

The three constants of system **T** correspond to the first, second and fifth Peano axioms.[2] The reduction rules for **R**$_\sigma$ are images of (10.1) and (10.2) under the contracting map b of Section 8.7, extended by b(**int** a) = **int**.

Note that the numerals (individuals) occurring in formulas get erased. What stays in after applying the erasing map are the *proofs* that given individuals are natural numbers. These proofs become the numbers of system **T**. Axiom P$_1$ becomes a constant **0** of type **int**, and axiom P$_2$ contracts to a one-argument constructor **s**. One can hardly name it otherwise than a *successor*.

So we have numerals also at the propositional level—denote by $\overline{n}$ the term $\mathbf{s}(\mathbf{s}(\cdots(\mathbf{s0})\cdots))$, where **s** occurs n times. But remember that $\overline{n}$ is *not* obtained by contracting the individual $\underline{n}$, but by contracting a proof, stating that $\underline{n}$ is a legal integer.

10.2.2. DEFINITION (Definable functions). A function $f : \mathbb{N}^k \to \mathbb{N}$ is *definable* in system **T** by a closed term F if and only if

1. ⊢ F : **int**^{k} → **int**;

2. $F\overline{m}_1\ldots\overline{m}_k =_T \overline{f(m_1,\ldots,m_k)}$, for all $m_1,\ldots,m_k \in \mathbb{N}$.

The above resembles very much Definition 3.7.2. The difference is that **int** is now a type constant rather than $(p \to p) \to p \to p$, cf. Exercise 10.9.

10.2.3. EXAMPLE. Here is a definition of addition.

$$\text{plus} \;=\; \lambda x^{\text{int}} y^{\text{int}}.\, \mathbf{R}_{\text{int}}(\lambda uv.\,\mathbf{s}v)xy$$

Indeed, plus $\overline{m}\,\mathbf{0} \twoheadrightarrow_T \mathbf{R}_{\text{int}}(\lambda uv.\,\mathbf{s}v)\,\overline{m}\,\mathbf{0} \twoheadrightarrow_T \overline{m}$, and

$$\begin{aligned}
\text{plus}\,\overline{m}(\mathbf{s}\overline{n}) \;&\twoheadrightarrow_T\; \mathbf{R}_{\text{int}}(\lambda uv.\,\mathbf{s}v)\,\overline{m}(\mathbf{s}\overline{n}) \\
&\twoheadrightarrow_T\; (\lambda uv.\,\mathbf{s}v)\,\overline{n}\,(\mathbf{R}_{\text{int}}(\lambda uv.\,\mathbf{s}v)\,\overline{m}\,\overline{n}) \\
&\twoheadrightarrow_T\; \mathbf{s}(\mathbf{R}_{\text{int}}(\lambda uv.\,\mathbf{s}v)\overline{m}\,\overline{n}) \\
&\twoheadrightarrow_T\; \mathbf{s}(\overline{m+n}) \\
&=\; \overline{m+n+1}
\end{aligned}$$

---

[2] We do not include constants representing Peano's third and fourth axioms, and other axioms about algebraic equations. The contracting map makes these axioms irrelevant.

10.2. SYNTAX

It is not difficult to see that the functions definable with the help of $\mathbf{R}_{\text{int}}$ are exactly the primitive recursive functions (Exercise 10.4). However, as the size of type τ increases, more and more functions become definable with the help of $\mathbf{R}_\tau$. For instance, to define the Ackermann's function in system $\mathbf{T}$ one only needs $\mathbf{R}_{\text{int}\to\text{int}}$ (Exercise 10.5).

We will see later (Theorem 10.4.10) that the class of functions definable in system $\mathbf{T}$ coincides with the class of provably total functions of PA.

RECURSION VS. ITERATION. One can consider a variant of system $\mathbf{T}$ with an *iterator* constant (one for every type) instead of the recursor:

$$\mathbf{I}_\sigma : (\sigma \to \sigma) \to \sigma \to \text{int} \to \sigma,$$

and with the reduction rules:

$$\mathbf{I}_\sigma MN\mathbf{0} \Rightarrow N;$$
$$\mathbf{I}_\sigma MN(\mathbf{s}P) \Rightarrow M(\mathbf{I}_\sigma MNP).$$

If we think of a recursor as a generalized **for**-loop, then the iterator corresponds to a **for**-loop that has no access to the control variable. Every execution of the body of the loop must then be identical.

One can say that the iterator is as powerful as the recursor, because their expressive power with respect to definable functions is the same (Exercise 10.11). However, a closer look reveals a difference. Consider the following definition of predecessor:

$$\text{pred}(u) = \mathbf{R}_{\text{int}}(\lambda uv.u)\mathbf{0}u.$$

This definition is uniform in that we not only have $\text{pred}(\mathbf{s}\overline{n}) \to_T \overline{n}$, for all n in one step, but we have also $\text{pred}(\mathbf{s}u) \to_T u$, for a variable u. The predecessor function can be defined with the help of iterator, using a trick similar to that used in the proof of Lemma 1.7.4. But this definition is not uniform. And no better definition is possible. It follows from Parigot [379] that every iterator-based definition of predecessor must be of at least linear time complexity (in terms of the number of reductions). In particular no such definition may satisfy $\text{pred}(\mathbf{s}u) =_T u$, for a variable u, because then $\text{pred}(\mathbf{s}\overline{n}) \twoheadrightarrow_T \overline{n}$ would be possible in a constant number of steps.

The type of the iterator can be seen as a contraction of a formula of arithmetic. This formula is Peano's fifth axiom of 1891 (Section 9.7):

$$\forall a(\varphi \to \varphi(\mathbf{s}a)) \to \varphi(0) \to \forall a(\text{int } a \to \varphi(a)).$$

The difference between the axioms is that the first quantifier may be restricted or not. The induction scheme corresponding to the iterator is weaker than that corresponding to the recursor, because it requires a stronger condition in the induction step.

10.3. Strong normalization

The basic properties of simply typed λ-calculus are preserved in system **T**. In particular, the extra expressibility does not come at the expense of loosing strong normalization. We show the strong normalization theorem for system **T** by the computability method of Tait, i.e. in a similar way as we did it for simply typed combinatory logic in Chapter 5. The slight additional complication is caused by the presence of abstraction and product types.

Another difference is that our system **T** is Church-style. In fact it is convenient to follow the "orthodox" Church-style, i.e. to assume that free variables in terms have fixed types.

As in Section 5.3, we begin with the definition of computable terms.

10.3.1. DEFINITION. For each type τ, we define the set $[\![\tau]\!]$ of *computable* terms of type τ.

- $[\![\mathbf{int}]\!] = \{M : \mathbf{int} \mid M \in \mathrm{SN}\};$
- $[\![\tau \to \sigma]\!] = \{M : \tau \to \sigma \mid \forall N (N \in [\![\tau]\!] \to MN \in [\![\sigma]\!])\};$
- $[\![\tau \wedge \sigma]\!] = \{M : \tau \wedge \sigma \mid \pi_1(M) \in [\![\tau]\!] \text{ and } \pi_2(M) \in [\![\sigma]\!]\}.$

In Chapter 5, to show that $G \in [\![\tau]\!]$ we had to consider applications of the form $GH_1 \ldots H_n$. Now we also have product types and we must deal with both applications and projections. We need a generalization.

10.3.2. DEFINITION. An *eliminator* is a term $E[x]$, with exactly one occurrence of x, of one of the following forms:

- $E[x] = x;$
- $E[x] = E'[x]Q$, where $E'[x]$ is an eliminator, Q is a computable term and $x \notin \mathrm{FV}(Q);$
- $E[x] = \pi_1(E'[x])$ or $E[x] = \pi_2(E'[x])$, where $E'[x]$ is an eliminator.

We write $E[M]$ for the term $E[x][x := M]$, and we call it an M-*eliminator*. Warning: An y-eliminator $E[y]$ does not have to be an eliminator.

The following easy lemmas collect some properties of eliminators.

10.3.3. LEMMA.

(i) *If* $E[M] : \sigma \to \tau$ *is an* M-*eliminator, and* $Q \in [\![\sigma]\!]$ *then* $E[M]Q$ *is an* M-*eliminator.*

(ii) *If* $E[M] : \sigma \wedge \tau$ *is an* M-*eliminator, then* $\pi_1(E[M])$ *and* $\pi_2(E[M])$ *are* M-*eliminators.*

(iii) *An* y-*eliminator* $E[y]$ *never reduces to an abstraction or pair.*

10.3. STRONG NORMALIZATION

10.3.4. LEMMA. *The following are equivalent for all terms M:*

(i) *The term M is computable.*

(ii) *All M-eliminators are computable.*

(iii) *All M-eliminators of type* **int** *are strongly normalizing.*

We shall now prove properties similar to Lemmas 5.3.3–5.3.5.

10.3.5. LEMMA. *For all types τ:*

(i) $[\![\tau]\!] \subseteq \text{SN}$;

(ii) *If y is a variable then any strongly normalizing y-eliminator of type τ is computable. In particular, variables of type τ are computable.*

PROOF. Induction with respect to τ. If $\tau = $ **int** then both parts are obvious. Part (i) for $\tau = \tau_1 \to \tau_2$ is shown as in the proof of Lemma 5.3.3. For $\tau = \tau_1 \wedge \tau_2$ note that $\pi_1(M) \in \text{SN}$ implies $M \in \text{SN}$.

In part (ii) we consider the case $\tau = \tau_1 \to \tau_2$ as an example. Let $E[y]$ be a strongly normalizing y-eliminator. For $M \in [\![\tau_1]\!]$, we want to show that $E[y]M \in [\![\tau_2]\!]$. This follows from the induction hypothesis for τ_2, because $E[y]M$ is a strongly normalizing y-eliminator, by Lemma 10.3.3(i, iii). □

A consequence of the above is that all eliminators are computable, in particular strongly normalizing. Indeed, once x is computable, so are all $E[x]$.

10.3.6. LEMMA.

(i) *If $M[z := N] \in [\![\tau]\!]$ and $N \in \text{SN}$, then $(\lambda z{:}\sigma. M)N \in [\![\tau]\!]$.*

(ii) *If M_1 is computable and M_2 is strongly normalizing then $\pi_1(\langle M_1, M_2 \rangle)$ and $\pi_2(\langle M_2, M_1 \rangle)$ are computable.*

(iii) *The constants* **0**, **s** *and* $\mathbf{R}_\sigma$ *are computable.*

PROOF. In part (i) we show a slightly stronger statement, namely that

$$E[M[z := N]] \in [\![\tau]\!] \text{ and } N \in \text{SN} \text{ implies } E[(\lambda z{:}\sigma. M)N] \in [\![\tau]\!],$$

for any eliminator $E[x]$. The proof is by induction with respect to τ. For $\tau = $ **int**, let $P_0 \to P_1 \to P_2 \to \cdots$ be a reduction sequence beginning with $P_0 = E[(\lambda z{:}\sigma. M)N]$. Since $M, N, E[x] \in \text{SN}$, for some m we must have

$$P_m = E'[(\lambda z{:}\sigma. M')N'] \to P_{m+1} = E'[M'[z := N']],$$

where $M \twoheadrightarrow M'$ and $N \twoheadrightarrow N'$ and $E[x] \twoheadrightarrow E'[x]$. But then we also have $E[M[z := N]] \twoheadrightarrow E'[M'[z := N']]$, so our reduction sequence terminates.

Now let $\tau = \tau_1 \to \tau_2$. To show $E[(\lambda z{:}\sigma.\,M)N] \in [\![\tau]\!]$, assume $Q \in [\![\tau_1]\!]$. Then $E[M[z := N]]Q$ is an $M[z := N]$-eliminator, and from the induction hypothesis for τ_2 we obtain $E[(\lambda z{:}\sigma.\,M)N]Q \in [\![\tau_2]\!]$. A similar argument applies when $\tau = \tau_1 \wedge \tau_2$.

Part (ii) is similar and left to the reader. We show part (iii). Of course **0** is a normal form, and if N : **int** is normal then $\mathbf{s}N$ is normal too. It follows that $\mathbf{0} \in [\![\mathbf{int}]\!]$ and $\mathbf{s} \in [\![\mathbf{int} \to \mathbf{int}]\!]$. To show that recursors are computable, we prove the following claim:

If M, N, Q are computable, then $\mathbf{R}_\sigma MNQ$ is computable. $\qquad(*)$

Since $Q \in [\![\mathbf{int}]\!]$, we know that $Q \in \text{SN}$. Let Q_0 be the normal form of Q. Then $Q_0 = \mathbf{s}^n(Q_0')$, for some $n \geq 0$, where Q_0' does not begin with $\mathbf{s}$. The claim $(*)$ is shown by induction with respect to n. By Lemma 10.3.4, it suffices to prove that $E[\mathbf{R}_\sigma MNQ] \in \text{SN}$, for every eliminator $E[x]$ of type **int**. Consider a reduction sequence beginning with $E[\mathbf{R}_\sigma MNQ]$. Since $M, N, Q \in \text{SN}$, this sequence terminates after a number of internal reductions, or eventually we have $E[\mathbf{R}_\sigma MNQ] \twoheadrightarrow_T E'[\mathbf{R}_\sigma M'N'Q']$, with $\mathbf{R}_\sigma M'N'Q'$ reduced in the next step. Suppose that $Q' = \mathbf{0}$. Then our term reduces to $E'[N']$, which is strongly normalizing by the computability of N.

Otherwise $Q' = \mathbf{s}Q''$ (because $\mathbf{R}_\sigma M'N'Q'$ is a redex) and the next step is $E'[M'Q''(\mathbf{R}_\sigma M'N'Q'')]$. Now Q'' : **int** is strongly normalizing (because Q' is strongly normalizing) and thus computable. Also, $\mathbf{R}_\sigma MNQ''$ is computable by the induction hypothesis, since the normal form of Q'' has fewer leading occurrences of $\mathbf{s}$ than Q_0. The conclusion now follows from the computability of M, because $E[MQ''(\mathbf{R}_\sigma MNQ'')] \twoheadrightarrow E'[M'Q''(\mathbf{R}_\sigma M'N'Q'')]$. $\qquad\square$

Let $\text{FV}(M) = \{x_1, \ldots, x_n\}$ and let variables $x_1, \ldots, x_n$ be respectively of types $\tau_1, \ldots, \tau_n$. A term of the form $M[x_1, \ldots, x_n := M_1, \ldots, M_n]$, where $M_i \in [\![\tau_i]\!]$ for $i = 1, \ldots, n$, is called a *computable instance* of M.

10.3.7. LEMMA. *A computable instance of an arbitrary term is computable.*

PROOF. Induction with respect to the length of M. The case of a variable is obvious and the case of a constant follows from Lemma 10.3.6(iii). Suppose that $M = \lambda x^\sigma N^\tau$. We show that every computable instance $M' = \lambda x^\sigma N'$ of M belongs to $[\![\sigma \to \tau]\!]$. Assume $Q \in [\![\sigma]\!]$. Then $N'[x := Q]$ is an instance of N (note that x is not free in the terms substituted in M to create M') and thus belongs to $[\![\tau]\!]$ by the induction hypothesis. By Lemma 10.3.6(i) we obtain $(\lambda x^\sigma N')Q \in [\![\tau]\!]$, as desired.

Now suppose that $M = \langle M_1, M_2 \rangle$ is of type $\sigma \wedge \tau$ and let $M' = \langle M_1', M_2' \rangle$ be an instance of M. By the induction hypothesis $M_1' \in [\![\sigma]\!]$ and $M_2' \in [\![\tau]\!]$, and thus $M' \in [\![\sigma \wedge \tau]\!]$ by Lemma 10.3.6(ii).

The cases of application and projection follow immediately from the induction hypothesis. $\qquad\square$

10.3. STRONG NORMALIZATION

10.3.8. THEOREM. *The relation $\to_T$ is strongly normalizing.*

PROOF. Immediate from Lemmas 10.3.5(i) and 10.3.7. Each term is a computable instance of itself, because variables are computable. □

The statement of Theorem 10.3.8 is expressible in the language of first-order arithmetic as follows: *"For each (Gödel number of a) term M there is a number n such that all reduction paths from M consist of at most n steps"*. Thus it is a meaningful question whether the proof above can also be formalized in arithmetic. Unfortunately, it cannot. The difficulty lies in expressing the notion of a computable term.

For instance, Lemma 10.3.5(i) asserts that every computable term (of any type) is strongly normalizing. One can try to express this statement in the language of arithmetic as

$$\forall M\, (computable(M) \to \mathrm{SN}(M)),$$

where M is understood as a Gödel number. As we have already observed, it is possible to write a formula $\mathrm{SN}(M)$ expressing strong normalization. But we do not know how to write a formula $computable(M)$, expressing that M is computable. Indeed, our definition of $[\![\tau]\!]$ is given by induction, and this essentially amounts to stating that the following set of postulates are to be satisfied by the set X of computable terms:

- *If $M : \mathbf{int}$ then $M \in X \leftrightarrow \mathrm{SN}(M)$.*
- *If $M : \tau_1 \to \tau_2$ then $M \in X \leftrightarrow \forall N (N : \tau_1 \wedge N \in X \to MN \in X)$.*
- *If $M : \tau_1 \wedge \tau_2$ then $M \in X \leftrightarrow \pi_1(M) \in X \wedge \pi_2(M) \in X$.*

Although these postulates determine X uniquely, they do not provide an explicit definition of X. We have no better way to formalize Lemma 10.3.5(i) than this: *For all terms M and all sets X, if X satisfies the postulates, and $M \in X$, then M is strongly normalizing.*

The above requires quantification over sets, and we conclude that the argument in Tait's proof is not arithmetical. A formalization of this proof is possible in *second-order arithmetic* (see Chapter 12), but not in the ordinary first-order arithmetic. We can say more: no proof of Theorem 10.3.8 can ever be formalized in PA (Corollary 10.4.11).

One should however note that the Turing-Prawitz technique we used in Section 3.5 for the simply typed λ-calculus can be formalized in PA. Needless to say, this technique does not generalize to system **T**.

10.3.9. REMARK. For any given type τ it is not difficult to write a separate formula $computable_\tau(M)$, stating that $M \in [\![\tau]\!]$. This does not solve the problem in the general case, because we do not have a *uniform* definition of computability. But sometimes we can settle for less, for instance when we are proving that a specific term M is strongly normalizing. Then the proof can be made arithmetical. Indeed, only finitely many types occur in M and we can formalize the argument using a finite number of formulas of the form $computable_\tau$. Similar circumstances occur when we deal with a single function $f : \mathbb{N}^k \to \mathbb{N}$, definable by a term $F : \mathbf{int}^k \to \mathbf{int}$, see Proposition 10.4.8.

We conclude this section with an expected consequence of Theorem 10.3.8.

10.3.10. THEOREM. *The relation $\to_T$ is Church-Rosser.*

PROOF. It is not difficult to see that $\to_T$ has the weak Church-Rosser property. As we have strong normalization, it is enough to apply Newman's Lemma 3.6.2. □

10.4. Modified realizability

We have not formally defined a lambda-calculus corresponding to first-order arithmetic, but we still want to state an appropriate correspondence between provability in HA and type assignment in system **T**. We will exhibit such a correspondence (in the form of Kreisel's *modified realizability*) to an extent sufficient to analyze proofs of termination. The modified realizability differs from Kleene's original realizability in that we use terms of system **T** as realizers rather than Gödel numbers of recursive functions.

The main result of this section (Theorem 10.4.7) is that every provable formula φ of arithmetic is *m-realized* by a term of type $\flat(\varphi)$, where $\flat$ is a forgetful map similar to the contracting map $\mathfrak{b}$ of Chapter 8. The difference is that the map $\flat$ is much more forgetful than $\mathfrak{b}$. It erases everything that is not relevant to the principal goal of our construction. Under the operator $\flat$, a large class of formulas gets erased to nothing, or more precisely to an artificial constant **1**. Just in order to simplify the definitions to follow, we assume the following:

10.4.1. DEFINITION. We extend the set of types of system **T** by a fresh constant **1** and we assume **i** : **1**, where **i** is another fresh constant. For each type σ over the extended language, we define a type $|\sigma|$, which is either **1**-free or equal to **1**. The type $|\sigma|$ is the normal form of σ with respect to the following rewrite rules:

$$\tau \wedge \mathbf{1} \Rightarrow \tau$$
$$\mathbf{1} \wedge \tau \Rightarrow \tau$$
$$\tau \to \mathbf{1} \Rightarrow \mathbf{1}$$
$$\mathbf{1} \to \tau \Rightarrow \tau$$

10.4. MODIFIED REALIZABILITY

The idea is of course to simplify e.g. $\tau \wedge \sigma$ to σ when $|\tau| = \mathbf{1}$. Sometimes it is also convenient to think of a type σ as if it was actually a product $\mathbf{1} \wedge \sigma$.

10.4.2. DEFINITION. For M of type $|\tau_1 \wedge \tau_2|$ define

$$\pi_i^{\tau_1 \tau_2}(M) = \begin{cases} \pi_i(M), & \text{if } |\tau_1|, |\tau_2| \neq \mathbf{1}; \\ \mathbf{i}, & \text{if } |\tau_i| = \mathbf{1}; \\ M, & \text{if } |\tau_i| \neq \mathbf{1} \text{ and } |\tau_{3-i}| = \mathbf{1}. \end{cases}$$

Similarly, if M is of type $|\tau_1 \to \tau_2|$, and N is of type $|\tau_1|$, then

$$(MN)^{\tau_1 \tau_2} = \begin{cases} MN, & \text{if } |\tau_1|, |\tau_2| \neq \mathbf{1}; \\ \mathbf{i}, & \text{if } |\tau_2| = \mathbf{1}; \\ M & \text{if } |\tau_1| = \mathbf{1} \text{ and } |\tau_2| \neq \mathbf{1}. \end{cases}$$

When τ_1 and τ_2 are known from the context, we skip the superscripts, and write $\pi_i(M)$ and MN instead of $\pi_i^{\tau_1 \tau_2}(M)$ and $(MN)^{\tau_1 \tau_2}$, respectively.

By Proposition 9.5.1(iii), disjunction is definable in HA. Without loss of generality, we can thus restrict our attention to disjunction-free formulas.

10.4.3. DEFINITION. The forgetful map $\flat$ from formulas of arithmetic to types of system **T** extended with **1** is defined by the following clauses:

$$\begin{aligned}
\flat(\bot) &= \mathbf{1}; \\
\flat(s = t) &= \mathbf{1}; \\
\flat(\exists a \varphi) &= |\mathbf{int} \wedge \flat(\varphi)|; \\
\flat(\forall a \varphi) &= |\mathbf{int} \to \flat(\varphi)|; \\
\flat(\varphi \wedge \psi) &= |\flat(\varphi) \wedge \flat(\psi)|; \\
\flat(\varphi \to \psi) &= |\flat(\varphi) \to \flat(\psi)|.
\end{aligned}$$

This definition is clearly inspired by the understanding of $\forall a$ and $\exists a$ as restricted quantifiers $\forall a^{\mathbf{int}}$ and $\exists a^{\mathbf{int}}$ ranging over natural numbers. Note that such restricted quantifiers are usually understood as abbreviations:

$$\begin{aligned}
\forall a^{\mathbf{int}}(\ldots) &= \forall a(\mathbf{int}\, a \to \ldots); \\
\exists a^{\mathbf{int}}(\ldots) &= \exists a(\mathbf{int}\, a \wedge \ldots).
\end{aligned}$$

The reader should easily observe that the map $\flat$ collapses to $\mathbf{1}$ everything that does not have an existential quantifier in a "target" position. Note also that $\flat$ ignores algebraic terms, in particular that $\flat(\varphi) = \flat(\varphi[a := t])$ holds.

10.4.4. DEFINITION. The conditions below describe all possible cases when a closed term M of type $\flat(\varphi)$ *m-realizes* a closed formula φ. It is assumed that M either equals $\mathbf{i}$ or is $\mathbf{i}$-free.

- **i** : **1** m-realizes $t = s$ iff t and s rewrite to the same numeral.

- $M : |\mathbf{int} \wedge \flat(\psi)|$ m-realizes $\exists a \psi$ iff $\pi_1(M) =_T \overline{n}$, for some n, and $\pi_2(M)$ m-realizes $\psi[a := \underline{n}]$.

- $M : |\mathbf{int} \to \flat(\psi)|$ m-realizes $\forall a \psi$ iff $M\overline{n}$ m-realizes $\psi[a := \underline{n}]$, for all n.

- $M : |\flat(\varphi) \wedge \flat(\psi)|$ m-realizes $\varphi \wedge \psi$ iff $\pi_1(M)$ m-realizes φ and $\pi_2(M)$ m-realizes ψ.

- $M : |\flat(\varphi) \to \flat(\psi)|$ m-realizes $\varphi \to \psi$ iff MN m-realizes ψ whenever N m-realizes φ.

- No term m-realizes $\bot$.

For example, the constant **0** m-realizes the formula $\exists a(a = 0)$. Indeed, according to our convention, we have $\pi_1^{\mathbf{int},\mathbf{1}}\mathbf{0} = \mathbf{0}$ and $\pi_2^{\mathbf{int},\mathbf{1}}\mathbf{0} = \mathbf{i}$. A term M is said to *m-realize* a judgement $\Gamma \vdash \varphi$, with $\Gamma = \{\gamma_1, \ldots, \gamma_n\}$, whenever it m-realizes the universal closure of $\gamma_1 \to \cdots \to \gamma_n \to \varphi$. Finally, we say that a formula or a judgement is *m-realizable* when it is m-realized by some term.

10.4.5. LEMMA.

(i) If M m-realizes $\varphi[a := t]$ and $t \approx s$ then M realizes $\varphi[a := s]$.

(ii) Let M and N be **i**-*free* and $M =_T N$. If M m-realizes φ then also N m-realizes φ.

PROOF. Induction with respect to φ. □

10.4.6. CONVENTION. Our realizers are always **i**-free terms, with the only exception of the constant **i** itself. For uniformity, it is however convenient sometimes to use "mixed" expressions, in which case we identify

$$\begin{array}{rcl} \langle \mathbf{i}, M \rangle & \text{with} & M; \\ \langle M, \mathbf{i} \rangle & \text{with} & M; \\ \lambda a\, \mathbf{i} & \text{with} & \mathbf{i}; \\ \lambda x^\sigma \mathbf{i} & \text{with} & \mathbf{i}; \\ \lambda x^\tau M & \text{with} & M, \end{array}$$

where $|\tau| = \mathbf{1}$ in the last clause. In addition to that, we follow the convention of Definition 10.4.2, identifying $M\mathbf{i}$ with M, etc.

10.4.7. THEOREM. *Every provable judgement of* HA *is m-realizable.*

10.4. MODIFIED REALIZABILITY

PROOF. First we check that all axioms of HA are m-realizable. For instance $\forall a(a = a)$ is m-realized by $\mathbf{i} : \mathbf{1}$ because $\mathbf{1} = |\mathbf{int} \to \mathbf{1}|$ and every application $\mathbf{i}\overline{n} = \mathbf{i}$ m-realizes the equation $\underline{n} = \underline{n}$. Also the axiom $\forall a \forall b(a = b \to b = a)$ is realized by $\mathbf{i} : \mathbf{1}$. Indeed, $\mathbf{1} = |\mathbf{int} \to \mathbf{int} \to \mathbf{1} \to \mathbf{1}|$ and if a term N m-realizes the equation $\underline{n} = \underline{m}$ then $n = m$ and $N = \mathbf{i}$. Thus $\mathbf{i} = \mathbf{i}\overline{n}\overline{m}N$ realizes also the formula $\underline{m} = \underline{n}$. The fourth Peano axiom $\forall a(sa = 0 \to \bot)$ is also realized by $\mathbf{i}$ because $s\underline{n} = 0$ is not m-realizable.

As the last example we consider induction, leaving the other axioms to the reader. If $|\varphi| = \mathbf{1}$, then the induction axiom is realized by $\mathbf{i}$, so let $|\varphi| \neq \mathbf{1}$. Assume for simplicity that the only free variable of $\varphi(a)$ is a. Let P m-realize the induction step $\forall a(\varphi(a) \to \varphi(sa))$ and let Q m-realize the base step $\varphi(0)$. Now, for an arbitrary $n > 0$ the term

$$P\overline{n-1}(P\overline{n-2}(\cdots(P\overline{1}(P\mathbf{0}Q))\cdots))$$

m-realizes $\varphi(\underline{n})$. It follows that the term $M = \lambda pqn.\mathbf{R}_{\flat(\varphi)}pqn$ m-realizes the induction axiom. Indeed, for every n, the application $MPQ\overline{n}$ realizes $\varphi(\underline{n})$.

To show that provable judgements $\Gamma \vdash \varphi$ are m-realizable we proceed by induction with respect to proofs. There are several cases depending on the last rule used. We only consider a few of them, leaving the others to the reader. In what follows we assume that Γ consists of k assumptions $\varphi_1, \ldots, \varphi_k$, with $|\varphi_i| = \sigma_i$, for $i = 1, \ldots, k$, and that the free individual variables in Γ and φ are $a_1, \ldots, a_r$ (abbreviated as $\vec{a}$).

If the last rule is $\wedge$-introduction, i.e. the proof ends with

$$\frac{\Gamma \vdash \psi, \quad \Gamma \vdash \vartheta}{\Gamma \vdash \psi \wedge \vartheta}$$

and terms M and N m-realize $\Gamma \vdash \psi$ and $\Gamma \vdash \vartheta$, respectively, then the the conclusion is m-realized by the term $\lambda \vec{x} \lambda \vec{y}.\langle M\vec{x}\vec{y}, N\vec{x}\vec{y}\rangle$, where $\vec{x}$ is a sequence of r variables of type $\mathbf{int}$ and $\vec{y}$ abbreviates $y_1 : \sigma_1, \ldots, y_k : \sigma_k$.

Suppose the proof ends with a $\forall$-elimination of the form

$$\frac{\Gamma \vdash \forall a \psi}{\Gamma \vdash \psi[a := t]}$$

where $a \in \mathrm{FV}(\psi)$. (The case $a \notin \mathrm{FV}(\psi)$ is easier.) Thus all free variables in t are among $\vec{a}$ and the term t defines an integer function, definable in $\mathbf{T}$ by a term $\lambda \vec{x}.T$. Assume that M is a term m-realizing $\Gamma \vdash \forall a \psi$. Since $T[x_1, \ldots, x_r := \overline{n}_1, \ldots, \overline{n}_r] \twoheadrightarrow_T \overline{m}$ whenever $t[a_1, \ldots, a_r := \underline{n}_1, \ldots, \underline{n}_r] \approx \underline{m}$, it follows from Lemma 10.4.5 that the conclusion is m-realized by $\lambda \vec{x} \vec{y}. M\vec{x}\vec{y}T$.

Now consider a proof ending with

$$\frac{\Gamma \vdash \psi[a := t]}{\Gamma \vdash \exists a \psi}$$

and assume that M m-realizes the premise. Let $\lambda \vec{x}.T$ define in system **T** the function induced by t. Then $\lambda \vec{x} \lambda \vec{y}. \langle T, M\vec{x} \rangle$ m-realizes the conclusion.

As the last example take a proof ending with an $\exists$-elimination:

$$\frac{\Gamma \vdash \exists a \psi \quad \Gamma, \psi \vdash \vartheta}{\Gamma \vdash \vartheta}$$

with the first premise m-realized by M and the second one by N. Then the conclusion is m-realized by $\lambda \vec{x} \lambda \vec{y}. N\vec{x}(\pi_1(M\vec{x}\vec{y}))\vec{y}(\pi_2(M\vec{x}\vec{y}))$. □

REMARK. The result above gives an alternative proof of the consistency of HA and of PA. Indeed, no term can m-realize $\bot$.

Observe that in the proof of Theorem 10.4.7 we actually construct a realizer for any proof in HA. In the case of a theorem of the form $\forall a \exists b \psi(a,b)$, with an atomic ψ, the realizer is a program computing a function f satisfying $\psi(n, f(n))$ for all n. The same can be achieved for arbitrary ψ with a slightly different definition of realizability (as in Exercise 10.15). In this way, modified realizability *realizes* the idea of program extraction (cf. Section 4.8).

EXPRESSSIVE POWER. Using modified realizability, we can characterize the definable functions of system **T**. The first step is to state an upper bound.

10.4.8. PROPOSITION. *All functions definable in* **T** *are provably total in* PA.

PROOF. Given a term $F : \text{int} \to \text{int}$, consider a Turing machine which halts for a given n if and only if $F\overline{n}$ is strongly normalizing, and then returns the normal form of $F\overline{n}$. Of course, our machine computes the function f definable by F in system **T**. For an appropriate primitive recursive t_f, proving the formula $\forall x \exists y \, (t_f(x,y) = 0)$ amounts thus to proving that all applications of the form $F\overline{n}$ are strongly normalizable. One can do it (with Tait's technique) in such a way that only finitely many types must be considered. As we have already observed (Remark 10.3.9), all this can be coded into arithmetic. □

10.4.9. THEOREM. *All functions provably total in first-order arithmetic*[3] *are definable in* **T**.

PROOF. Suppose HA $\vdash \forall a \exists b (t_f(a,b) = 0))$. From Theorem 10.4.7 there is a term M which m-realizes this formula. This means that for an arbitrary n the application $M\overline{n}$ m-realizes $\exists b(t_f(\underline{n}, b) = 0))$. Thus $\pi_1(M\overline{n}) =_T \overline{m}$ for some m, and $\pi_2(M\overline{n})$ m-realizes $t_f(\underline{n}, \underline{m}) = 0$. Since the latter equation is an m-realized atomic formula, it follows that $t_f(n,m) = 0$ must actually hold, and thus $f(n)$ is the left component of m. It follows that f is defined by $\lambda x \, L(\pi_1(Mx))$, where L defines the left converse of the pairing function. □

---

[3] By Corollary 9.5.7 it does not matter whether we mean PA or HA in this context.

Together with Proposition 10.4.8, this gives:

10.4.10. THEOREM. *The functions definable in system* **T** *are exactly the provably total functions of first-order arithmetic.* □

10.4.11. COROLLARY. *The strong normalization theorem for system* **T** *is a statement independent from* PA.

PROOF. Exercise 10.13 □

10.5. Notes

A standard reference for system **T** is Gödel's original article [194] of 1958. But the system T of "computable functions of finite types" as defined in [194] is a less precise and more general theory than what is usually understood today as "Gödel's system **T**." In particular, the paper mentions neither any recursor constant nor the issue of normalization. What we now call system **T** can be seen merely as a "term model" of Gödel's original T [20, 481], while for instance Tait's infinitely long terms of [466] make a different model.

System **T** (initially called Σ) is also much older than 1958. Gödel's lecture notes [192] of 1941 contain an exposition of the ideas of [194], where a higher-order recursion scheme is explicitly mentioned. According to [480], in one of his lectures in 1941, Gödel even discussed possible techniques for a normalization proof, one of these resembling the computability method later introduced by Tait.

The main result of [194] is a proof of consistency of arithmetic by means of what became known as the *Dialectica* interpretation, after the name of the journal where the paper was published. Roughly speaking, the idea is as follows. Every formula φ in the language of arithmetic can be translated to a "λ-formula," that is, an expression of the form $\varphi^D = \exists u^\sigma \forall v^\tau \varphi_D$, where φ_D is a term of system **T** of type **int**. It is shown that HA $\vdash \varphi$ implies the following property: There exists a term M of type σ (the realizer) with $v \notin \mathrm{FV}(M)$ such that for all closed N of type τ one has $\varphi_D[u := M][v := N] =_T \bar{1}$. Since $\bot_D$ is defined as 0, it follows that $\bot$ is not provable. In addition, a realizer of a formula of the form $\varphi = \forall x \, \exists y \, (t_f(x, y) = 0)$ turns out to be a definition of f in system **T**, and this gives an alternative proof of Theorem 10.4.10.[4]

Applications of the *Dialectica* interpretation to various extensions of first-order arithmetic were investigated by Kreisel [277], Spector [451], Girard [179] and others. Kreisel's paper [277] also contains a first hint of the idea of modified realizability, introduced later in [279].[5] Our proof of Theorem 10.4.10 is inspired by [435].

The papers of Grzegorczyk [210] and Dragalin [132] contributed to the modern style of presenting system **T** as a λ-calculus with integers and recursor, which became customary after the books [239, 459, 482].

Many expositions of system **T** do not include product types, see for instance [241] or [189]. We do this, because it is convenient (products correspond directly to conjunctions), and because this extension does not change the expressive power of the system (Theorem 10.4.10 remains true without products) as shown in [210].

---

[4] See [443] for yet another proof, an early application of the reducibility approach.
[5] Modified realizability was independently discovered by Dragalin [132, 344].

Stenlund's book [459] was probably the first to contain a Tait-style *strong* normalization proof for system **T**, based on Tait's [467], where the computability method was introduced, but only "weak" normalization was shown. Other strong normalization proofs [246, 421] used various forms of transfinite induction.

The integers of system **T** can be seen as a simple example of an inductive type. Calculi with inductive types are presently studied by many authors for their relevance to programming. Our strong normalization proof resembles to some extent the more sophisticated arguments used with arbitrary inductive types [2, 430].

To learn more about system **T**, program extraction from arithmetical proofs, and related subjects, see [430, 444, 482], and volumes II and III of [195]. For the *Dialectica* interpretation, see [20, 241, 443].

10.6. Exercises

10.1. Extend $\lambda_\to$ by a new constant $\mathbf{D}^\varphi$, and the reductions $\mathbf{D}^\varphi MN \to_D N$ and $\mathbf{D}^\varphi MN \to_D M(\mathbf{D}^\varphi MN)$, for $M : \varphi \to \varphi$ and $N : \varphi$. Show that $M =_{\beta D} N$, for all terms M, N of the same type.

10.2. Show that multiplication, exponentiation, subtraction, and all your favourite integer functions are definable in system **T**.

10.3. Here are our favourite integer functions: let f_0 be the successor function, and let $f_{k+1}(n) := f_k^n(n)$ (apply f_k to n exactly n times). Show that all functions f_k are definable in system **T**.

10.4. Show that a function is definable in system **T** with the help of $\mathbf{R}_{\text{int}}$ if and only if it is primitive recursive.

10.5. The function f_ω is given by the diagonal equation $f_\omega(n) = f_n(n)$, where f_n is as in Exercise 10.3. (This is essentially the *Ackermann function*, also called *Ackermann-Sudan function*.) Prove that f_ω is not primitive recursive. *Hint:* Show that if f is a primitive recursive function then there is k such that $f(\vec{n}) \leq f_k(\max \vec{n})$ holds whenever $\max \vec{n} \geq 2$. In other words, every primitive recursive function is majorized by one of the f_k's.

10.6. Show that the Ackermann function (Exercise 10.5) is definable in system **T**.

10.7. Show that all functions $f_{\omega+k+1}(n) = f_{\omega+k}^n(n)$ are definable in system **T**, as well as the function $f_{\omega \cdot 2}(n) = f_{\omega+n}(n)$.

10.8. Go ahead, define even faster growing functions, all definable in system **T**. Will this ever come to an end?

10.9. Show that all the functions definable in $\lambda_\to$ are definable in system **T**.

10.10. Without appealing to Theorem 10.4.8 prove that there are recursive functions not definable in system **T**.

10.11. Show that every function definable in system **T** is also definable in a variant of system **T** where recursors are replaced by iterators.

10.12. Let $\{F_n\}_{n\in\mathbb{N}}$ be an effective enumeration of all closed terms of system **T** of type $\text{int} \to \text{int}$. Let $g(n,k)$ be the length of the longest reduction sequence beginning with $F_n\overline{k}$. By a diagonal argument show that g is not provably total in PA.

10.13. Prove Corollary 10.4.11 (that SN for system **T** is independent from PA). *Hint:* Use Exercise 10.12.

10.14. Prove strong normalization for system **T** extended with sum types. *Hint:* Define $A \vee B$ as $\text{int} \wedge (A \wedge B)$ with $\mathbf{in}_1(M^A) = \langle 0, \langle M, c^B \rangle \rangle$, $\mathbf{in}_2(M^B) = \langle 1, \langle c^A, M \rangle \rangle$.

10.6. EXERCISES

10.15. Define a notion of modified ⊢-realizability, by adding provability conditions to Definition 10.4.4 in a similar way as in the proof of Proposition 9.6.6. Prove that a closed formula is ⊢-m-realizable if and only if it is provable in HA.

10.16. One can define the notion of a function representable and strongly representable in HA in exactly the same way as in PA (cf. Definition 9.4.1). Show that every function strongly representable in HA is provably recursive in HA. *Hint:* Use Exercise 10.15.

Chapter 11

Second-order logic and polymorphism

In first-order logic quantifiers range over individual objects (see Chapter 8). In contrast, by *second-order logic* one usually means a logic involving quantification over sets or relations. In *classical* second-order logics[1] it is natural to assume that a set variable ranges over all possible subsets of a domain. Constructively, it is more adequate to assume that quantifiers range over definable sets or relations (i.e. predicates expressible in the language) rather than over all sets or relations.

Typically, second-order logic is defined as an extension of first-order logic (and then it may be called second-order *predicate* logic), but quantification over propositions (nullary predicates) makes perfect sense also in the absence of individuals. The *second-order propositional logic* is obtained by adding second-order quantifiers, binding propositional variables, directly to propositional calculus.

In the next chapter we consider second-order predicate logic (though we do not study it in depth). Now we concentrate on the simpler formalism of the intuitionistic second-order propositional logic.

Although the language of second-order predicate logic is much richer than that of the propositional fragment, many properties of the former can be studied in the simplified setting of the latter. In the presence of second-order quantification, first-order features turn out to be somehow less distinctive than it might be expected. For instance, consider the second-order induction axiom for natural numbers written as

$$\forall R(\forall a(a \in R \to sa \in R) \to 0 \in R \to \forall a(a \in R)).$$

The second-order propositional formula corresponding to this axiom is

$$\forall r((r \to r) \to r \to r).$$

---
[1]Some of these have gained significant attention in computer science, see e.g. [477].

Under the Curry-Howard isomorphism, this formula becomes a type assigned to all the Church numerals in the *polymorphic lambda-calculus* (also known as λ2, or *Girard's system* **F**). This calculus is an excellent example of the Curry-Howard correspondence, and provides undeniable evidence for the intimate relation between computer science and logic. First introduced in 1970 by the logician Jean-Yves Girard, it was re-invented in 1974 by the computer scientist John Reynolds, for an entirely different purpose.

11.1. Propositional second-order logic

We extend the language of propositional logic by quantifiers binding propositional variables.

11.1.1. DEFINITION. Assuming an infinite set PV of propositional variables, we define the *second-order propositional formulas*:

- Each propositional variable is a formula.

- The constant $\bot$ is a a formula.

- If φ and ψ are formulas then $(\varphi \to \psi)$, $(\varphi \vee \psi)$, $(\varphi \wedge \psi)$ are formulas.

- If φ is a formula and $p \in \text{PV}$ then $\forall p \varphi$ and $\exists p \varphi$ are formulas.

The set of all propositional second-order formulas built over a set PV of propositional variables is denoted by $\Phi_2(\text{PV})$ or simply by Φ_2, when PV is known. The variable p is bound in $\forall p \varphi$ and $\exists p \varphi$; that is, we have the following definition of free type variables $\text{FV}(\varphi)$ in a formula φ.

- $\text{FV}(\bot) = \varnothing$ and $\text{FV}(p) = \{p\}$;

- $\text{FV}(\varphi \circ \psi) = \text{FV}(\varphi) \cup \text{FV}(\psi)$, for $\circ \in \{\to, \vee, \wedge\}$;

- $\text{FV}(\nabla p \varphi) = \text{FV}(\varphi) - \{p\}$, for $\nabla \in \{\forall, \exists\}$.

Of course $\text{FV}(\Gamma)$, for a set Γ of formulas, stands for the union of all $\text{FV}(\varrho)$, for $\varrho \in \Gamma$. Alpha-conversion is assumed as usual; we identify formulas which differ only in their bound variables. Also the definition of substitution should now be considered routine:

- $p[p := \vartheta] = \vartheta$ and $q[p := \vartheta] = q$, when $q \neq p$;

- $(\varphi \circ \psi)[p := \vartheta] = \varphi[p := \vartheta] \circ \psi[p := \vartheta]$ for $\circ \in \{\to, \vee, \wedge\}$;

- $(\nabla q \varphi)[p := \vartheta] = \nabla q \varphi[p := \vartheta]$, for $\nabla \in \{\forall, \exists\}$ and $q \notin \text{FV}(\vartheta) \cup \{p\}$.

11.1. PROPOSITIONAL SECOND-ORDER LOGIC

The intended understanding of $\forall p \varphi(p)$ is that $\varphi(p)$ holds for all possible meanings of p. The understanding of $\exists p \varphi(p)$ is that $\varphi(p)$ holds for some meaning of p. Classically, there are just two possible such meanings: the two truth values. Thus, the universal formula $\forall p \varphi(p)$ is classically equivalent to $\varphi(\top) \wedge \varphi(\bot)$, and $\exists p \varphi(p)$ is equivalent to $\varphi(\top) \vee \varphi(\bot)$. Therefore, every property expressible with quantifiers can be also expressed without them.[2] With intuitionistic logic, the situation is different (see Exercises 11.9–11.10). There is no finite set of truth-values, and the propositional quantifiers should be regarded as ranging over some infinite space of propositions.

The intuitive meaning of quantified expressions is best explained by means of the Brouwer-Heyting-Kolmogorov interpretation.

- A *construction of* $\forall p \varphi(p)$ *is a method (function) transforming any proposition* **P** *into a proof of* $\varphi(\mathbf{P})$.

- A *construction of* $\exists p \varphi(p)$ *consists of a proposition* **P** *and a construction of* $\varphi(\mathbf{P})$.

We formalize the principles above by assuming that a "proposition" is an arbitrary formula—the second-order quantifiers range over the class of *all* formulas. This is an important assumption, called *full comprehension*, which does not have to be accepted in all systems of second-order logic (cf. *ramified* second-order logic in e.g. [307, 403]). Sometimes the principle of full comprehension is expressed by the following schemes, where $p \notin \mathrm{FV}(\psi)$:

$$\forall p \varphi \to \varphi[p := \vartheta];$$
$$\exists p (p \leftrightarrow \psi).$$

One has to be aware that the full comprehension principle has the following consequence, called *impredicativity* of second-order logic. The meaning of a formula $\forall p \varphi$ is determined by the meanings of all formulas $\varphi[p := \vartheta]$, including the cases where ϑ is equally or more complex than $\forall p \varphi$ itself. Unlike the first-order case, an instance of a universal formula can thus by no means be considered simpler than the formula itself, and the ordinary well-founded hierarchy of formulas is no longer available. This means in particular that many proof methods based on induction must fail for second-order logic.

The Brouwer-Heyting-Kolmogorov interpretation can be seen as a justification of the following natural deduction rules, where the informal notion of "proposition" is replaced by "formula".

11.1.2. DEFINITION. The natural deduction system for intuitionistic second-order propositional logic consists of the ordinary rules for propositional connectives (Figure 2.1) plus the following rules for quantifiers, very similar to the first-order quantifier rules of Section 8.3.

---

[2]But at a certain cost: The size of formulas grows exponentially.

$$(\forall \text{I}) \frac{\Gamma \vdash \varphi}{\Gamma \vdash \forall p \varphi} \ (p \notin \text{FV}(\Gamma)) \qquad (\forall \text{E}) \frac{\Gamma \vdash \forall p \varphi}{\Gamma \vdash \varphi[p := \psi]}$$

$$(\exists \text{I}) \frac{\Gamma \vdash \varphi[p := \psi]}{\Gamma \vdash \exists p \varphi} \qquad (\exists \text{E}) \frac{\Gamma \vdash \exists p \varphi \quad \Gamma, \varphi \vdash \psi}{\Gamma \vdash \psi} \ (p \notin \text{FV}(\Gamma, \psi))$$

To obtain a Hilbert-style proof system or sequent calculus, we proceed in the same way as we did for first-order logic in Chapter 8.

SEMANTICS. We begin, as usual, with the algebraic approach, based on Heyting algebras, and then we generalize the notion of a Kripke model.

11.1.3. DEFINITION. Let $v : \text{PV} \to \mathcal{H}$ be a valuation of propositional variables in a complete Heyting algebra $\mathcal{H}$. Then

$$\begin{aligned}
[\![p]\!]_v &= v(p); \\
[\![\bot]\!]_v &= 0; \\
[\![\varphi \vee \psi]\!]_v &= [\![\varphi]\!]_v \sqcup [\![\psi]\!]_v; \\
[\![\varphi \wedge \psi]\!]_v &= [\![\varphi]\!]_v \sqcap [\![\psi]\!]_v; \\
[\![\varphi \to \psi]\!]_v &= [\![\varphi]\!]_v \Rightarrow [\![\psi]\!]_v; \\
[\![\forall p \varphi]\!]_v &= \inf\{[\![\varphi]\!]_{v(p \mapsto a)} \mid a \in \mathcal{H}\}; \\
[\![\exists p \varphi]\!]_v &= \sup\{[\![\varphi]\!]_{v(p \mapsto a)} \mid a \in \mathcal{H}\}.
\end{aligned}$$

We use the symbol $\models$ in the obvious way, remembering that we deal exclusively with complete algebras. For instance, we write $\models \varphi$ (and we say that φ is a tautology) iff φ has the value 1 under all valuations in all complete Heyting algebras.

11.1.4. THEOREM (Heyting completeness). *The conditions $\Gamma \models \varphi$ and $\Gamma \vdash \varphi$ are equivalent.*

PROOF. Similar to the proof of Theorem 8.5.6. Can be found in [172]. □

Unfortunately, as in the first-order case, Euclidean spaces do not define a complete semantics for second-order propositional logic (Exercise 11.3).

The definition of a Kripke model below is motivated by the following idea: The meaning of a proposition is the set of states where this proposition is satisfied. Because of monotonicity, this set is always upward-closed. Thus propositional variable should be interpreted as upward-closed set of states.

11.1.5. DEFINITION.

(i) A second-order *Kripke model* is a tuple of the form

$$\mathcal{C} = \langle C, \leq, \{D_c \mid c \in C\}\rangle,$$

11.1. PROPOSITIONAL SECOND-ORDER LOGIC

where C is a non-empty set, $\leq$ is a partial order in C, and each D_c is a non-empty collection of upward-closed subsets of C, satisfying

$$\text{if } c \leq c' \text{ then } D_c \subseteq D_{c'}.$$

The intuition is that D_c is the family of predicates meaningful or understood at state c. Expanding the knowledge means to assert more propositions but also to *comprehend* more statements.

(ii) A *valuation* in $\mathcal{C}$ assigns upward-closed subsets of C to propositional variables. We say that a valuation v is *admissible* for a state c and a formula φ iff $v(p) \in D_c$, for all $p \in \mathrm{FV}(\varphi)$. Clearly, a valuation is admissible for c and φ is also admissible for c' and φ, whenever $c' \geq c$.

(iii) The forcing relation $c, v \Vdash \varphi$ is defined when v is admissible for c and φ:

- $c, v \Vdash p$ iff $c \in v(p)$;
- $c, v \Vdash \varphi \vee \psi$ iff $c, v \Vdash \varphi$ or $c, v \Vdash \psi$;
- $c, v \Vdash \varphi \wedge \psi$ iff $c, v \Vdash \varphi$ and $c, v \Vdash \psi$;
- $c, v \Vdash \varphi \to \psi$ iff $c', v \Vdash \psi$, for all $c' \geq c$ with $c', v \Vdash \varphi$;
- $c, v \Vdash \bot$ never happens;
- $c, v \Vdash \exists p \varphi$ iff $c, v(p \mapsto x) \Vdash \varphi$, for some $x \in D_c$;
- $c, v \Vdash \forall p \varphi$ iff $c', v(p \mapsto x) \Vdash \varphi$, for all $c' \geq c$, and all $x \in D_{c'}$.

One writes $c, v \Vdash \Gamma$ when $c, v \Vdash \gamma$ for all $\gamma \in \Gamma$. The notation $\mathcal{C} \Vdash \varphi$ means that $c, v \Vdash \varphi$ whenever v is admissible for c and φ.

As usual, the meaning of a formula depends only on its free variables.

11.1.6. LEMMA. *Let v_1, v_2 be admissible for c and φ and let $v_1(p) = v_2(p)$ for all $p \in \mathrm{FV}(\varphi)$. Then*

$$c, v_1 \Vdash \varphi \quad \text{if and only if} \quad c, v_2 \Vdash \varphi.$$

PROOF. Routine induction with respect to the length of φ. □

It is easy to see that Kripke semantics is monotone in that $c, v \Vdash \varphi$ implies $c', v \Vdash \varphi$, for all $c' \geq c$. In other words, the "meaning" of φ, i.e. the set $\{c \mid c, v \Vdash \varphi\}$ is upward-closed, for each φ and v. However it does not a priori have to be a member of D_c, for any given c. Because of this, the semantics based on arbitrary Kripke models would be unsound (Lemma 11.1.8), and we need the following restriction (note that in state c only the states $c' \geq c$ must be taken into account).

11.1.7. DEFINITION.

(i) An element $x \in D_c$ *represents* a formula φ in state c under a valuation v if the following equivalence holds for all $c' \geq c$:

$$c' \in x \quad \text{if and only if} \quad c', v \Vdash \varphi.$$

A Kripke model is *full for* PV (or just *full* when PV is fixed) iff for every formula $\varphi \in \Phi_2(\text{PV})$, every c and every v, admissible for c and φ, there exists $x \in D_c$ which represents φ in c under v. (If we understand the meaning of propositional variables in φ then we should understand φ.)

(ii) We write $\Gamma \Vdash \varphi$ if $c, v \Vdash \Gamma$ implies $c, v \Vdash \varphi$ for every full Kripke model $\mathcal{C}$, every $c \in \mathcal{C}$ and every valuation v admissible for c and all formulas in $\Gamma \cup \{\varphi\}$.

11.1.8. LEMMA. *A model $\mathcal{C}$ is full for* PV *if and only if* $\mathcal{C} \Vdash \exists p(p \leftrightarrow \varphi)$, *for every* $\varphi \in \Phi_2(\text{PV})$ *and* $p \notin \text{FV}(\varphi)$.

PROOF. Let v be admissible for c and φ. The statement $c, v \Vdash \exists p(p \leftrightarrow \varphi)$ means that there is an $x \in D_c$ such that $c, v(p \mapsto x) \Vdash p \leftrightarrow \varphi$. This is the same as saying that the conditions $c', v(p \mapsto x) \Vdash p$ and $c', v(p \mapsto x) \Vdash \varphi$ are equivalent, for all $c' \geq c$. By Lemma 11.1.6 and the definition of forcing, we can reformulate the latter equivalence as: "$c' \in x$ iff $c', v \Vdash \varphi$". □

Our next lemma is similar to Lemma 8.6.4:

11.1.9. LEMMA. *Assume that $x \in D_c$ represents a formula ψ in state c under a valuation v, admissible for c and $\varphi[p := \psi]$. Then*

$$c, v \Vdash \varphi[p := \psi] \quad \text{if and only if} \quad c, v(p \mapsto x) \Vdash \varphi.$$

PROOF. Induction with respect to the length of φ, using Lemma 11.1.6. □

The proof of Kripke completeness is similar to the proof of Theorem 8.6.7 and uses the following analogue of Henkin's Lemma 8.6.6.

11.1.10. DEFINITION. Let $\Gamma \subseteq \Phi_2(\text{PV})$ be such that the following holds for all formulas $\varphi, \psi \in \Phi_2(\text{PV})$:

(i) If $\Gamma \vdash \varphi \vee \psi$ then either $\Gamma \vdash \varphi$ or $\Gamma \vdash \psi$.

(ii) If $\Gamma \vdash \exists p \varphi(p)$ then $\Gamma \vdash \varphi(q)$, for some $q \in \text{PV}$.

Then we say that Γ is *prime with respect to* PV.

11.1. PROPOSITIONAL SECOND-ORDER LOGIC

11.1.11. LEMMA. *Assume that* $\Gamma \nvdash \varphi$ *for some* $\Gamma, \varphi \subseteq \Phi_2(\text{PV})$. *Let* PV' *be a countably infinite set of new propositional variables. Then there exists a set* $\Gamma' \subseteq \Phi_2(\text{PV} \cup \text{PV}')$, *which is prime with respect to* $\text{PV} \cup \text{PV}'$, *and such that* $\Gamma \subseteq \Gamma'$ *and* $\Gamma' \nvdash \varphi$.

PROOF. A routine adaptation of the proof of Lemma 8.6.6. □

11.1.12. THEOREM (Kripke completeness). *The conditions* $\Gamma \Vdash \varphi$ *and* $\Gamma \vdash \varphi$ *are equivalent.*

PROOF. The soundness verification is routine (use Lemma 11.1.9). Observe that the restriction to full models is needed in cases ($\forall$E) and ($\exists$I).

The completeness proof is similar to that of Theorem 8.6.7. Take an infinite sequence $\text{PV} = \text{PV}_0 \subseteq \text{PV}_1 \subseteq \text{PV}_2 \subseteq \cdots$ of sets of propositional variables, such that each $\text{PV}_{n+1} - \text{PV}_n$ is countably infinite. Consider a Kripke model $\mathcal{C} = \langle C, \leq, \{D_c \mid c \in C\}\rangle$, where C consists of all pairs $\langle \Delta, \text{PV}_n \rangle$, where $\Delta \subseteq \Phi_2(\text{PV}_n)$ is prime with respect to PV_n. As in the first-order case, we define $\langle \Delta, \text{PV}_n \rangle \leq \langle \Delta', \text{PV}_m \rangle$ when $\Delta \subseteq \Delta'$ and $n \leq m$.

Let $v(p) = \{\langle \Delta, \text{PV}_n \rangle \in C \mid \Delta \vdash p\}$, for all $p \in \bigcup_{n \in \mathbb{N}} \text{PV}_n$, and define $D_c = \{v(p) \mid p \in \text{PV}_n\}$, when $c = \langle \Delta, \text{PV}_n \rangle$. By induction with respect to the length of $\psi \in \Phi_2(\text{PV}_n)$ one proves that

$$\langle \Delta, \text{PV}_n \rangle, v \Vdash \psi \quad \text{if and only if} \quad \Delta \vdash \psi. \qquad (*)$$

If ψ is a variable then $(*)$ holds by definition. The cases of propositional connectives are handled as in the first-order case, so assume that $\psi = \exists p \psi_1$ and let $c = \langle \Delta, \text{PV}_n \rangle$ be such that $c, v \Vdash \psi$. Then there exists an element $v(q) \in D_c$, where $q \in \text{PV}_n$, such that $c, v(p \mapsto v(q)) \Vdash \psi_1$. From Lemma 11.1.9 it follows that $c, v \Vdash \psi_1[p := q]$, so by the induction hypothesis we have $\Delta \vdash \psi_1[p := q]$, and thus also $\Delta \vdash \psi$. The implication from right to left in the case of $\psi = \exists p \psi_1$ is a consequence of primeness.

Suppose now that $\psi = \forall p \psi_1$. For the implication from left to right assume that $\Delta \nvdash \forall p \psi_1$. We have $\Delta \nvdash \psi_1[p := q]$, for $q \in \text{PV}_{n+1} - \text{PV}_n$. Expand Δ to a Δ', such that $\Delta' \nvdash \psi_1[p := q]$ and prime with respect to PV_{n+2}. Then $\langle \Delta', \text{PV}_{n+2} \rangle, v \nVdash \psi_1[p := q]$ by the induction hypothesis. It follows that $\langle \Delta', \text{PV}_{n+2} \rangle, v(p \mapsto v(q)) \nVdash \psi_1$ and thus $\langle \Delta, \text{PV}_n \rangle, v \nVdash \psi$.

For the other implication, assume $\Delta \vdash \psi$ and $\langle \Delta, \text{PV}_n \rangle \leq c' = \langle \Delta', \text{PV}_m \rangle$ and take any $q \in \text{PV}_m$. We have to show that $c', v(p \mapsto v(q)) \Vdash \psi_1$, i.e. that $c', v \Vdash \psi_1[p := q]$. But that is a consequence of $\Delta' \vdash \psi_1[p := q]$ and the induction hypothesis.

The property $(*)$ has been shown, so we can begin the main part of the proof. We prove completeness for formulas in $\Phi_2(\text{PV})$. Let $\Gamma, \varphi \subseteq \Phi_2(\text{PV})$ and suppose that $\Gamma \nvdash \varphi$. By $(*)$ and Lemma 11.1.11 we have a set Δ, prime with respect to PV_1 and such that $\langle \Delta, \text{PV}_1 \rangle, v \Vdash \Gamma$ but $\langle \Delta, \text{PV}_1 \rangle, v \nVdash \varphi$. To conclude that $\Gamma \nVdash \varphi$ we only need to observe that the model is full for PV.

This follows from (∗) and Lemmas 11.1.8 and 11.1.9, because all the formulas $\exists p(p \leftrightarrow \psi)$, where $\psi \in \Phi_2(\mathrm{PV})$ and $p \notin \mathrm{FV}(\psi)$, are forced in every state. Indeed, $c, w \Vdash \vartheta$ iff $c, v \Vdash \vartheta'$ for a substitution instance ϑ' of ϑ. □

11.1.13. REMARK. An additional axiom scheme of Gabbay [156, 157] is the second-order version of the Grzegorczyk scheme:

$$\forall p(\psi \vee \varphi(p)) \to \psi \vee \forall p \varphi(p), \quad p \notin \mathrm{FV}(\psi).$$

This is a classical second-order tautology, but not an intuitionistic tautology. The class of models corresponding to propositional second-order intuitionistic logic extended with Gabbay's axiom is obtained by postulating that all D_c are equal to each other (models with constant domains), quite like in the first-order case (Exercise 11.5).

11.1.14. REMARK. It is tempting to consider Kripke models with all D_c equal to the family of all upward-closed subsets of C (*principal* Kripke models). But it turns out that the class of all formulas valid in principal models is not recursively enumerable. In particular, our natural deduction system (and any other finitary proof system) is not complete with respect to the principal model semantics. This result is due to Skvortsov [445] and Kremer [282].

11.2. Polymorphic lambda-calculus (system F)

The polymorphic lambda-calculus, often referred to as "system **F**", serves both as a proof notation for the intuitionistic second-order propositional logic and as a prototype polymorphic programming language. We begin with a Church-style variant.

11.2.1. DEFINITION.

(i) Second-order propositional formulas built with $\to$ and $\forall$ only are called *polymorphic types* or *second-order types*.

(ii) A *raw term* of system **F** is either a variable, an ordinary application or abstraction, or it is

- a *polymorphic abstraction*, written $(\Lambda p M)$, where M is a raw term and p is a type variable, or
- a *type application*, written $(M\tau)$, where M is a raw term and τ is a type.

In summary, we have this grammar for raw terms:

$$M ::= x \mid (MM) \mid (\lambda x{:}\tau\, M) \mid (\Lambda p\, M) \mid (M\tau)$$

The usual notational conventions apply.

11.2. POLYMORPHIC LAMBDA-CALCULUS (SYSTEM F)

(iii) Well-typed lambda-terms (Church style) are defined by the type inference rules of Figure 11.1. An environment is again a finite set of declarations $(x : \tau)$, for different variables (i.e. finite partial function from variables to types). For simplicity we write $\mathrm{FV}(\Gamma)$ for $\mathrm{FV}(\mathrm{rg}(\Gamma))$.

$$
\begin{array}{ll}
(\text{Var}) & \Gamma, x{:}\tau \vdash x : \tau \\[1em]
(\text{Abs}) & \dfrac{\Gamma, x{:}\sigma \vdash M : \tau}{\Gamma \vdash (\lambda x{:}\sigma\, M) : \sigma \to \tau} \\[1em]
(\text{App}) & \dfrac{\Gamma \vdash M : \sigma \to \tau \quad \Gamma \vdash N : \sigma}{\Gamma \vdash (MN) : \tau} \\[1em]
(\text{Gen}) & \dfrac{\Gamma \vdash M : \sigma}{\Gamma \vdash (\Lambda p\, M) : \forall p\, \sigma} \quad (p \notin \mathrm{FV}(\Gamma)) \\[1em]
(\text{Inst}) & \dfrac{\Gamma \vdash M : \forall p\, \sigma}{\Gamma \vdash (M\tau) : \sigma[p := \tau]}
\end{array}
$$

FIGURE 11.1: CHURCH-STYLE SYSTEM F

The axiom and rules for $\to$ are as usual, and we also have the *generalization* (Gen) and *instantiation* (Inst) rules corresponding to natural deduction rules ($\forall$I) and ($\forall$E). The intuitive meaning of $\Lambda p\, M$ is that the term M, which may refer to a free type variable p, is taken as a polymorphic procedure with a type parameter p. Under this interpretation, the restriction in (Gen) states that a type parameter is a *local* identifier. Type application corresponds to a call to such a generic procedure with an actual type parameter. This is an *explicit* form of polymorphism (type as parameter), known for instance from the programming language Ada, as opposed to *implicit* polymorphism, which is present in the Curry-style variant of system F (Section 11.4) or in the programming language ML.

As in the first-order case, the universal type $\forall p\, \sigma(p)$ corresponds to a product construction of the form $\Pi\{\sigma(\tau) \mid \tau \text{ a type}\}$. This responds very well to the idea that a proof of $\forall p\, \sigma(p)$ is a function translating proofs of τ into proofs of $\sigma(\tau)$.

It is often said that a Church-style term contains all the type information (as opposed to Curry style). Strictly speaking, this is not true. In our formulation (and many similar presentations), types of free variables are *not* part of a term. Surprisingly, it is undecidable [428] if a given raw Church-style term is typable, i.e. if it can be assigned a type in some environment. (This should not be confused with Theorem 11.4.7 stating that the typability problem for Curry-style system **F** is also undecidable.)

11.2.2. CONVENTION. As usual, whenever we talk about a Church-style term, we always mean a term together with a certain environment that assigns types to all its free variables. We sometimes stress this fact by placing (informally) an upper index, as for instance in $x^{\forall p(\sigma \to \varrho)} \tau y^{\sigma[p:=\tau]}$. As usual, we may also write upper indices to mark types of certain (sub)terms, or to write e.g. $\lambda x^\tau M$, rather than $\lambda x{:}\tau M$, to improve readability.

11.2.3. EXAMPLE. Here are some well-typed Church-style terms:

$$\vdash \Lambda p \lambda f^{p \to p} \lambda x^p . f(fx) : \forall p((p \to p) \to p \to p);$$

$$\vdash \Lambda q \lambda x^{\forall p(p \to p)} . x(q \to q)(xq) : \forall q(\forall p(p \to p) \to q \to q);$$

$$\vdash \lambda f^{\forall p(p \to q \to p)} \Lambda p \lambda x^p . f(q{\to}p)(fpx) : \forall p(p \to q \to p) \to \forall p(p \to q \to q \to p).$$

11.2.4. THEOREM (Curry-Howard isomorphism).

(i) *If $\Gamma \vdash M : \tau$ in the polymorphic lambda calculus then $\mathrm{rg}(\Gamma) \vdash \tau$ in the $\{\forall, \to\}$-fragment of the second-order intuitionistic propositional logic.*

(ii) *If $\Gamma \vdash \tau$ is provable then $\Delta \vdash M : \tau$ for some term M and some Δ with $\mathrm{rg}(\Delta) = \Gamma$.*

PROOF. A routine induction. □

In Church-style system **F** there are two sorts of free variables (which we assume to belong to disjoint sets): object variables $x, y, \ldots$ and type (propositional) variables $p, q, r, \ldots$ There are also two binding operators: the big lambda and the small lambda. Thus, the following definition of free variables must account for both. (Warning: binding an object variable x^ϱ does not mean binding type variables in ϱ.)

11.2.5. DEFINITION. By FV(M) we denote the set of all object and type variables free in M.

$$\begin{aligned}
\mathrm{FV}(x) &= \{x\}; \\
\mathrm{FV}(\lambda x{:}\varphi . M) &= \mathrm{FV}(\varphi) \cup (\mathrm{FV}(M) - \{x\}); \\
\mathrm{FV}(MN) &= \mathrm{FV}(M) \cup \mathrm{FV}(N); \\
\mathrm{FV}(\Lambda p\, M) &= \mathrm{FV}(M) - \{p\}; \\
\mathrm{FV}(M\sigma) &= \mathrm{FV}(M) \cup \mathrm{FV}(\sigma).
\end{aligned}$$

As usual, we assume α-conversion on terms. The definition is routine, but it must account for the two sorts of bindings. We must also define two forms of substitution, one of the form $M[x := N]$ and another of the form $M[p := \tau]$.

11.2. POLYMORPHIC LAMBDA-CALCULUS (SYSTEM F)

11.2.6. DEFINITION. Substitution is defined as it can be expected:

$$\begin{aligned}
x[x := P] &= P; \\
y[x := P] &= y; \\
(MN)[x := P] &= M[x := P]N[x := P]; \\
(\lambda y{:}\varphi.\,M)[x := P] &= \lambda y{:}\varphi.\,M[x := P], \text{ where } y \notin \mathrm{FV}(P) \cup \{x\}; \\
(M\tau)[x := P] &= M[x := P]\tau; \\
(\Lambda p M)[x := P] &= \Lambda p M[x := P], \text{ where } p \notin \mathrm{FV}(P).
\end{aligned}$$

$$\begin{aligned}
x[p := \sigma] &= x; \\
(MN)[p := \sigma] &= M[p := \sigma]N[p := \sigma]; \\
(\lambda x{:}\varphi.\,M)[p := \sigma] &= \lambda x{:}\varphi[p := \sigma].\,M[p := \sigma]; \\
(M\tau)[p := \sigma] &= M[p := \sigma]\tau[p := \sigma]; \\
(\Lambda q M)[p := \sigma] &= \Lambda q M[p := \sigma], \text{ where } q \notin \mathrm{FV}(\sigma) \cup \{p\}.
\end{aligned}$$

By $\Gamma[p := \sigma]$ we denote Γ, after replacing each $x : \tau$ by $x : \tau[p := \sigma]$.

The result of substitution $x[p := \tau]$ is of course x. But if x is of type σ then the substitution $x[p := \tau]$ may be better explained by the informal $x^\sigma[p := \tau] = x^{\sigma[p:=\tau]}$. Indeed, if $\Gamma \vdash x : \sigma$ then $\Gamma[p := \tau] \vdash x : \sigma[p := \tau]$, and in general we have the following lemma.

11.2.7. LEMMA. *If $\Gamma \vdash M : \varphi$ then $\Gamma[p := \tau] \vdash M[p := \tau] : \varphi[p := \tau]$.*

PROOF. Induction with respect to the length of M. If M is a type abstraction, then an additional application of the induction hypothesis may be needed to rename the type variable to be quantified. □

The lemma above, together with the next one (similar to Lemma 3.1.5(iii)), is crucial for the proper handling of alpha-equivalent terms.

11.2.8. LEMMA.

(i) *If $\Gamma \vdash \lambda x{:}\tau_1.M : \tau_1 \to \tau_2$ then $\Gamma, x{:}\tau_1 \vdash M : \tau_2$, provided $x \notin \mathrm{dom}(\Gamma)$.*

(ii) *If $\Gamma \vdash \Lambda p M : \forall p \tau$ and $p \notin \mathrm{FV}(\Gamma)$ then $\Gamma \vdash M : \tau$.*

PROOF. The proof is similar to the proof of Lemma 3.1.5(iii). Part (i) requires an analogue of Lemma 3.1.4(ii) to be shown first. In part (ii) one uses Lemma 11.2.7. □

Now we have an analogue of Lemma 3.1.6.

11.2.9. LEMMA.

(i) *If $\Gamma \vdash M : \sigma$ and $\Gamma(x) = \Gamma'(x)$ for all $x \in \mathrm{FV}(M)$ then $\Gamma' \vdash M : \sigma$.*

(ii) *If $\Gamma, x : \tau \vdash M : \varphi$ and $\Gamma \vdash P : \tau$ then $\Gamma \vdash M[x := P] : \varphi$.*

PROOF. Induction with respect to the length of M, using Lemma 11.2.8. □

11.2.10. DEFINITION (Beta-reduction). There are two beta-reduction rules:

$$(\lambda x{:}\tau.M)N \to_\beta M[x := N];$$
$$(\Lambda p M)\tau \to_\beta M[p := \tau].$$

11.2.11. PROPOSITION. *If* $\Gamma \vdash M : \tau$ *and* $M \twoheadrightarrow_\beta N$ *then also* $\Gamma \vdash N : \tau$.

PROOF. Easy induction based on Lemmas 11.2.7 and 11.2.9. □

By inspecting the possible interaction between redexes one can see that no critical pair (Remark 6.3.10) occurs and thus the following holds.

11.2.12. PROPOSITION. *System* **F** *has the weak Church-Rosser property.*

The Church-Rosser property (Corollary 11.5.8) will be shown later.

11.3. Expressive power

In classical logic, the connectives $\neg$, $\vee$ and $\wedge$ can be defined by means of $\bot$ and $\to$. The quantifier $\exists$ is expressible by means of $\forall$ via the De Morgan law, so that $\bot$, $\to$ and $\forall$ make a sufficient set of operators. This is not the case in intuitionistic logic, neither propositional nor first-order. However, in second-order logic it is possible. Using $\to$ and $\forall$ one can express the other connectives and also the constant $\bot$.

11.3.1. DEFINITION (Absurdity or the empty type). It is absurd (contradictory) to claim that all propositions hold. With $\bot := \forall p\, p$, and $\varepsilon_\tau(M) = M\tau$, we have the following derived rule:

$$\frac{\Gamma \vdash M : \bot}{\Gamma \vdash \varepsilon_\tau(M) : \tau}$$

11.3.2. DEFINITION (Disjunction or co-product). A disjunction $\tau \vee \sigma$ is the weakest proposition implied by τ and also implied by σ. In other words, it holds when all such propositions hold. This motivates the definition:

$$\tau \vee \sigma := \forall p((\tau \to p) \to (\sigma \to p) \to p),$$

where $p \notin \mathrm{FV}(\sigma) \cup \mathrm{FV}(\tau)$. We define injections and case eliminator this way:

- $\mathbf{in}_1(M^\tau) := \Lambda p \lambda u^{\tau \to p} \lambda v^{\sigma \to p}.uM;$
- $\mathbf{in}_2(M^\sigma) := \Lambda p \lambda u^{\tau \to p} \lambda v^{\sigma \to p}.vM;$
- **case** $L^{\tau \vee \sigma}$ **of** $[x^\tau]M^\varrho$ **or** $[y^\sigma]N^\varrho := L\varrho(\lambda x^\tau M)(\lambda y^\sigma N).$

11.3. EXPRESSIVE POWER

The reader can easily check the correctness of these rules:

$$\frac{\Gamma \vdash M : \tau}{\Gamma \vdash \mathbf{in}_1(M) : \tau \vee \sigma} \qquad \frac{\Gamma \vdash M : \sigma}{\Gamma \vdash \mathbf{in}_2(M) : \tau \vee \sigma}$$

$$\frac{\Gamma \vdash L : \tau \vee \sigma \quad \Gamma, x{:}\tau \vdash M : \varrho \quad \Gamma, y{:}\sigma \vdash N : \varrho}{\Gamma \vdash \mathbf{case}\ L\ \mathbf{of}\ [x]M\ \mathbf{or}\ [y]N : \varrho}$$

as well as the correctness of beta-reduction:

$$\mathbf{case}\ \mathbf{in}_1^{\tau \vee \sigma}(N)\ \mathbf{of}\ [x]P\ \mathbf{or}\ [y]Q \twoheadrightarrow_\beta P[x := N];$$
$$\mathbf{case}\ \mathbf{in}_2^{\tau \vee \sigma}(N)\ \mathbf{of}\ [x]P\ \mathbf{or}\ [y]Q \twoheadrightarrow_\beta Q[y := N].$$

11.3.3. DEFINITION (Conjunction or product). Define:

$$\tau \wedge \sigma := \forall p((\tau \to \sigma \to p) \to p),$$

where p is free in neither τ nor σ. That is, $\tau \wedge \sigma$ holds iff everything holds that can be derived from $\{\tau, \sigma\}$. We define pairs and projections as follows:

$$\begin{aligned}
\langle P^\tau, Q^\sigma \rangle &:= \Lambda p \lambda z^{\tau \to \sigma \to p}. zPQ; \\
\pi_1(M^{\tau \wedge \sigma}) &:= M\tau(\lambda x^\tau \lambda y^\sigma . x); \\
\pi_2(M^{\tau \wedge \sigma}) &:= M\sigma(\lambda x^\tau \lambda y^\sigma . y).
\end{aligned}$$

It is left to the reader to check that the term assignment to the $\wedge$-related rules of natural deduction, described in Section 4.5, is correct:

$$\frac{\Gamma \vdash M : \tau \quad \Gamma \vdash N : \sigma}{\Gamma \vdash \langle M, N \rangle : \tau \wedge \sigma} \qquad \frac{\Gamma \vdash M : \tau \wedge \sigma}{\Gamma \vdash \pi_1(M) : \tau} \qquad \frac{\Gamma \vdash M : \tau \wedge \sigma}{\Gamma \vdash \pi_2(M) : \sigma}$$

With our definitions we have $\pi_i(\langle P_1, P_2 \rangle) \twoheadrightarrow_\beta P_i$, so we can say that beta-reduction is correctly implemented (but with $\to_\beta$ replaced by $\twoheadrightarrow_\beta$). Note however that eta-reduction is *not* implemented: if $y^{\tau \wedge \sigma}$ is a variable then $\langle \pi_1(y), \pi_2(y) \rangle$ is a normal form.

THE EXISTENTIAL QUANTIFIER. Here is a term assignment for the natural deduction rules for the existential second-order quantifier.

$$(\exists \mathrm{I}) \quad \frac{\Gamma \vdash M : \sigma[p := \tau]}{\Gamma \vdash [\tau, M]_{\exists p \sigma} : \exists p \sigma}$$

$$(\exists \mathrm{E}) \quad \frac{\Gamma \vdash M : \exists p \sigma \quad \Gamma, x{:}\sigma \vdash N : \varrho}{\Gamma \vdash \mathbf{let}\ [p, x{:}\sigma] = M\ \mathbf{in}\ N : \varrho} \quad (p \notin FV(\Gamma, \varrho))$$

As in the first-order case, existential quantification corresponds to data abstraction. An existential type of the form $\exists p\sigma$ can be seen as a partial type specification, where type p is "private" and not accessible for the user. For instance, one can consider a type of push-down stores (with the *push* and *pop* operations on data of type τ) defined as

$$\tau\text{-}pds := \exists p(p \wedge (\tau \to p \to p) \wedge (p \to p \wedge \tau)).$$

A user can operate on such a pds without knowing the actual type used to implement it. A generic pds type may now be defined this way:

$$generic\text{-}pds := \forall q \exists p(p \wedge (q \to p \to p) \wedge (p \to p \wedge q)).$$

Here is the beta-reduction rule for existential types:

let $[p, x{:}\sigma] = [\tau, M]_{\exists p\sigma}$ **in** $N \to N[p := \tau][x := M]$.

This corresponds to the use of an encapsulated object at run time. The contents is "unpacked" and the actual implementation is now accessed.

Existential quantification can be represented in system **F** as follows:

11.3.4. Definition. Assuming $q \notin \text{FV}(\sigma)$, we define

$$\exists p\sigma := \forall q(\forall p(\sigma \to q) \to q).$$

The packing and unpacking terms are these:

- $[\tau, M]_{\exists p\sigma} = \Lambda q \, \lambda x^{\forall p(\sigma \to q)}.x\tau M$;

- **let** $[p, x{:}\sigma] = M$ **in** $N^\varrho = M\varrho(\Lambda p \lambda x^\sigma N)$.

Compare the above with Definition 11.3.2. We now have the weakest common consequence of all $\sigma(p)$, where p ranges over all types. This supports the understanding of existential quantifier as infinite disjunction. But note also that there is a similarity to De Morgan's law here: take $\bot$ instead of q and we obtain $\neg \forall p \neg \sigma$. We leave to the reader the verification of beta-reduction.

DATA TYPES. Various data types can be implemented in system **F**. For instance, **bool** can be interpreted as $\forall p(p \to p \to p)$, with true $= \Lambda p \lambda x^p y^p.x$ and false $= \Lambda p \lambda x^p y^p.y$. Natural numbers are represented as the polymorphic Church numerals

$$\mathbf{c}_n = \Lambda p \lambda f^{p \to p} x^p. f(\cdots f(x) \cdots)$$

of type

$$\omega = \forall p((p \to p) \to p \to p).$$

Once we have defined numerals we can generalize the notion of a *definable* integer function in the obvious way.

11.3. EXPRESSIVE POWER   283

11.3.5. DEFINITION. We say that $f : \mathbb{N}^k \to \mathbb{N}$ is *definable* in system **F** iff there is a closed M of type $\omega^k \to \omega$ such that $M\mathbf{c}_{n_1} \ldots \mathbf{c}_{n_k} =_\beta \mathbf{c}_{f(n_1,\ldots,n_k)}$ holds for all $n_1, \ldots, n_k \in \mathbb{N}$.

Clearly, all functions definable in simple types can also be defined in system **F** using essentially the same definitions. For instance, we define the successor function as:

$$\text{succ} := \lambda n^\omega \Lambda p \lambda f^{p \to p} x^p . f(npfx).$$

But one can do much more, for instance the function $n \mapsto n^n$ is defined by:

$$\lambda n^\omega . n(p \to p)(np).$$

Note that this trick uses polymorphism in an essential way. We can generalize it to represent primitive recursion. Indeed, system **T** (as an equational theory) can be embedded into system **F**.

11.3.6. PROPOSITION. *For any given type σ, let the* recursor *over σ be the term* $\mathbf{R}_\sigma : (\omega \to \sigma \to \sigma) \to \sigma \to \omega \to \sigma$ *defined as:*

$$\lambda f^{\omega \to \sigma \to \sigma} \lambda y^\sigma \lambda n^\omega . \pi_1(n(\sigma \wedge \omega)(\lambda v^{\sigma \wedge \omega} \langle f(\pi_2(v))(\pi_1(v)), \text{succ}(\pi_2(v))\rangle)\langle y, \mathbf{c}_0\rangle).$$

Then $\mathbf{R}_\sigma MN(\mathbf{c}_0) =_\beta N$ *and* $\mathbf{R}_\sigma MN(\mathbf{c}_{n+1}) =_\beta M\mathbf{c}_n(\mathbf{R}_\sigma MN\mathbf{c}_n)$, *for all n.*

PROOF. Left to the reader. □

11.3.7. COROLLARY. *The class of integer functions definable in system* **F** *includes all functions provably total in first-order arithmetic.*

We will see in the next chapter that system **F** defines much more than system **T**, namely all provably total functions of *second-order* arithmetic.

Our last example is the type of lists of natural numbers. Recall that a *list* is either the empty list nil or it is obtained by adding a number n in front of another list ℓ, and then it is written as $n :: \ell$. One possible way to represent lists in system **F** is to define

$$\text{list} = \forall p((\omega \to p \to p) \to p \to p),$$

with nil $= \Lambda p \lambda f x . x$ and $n :: \ell = \Lambda p \lambda f x . f \mathbf{c}_n(\ell p f x)$. Then, for instance $1 :: 2 :: 3 ::$ nil is represented by $\Lambda p \lambda f x . f\mathbf{3}(f\mathbf{2}(f\mathbf{1}x))$.

Integers and lists are examples of *inductive types*, commonly used in programming. Various other inductive types can be implemented in system **F** in a similar fashion (Exercise 11.12). We will see a more general approach to define data types in Section 12.3.

$$
\begin{array}{ll}
\text{(Var)} & \Gamma, x:\tau \vdash x:\tau \\[1em]
\text{(Abs)} & \dfrac{\Gamma, x:\sigma \vdash M:\tau}{\Gamma \vdash (\lambda x M):\sigma \to \tau} \\[1em]
\text{(App)} & \dfrac{\Gamma \vdash M:\sigma \to \tau \quad \Gamma \vdash N:\sigma}{\Gamma \vdash (MN):\tau} \\[1em]
\text{(Gen)} & \dfrac{\Gamma \vdash M:\sigma}{\Gamma \vdash M:\forall p\sigma} \quad (p \notin \mathrm{FV}(\Gamma)) \\[1em]
\text{(Inst)} & \dfrac{\Gamma \vdash M:\forall p\sigma}{\Gamma \vdash M:\sigma[p := \tau]}
\end{array}
$$

FIGURE 11.2: CURRY-STYLE SYSTEM **F**

11.4. Curry-style polymorphism

The Curry-style variant of system **F** is defined by the type assignment rules in Figure 11.2. (The notion of a type and an environment remains the same.)

The rules for abstraction and application are the same as for the simply typed lambda-calculus à la Curry. The rules for *generalization* and *instantiation* reflect the idea of *implicit* polymorphism: to have the universal type $\forall p\tau$ means to have all possible instances of this type (all types $\tau[p := \sigma]$).

11.4.1. DEFINITION. The erasing map $|\cdot|$ from terms of the Church-style system **F** to pure lambda-terms is defined by:

$$
\begin{aligned}
|x| &= x \\
|MN| &= |M||N| \\
|\lambda x{:}\sigma.M| &= \lambda x|M| \\
|\Lambda p M| &= |M| \\
|M\tau| &= |M|
\end{aligned}
$$

In a given environment, a Church-style term M can be seen as a type derivation for the Curry-style term $|M|$. In particular, exactly the same types are inhabited, and Church-style reductions translate to Curry-style reductions.

11.4. CURRY-STYLE POLYMORPHISM

11.4.2. PROPOSITION.

(i) *For a pure lambda-term M, we have $\Gamma \vdash M : \tau$ if and only if there is a Church-style term M_0 with $|M_0| = M$, such that $\Gamma \vdash M_0 : \tau$.*

(ii) *Erasing commutes with substitutions, i.e. for all M, N, and σ, we have $|M[p := \sigma]| = |M|$ and $|M[x := N]| = |M|[x := |N|]$.*

(iii) *If $M \to_\beta N$ for Church-style terms M and N then $|M| \to_\beta |N|$ in case of object reduction and $|M| = |N|$ in case of type reduction.*

PROOF. Easy. Part (iii) uses part (ii). □

It might seem at first sight that the subject reduction à la Curry:

If $\Gamma \vdash M : \tau$, and $M \twoheadrightarrow_\beta N$ then also $\Gamma \vdash N : \tau$,

is an easy consequence of Propositions 11.2.11 and 11.4.2. Indeed, every beta-reduction performed in a typable pure lambda-term should correspond to a reduction in the corresponding Church-style term. This is true but not so obvious. (Exercise 11.19 shows that an analogous property does not hold for eta-reduction.) For the proof of the subject reduction property, we need a few lemmas, the first one easily translated from Church style.

11.4.3. LEMMA.

(i) *If $\Gamma \vdash M : \sigma$ and $\Gamma(x) = \Gamma'(x)$, for all $x \in \mathrm{FV}(M)$, then $\Gamma' \vdash M : \sigma$.*

(ii) *If $\Gamma, x : \tau \vdash M : \varphi$ and $\Gamma \vdash P : \tau$ then $\Gamma \vdash M[x := P] : \varphi$.*

(iii) *If $\Gamma \vdash M : \varphi$ then $\Gamma[p := \sigma] \vdash M : \varphi[p := \sigma]$.*

PROOF. By Lemmas 11.2.7, 11.2.9, and 11.4.2(i),(ii). □

For two types σ and τ, we write $\sigma \preceq \tau$ when $\sigma = \forall \vec{p}\sigma'$, and $\tau = \forall \vec{q}\sigma'[\vec{p} := \vec{\varrho}]$, for some sequence of types $\vec{\varrho}$, and some sequence $\vec{q}$ of type variables, not free in σ. For instance, $\forall pq(p \to q \to r) \preceq \forall p((p \to p) \to p \to r)$. Thus, $\sigma \preceq \tau$ implies that τ can be obtained from σ by a sequence of instantiation steps followed by a sequence of generalization steps. But note that not every correct sequence of instantiations and generalizations corresponds to an application of $\preceq$. For instance, the judgement $\vdash \mathbf{I} : p \to p$ generalizes to $\vdash \mathbf{I} : \forall p(p \to p)$, while $p \to p \not\preceq \forall p(p \to p)$. The following easy lemma puts together the main properties of the relation $\preceq$.

11.4.4. LEMMA.

(i) *The relation $\preceq$ is a quasi-order, i.e. it is reflexive and transitive. But it is not anti-symmetric, and thus not a partial order.*

(ii) *If* $\Gamma \vdash M : \sigma$ *and* $\sigma \preceq \tau$ *then* $\Gamma \vdash M : \tau$.

(iii) *If* $\Gamma \vdash M : \tau$ *is derived from* $\Gamma \vdash M : \sigma$ *by a sequence of generalization and instantiation steps then* $\forall \vec{p} \sigma \preceq \forall \vec{q} \tau$, *where* $\vec{p} = \text{FV}(\sigma) - \text{FV}(\Gamma)$ *and* $\vec{q} = \text{FV}(\tau) - \text{FV}(\Gamma)$.

11.4.5. LEMMA (Generation lemma).

(i) *If* $\Gamma \vdash x : \tau$ *then* $\Gamma(x) \preceq \tau$.

(ii) *If* $\Gamma \vdash \lambda x M : \tau$ *and* $x \notin \text{dom}(\Gamma)$ *then* $\tau = \forall \vec{p}(\varrho \to \sigma)$, *where* ϱ *and* σ *are such that* $\Gamma, x:\varrho \vdash M : \sigma$, *and the variables* $\vec{p}$ *are not free in* Γ.

(iii) *If* $\Gamma \vdash MN : \tau$ *then* $\Gamma \vdash M : \varrho \to \sigma$ *and* $\Gamma \vdash N : \varrho$, *for some* ϱ *and* σ *satisfying* $\forall \vec{p} \sigma \preceq \tau$, *where* $\vec{p} = \text{FV}(\sigma) - \text{FV}(\Gamma)$.

PROOF. We begin with part (i). If $\Gamma \vdash x : \tau$ is an axiom, then $\Gamma(x) \preceq \tau$ is trivially satisfied. Otherwise, the type assignment is obtained by an application of rule (Var) followed by a sequence of generalization and instantiation steps. In this case we apply Lemma 11.4.4(iii).

In part (iii) the obvious case is that of a derivation ending with (App). Otherwise we proceed as in part (i).

The proof of part (ii) is slightly more delicate. A type assignment to an abstraction is obtained by an application of (Abs), followed by a sequence of generalization and instantiation steps. We show part (ii) by induction with respect to the number of these steps. The base case when $\Gamma \vdash \lambda x M : \tau$ is obtained by (Abs) is handled as in the proof of Lemma 3.1.5 (use Lemma 11.4.3(i),(ii) to derive an analogue of Lemma 11.2.8(i)). If the last rule is (Gen), then all we need is the induction hypothesis. Let the last rule be (Inst). By the induction hypothesis, $\tau = \forall \vec{p}(\varrho[p := \mu] \to \sigma[p := \mu])$, where $\Gamma, x:\varrho \vdash M : \sigma$. Now we use Lemma 11.4.3(iii) to conclude that $\Gamma, x:\varrho[p := \mu] \vdash M : \sigma[p := \mu]$. □

The lemma above can be read as follows. The type assignment rules in Figure 11.2 are equivalent to the following *syntax-oriented* rules:

$$\Gamma \vdash x : \tau \qquad \text{if } (x : \sigma) \text{ is in } \Gamma \text{ and } \sigma \preceq \tau;$$

$$\frac{\Gamma, x:\varrho \vdash M : \sigma}{\Gamma \vdash (\lambda x M) : \forall \vec{p}(\varrho \to \sigma)} \qquad \text{if } \vec{p} \notin \text{FV}(\Gamma);$$

$$\frac{\Gamma \vdash M : \varrho \to \sigma \quad \Gamma \vdash N : \varrho}{\Gamma \vdash (MN) : \tau} \qquad \text{if } \forall \vec{p} \sigma \preceq \tau, \text{ for } \vec{p} = \text{FV}(\sigma) - \text{FV}(\Gamma).$$

11.4.6. THEOREM (Subject reduction). *Let* $\Gamma \vdash M : \tau$ *for a pure lambda term* M. *Then* $M \to_\beta M'$ *implies* $\Gamma \vdash M' : \tau$.

11.5. STRONG NORMALIZATION

PROOF. The proof is by induction with respect to the definition of beta-reduction. Most cases are routine. We consider only the case of a beta redex. Let $\Gamma \vdash (\lambda x\, P)Q : \tau$. By Lemma 11.4.5, there are types ϱ and σ such that

$$\Gamma, x : \varrho \vdash P : \sigma, \qquad \Gamma \vdash Q : \varrho, \qquad \forall \vec{p}\,\sigma \preceq \tau,$$

where $\vec{p} = \mathrm{FV}(\sigma) - \mathrm{FV}(\Gamma)$. From Lemma 11.4.3 it follows that

$$\Gamma \vdash P[x := Q] : \sigma.$$

Since the variables in $\vec{p}$ are not free in Γ, we have $\Gamma \vdash P[x := Q] : \forall \vec{p}\,\sigma$, and thus also $\Gamma \vdash P[x := Q] : \tau$, by Lemma 11.4.4. □

POLYMORPHIC TYPABILITY. With polymorphism, one can assign types to many pure lambda terms which are untypable in simple types. A prominent example is $\lambda x.xx$, and another one is **2K**. By Theorem 11.5.7 below, only strongly normalizing terms are typable. But there are strongly normalizing terms untypable in **F**, like this one:

$$(\lambda zy.y(z\mathbf{I})(z\mathbf{K}))(\lambda x.xx).$$

The essential thing here is that we cannot find *one* type for $\lambda x.xx$ that could be applied to both **I** and **K**. Another such example is **22K** (Exercise 11.23). Compare this to the typable term **2(2K)**.

Unlike $\lambda_\rightarrow$, the polymorphic type assignment does not have the principal type property (see Definition 3.2.4). For instance, the term $\lambda x.xx$ has no principal type, and the set of all types of this term is not so easy to describe. This makes type reconstruction difficult. It was long an open question whether the typability and type checking problems for system **F** are decidable. The answer is negative [506]; the complex proof is omitted.

11.4.7. THEOREM (Wells, 1993). *Typability and type checking in the second-order λ-calculus are both undecidable.*

11.5. Strong normalization

Before we begin the strong normalization proof for system **F**, let us recall the proof method used in Chapter 5 to show the analogous results for combinatory terms. (A variant of this proof was also used for Gödel's system **T**.) For each type τ we defined a set $[\![\tau]\!]$ of terms *computable* for τ, so that

$$M \in [\![\tau \to \sigma]\!] \quad \text{iff} \quad N \in [\![\tau]\!] \text{ implies } MN \in [\![\sigma]\!].$$

Then we proved three crucial lemmas:

(i) Each computable term is strongly normalizable.

(ii) Each term of the form $xN_1\ldots N_k$, where $N_1,\ldots,N_k \in \mathrm{SN}$, is computable. In particular variables are computable.

(iii) If $M \to_\beta N$ by applying a "head reduction," and N is computable, then also M is computable.

With the help of properties (ii) and (iii) one then shows that every term is computable. The conclusion follows from property (i).

The above *computability* or *reducibility* proof method, known also as *Tait's method*, generalizes easily to many typed systems and can also be used to prove properties other than strong normalization (like the Church-Rosser property). For example, in the system of *intersection types*, the type constructor $\cap$ has the property

$$\Gamma \vdash M : \tau \cap \sigma \quad \text{iff} \quad \Gamma \vdash M : \tau \text{ and } \Gamma \vdash M : \sigma.$$

Although the system is of enormous expressive power (all strongly normalizable terms are typable, see [8]) the SN proof for it requires only a minor modification of the computability method. Just define

$$[\![\tau \cap \sigma]\!] = [\![\tau]\!] \cap [\![\sigma]\!],$$

and everything works so well, that for system **F** one may naively try this:

$$[\![\forall p\tau]\!] = \bigcap\{[\![\tau[p:=\sigma]]\!] \mid \text{all types } \sigma\}.$$

Unfortunately, this definition is circular: To define $[\![\forall p\tau]\!]$ we need all sets $[\![\tau[p:=\sigma]]\!]$ to be defined, including the case $\sigma = \forall p\tau\ldots$

We thus use Girard's *method of candidates*. We give up the idea of an "absolute" definition of a computable term. Instead, we accept any choice of sets $[\![\tau]\!]$, as long as they have the properties we need. In particular, the conditions (i)–(iii) listed above should be satisfied.

11.5.1. DEFINITION. A set X of lambda-terms is a *candidate* when

(i) $X \subseteq \mathrm{SN}$;

(ii) If $M_1, \ldots M_n \in \mathrm{SN}$ then $xM_1\ldots M_n \in X$;

(iii) If $M_0[x := M_1]M_2\ldots M_n \in X$ then $(\lambda x.M_0)M_1M_2\ldots M_n \in X$, provided $M_1,\ldots,M_n \in \mathrm{SN}$.

We use the notation $\mathbb{C}$ for the family of all candidates.

11.5.2. DEFINITION. A *valuation* in $\mathbb{C}$ is any map $\xi : \mathrm{PV} \to \mathbb{C}$. For each valuation ξ in $\mathbb{C}$ and each type σ, the set $[\![\sigma]\!]_\xi$ is defined by:

$$\begin{aligned}
[\![p]\!]_\xi &= \xi(p); \\
[\![\sigma \to \tau]\!]_\xi &= \{M \mid N \in [\![\tau]\!]_\xi \text{ implies } MN \in [\![\sigma]\!]_\xi\}; \\
[\![\forall p\sigma]\!]_\xi &= \bigcap\{[\![\sigma]\!]_{\xi(p \mapsto X)} \mid X \in \mathbb{C}\}.
\end{aligned}$$

A term M is *computable* iff $M \in [\![\sigma]\!]_\xi$, for some σ and ξ.

11.5. STRONG NORMALIZATION

11.5.3. LEMMA.

(i) *If ξ and ξ' coincide on $\mathrm{FV}(\sigma)$ then $[\![\sigma]\!]_\xi = [\![\sigma]\!]_{\xi'}$.*

(ii) *$[\![\sigma[p := \tau]]\!]_\xi = [\![\sigma]\!]_{\xi(p \mapsto [\![\tau]\!]_\xi)}$.*

PROOF. Routine induction with respect to σ. □

11.5.4. LEMMA.

(i) *$\mathrm{SN} \in \mathbb{C}$, in particular $\mathbb{C}$ is non-empty.*

(ii) *For each ξ and σ, we have $[\![\sigma]\!]_\xi \in \mathbb{C}$.*

PROOF. Part (i) is analogous to the base case in Lemmas 5.3.3 and 10.3.6(i). Case $\sigma = \sigma_1 \to \sigma_2$ of part (ii) is shown quite like the induction step in Lemma 5.3.3. For case $\sigma = \forall p \sigma'$ observe that an intersection of candidates is a candidate. □

11.5.5. DEFINITION. Let $\Gamma \vdash M : \tau$, where $\mathrm{dom}(\Gamma) = \{x_1, \ldots, x_k\}$, and let ξ be a valuation. A term of the form $M[x_1, \ldots, x_n := N_1, \ldots, N_k]$, where $N_i \in [\![\Gamma(x_i)]\!]_\xi$, for $i = 1, \ldots, k$, is called a *computable instance* of M.

As in Chapter 10, we have the following main lemma.

11.5.6. LEMMA. *A computable instance of a typable term is computable.*

PROOF. Of course the argument is similar to the proof of Lemma 10.3.7, but now we proceed by induction with respect to the type assignment. We consider the two essentially new cases when the last rule used concerns the universal quantifier.

First suppose that $\Gamma \vdash M : \forall p \sigma$ is obtained by a generalization step from $\Gamma \vdash M : \sigma$, with $p \notin \mathrm{FV}(\Gamma)$. Let $M' = M[x_1, \ldots, x_k := N_1 \ldots, N_k]$, where $N_i \in [\![\Gamma(x_i)]\!]_\xi$, for $i = 1, \ldots, k$. By the induction hypothesis, $M' \in [\![\sigma]\!]_\nu$ for all ν, such that $N_i \in [\![\Gamma(x_i)]\!]_\nu$, for $i = 1, \ldots, k$. Since $p \notin \mathrm{FV}(\Gamma(x_i))$, we have $[\![\Gamma(x_i)]\!]_\xi = [\![\Gamma(x_i)]\!]_{\xi(p \mapsto X)}$, for arbitrary i and X. Thus $M' \in [\![\sigma]\!]_{\xi(p \mapsto X)}$ for each $X \in \mathbb{C}$, so that $M' \in [\![\forall p \sigma]\!]_\xi$.

Now assume that $\Gamma \vdash M : \sigma[p := \tau]$ is obtained from $\Gamma \vdash M : \forall p \sigma$. By the induction hypothesis, $M[x_1, \ldots, x_k := N_1, \ldots, N_k] \in [\![\forall p \sigma]\!]_\xi$, thus also $M[x_1, \ldots, x_k := N_1, \ldots, N_k] \in [\![\sigma]\!]_{\xi(p \mapsto [\![\tau]\!]_\xi)} = [\![\sigma[p := \tau]]\!]_\xi$, as needed. □

11.5.7. THEOREM (Girard). *All typable terms of system $\mathbf{F}$ are strongly normalizing.*

PROOF. Every typable term is a computable instance of itself. □

The strong normalization theorem for Church-style system $\mathbf{F}$ is a consequence of the above (Exercise 11.27). Using Proposition 11.2.12 we obtain:

11.5.8. COROLLARY. *System* **F** *has the Church-Rosser property.* □

The proof of Theorem 11.5.7 differs in a substantial way from Tait's strong normalization proof for simple types or for system **T**. In the previous case we defined by induction a unary predicate *computable*(M) over (Gödel numbers of) terms. To turn this induction into an explicit definition we needed to quantify over such unary predicates (or sets).

Now the inductive definition of "computable" involves a valuation ξ as an additional parameter. We cannot encode ξ by a single number, and the best we can do is to represent it as a set of numbers, say $\{n \mid r(n) \in \xi(\ell(n))\}$. This means that we deal with an inductive definition of a predicate over sets. To express this definition in an explicit way we must quantify over predicates on sets (or equivalently over sets of sets).

In other words, our proof cannot be formalized even if we extend the language of arithmetic by quantification over sets.

However, suppose all we ask for is strong normalization for terms typable using a fixed finite set of types. Then we can avoid induction by expressing the statements $M \in [\![\tau]\!]_\xi$ for each type τ separately (cf. Remark 10.3.9). For a proof of Theorem 11.5.7 restricted to this special case, it is enough to use quantification over numbers and sets of numbers (we still must quantify over candidates), but not over sets of sets. Thus, such a restricted strong normalization result can be formalized in the language of second-order arithmetic.

11.6. The inhabitation problem

The strong normalization of system **F** has the following important consequence: If a type is inhabited, then there is an inhabitant in normal form. As in the simply typed case we can also require that the inhabitant is in long normal form. We generalize Definition 4.2.1 and Lemma 4.2.2 as follows:

11.6.1. DEFINITION. The *long normal forms* of Church style **F** are as follows:

- If x is a variable, p is a type variable, and $\vec{\Delta}$ is a sequence of types and terms in long normal forms, such that $x\vec{\Delta}$ has type p, then $x\vec{\Delta}$ is in long normal form.

- If M^σ is in long normal form then so are $(\lambda x{:}\tau.M)^{\tau \to \sigma}$ and $(\Lambda p M)^{\forall p \sigma}$.

11.6.2. LEMMA. *If* $\Gamma \vdash M : \tau$ *then also* $\Gamma \vdash M' : \tau$, *for some* M' *in long normal form.*

PROOF. Left to the reader. □

In Chapter 4, the restriction to long normal forms helped us to define an inhabitation algorithm (Lemma 4.2.3). Now we will use long normal forms to

11.6. THE INHABITATION PROBLEM

show that polymorphic inhabitation is undecidable. We work with Church-style system **F**. The proof uses the analogous result for the first-order case.

Recall that Theorem 8.8.2 states the undecidability of first-order consequences of the form $\Gamma \vdash \mathbf{ff}$, where $\mathbf{ff}$ is a distinguished nullary relation symbol and Γ is a set of special formulas over a function-free signature with at most binary relation symbols. That is, Γ consists of three types of assumptions:

(i) Atomic formulas, different than $\mathbf{ff}$;

(ii) Closed formulas of the form $\forall \vec{a}(\alpha_1 \to \cdots \to \alpha_n \to \beta)$;

(iii) Closed formulas of the form $\forall a(\forall b(pab \to \mathbf{ff}) \to \mathbf{ff})$.

Here, α_i, β, and pab are atomic formulas, and in addition all α_i are different from $\mathbf{ff}$. Each variable occurring in β must also occur in some α_i. Without loss of generality we can assume in addition that all relation symbols in the signature of Γ (except $\mathbf{ff}$) are binary.

We reduce the problem above to provability in second-order propositional intuitionistic logic (or, equivalently, to inhabitation in system **F**). The basic idea of the translation is simple. Assign a type variable p_a to each individual variable a, and introduce a type variable $p_{\mathbf{ff}}$ to represent $\mathbf{ff}$. Encode an atomic formula rab as a certain type $\overline{r}(p_a, p_b)$ with free type variables p_a, p_b. A first-order quantifier $\forall a$ is represented by the second-order $\forall p_a$ and every first-order formula γ translates in this way to a polymorphic type $\overline{\gamma}$. Then $\Gamma \vdash \mathbf{ff}$ should hold in first-order logic if and only if $\{x_\gamma : \overline{\gamma} \mid \gamma \in \Gamma\} \vdash M : p_{\mathbf{ff}}$ is derivable in system **F** for some M.

It is quite easy to define the translation so that the left-to-right part of the equivalence above holds. The difficult part is the converse. The problem is that types used in the derivation for M do not a priori have to represent first-order formulas. In other words, M can be an "ad hoc" term, not corresponding to any first-order proof. We must make sure that this cannot happen.

To simplify the exposition we abuse the notation in various ways. For instance, we can write $\Gamma \vdash \varphi$ meaning that $\Gamma \vdash M : \varphi$ for some M. Also, we skip variable names in environments if these are irrelevant.

Further, we assume that all individual variables and relation symbols (including $\mathbf{ff}$) can be used as type variables. That is, we identify p_a with a and $p_{\mathbf{ff}}$ with $\mathbf{ff}$. In addition, our encoding uses a number of distinguished type variables. Here is a full list of all type variables we use:

(i) Individual variables $a, b, \ldots$;

(ii) The distinguished variable $\mathbf{ff}$;

(iii) Three variables: p, p_1 and p_2, for each binary relation symbol p;

(iv) And four more variables: $\bullet, \circ_1, \circ_2$ and $\star$.

11.6.3. DEFINITION. For an arbitrary type A we write $A^\bullet$ for $A \to \bullet$. If p is a binary relation symbol, and A, B are arbitrary types, then we define

$$p_{AB} = (A^\bullet \to p_1) \to (B^\bullet \to p_2) \to p;$$
$$p(A, B) = p_{AB} \to \star$$

For every type A, let $\mathcal{U}(A)$ be the set of all types of the form

$$(A^\bullet \to p_i) \to \circ_1 \quad \text{and} \quad A^\bullet \to \circ_2,$$

where $i = 1, 2$. We write $\Gamma \vdash \mathcal{U}(A)$ to mean that $\Gamma \vdash \theta$, for all $\theta \in \mathcal{U}(A)$. An expression of the form $\mathcal{U}(A) \to \xi$ stands for $\theta_1 \to \cdots \to \theta_n \to \xi$, where $\theta_1, \ldots, \theta_n$ are all members of $\mathcal{U}(A)$. Finally $\mathcal{U}(\vec{A})$ is the union of all $\mathcal{U}(A)$, for A in $\vec{A}$.

The intended meaning of $\mathcal{U}$ is to define the universe of individuals. Under the assumptions $\mathcal{U}(a)$, for each individual a, any type A satisfying $\mathcal{U}(A)$ behaves (to a sufficient extent) like an individual variable. First-order quantifiers $\forall a$ can now be encoded as second-order quantifiers relativized to $\mathcal{U}(a)$.

11.6.4. DEFINITION. Any formula φ of the $(\forall, \to)$-fragment of first-order logic is encoded as a type $\overline{\varphi}$:

(i) $\overline{pab} = p(a, b)$ and $\overline{\mathbf{ff}} = \mathbf{ff}$.

(ii) $\overline{\varphi} = \overline{\vartheta} \to \overline{\psi}$, for $\varphi = \vartheta \to \psi$.

(iii) $\overline{\varphi} = \forall \vec{a}(\mathcal{U}(\vec{a}) \to \overline{\psi})$, for $\varphi = \forall \vec{a}\psi$, with ψ not beginning with $\forall$.

If Γ is a set of first-order formulas then $\overline{\Gamma} = \{\overline{\gamma} \mid \gamma \in \Gamma\} \cup \mathcal{U}(\vec{a})$, where $\vec{a}$ are all the free individual variables in Γ.

Every first-order proof can now easily be translated into a type derivation.

11.6.5. LEMMA. *If* $\Gamma \vdash \varphi$ *in first-order logic and* $\vec{a} = \mathrm{FV}(\varphi) - \mathrm{FV}(\Gamma)$ *then* $\overline{\Gamma} \cup \mathcal{U}(\vec{a}) \vdash M : \overline{\varphi}$, *for some term M of system* **F**.

PROOF. Easy induction. □

The converse of the lemma above will work only under certain additional assumptions we now begin to identify.

11.6.6. DEFINITION. An environment $\Sigma = \overline{\Gamma} \cup \Delta \cup \mathcal{U}(\vec{b})$ is *simple* when

(i) Γ is a set of special formulas;

(ii) All types in Δ are of the forms p_{AB} or $p(A, B)$;

(iii) If $p(A, B)$ or p_{AB} occurs in Σ then $\Sigma \vdash \mathcal{U}(A)$ and $\Sigma \vdash \mathcal{U}(B)$.

11.6. THE INHABITATION PROBLEM

A simple environment is *good* if it contains no declarations of the form p_{AB}. In a type of the form $\forall \vec{p}(\alpha_1 \to \cdots \to \alpha_n \to q)$ the variable q is the *target*.

Observe that in a simple environment targets of all formulas are free, and that individual variables, and variables p_i and $\bullet$, do not occur as targets.

11.6.7. DEFINITION. An individual variable a *represents* a type A in an environment Γ iff the conditions

$$\Gamma, A^\bullet \vdash a^\bullet,$$
$$\Gamma, A^\bullet \to p_i \vdash a^\bullet \to p_i,$$

hold for every binary relation symbol p and every $i \in \{1,2\}$.

The intuition to be associated with this definition is that types A and a are "nearly equivalent" in some contexts, which means that a can be used instead of A. Under some circumstances, such a is uniquely determined by A.

11.6.8. LEMMA. *Assume that all formula targets in Γ are free, no individual variable nor $\bullet$ occurs as a target in Γ, and that*

$$\Gamma, A^\bullet \vdash a^\bullet,$$
$$\Gamma, A^\bullet \to p_i \vdash b^\bullet \to p_i,$$

hold for some p and i. If p_i is not a target in Γ then $a = b$.

PROOF. From $\Gamma, A^\bullet \vdash a^\bullet$ we obtain $\Gamma, a^\bullet \to p_i \vdash A^\bullet \to p_i$. It follows that $\Gamma, a^\bullet \to p_i \vdash b^\bullet \to p_i$. Thus $\Gamma, x\!:\!a^\bullet \to p_i, y\!:\!b^\bullet \vdash N : p_i$, for some long normal form N. There is only one declaration with target p_i, and we must have $N = xN'$, for some N' with $\Gamma, x\!:\!a^\bullet \to p_i, y\!:\!b^\bullet \vdash N' : a^\bullet$. It follows that $\Gamma, a^\bullet \to p_i, b^\bullet, a \vdash \bullet$. We can again ask for a term in long normal form and conclude that $\Gamma, a^\bullet \to p_i, b^\bullet, a \vdash b$ must hold, as there is only one assumption with target $\bullet$. The latter can only happen when $a = b$. □

11.6.9. LEMMA. *Let Σ be a simple environment and suppose that $\Sigma \vdash \mathcal{U}(A)$, for some type A. Then there is a unique[3] individual variable a representing A in Σ. In addition, Σ must contain the assumptions $\mathcal{U}(a)$.*

PROOF. Since $\Sigma \vdash \mathcal{U}(A)$, we have $\Sigma \vdash A^\bullet \to o_2$, that is, $\Sigma, A^\bullet \vdash o_2$. The variable o_2 may occur in $\Sigma, A^\bullet$ only as target of some $\mathcal{U}(a)$, so we must have

$$\Sigma, A^\bullet \vdash N : a^\bullet,$$

for some term N. On the other hand, we also have $\Sigma \vdash (A^\bullet \to p_i) \to o_1$, that is, $\Sigma, A^\bullet \to p_i \vdash o_1$. Again, o_1 may occur in Σ only as target of some $\mathcal{U}(b)$,

---
[3]Note that a variable represents itself.

and we get $\Sigma, A^\bullet \to p_i \vdash b^\bullet \to q_j$, for some relation symbol q and some $j \in \{1,2\}$. Since q_j is not a target in Σ, it must be that $q_j = p_i$, that is, we have $\Sigma, A^\bullet \to p_i \vdash b^\bullet \to p_i$. Thus $b = a$, by Lemma 11.6.8, so that

$$\Sigma, A^\bullet \to p_i \vdash a^\bullet \to p_i.$$

This holds for all p and i. Uniqueness follows from Lemma 11.6.8. □

11.6.10. DEFINITION. Let $\Sigma = \overline{\Gamma} \cup \Delta \cup \mathcal{U}(\vec{b})$ be a simple environment. Then $|\Sigma| = \Gamma \cup \{pab \mid p(A,B) \text{ occurs in } \Delta \text{ and } a,b \text{ represent } A,B \text{ in } \Sigma\}$.

Observe that by Lemma 11.6.9, to every $p(A,B)$ in Σ we can assign exactly one pab in $|\Sigma|$.

11.6.11. LEMMA. *Let Σ be a simple environment such that $\Sigma \vdash p_{AB}$, for some types A,B with $\Sigma \vdash \mathcal{U}(A)$ and $\Sigma \vdash \mathcal{U}(B)$. Then Σ contains a declaration p_{CD} such that types A and C are represented in Σ by the same variable, and the same holds for B and D.*

PROOF. We have $\Sigma, x : A^\bullet \to p_1, y : B^\bullet \to p_2 \vdash N : p$, for some term N in long normal form. The target p can occur in Σ only in a declaration of the form $z : p_{CD}$. Thus we have $N = zN'N''$, where

$$\Sigma, \ x : A^\bullet \to p_1, \ y : B^\bullet \to p_2 \vdash N' : C^\bullet \to p_1$$
$$\Sigma, \ x : A^\bullet \to p_1, \ y : B^\bullet \to p_2 \vdash N'' : D^\bullet \to p_2.$$

By Lemma 11.6.9, types A,B,C,D are represented in Σ by some variables a,b,c,d. We have $\Sigma, C^\bullet \to p_1 \vdash c^\bullet \to p_1$, and thus

$$\Sigma, \ A^\bullet \to p_1, \ B^\bullet \to p_2 \vdash c^\bullet \to p_1.$$

By Lemma 11.6.8, we have $c = a$. In a similar way we obtain $b = d$. □

11.6.12. LEMMA. *Let Σ be a good environment with $\Sigma \vdash \mathcal{U}(A)$ and $\Sigma \vdash \mathcal{U}(B)$, and let a,b represent A,B in Σ. If $\Sigma \vdash p(A,B)$ then $|\Sigma| \vdash pab$.*

PROOF. If Σ is good then $\Sigma, x : p_{AB}$ is simple, and $|\Sigma, x : p_{AB}| = |\Sigma|$. Also, a,b represent A,B in $\Sigma, x : p_{AB}$. Therefore it suffices to prove the following:

If $\Sigma \vdash M : \star$, where Σ is a simple environment, then there is a declaration $x : p_{AB}$ in Σ such that $|\Sigma| \vdash pab$, where a,b represent A,B in Σ.

The proof is by induction with respect to M, assuming that M is a long normal form. Let z be the head variable of M. We have two cases:
CASE 1: The type of z is of the form $p(A,B)$. Let a,b represent A,B. Then $pab \in |\Sigma|$, in particular $|\Sigma| \vdash pab$.

11.6. THE INHABITATION PROBLEM

CASE 2: The type of z is of the form $\overline{\varphi}$, where φ is a universal formula, i.e.

$$z : \forall \vec{a}(\mathcal{U}(\vec{a}) \to \overline{\alpha}_1 \to \cdots \to \overline{\alpha}_n \to p(a_u a_v)),$$

where $\vec{a} = a_1, \ldots, a_m$ and $u, v \in \{1, \ldots, m\}$. We must have $M = z\vec{A}\vec{U}\vec{N}L$, where $\vec{A} = A_1, \ldots, A_m$ is a sequence of types substituted for $\vec{a}$, the symbol $\vec{U}$ denotes a sequence of proofs for the components of $\mathcal{U}(\vec{A})$, and $\vec{N}$ are terms of types $\overline{\alpha}_i[\vec{a} := \vec{A}]$. Finally, L has type $p_{A_u A_v}$.

Take any $j \in \{1, \ldots, n\}$, and suppose that $\overline{\alpha}_j = q(a_l, a_r)$. Then N_j must be an abstraction of the form $N_j = \lambda y{:}q_{A_l A_r}.N'_j$. By the induction hypothesis applied to N'_j and the environment Σ, $y : q_{A_l A_r}$, there is a declaration $x : p_{AB}$ in Σ, $y : q_{A_l A_r}$ such that $|\Sigma| \vdash pb_l b_r$, where $b_1, \ldots, b_m$ is the unique sequence of variables representing types $\vec{A}$ in Σ.

CASE 2A: The declaration $x : p_{AB}$ occurs in Σ. If this happens for at least one number $j \in \{1, \ldots, n\}$ then we are done.

CASE 2B: Otherwise, we have $x = y$, that is $|\Sigma| \vdash qb_l b_r$, and we may assume that a similar situation takes place for all $j \in \{1, \ldots, n\}$. In other words, $|\Sigma| \vdash \alpha_j[\vec{a} := \vec{b}]$ holds for every j. Now recall that the universal formula $\forall \vec{a}(\alpha_1 \to \cdots \to \alpha_n \to p a_u a_v)$ is in $|\Sigma|$, and an immediate consequence of this is $|\Sigma| \vdash pb_u b_v$. In addition, we have $\Sigma \vdash L : p_{A_u A_v}$, so by Lemma 11.6.11, a type p_{CD} must occur in Σ, for some C, D represented by the same variables as A_u, A_v, namely by b_u, b_v. □

11.6.13. LEMMA. *If Σ is a good environment and $\Sigma \vdash M : \mathbf{ff}$ then $|\Sigma| \vdash \mathbf{ff}$.*

PROOF. We proceed by induction with respect to M, assumed to be in long normal form. Then M must be an application of a variable z to a sequence of terms and types, and we have two cases.

CASE 1: The type of z is of the form $\forall \vec{a}(\mathcal{U}(\vec{a}) \to \overline{\alpha}_1 \to \cdots \to \overline{\alpha}_n \to \mathbf{ff})$. Then $M = z\vec{A}\vec{U}\vec{N}$, where $\vec{A}$ is a sequence of types, each N_j has type $\overline{\alpha}_j[\vec{a} := \vec{A}]$ and (slightly abusing notation) we can write $\vec{U} : \mathcal{U}(\vec{A})$.

If $\alpha_j = q a_l a_r$ then $\overline{\alpha}_j[\vec{a} := \vec{A}] = q(A_l, A_r)$. Lemma 11.6.12 implies that $|\Sigma| \vdash q b_l b_r$, where the variables $\vec{b}$ represent $\vec{A}$. That is, $|\Sigma| \vdash \alpha_j[\vec{a} := \vec{b}]$ holds for all j. It follows that $|\Sigma| \vdash \mathbf{ff}$, because $\forall \vec{a}(\alpha_1 \to \cdots \to \alpha_n \to \mathbf{ff})$ is in $|\Sigma|$.

CASE 2: The type of z has the form $\forall a(\mathcal{U}(a) \to \forall b(\mathcal{U}(b) \to p(a,b) \to \mathbf{ff}) \to \mathbf{ff})$. Then $M = zA\vec{U}L$, where A is a type and the meaning of $\vec{U}$ can be informally stated as $\vec{U} : \mathcal{U}(A)$. In addition, we have

$$\Sigma \vdash L : \forall b(\mathcal{U}(b) \to p(A,b) \to \mathbf{ff}).$$

Clearly, $L = \Lambda b \lambda \vec{x}{:}\mathcal{U}(b) \lambda y{:}p(A,b).N$, where

$$\Sigma, \vec{x} : \mathcal{U}(b), y : p(A,b) \vdash N : \mathbf{ff}.$$

We apply induction to N and obtain $|\Sigma|, pcb \vdash \mathbf{ff}$, where c represents type A. The variable b does not occur in $|\Sigma|$ so we have $|\Sigma| \vdash \forall b(pcb \to \mathbf{ff})$. Since the formula $\forall a(\forall b(pab \to \mathbf{ff}) \to \mathbf{ff})$ is in $|\Sigma|$, we conclude that $|\Sigma| \vdash \mathbf{ff}$. □

11.6.14. THEOREM. *The inhabitation problem for system* **F** *is undecidable. In other words, it is undecidable whether a given formula* $\varphi \in \Phi_2$ *is a theorem of second-order propositional intuitionistic logic.*

PROOF. It is enough to show the equivalence

$$\Gamma \vdash \mathbf{ff} \quad \text{iff} \quad \overline{\Gamma} \vdash \mathbf{ff},$$

and apply Theorem 8.8.2. The implication from left to right follows by Lemma 11.6.5. The other implication is a special case of Lemma 11.6.13. Indeed, if $\Sigma = \overline{\Gamma}$ then $|\Sigma| = \Gamma$. □

11.7. Higher-order polymorphism

A type τ with a free type variable p can be seen as an operator $\lambda p \tau$ which, applied to a type σ, yields $\tau[p := \sigma]$. Since the result is a type, we may say that $\lambda p \tau$ is a *type constructor*.

The notion of a function from types to types can be useful. For example, each of the following two types can be assigned to the Church numeral **2**:

$$\forall p(p \to q \to p) \to \forall p(p \to q \to q \to p);$$
$$\forall p(p \to q \to q \to p) \to \forall p(p \to q \to q \to q \to p),$$

and the reader will easily observe that both obey a common pattern

$$\forall p(p \to \alpha p) \to \forall p(p \to \alpha(\alpha p)).$$

Just a little imagination and we can write this type assignment

$$\mathbf{2} : \forall \alpha (\forall p(p \to \alpha p) \to \forall p(p \to \alpha(\alpha p))),$$

where α ranges over type constructors. Instantiating the variable α in two different ways, one can now assign a type to the term **22K**, which is not typable in **F** (cf. Exercise 11.23).

The type assignment system $\mathbf{F}_\omega$ makes this trick possible by adding type constructors to system **F**. There is of course no reason to consider only unary constructors, and once we agree to consider functions from types to types, we should have no objections against operations on constructors, i.e. constructors of higher orders. This makes it necessary to classify constructors in a similar way as we classify terms in the simply typed lambda calculus.

11.7.1. DEFINITION. The set of *kinds* of $\mathbf{F}_\omega$ is the least set containing the kind $*$ of types and such that whenever κ, κ' are kinds then also $\kappa \Rightarrow \kappa'$ is a kind. Now assume an infinite supply of *constructor variables* of every kind (variables of kind $*$ called *type variables*) and for each kind κ define *constructors* of kind κ (called *types* when $\kappa = *$):

11.7. HIGHER-ORDER POLYMORPHISM

- A constructor variable of kind κ is a constructor of kind κ.

- If φ is a constructor of kind $\kappa_1 \Rightarrow \kappa_2$ and τ is a constructor of kind κ_1, then $\varphi\tau$ is a constructor of kind κ_2.

- If α is a constructor variable of kind κ_1 and τ is a constructor of kind κ_2, then $\lambda\alpha\tau$ is a constructor of kind $\kappa_1 \Rightarrow \kappa_2$.

- If α is a variable of any kind, and τ is a type, then $\forall\alpha\tau$ is a type.

- If τ and σ are types then $\tau \to \sigma$ is a type.

Thus, for instance, $\lambda p(q \to p)$, where p and q are type variables, is a constructor of kind $* \Rightarrow *$. Both quantifiers and lambda-abstractions bind variables and we assume alpha-conversion as usual. By convention, we often write α^κ, meaning that α is a constructor variable of kind κ.

Beta-reduction on constructors is defined as expected, i.e. as the compatible closure of the rule (where α and ψ must be of the same kind):

$$(\lambda\alpha\varphi)\psi \to \varphi[\alpha := \psi].$$

A Curry-style type assignment system $\mathbf{F}_\omega$ is given by the rules in Figure 11.3, where α can be of any kind. Judgements are of the form $\Gamma \vdash M : \tau$, where τ must be a type. Note that, by rule (Conv), beta-equal types are in fact identified (and thus can be always assumed normal, cf. Exercise 11.30).

System $\mathbf{F}_\omega$ has the strong normalization property. The proof uses a natural generalization of the method of candidates. We sketch this proof below, omitting all technicalities related to generation, subject reduction, etc.

11.7.2. DEFINITION. For every kind κ, we define a set $\mathbb{C}_\kappa$ by induction. First we take $\mathbb{C}_* = \mathbb{C}$, the family of all candidates (see Definition 11.5.1) and then $\mathbb{C}_{\kappa_1 \Rightarrow \kappa_2}$ is the set of all functions from $\mathbb{C}_{\kappa_1}$ to $\mathbb{C}_{\kappa_2}$.

A *valuation* is now a function ξ, which assigns a set $\xi(\alpha) \in \mathbb{C}_\kappa$ to every constructor variable α of kind κ. For any $\varphi : \kappa$ define $[\![\varphi]\!]_\xi$ by

$$\begin{aligned}
[\![\alpha]\!]_\xi &= \xi(\alpha); \\
[\![\psi(\tau)]\!]_\xi &= [\![\psi]\!]_\xi([\![\tau]\!]_\xi); \\
[\![\lambda\alpha^\kappa\tau]\!]_\xi(X) &= [\![\tau]\!]_{\xi(\alpha \mapsto X)}; \\
[\![\sigma \to \tau]\!]_\xi &= \{M \mid \forall N(N \in [\![\tau]\!]_\xi \to MN \in [\![\sigma]\!]_\xi)\}; \\
[\![\forall\alpha^\kappa\sigma]\!]_\xi &= \bigcap\{[\![\sigma]\!]_{\xi(\alpha \mapsto X)} \mid X \in \mathbb{C}_\kappa\}.
\end{aligned}$$

The reader can easily state and prove an analogue of Lemma 11.5.3, as well as the following version of Lemma 11.5.4:

11.7.3. LEMMA. *If σ is a constructor of type κ then $[\![\sigma]\!]_\xi \in \mathbb{C}_\kappa$.*

$$
\begin{array}{ll}
\text{(Var)} & \Gamma, x:\tau \vdash x:\tau \\[1em]
\text{(Abs)} & \dfrac{\Gamma, x:\sigma \vdash M:\tau}{\Gamma \vdash (\lambda x M):\sigma \to \tau} \\[1em]
\text{(App)} & \dfrac{\Gamma \vdash M:\sigma \to \tau \quad \Gamma \vdash N:\sigma}{\Gamma \vdash (MN):\tau} \\[1em]
\text{(Gen)} & \dfrac{\Gamma \vdash M:\sigma}{\Gamma \vdash M:\forall \alpha \sigma} \quad (\alpha \notin \mathrm{FV}(\Gamma)) \\[1em]
\text{(Inst)} & \dfrac{\Gamma \vdash M:\forall \alpha^\kappa \sigma}{\Gamma \vdash M:\sigma[p := \varphi]} \quad (\varphi \text{ is of kind } \kappa) \\[1em]
\text{(Conv)} & \dfrac{\Gamma \vdash M:\sigma}{\Gamma \vdash M:\tau} \, (\sigma =_\beta \tau)
\end{array}
$$

FIGURE 11.3: CURRY-STYLE SYSTEM $\mathbf{F}_\omega$

Finally, Definition 11.5.5 of a computable instance of a term is adopted without any modification, and we state the main lemma.

11.7.4. LEMMA. *A computable instance of a typable term is computable.*

PROOF. The proof is almost the same as that of Lemma 11.5.6. In the case of (Conv) we use an easy additional claim: If $\tau =_\beta \sigma$ then $[\![\tau]\!]_\xi = [\![\sigma]\!]_\xi$. □

11.7.5. THEOREM. *System $\mathbf{F}_\omega$ has the strong normalization property.*

PROOF. Immediate from Lemma 11.7.4. □

11.7.6. COROLLARY. *System $\mathbf{F}_\omega$ has the Church-Rosser property.*

PROOF. By the standard argument based on Newman's lemma. □

The goal of this section was to give the reader a flavour of higher-order polymorphism rather than a full exposition. We stop here; in particular we omit the definition of a Church-style version of $\mathbf{F}_\omega$, which will later occur as a subsystem of Barendregt's cube in Chapter 14.

11.8. Notes

The idea of a propositional quantifier dates from the beginning of the 20th century. Church [75] mentions Russell, Leśniewski, Łukasiewicz, and Tarski in this respect. Then there were the modal logics studied by Bull, Fine, Kaplan, and others in the late 60's, and finally the 1970 paper [179] by Jean-Yves Girard, introducing the second-order propositional intuitionistic logic. The two-valued *classical* propositional quantifiers regained popularity after the QBF problem was shown PSPACE-complete [460] in 1973.

Girard's [179] contains the first syntactic proof of *Takeuti's conjecture* concerning cut-elimination for higher-order logic. This conjecture was initially posed in 1953 for classical logic, and various semantical solutions (building on earlier work of Schütte) were given in the late 60's by Tait, Takahashi and Prawitz (see e.g. [181, 404]). Strictly speaking, these proofs were showing completeness of the cut-free fragment, and were less informative than a syntactic cut-elimination procedure. Girard's novel approach was to use a generalized form of Gödel's *Dialectica* interpretation and a normalization result for terms of system **F**. The results of [179] became the basis of Girard's *Thèse d'État* [180] of 1972. In this habilitation thesis further issues, some of which were briefly announced in [179], were elaborated. This included strong normalization and extension of the approach to higher-order logic.

Later, many authors obtained numerous variants and simplifications of the candidates proof. A proof of strong normalization for second-order intuitionistic logic, due to Prawitz [405] and based on Girard's technique of candidates, was included in the very book [148] in which Girard's original paper was published. Also, Martin-Löf's simplified normalization proof [324] was eventually included in the same proceedings. Another strong normalization proof appeared soon in the book [459] by Stenlund. Further simplifications were obtained by Tait [468] and Mitchell. Our presentation is based on handwritten notes by Val Breazu-Tannen from a lecture given by John Mitchell in 1986. See [163] for subtleties concerning candidates, and [330] for simple proofs of strong normalization for some extensions of system **F**.

Algebraic semantics for second- and higher-order predicate intuitionistic logic was studied e.g. by Dragalin [134]. Various completeness results for propositional higher-order logics can be found in Geuvers' [172]. In particular, the undecidability of inhabitation in $\mathbf{F}_\omega$ is obtained there from the conservativity over **F**.

Kripke semantics for second-order propositional formulas was considered by several authors. There are various sorts of models and different variants of the logics under consideration. Prawitz [404] first proved a completeness theorem for second-order intuitionistic *predicate* logic with respect to a class of *Beth models*, structures similar in spirit to Kripke models. Then Gabbay [156, 157] showed completeness for his extended logic (see Remark 11.1.13) over Kripke models with constant domains. This result was adjusted to the general case by Sobolev [447].

The first proof of undecidability of second-order propositional intuitionistic logic was given by Gabbay [156, 157]. But this (completeness-based) proof applies to the logic extended by the Gabbay-Grzegorczyk axiom, and it does not automatically generalize to the pure intuitionistic case, which was later treated in [447].

In the meantime, M.H. Löb [313] published another proof of Theorem 11.6.14. The main result of [313], which implies undecidability, is an effective translation from first-order classical logic to second-order propositional intuitionistic logic. Our proof of Theorem 11.6.14, inspired by [313], is a corrected version of the one in [494].

As we mentioned at the beginning of this chapter, the polymorphic lambda-calculus was actually invented twice: first by Girard, and then by Reynolds [415]. But they arrived at essentially the same target from two different sides. Girard needed a proof notation for his work on second-order logic and Takeuti's conjecture. Reynolds' goal was a polymorphic type system for programming, in the line of work initiated by Strachey. Although Reynolds did not know at that time about Girard's work, the resulting calculi (after dissolving the syntactic sugar) were essentially the same. Some results of Girard were even re-invented by other authors.

The classical reference to definability of inductive data types in system **F** is Böhm and Berarducci [52]. The relationship between the type ω and second-order induction axiom was noted by Leivant [302] and Krivine and Parigot [288]. See also [285, 379, 382]. The use of existential quantifiers in a polymorphic type system was first discussed by Mitchell and Plotkin in 1985, see Mitchell's book [349] for more details. The old paper [66] can still explain a lot.

The idea of implicit (Curry-style) polymorphism goes back to Hindley's principal types [234], which later returned in Milner's approach to ML [342]. The first example of a strongly normalizing term untypable in **F** was shown by Ronchi Della Rocca and Giannini [177], confirming a conjecture of Honsell. The full proof of the undecidability result of Wells (Theorem 11.4.7) can be found in [506]. For information on system **F** extending the scope of this book, one can consult [8, 18, 182, 189, 285, 349]. The reader is also referred to [172, 330, 494, 496] for more on system $\mathbf{F}_\omega$, and to [157, 394, 399, 400] for more on second-order propositional logic.

11.9. Exercises

11.1. Show that the following formulas are intuitionistically valid.

(i) $\neg\exists p \neg(p \vee \neg p)$;

(ii) $\forall p(\neg\neg p \leftrightarrow \exists q(p \leftrightarrow (\neg q \vee q)))$;

(iii) $(p \leftrightarrow q) \to \varphi \to \varphi[p := q]$;

(iv) $\varphi \leftrightarrow ((\varphi \to p) \to q)$, where $\varphi = \forall r((((r \to p) \to q) \to r) \to r)$;

(v) $(((s \to r) \to q) \to p) \leftrightarrow \forall t(((t \to r) \to q) \to (t \to s) \to p)$.

11.2. The purpose of this exercise is to prove a lemma necessary for our solution to Exercise 11.3. Let $A \subseteq \mathbb{R}$ be an open set and let Z be the closure of the set $\sim\sim A - A$. Let T be a maximal subset of A such that $|x - y| \geq \frac{1}{2}(|x - Z| + |y - Z|)$ holds for all distinct $x, y \in T$, where $|u - Z|$ abbreviates $\inf_{z \in Z} |u - z|$. Prove that Z is the set of all the accumulation points of T.

11.3. (After [399, 425].) Show that the formulas

(i) $\neg\forall p(p \vee \neg p)$;

(ii) $\neg\neg p \to \exists q((p \to \neg q \vee q) \to p)$,

are valid in the Heyting algebra $\mathcal{O}(\mathbb{R})$, but not intuitionistically valid in general.

11.4. (Zdanowski) Let φ be a closed formula built from the connectives $\to, \vee, \wedge, \bot$, and $\exists$. Prove that if $\mathcal{O}(\mathbb{R}) \models \varphi$ then φ is a classical tautology.

11.5. Prove that Gabbay's axiom of Remark 11.1.13 holds in all Kripke models with constant domains, but not in all full Kripke models.

11.9. EXERCISES

11.6. Show that in full Kripke models, the semantics of the defined connectives (see Section 11.3) coincides with the semantics of their definitions. Does this hold in non-full models?

11.7. Show that in Lemma 11.1.6 the assumption $v_1(p) = v_2(p)$ can be replaced by a weaker one: $v_1(p) \cap \{c' \mid c' \geq c\} = v_2(p) \cap \{c' \mid c' \geq c\}$.

11.8. Will Theorem 11.1.12 remain true under this definition of a full model?

For every formula $\varphi \in \Phi_2(\text{PV})$, every c and every v, admissible for c and φ, the set $\{c' \geq c \mid c', v \Vdash \varphi\}$ belongs to D_c.

11.9. Prove that the intuitionistic propositional existential quantifier cannot be defined from the connectives $\to$, $\vee$, $\wedge$ and $\bot$. *Hint:* Use Exercise 11.3(ii).

11.10. Prove that the intuitionistic propositional universal quantifier cannot be defined from the connectives $\to$, $\vee$, $\wedge$, $\bot$ and $\exists$. *Hint:* Use Exercises 11.3(i) and 11.4.

11.11. Assuming type **list** is defined as in Section 11.3, write a term M of type **list** $\to$ **int** such that M applied to a list returns the product of all its entries.

11.12. Implement in system **F** the types of:

- words over a fixed finite alphabet;
- finite binary trees,

with the appropriate basic operations on these types.

11.13. Will Proposition 11.3.6 remain true with $\twoheadrightarrow_\beta$ in place of $=_\beta$?

11.14. Show that all normal forms are typable in Curry-style system **F**.

11.15. Show that $(\lambda x.xx)(\lambda z.zyz)$ is typable in Curry-style system **F**.

11.16. Prove that $(\lambda x.xx)(\lambda y.yy)$ is untypable in system **F**, not using the fact that typable terms can be normalized.

11.17. Consider a variant of Curry-style system **F** with rule (Inst) in Figure 11.2 restricted so that τ must be quantifier-free. Does this restriction reduce the class of typable terms?

11.18. Suppose that pure λ-terms M^0 and M^1 are β-equal and that both can be assigned type σ in an environment Γ. Can you find β-equal Church-style terms N^0 and N^1 of type σ in Γ, such that $|N^0| = M^0$ and $|N^1| = M^1$?

11.19. Show an example of pure terms M, N such that $M \twoheadrightarrow_\eta N$ and $\Gamma \vdash M : \tau$ in Curry-style system **F**, but $\Gamma \nvdash N : \tau$.

11.20. Show an example of a pure lambda-term N, untypable in Curry-style system **F**, and such that $M \twoheadrightarrow_\eta N$, for some typable M.

11.21. (Mitchell [348]) Let $\sqsubseteq$ be the least reflexive and transitive relation between types, satisfying the axioms

$$\forall p\tau \sqsubseteq \tau[p := \sigma], \qquad \forall p(\tau \to \sigma) \sqsubseteq \forall p\tau \to \forall p\sigma, \qquad \rho \sqsubseteq \forall p\rho,$$

where $p \notin \text{FV}(\rho)$, and closed under the rules

$$\frac{\tau' \sqsubseteq \tau \quad \sigma \sqsubseteq \sigma'}{\tau \to \sigma \sqsubseteq \tau' \to \sigma'} \qquad \frac{\tau \sqsubseteq \tau'}{\forall p\tau \sqsubseteq \forall p\tau'}$$

Prove that the Curry-style system **F** extended by either of the following rules derives the same judgements. (The type assignment with $\sqsubseteq$ is closed under eta-reductions.)

$$\frac{\Gamma \vdash M : \tau}{\Gamma \vdash M : \sigma} \ (\tau \sqsubseteq \sigma) \qquad \frac{\Gamma \vdash M : \tau}{\Gamma \vdash N : \tau} \ (M \twoheadrightarrow_\eta N)$$

11.22. It seems an appealing idea that existential data abstraction might be done with only implicit typing discipline. Look, how beautiful rules we may have, if we add existential quantification to system **F** à la Curry:

$$(\exists\mathrm{I})\ \frac{\Gamma \vdash M : \varphi[p := \sigma]}{\Gamma \vdash M : \exists p \varphi} \qquad (\exists\mathrm{E})\ \frac{\Gamma \vdash M : \exists p \varphi \quad \Gamma, v : \varphi \vdash N : \psi}{\Gamma \vdash N[v := M] : \psi}\ (p \notin \mathrm{FV}(\Gamma, \psi))$$

Does the resulting calculus satisfy the subject-reduction property?

11.23. Prove that the terms **22K** and $(\lambda zy.y(z\mathbf{I})(z\mathbf{K}))(\lambda x.xx)$ are untypable in Curry-style system **F**.

11.24. The eight problems of Section 3.2.8 make sense for Curry-style **F** and $\mathbf{F}_\omega$. Prove that for both systems:

(i) Problems (i) and (ii) reduce to each other and to all the problems (iii)–(vi).

(ii) Problems (iii)–(vi) reduce to each other.

(iii) Problems (vii) and (viii) reduce to each other.

11.25. Do you agree with the statement: *It is a decidable problem whether two given terms typable in Curry-style system* **F** *are beta-equal?*

11.26. Show that $\vdash C : \omega \to \omega \to \omega$, where C is the term defined in Exercise 1.17. Is C typable in system **F**? Is the function c definable in system **F**?

11.27. Assume strong normalization for Curry-style system **F**. Derive strong normalization for Church-style system **F**.

11.28. Complete the proof of Corollary 1.7.7 (cf. Exercise 1.21) by showing that every primitive recursive function is definable by a term F such that every application $F\mathbf{c}_{n_1}\ldots\mathbf{c}_{n_m}(\lambda v.\mathbf{false})\mathbf{true}$ is strongly normalizing.

11.29. From strong normalization of system **F** derive strong normalization of the extended lambda-calculus with products and sums (Theorem 4.5.4).

11.30. Prove that constructors in $\mathbf{F}_\omega$ are strongly normalizing.

11.31. Prove that the inhabitation problem for $\mathbf{F}_\omega$ is undecidable. *Hint:* Adopt the proof of Theorem 8.8.2, using free constructor variables in place of relation symbols.

Chapter 12

Second-order arithmetic

In this chapter we show that the number-theoretic functions definable in System **F** coincide with the provably total functions of *second-order arithmetic*. This fact is a further support for the slogan *"proofs into programs"*. If an algorithm can be shown to terminate by means of second-order arithmetic (i.e. in *analysis,* as they say) then a polymorphic program implementing this algorithm can be extracted from the proof.

The statement of this result requires a definition of second-order arithmetic and second-order predicate logic, to which we have referred until now only informally. We briefly discuss these systems in the following few sections, without attempting an exhaustive study.

Our strategy for the main proof is thus: First we show that provably total functions of classical and intuitionistic second-order arithmetic coincide (Theorem 12.5.6). Then we introduce a proof notation for second-order intuitionistic arithmetic and show strong normalization. Finally we apply a contracting map to a proof of totality of a function f to discover its computational contents: a polymorphic term representing f in system **F**.

12.1. Second-order syntax

We begin with the syntax of second-order predicate logic. As in the case of first-order logic, the syntax depends on the choice of a signature Σ (cf. Section 8.1.1). In addition to individual variables, we assume an infinite set $\mathcal{R}$ of *relation variables* (different from signature symbols). Since we need to distinguish between relations of various arities, we assume that the set $\mathcal{R}$ splits into a family of disjoint infinite subsets $\mathcal{R}_k$, for all $k \in \mathbb{N}$, and we assume that elements of $\mathcal{R}_k$ are of *arity* k.

The definition of a formula of second-order logic is very similar to Definition 8.1.1, but consists of a few more clauses.

12.1.1. DEFINITION.

1. An *atomic formula* is either the constant $\bot$ or an expression of the form $(rt_1 \ldots t_n)$, where r is an n-ary relation symbol in Σ, or an n-ary relation variable, and $t_1, \ldots, t_n$ are algebraic terms over Σ.

2. The set Φ_Σ^2 of *second-order formulas* over Σ, is the least set such that:
    - All atomic formulas are in Φ_Σ^2;
    - If $\varphi, \psi \in \Phi_\Sigma^2$ then $(\varphi \to \psi), (\varphi \vee \psi), (\varphi \wedge \psi) \in \Phi_\Sigma^2$;
    - If $\varphi \in \Phi_\Sigma^2$ and a is an individual variable, then $\forall a \varphi, \exists a \varphi \in \Phi_\Sigma^2$.
    - If $\varphi \in \Phi_\Sigma^2$ and R is a relation variable, then $\forall R \varphi, \exists R \varphi \in \Phi_\Sigma^2$.

All ordinary abbreviations and parentheses-avoiding conventions extend to the second-order syntax in a natural way. For instance, if R is binary then the formula $\forall R. \forall ab (Rab \to Rba) \to \neg \exists ab \, Rab$ states that all symmetric relations are empty. A second-order quantifier $\forall R$ or $\exists R$ binds the variable R in the body of a formula. Of course, we assume alpha-conversion as usual.

12.1.2. DEFINITION. The set $\mathrm{FV}(\varphi)$ of free variables of a second-order formula φ is defined as follows. (Of course, the free variables belong to two disjoint domains: the individual variables and the relation variables.)

- $\mathrm{FV}(rt_1 \ldots t_n) = \mathrm{FV}(t_1) \cup \ldots \cup \mathrm{FV}(t_n)$, if r is a relation symbol in Σ;
- $\mathrm{FV}(Rt_1 \ldots t_n) = \{R\} \cup \mathrm{FV}(t_1) \cup \ldots \cup \mathrm{FV}(t_n)$, if R is a relation variable;
- $\mathrm{FV}(\varphi \to \psi) = \mathrm{FV}(\varphi \vee \psi) = \mathrm{FV}(\varphi \wedge \psi) = \mathrm{FV}(\varphi) \cup \mathrm{FV}(\psi)$;
- $\mathrm{FV}(\forall a \varphi) = \mathrm{FV}(\exists a \varphi) = \mathrm{FV}(\varphi) - \{a\}$;
- $\mathrm{FV}(\forall R \varphi) = \mathrm{FV}(\exists R \varphi) = \mathrm{FV}(\varphi) - \{R\}$.

The meaning of $\varphi[a := t]$ should be entirely obvious, just add to Definition 8.1.3 the clauses:

- $(Rt_1 \ldots t_n)[a := t] = Rt_1[a := t] \ldots t_n[a := t]$;
- $(\forall R \varphi)[a := t] = \forall R \varphi[a := t]$;
- $(\exists R \varphi)[a := t] = \exists R \varphi[a := t]$.

In order to define the notion of a substitution for a relation variable we need to determine first *what* should be substituted. We can express relations as formulas, but a substitution $\varphi[R := \psi]$, for $R \in \mathcal{R}_k$, would be ill-formed as long as we do not specify which variables of ψ should correspond to the k arguments of R. What we need is a syntactic notation for a formula with distinguished parameters. As we somehow like the Greek letter λ, we will use it again.

12.1.3. DEFINITION.

(i) If ψ is a formula and $a_1, \ldots a_k$ are distinct individual variables, then the expression "$\lambda a_1 \ldots a_k.\psi$" is called a *species* of arity k. (Note that the variables $a_1, \ldots a_k$ do not have to occur in ψ.) Of course we define $\mathrm{FV}(\lambda a_1 \ldots a_k.\psi) = \mathrm{FV}(\psi) - \{a_1, \ldots, a_k\}$. If Q is a k-ary relation variable, we abbreviate $\lambda a_1 \ldots a_k.Q a_1 \ldots a_k$ by just Q.

(ii) The substitution of a k-ary species $\lambda a_1 \ldots a_k.\psi$ for a k-ary relation variable in a formula φ is denoted by $\varphi[R := \lambda a_1 \ldots a_k.\psi]$ and defined by the following clauses, where we take $\circ \in \{\to, \vee, \wedge\}$ and $\nabla \in \{\forall, \exists\}$ to reduce the number of cases.

- $\bot[R := \lambda a_1 \ldots a_k.\psi] = \bot$;
- $(rt_1 \ldots t_n)[R := \lambda a_1 \ldots a_k.\psi] = (rt_1 \ldots t_n)$, if r is either a signature symbol, or a variable other than R;
- $(Rt_1 \ldots t_k)[R := \lambda a_1 \ldots a_k.\psi] = \psi[a_1, \ldots, a_k := t_1, \ldots, t_k]$;
- $(\varphi \circ \psi)[R := \lambda a_1 \ldots a_k.\psi] = \varphi[R := \lambda a_1 \ldots a_k.\psi] \circ \psi[R := \lambda a_1 \ldots a_k.\psi]$;
- $(\nabla a\, \varphi)[R := \lambda a_1 \ldots a_k.\psi] = \nabla a\, \varphi[R := \lambda a_1 \ldots a_k.\psi]$, where a is chosen so that $a \notin \mathrm{FV}(\lambda a_1 \ldots a_k.\psi)$;
- $(\nabla P\, \varphi)[R := \lambda a_1 \ldots a_k.\psi] = \nabla P\, \varphi[R := \lambda a_1 \ldots a_k.\psi]$, where P is chosen so that $P \neq R$ and $P \notin \mathrm{FV}(\psi)$.

(iii) Sometimes we write e.g. $\varphi(R)$ if φ is (informally) meant to express a property of R. We then also write $\varphi(\lambda \vec{a}.\psi)$ instead of $\varphi[R := \lambda \vec{a}.\psi]$.

12.1.4. REMARK. An alternative syntax for second-order logic may involve also function variables of arbitrary arity (see [307, 337]).

12.2. Classical second-order logic

The *standard* or *Tarskian* semantics of second-order classical logic is obtained as a straightforward extension of the first-order semantics as described in Section 8.4. The notion of a model remains the same, but in order to interpret formulas we must provide meaning for relation variables. That is, we now define the value of a formula φ in a model $\mathcal{A}$ (cf. Definition 8.4.5) relative to a valuation which assigns

- elements of $\mathcal{A}$ to the individual variables;
- relations over $\mathcal{A}$ (of appropriate arities) to the relation variables.

12.2.1. DEFINITION. Given a valuation ϱ, the value $[\![\varphi]\!]_\varrho$ of φ is defined along the lines of Definition 8.4.5. In addition, we have the following clauses (where R is a n-ary relation variable):

$$\begin{aligned}
{[\![Rt_1\ldots t_n]\!]_\varrho} &= \varrho(R)([\![t_1]\!]_\varrho,\ldots,[\![t_n]\!]_\varrho); \\
{[\![\forall R\,\varphi]\!]_\varrho} &= \inf\{[\![\varphi]\!]_{\varrho(R\mapsto \mathbf{R})} \mid \mathbf{R} \subseteq \mathcal{A}^n\}; \\
{[\![\exists R\,\varphi]\!]_\varrho} &= \sup\{[\![\varphi]\!]_{\varrho(R\mapsto \mathbf{R})} \mid \mathbf{R} \subseteq \mathcal{A}^n\}.
\end{aligned}$$

Note that in the last two clauses, the value associated to R ranges over the set of all relations. This gives the language an enormous expressive power.

12.2.2. EXAMPLE.

(i) The formula $\forall R(Ra \to Rb)$ is satisfied by elements $\mathbf{a}$ and $\mathbf{b}$ of an arbitrary model $\mathcal{A}$ if and only if $\mathbf{a} = \mathbf{b}$. That is, equality $a = b$ is definable. (Equality defined this way is called *Leibniz equality*.)

(ii) The formula $\forall \vec{a}(P\vec{a} \to R\vec{a})$ is satisfied by two relations $\mathbf{P}$ and $\mathbf{Q}$ (of the same arity) if and only if $\mathbf{P} \subseteq \mathbf{Q}$. Thus, inclusion and equality of relations is definable too.

(iii) The formula $\mathsf{Int}(a) = \forall R(\forall a(Ra \to R(sa)) \to R0 \to Ra)$ is a second-order definition of a natural number based on Peano's fifth axiom, as discussed in Section 9.7. In an arbitrary model of arithmetic, this formula defines exactly the set of "standard numbers" $0, 1, 2, \ldots$ that can be obtained from zero by a finite number of applications of the sucessor function. It follows that the standard model of arithmetic is definable in second-order logic up to isomorphism (Exercise 12.3), although it is not the case for first-order logic (Theorem 9.1.2).

(iv) Consider the least ternary relation $\mathbf{R}$ over $\mathbb{N}$, such that $(n, 0, n) \in \mathbf{R}$ for all n, and whenever $(m, n, k) \in \mathbf{R}$ then also $(m, n+1, k+1) \in \mathbf{R}$. Clearly, $(m, n, k) \in \mathbf{R}$ if and only if $m + n = k$, i.e. the following formula states that $a + b = c$.

$$\forall R(\forall a\,(Ra0a) \to \forall abc\,(Rabc \to Ra(sb)(sc)) \to Rabc).$$

That is, integer addition is *explicitly* definable in second-order logic from zero and successor. In a similar way one can define multiplication and any other primitive recursive function. Compare this to the *inductive* definability in Chapter 9.

(v) Since addition and multiplication in $\mathbb{N}$ are definable, all arithmetical relations, in particular all partial recursive functions (cf. Corollary 9.1.6), are definable from zero and successor.

12.2. CLASSICAL SECOND-ORDER LOGIC

(vi) The technique used in part (iv) can be generalized to other recursive definitions (i.e. to all relations defined as least fixed points of monotone maps.) For instance, if R is unary then a formula $\varphi(R, a)$ defines an operation on unary relations, informally written as $R \mapsto \lambda a. \varphi(R, a)$. Suppose that this function is monotone in a model $\mathcal{A}$, i.e. assume that

$$\mathcal{A} \models \forall RS(R \subseteq S \to \varphi(R) \subseteq \varphi(S)),$$

where $\varphi(R)$ abbreviates the species $\lambda a. \varphi(R, a)$, and inclusion is defined as in (ii). By the fixed-point Theorem A.1.2, the formula

$$\forall R((\varphi(R) \subseteq R) \to Ra)$$

defines the least fixed point of the map given by φ, i.e. the relation r given by the recursive definition $ra \leftrightarrow \varphi(r, a)$.

(vii) Another definition of the least fixed point is suggested by Remark A.1.3:

$$\forall R(\forall P(P \subseteq R \to \varphi(P) \subseteq R) \to Ra).$$

INCOMPLETENESS. A natural way to design a proof system for classical second-order logic is to expand any of the first-order proof systems with a mechanism to handle second-order quantifiers. For instance, one can add the following axioms to the Hilbert-style proof system of Section 8.3.

(Q5) $\forall R \varphi(R) \to \varphi(\lambda \vec{a}. \psi)$;

(Q6) $\varphi(\lambda \vec{a}. \psi) \to \exists R \varphi(R)$;

(Q7) $\forall R(\varphi \to \psi) \to \varphi \to \forall R \psi$, where $R \notin \mathrm{FV}(\varphi)$;

(Q8) $\forall R(\varphi \to \psi) \to \exists R \varphi \to \psi$, where $R \notin \mathrm{FV}(\psi)$.

Then add a new form of the generalization rule

$$\frac{\varphi}{\forall R \varphi}$$

applicable when R is not free in the assumptions.

There should be no doubt that the proof system above is sound with respect to the standard semantics. However, here are the bad news. From part (iii) of Example 12.2.2 it follows that the set of tautologies of classical second-order logic is not recursively enumerable, see Exercise 12.4. An important consequence is that our proof system is not complete for the standard semantics, and no other finitary proof system can be complete. The incompleteness is the price we pay for the expressive power. One can recover completeness by changing the semantics to *Henkin models*, see Section 12.8.

12.3. Intuitionistic second-order logic

The natural deduction proof system for intuitionistic first-order logic (see Section 8.3) can be extended in a natural way by the following rules for second-order quantifiers.

$$(\forall^2 I) \ \frac{\Gamma \vdash \varphi}{\Gamma \vdash \forall R\, \varphi} \ (R \notin \mathrm{FV}(\Gamma)) \qquad (\forall^2 E) \ \frac{\Gamma \vdash \forall R\, \varphi}{\Gamma \vdash \varphi[R := \lambda \vec{a}.\psi]}$$

$$(\exists^2 I) \ \frac{\Gamma \vdash \varphi[R := \lambda \vec{a}.\psi]}{\Gamma \vdash \exists R\, \varphi} \qquad (\exists^2 E) \ \frac{\Gamma \vdash \exists R\, \varphi \quad \Gamma, \varphi \vdash \psi}{\Gamma \vdash \psi} \ (R \notin \mathrm{FV}(\Gamma, \psi))$$

As it can be expected, exactly the same rules added to the classical first-order natural deduction system make a proof system for second-order classical logic, equivalent to that of the previous section. Conversely, the axioms (Q5) to (Q8) together with the second-order generalization can be used to build a Hilbert-style system for the second-order intuitionistic logic.

The Brouwer-Heyting-Kolmogorov interpretation also extends to second-order logic in a natural way.

- A construction of $\forall R\, \varphi(R)$ is a method (function) transforming every predicate $\mathbf{R}$ into a proof of $\varphi(\mathbf{R})$.

- A construction of $\exists R\, \varphi(R)$ consists of a predicate $\mathbf{R}$ and a construction of $\varphi(\mathbf{R})$.

Our understanding of "every predicate" is "every property expressible by a formula". Indeed, the principle of full comprehension, expressed by

$$\exists R \forall \vec{a}(R\vec{a} \leftrightarrow \psi),$$

is intuitionistically provable (for arbitrary ψ with $\mathrm{FV}(\psi) = \vec{a}$).

As in the case of second-order propositional logic (cf. Section 11.1), a consequence of full comprehension is impredicativity: The meaning of a universal formula can only be explained by a reference to a collection (the set of all formulas) to which the formula in question itself belongs.

On the other hand, the assumption that quantifiers range over *definable* propositions only is a sharp restriction compared to Tarskian semantics. However, from a constructive point of view, a *proof* and not a *model* is the ultimate criterion, and this restriction is fully justified.

A relationship between the classical and intuitionistic second-order proof systems can be again formulated in the form of Kolmogorov's translation, extended as follows.

- $k(\forall R\, \varphi) := \neg\neg \forall R\, k(\varphi)$;
- $k(\exists R\, \varphi) := \neg\neg \exists R\, k(\varphi)$.

12.3. INTUITIONISTIC SECOND-ORDER LOGIC

12.3.1. PROPOSITION. *A formula φ is a classical theorem iff $k(\varphi)$ is an intuitionistic theorem.*

PROOF. Similar to the proof of Proposition 8.3.1. Use the intuitionistically provable equivalence $k(\varphi[R := \lambda \vec{a}.\psi]) \leftrightarrow k(\varphi)[R := \lambda \vec{a}.k(\psi)]$. □

The reader must remember however that now the translation works only for provable *theorems* of classical logic, and not for all valid *tautologies*.

CONTRACTING MAP. Propositional second-order logic can be seen as a fragment of the second-order language over an empty signature with only nullary relation variables. Alternatively, propositional second-order formulas are obtained from predicate second-order formulas by applying a contracting map.

12.3.2. CONVENTION. We extend Notation 8.7.5 by assuming that a fixed propositional variable p_R is associated to every relation variable R, and that this association is injective. In order to simplify notation we write $p, q, r, \ldots$ for $p_P, p_Q, p_R, \ldots$

12.3.3. DEFINITION. We extend the map $\flat$ of Definition 8.7.6 to second-order formulas, by adding these clauses:

- $\flat(Rt_1 \ldots t_n) = r$, the propositional variable associated to R.
- $\flat(\nabla R \, \varphi) = \nabla r \, \flat(\varphi)$, for $\nabla \in \{\exists, \forall\}$.

FROM SECOND-ORDER LOGIC TO SYSTEM **F**. We have seen in Chapter 10 that the contracting map applied to the induction scheme yields the type of recursor in system **T**. As another example, the formula $\mathsf{Int}(a)$ of Example 12.2.2(iii), which states the second-order definition of a natural number, translates under $\flat$ to ω, the type of polymorphic Church numerals. Let us see how the contracting map works in a few other cases.

Consider the following definition of the union of two sets, as the least set containing both the components:

$$x \in P \cup Q \quad \text{iff} \quad \forall R(P \subseteq R \to Q \subseteq R \to x \in R).$$

Replacing the set variables P, Q and R by variables ranging over unary predicates, we rewrite this definition as

$$Px \vee Qx \Leftrightarrow \forall R(\forall y(Py \to Ry) \to \forall y(Qy \to Ry) \to Rx).$$

This can be used as a definition of disjunction in second-order predicate logic. The map $\flat$ turns it into the propositional variant, as in Definition 11.3.2:

$$p \vee q \Leftrightarrow \forall r((p \to r) \to (q \to r) \to r),$$

The first-order aspect could simply be disposed of without loss of generality.

Similarly, the definition of $\bot$ in system **F** (cf. Definition 11.3.2) can be seen as $\flat(\forall P.Px)$, that is, as a contraction of a definition of the empty set:

$$x \in \varnothing \quad \text{iff} \quad \forall P(x \in P).$$

Finally, to understand why the formula $\forall r \left((\tau \to \sigma \to r) \to r \right)$ is taken as a definition of $\tau \wedge \sigma$ (Definition 11.3.3) one should observe that

$$x \in P \cap Q \quad \text{iff} \quad \forall R(\forall z (z \in P \to z \in Q \to z \in R) \to x \in R).$$

INDUCTIVE TYPES. Let us have a look now at the second-order definitions of the least fixed point of a monotone map, as given by parts (vi) and (vii) of Example 12.2.2. Before discussing the general case, let us first consider something more familiar. For instance, we may want to implement a list of the form $n :: \ell$ as a pair $\langle n, \ell \rangle$. Assuming that nil is the only element of a certain "unit" type **1**, we may thus postulate the following equation:

$$\textbf{list} = \textbf{1} \vee (\omega \wedge \textbf{list}). \tag{$*$}$$

Of course, in system **F** there is no type satisfying this equation. We can however define a decent representation of lists if we settle for a little less than the exact solution. Consider the following definition, inspired by Example 12.2.2(vii), where $\tau(p)$ stands for $\textbf{1} \vee (\omega \wedge p)$:

$$\textbf{list} = \forall q \left(\forall p \left((p \to q) \to \tau(p) \to q \right) \to q \right). \tag{$**$}$$

This type does not satisfy the equation $(*)$, but we can define a closed term

$$\text{in} : \tau(\textbf{list}) \to \textbf{list},$$

which works as an "embedding" from $\tau(\textbf{list})$ into **list**. If $n : \omega$ and $\ell :$ **list** then $\text{in}(\langle n, \ell \rangle)$ is of type **list**. Also in(nil) is a list. A possible definition of such an embedding is:

$$\text{in} = \lambda x^{\tau(\textbf{list})} \Lambda q \lambda z^{\forall p \left((p \to q) \to \tau(p) \to q \right)} . z \, \textbf{list}(\lambda y^{\textbf{list}}.yqz)x.$$

Each list (a sequence of numbers) can thus be interpreted as a closed term of type **list**, for instance the term $\text{in}\langle 1, \text{in}\langle 3, \text{in}\langle 3, \text{in nil}\rangle\rangle\rangle$, represents the list $1 :: 3 :: 3 ::$ nil.

The type **list** in $(**)$ has been chosen so that it supports definitions by induction over lists. Suppose we have a type σ and an operator

$$Z : \forall p \left((p \to \sigma) \to \tau(p) \to \sigma \right).$$

We can think of Z as of a (generic) method to translate a function of type $p \to \sigma$ into a function of type $\tau(p) \to \sigma$. Every such operator induces an inductive definition of a function from **list** to σ. For instance, if $\sigma = \omega$ and

$$Z = \Lambda p \lambda f^{p \to \omega} \lambda z^{\tau(p)}. \textbf{case } z \textbf{ of } [x]\, \textbf{0} \textbf{ or } [y]\, \text{add}(\pi_1(y))(f(\pi_2(y)),$$

12.4. Second-order Peano Arithmetic   311

then Z can be used to define the sum of all entries of a list, by setting

$$\text{sum} = \lambda \ell^{\text{list}}.\ell \omega Z.$$

Indeed, $\text{sum}(\text{in nil}) =_\beta \mathbf{0}$ and $\text{sum}(\text{in}\langle n, \ell\rangle) =_\beta \text{add}\, n\, \ell$. In general, we have $\text{sum}(\text{in}\, M) =_\beta Z\, \textbf{list}\, \text{sum}\, M$. The function sum can be seen as obtained from Z by application of the *list iterator*

$$\text{It} = \Lambda q \lambda z^{\forall p\, ((p \to q) \to \tau(p) \to q)} \lambda \ell^{\text{list}}.\ell q z.$$

In general, the induction principle for lists can be written as the equation

$$\text{It}\,\sigma\, Z(\text{in}\, M) =_\beta Z\,\textbf{list}(\text{It}\,\sigma\, Z)M.$$

12.3.4. REMARK. The construction above generalizes to any type τ. Define

$$\mu = \forall q\, (\forall p\, ((p \to q) \to \tau(p) \to q) \to q),$$

and let $\text{in} : \tau(\mu) \to \mu$ and $\text{It} : \forall q\, (\forall p\, ((p \to q) \to \tau(p) \to q) \to \mu \to q)$ be defined as above with **list** replaced by μ:

$$\text{in} = \lambda x^{\tau(\mu)} \Lambda q \lambda z^{\forall p\, ((p\to q)\to\tau(p)\to q)}.\, z\mu(\lambda y^\mu. yqz)x;$$
$$\text{It} = \Lambda q \lambda z^{\forall p\, ((p\to q)\to\tau(p)\to q)} \lambda \ell^\mu.\ell q z.$$

Then the induction principle $\text{It}\,\sigma\, Z(\text{in}\, M) =_\beta Z\mu(\text{It}\,\sigma\, Z)M$ holds for all Z and M of appropriate types.

The definition of μ is obtained by contracting the formula in Example 12.2.2(vii). Part (vi) of Example 12.2.2 gives another way to implement fixed points. See Exercise 12.9.

12.4. Second-order Peano Arithmetic

If we add the signature of arithmetic to second-order logic we obtain a language capable of expressing second-order concepts concerning the algebra of natural numbers. This extension can be understood literally, that is, we can include the symbols 0, s, $+$, $\cdot$, $=$ in the vocabulary, and postulate all first-order axioms of Peano Arithmetic. However, we can now simplify the presentation using a true *induction axiom* rather than *induction scheme*. This axiom can be formulated as

$$\forall R(\forall a(Ra \to R(sa)) \to R0 \to \forall a Ra),$$

but it will be more convenient to use an equivalent universal statement

$$\forall a\, \mathsf{Int}(a),$$

where $\mathsf{Int}(a)$ is the definition of a standard number, as in Example 12.2.2:

$$\mathsf{Int}(a) = \forall R(\forall a(Ra \to R(sa)) \to R0 \to Ra).$$

We also know from Example 12.2.2 that equality, as well as addition and multiplication, are definable in second-order logic. One can thus make further simplifications in the syntax and axiomatization. We will indeed consider such simplified syntax in Section 12.6. However, as in Chapter 9, it will be convenient for a while to have more symbols in the signature, because it is easier to deal with atomic formulas than with complex ones. Thus, for our initial presentation we keep all the old symbols, and we even expand the signature with names of all primitive recursive functions, exactly as we did for first-order arithmetic in Chapter 9.

For a similar reason it will be convenient to assume that the standard logical symbols $\bot$, $\lor$, $\land$ and $\exists$ are still available in the language as primitives, rather than as defined abbreviations.

12.4.1. DEFINITION. The *second-order Peano Arithmetic* (PAS) is a classical second-order theory (cf. Definition 8.3.11) in the signature consisting of the symbols 0, s, $=$, and of names for all primitive recursive functions. The axioms of PAS include the axioms of equality

(E_1) $\forall a(a = a)$;

(E_2) $\forall ab(a = b \to b = a)$;

(E_3) $\forall a \forall b \forall c(a = b \to b = c \to a = c)$;

(E_4) $\forall ab(a = b \to sa = sb)$,

the three Peano axioms

(P_3) $\forall a \forall b(sa = sb \to a = b)$;

(P_4) $\forall a(sa = 0 \to \bot)$;

(P_5) $\forall a\, \mathsf{Int}(a)$,

and the defining equations of all primitive recursive functions (including addition and multiplication).

The properties of second-order Peano Arithmetic are similar to the properties of PA, but the new system is of much larger expressive power. For instance, the proof of strong normalization for System **T** (Theorem 10.3.8) can be formalized in PAS, but not in PA (cf. Corollary 10.4.11 and the discussion following Theorem 10.3.8). It follows that there are functions provably total in PAS but not in PA.

12.4. SECOND-ORDER PEANO ARITHMETIC

12.4.2. PROPOSITION. *The class of provably total functions of* PAS *properly extends the class of functions provably total in* PA.

PROOF. Exercise 12.6. □

As we have already noticed in Chapter 11, the proof of strong normalization for system **F** (Theorem 11.5.7) requires the use of quantifiers ranging over sets of sets. This obviously leads to the idea of *third-order* and generally *higher-order* arithmetic (related to $\mathbf{F}_\omega$ in a similar way as second-order arithmetic is related to system **F**). We do not discuss these stronger systems here. For our purposes second-order logic is enough.

12.4.3. PROPOSITION. *Every function definable in system* **F** *is provably total in second-order arithmetic.*

PROOF. If our k-ary function f is defined by a term $F : \omega^k \to \omega$ then every application $F\mathbf{c}_{n_1}\ldots\mathbf{c}_{n_k}$ is typable using a finite set of types. It follows that strong normalization of all $F\mathbf{c}_{n_1}\ldots\mathbf{c}_{n_k}$ (and thus totality of f) can be shown using a simplified version of the candidates argument used in the proof of Theorem 11.5.7. As observed at the end of Section 11.5, one can write a finite number of second-order formulas $computable_\tau(M,\xi)$, expressing that $M \in [\![\tau]\!]_\xi$, with ξ represented as a set variable. Using these formulas, the proof in our special case can be formulated so that quantification over sets of sets is not necessary, and the whole proof can be formalized in PAS. □

We conclude this section with an observation which later will help us prove Theorem 12.7.11, the converse of Proposition 12.4.3. We show that the class of provably total functions does not change if we choose a slightly weaker version of PAS, which we denote by PAs. We need this weaker version, because in Section 12.7 we want to apply a contracting map (extending the map $\mathfrak{b}$ of Definition 12.3.3) to formulas of second-order arithmetic. Unfortunately, if we apply $\mathfrak{b}$ to Peano's fourth axiom $\forall a(sa = 0 \to \bot)$ we obtain a non-inhabited type (because of the occurrence of absurdity), and we would rather prefer to avoid this.

12.4.4. DEFINITION. The theory PAs is obtained from PAS by replacing axiom (P_4) by

(P_4') $\forall ab(sa = 0 \to b = 0)$.

The idea of this definition (following Leivant [304, 308]) is to replace $\bot$ by $\forall b\,(b = 0)$, which is *almost* as devastating as the absurdity itself, at least from the point of view of arithmetic. Although PAs is a weaker theory (Exercise 12.13) it is strong enough for our specific purpose.

12.4.5. LEMMA. *If a function is provably total in* PAS *then it is also provably total in* PAs.

PROOF. Let φ denote the formula $\forall a \exists b \, t_f(a,b) = 0$ and suppose that φ has a proof in PAS. It is not hard to see that from P_4' one can (classically) derive the disjunction $P_4 \vee \forall b (b = 0)$. From either of the components we can derive φ. Indeed, with P_4 we regain the power of the full PAS, and in the other case everything is zero anyway. Therefore φ has a proof in PAs. $\square$

12.5. Second-order Heyting Arithmetic

The language of second-order intuitionistic arithmetic is the same as the language of second-order classical arithmetic, and the difference is only in the underlying logic. We can of course choose several variants of the syntax, corresponding to analogous choices for the classical theory.

12.5.1. DEFINITION. *Second-order Heyting Arithmetic* (HAS) is an intuitionistic second-order theory over the signature of PAS and with the same axioms as PAS. By HAs we denote a fragment of HAS with axiom (P_4) replaced by (P_4').

We now discuss the relationship between classical and intuitionistic second-order arithmetic. Our goal is to show that all provably recursive functions of PAS can be shown total in HAs. The first step is that the Kolmogorov translation k for second-order logic (see Proposition 12.3.1) interprets PAS in HAS and PAs in HAs.

12.5.2. PROPOSITION. *For all φ, the condition* PAS $\vdash \varphi$ *is equivalent to* HAS $\vdash k(\varphi)$ *and the condition* PAs $\vdash \varphi$ *is equivalent to* HAs $\vdash k(\varphi)$.

PROOF. Similar to the proof of Proposition 9.5.2. See Exercise 12.11. $\square$

12.5.3. DEFINITION. Let ϱ be a fixed formula. As in Chapter 9, we denote by φ^ϱ the formula obtained from φ by replacing every atomic subformula ν (including $\bot$) by the disjunction $\nu \vee \varrho$. (It is assumed that free variables of ϱ are not bound in φ.)

12.5.4. LEMMA.

(i) *For every φ, the formula $\varrho \to \varphi^\varrho$ is an intuitionistic theorem.*

(ii) *For every φ and t, if $x \notin \mathrm{FV}(\varrho)$ then $(\varphi[x := t])^\varrho = \varphi^\varrho[x := t]$.*

(iii) *Let R and $\lambda \vec{a}.\psi$ be of the same arity, and let $R, \vec{a} \notin \mathrm{FV}(\varrho)$. Then, for every φ, the formula $(\varphi[R := \lambda \vec{a}.\psi])^\varrho \leftrightarrow \varphi^\varrho[R := \lambda \vec{a}.\psi^\varrho]$ is an intuitionistic theorem.*

(iv) *If $\Gamma \vdash \varphi$ then $\Gamma^\varrho \vdash \varphi^\varrho$, where $\Gamma^\varrho = \{\gamma^\varrho \mid \gamma \in \Gamma\}$.*

PROOF. The proof of parts (i) and (ii) is by a routine induction with respect to φ. Both (i) and (ii) are used in the proof of (iii), which is also an easy induction on φ. Part (iv) goes by induction with respect to $\Gamma \vdash \varphi$, using parts (ii) and (iii) in cases when the proof involves instantiation. □

12.5.5. LEMMA.

(i) *If φ is an axiom of* HAs *then* HAs $\vdash \varphi^\varrho$.

(ii) *If* HAs $\vdash \varphi$ *then* HAs $\vdash \varphi^\varrho$.

PROOF. For part (i) observe that most of the axioms are universal formulas of the form $\varphi = \forall \vec{a}(\alpha_1 \to \cdots \to \alpha_n \to \beta)$, where α_i and β are equations. Then φ^ϱ is $\forall \vec{a}(\alpha_1 \vee \varrho \to \cdots \to \alpha_n \vee \varrho \to \beta \vee \varrho)$ and this formula is easily provable from φ. The only axiom not of the form above is P_5. But the translation $\forall a \forall R(\forall a(Ra \vee \varrho \to R(sa) \vee \varrho) \to R0 \vee \varrho \to Ra \vee \varrho)$ of P_5 is essentially a special case of P_5. Part (ii) is an immediate consequence of part (i) and Lemma 12.5.4(iv). □

The following is a second-order version of Corollary 9.5.7.

12.5.6. THEOREM (Spector). *If a function f is provably total in* PAS *then it is also provably total in* HAs.

PROOF. From Lemma 12.4.5 we know that f is provably total in PAs. Thus PAs proves $\exists b\, t_f(\vec{a}, b) = 0$. By Lemma 12.5.2, we have

$$\text{HAs} \vdash \neg\neg \exists b\, (\neg\neg t_f(\vec{a}, b) = 0).$$

To this formula we apply Friedman's translation, where ϱ is $\exists b\, (t_f(\vec{a}, b) = 0)$, and (after simplifying $\bot \vee \varrho$ to ϱ) we obtain:

$$\text{HAs} \vdash (\exists b\, [((t_f(\vec{a}, b) = 0 \vee \varrho) \to \varrho) \to \varrho] \to \varrho) \to \varrho.$$

The formula $(t_f(\vec{a}, b) = 0 \vee \varrho) \to \varrho$ is provable, and thus

$$\text{HAs} \vdash ((\exists b\, \varrho) \to \varrho) \to \varrho.$$

But of course $\exists b\, \varrho$ is equivalent to ϱ, so we obtain HAs $\vdash (\varrho \to \varrho) \to \varrho$ and finally HAs $\vdash \varrho$. □

12.6. Simplified syntax

As observed at the beginning of this chapter, the definitional strength of the second-order language makes it possible to eliminate certain primitives from the language of second-order arithmetic. We now use this opportunity in order to reduce the syntax of second-order arithmetic to a minimum. First we get rid of the whole vocabulary of primitive recursive functions. It is not difficult to see the following.

12.6.1. LEMMA. *For every primitive recursive f of k arguments there exists a formula φ_f with $k+1$ free individual variables, and with no symbols of primitive recursive function, such that* $\mathrm{HAs} \vdash f(\vec{a}) = b \leftrightarrow \varphi_f(\vec{a}, b)$.

In other words, all primitive recursive functions are explicitly definable in HAs. Therefore, HAs is conservative over its fragment HAs°, where the special symbols for primitive recursive functions are not available (cf. [489]). From now on, we identify HAs and HAs°, as we did in Chapter 9 in case of first-order arithmetic. Now recall that equality defined as in part (i) of Example 12.2.2 satisfies the axioms (E_1) to (E_4) in Definition 12.4.1. To be more precise, if we replace equality by its second-order definition (Leibniz equality), these axioms become provable in second-order intuitionistic logic (Exercise 12.7).

Similarly, as shown in Section 11.3, the propositional connectives are definable in terms of $\to$ and $\forall$. The existential quantifiers can also be defined:

- $\exists a \varphi$ unfolds to $\forall R(\forall a(\varphi \to R) \to R))$;

- $\exists P \varphi$ unfolds to $\forall R(\forall P(\varphi \to R) \to R))$.

Formulas written exclusively with the logical symbols $\to$ and $\forall$ (first- or second-order) and the function symbols 0 and s are referred to as formulas over the *simplified syntax*.

12.6.2. DEFINITION. Let φ be a formula in the language of HAs. By φ_L we denote the formula (over the simplified syntax) obtained from φ by replacing all occurrences of $\bot, \vee, \wedge, \exists$ and $=$ by their second-order definitions. The notation Γ_L (in particular HAs_L) is used accordingly for sets of formulas.

12.6.3. LEMMA. *If $\Gamma \vdash \varphi$ then $\Gamma_L \vdash \varphi_L$.*

PROOF. Left to the reader. □

Whenever we refer to the simplified syntax, any use of the eliminated symbols should be understood as an abbreviation representing the appropriate definition.

Replacing the symbol $=$ by Leibniz equality makes the axioms of equality provable in second-order logic, so that we can safely make the identification

$$\mathrm{HAs}_L = \{(P_3)_L, (P_4')_L, P_5\}.$$

In a sense, we can also eliminate P_5 or, more precisely, replace the use of P_5 by imposing the condition $\mathsf{Int}(a)$ on all individuals under consideration. This requires relativizing individual quantifiers as follows. We abbreviate $\forall a(\mathsf{Int}(a) \to \cdots)$ by $\forall a^{\mathsf{Int}}(\cdots)$.

12.6. SIMPLIFIED SYNTAX

12.6.4. DEFINITION. For a formula φ in the simplified syntax, let φ^{Int} be obtained from φ by relativizing all quantifiers, i.e. replacing each $\forall$ by $\forall^{\mathsf{Int}}$. If Γ is a set of formulas in the simplified syntax then Γ^{Int} is obtained by:

- relativizing all formulas in Γ;
- adding assumptions $\mathsf{Int}(a)$, for all $a \in \mathrm{FV}(\Gamma)$;
- adding the axioms $(P_3)_L$ and $(P'_4)_L$.

The expression $\mathsf{Int}(\vec{a})$ abbreviates a conjunction of all $\mathsf{Int}(a)$ for $a \in \vec{a}$.

Note that $(\exists a \varphi)^{\mathsf{Int}} = \forall R(\forall a(\mathsf{Int}(a) \to \varphi^{\mathsf{Int}} \to R) \to R)$, which is the same as $\exists a(\mathsf{Int}(a) \wedge \varphi^{\mathsf{Int}})$.

12.6.5. LEMMA. *For formulas φ in the simplified syntax:*

(i) $\mathrm{HAs}_L \vdash \varphi^{\mathsf{Int}} \leftrightarrow \varphi$.

(ii) $(\varphi[a := t])^{\mathsf{Int}} = \varphi^{\mathsf{Int}}[a := t]$ and $(\varphi[R := \lambda \vec{a}.\psi])^{\mathsf{Int}} = \varphi^{\mathsf{Int}}[R := \lambda \vec{a}.\psi^{\mathsf{Int}}]$.

(iii) *If* $\Gamma \vdash \mathsf{Int}(a) \to \varphi$ *and* $a \notin \mathrm{FV}(\Gamma, \varphi)$ *then* $\Gamma \vdash \varphi$.

PROOF. Parts (i) and (ii) are shown by induction with respect to φ. To show part (iii), generalize over a and instantiate it to zero. □

12.6.6. LEMMA. *Let φ be a formula in the simplified syntax with $\mathrm{FV}(\varphi) = \vec{a}$. Then $\mathrm{HAs}_L, \Gamma \vdash \varphi$ iff $\Gamma^{\mathsf{Int}}, \mathsf{Int}(\vec{a}) \vdash \varphi^{\mathsf{Int}}$.*

PROOF. The proof from left to right is by induction with respect to the derivation of φ. Among the axioms of HAs_L only induction is non-obvious. We need to find a derivation for the following judgement:

$$\Gamma^{\mathsf{Int}}, \mathsf{Int}(a) \vdash \forall R \left(\forall b(\mathsf{Int}(b) \to Rb \to R(sb)) \right) \to R0 \to Ra).$$

This can be done by instantiating the $\forall R$ in $\mathsf{Int}(a)$ by $\lambda b.(\mathsf{Int}(b) \wedge Rb)$. The other cases in the proof are essentially routine, but Lemma 12.6.5 is needed.

For the other direction we use Lemma 12.6.5(i). Since HAs_L, Γ proves all formulas in $\Gamma^{\mathsf{Int}}, \mathsf{Int}(\vec{a})$, we have $\mathrm{HAs}_L, \Gamma \vdash \varphi^{\mathsf{Int}}$, and thus $\mathrm{HAs}_L, \Gamma \vdash \varphi$. □

We summarize the results of this section as follows.

12.6.7. LEMMA. *If φ is closed and $\mathrm{HAs} \vdash \varphi$ then $(P_3)_L, (P'_4)_L \vdash (\varphi_L)^{\mathsf{Int}}$.*

12.7. Lambda-calculus

As we know, formulas of various languages can be seen as types of their proofs, and the proofs may be identified with terms in certain lambda-calculi. This time we want to design an appropriate calculus corresponding to intuitionistic second-order arithmetic. The corresponding lambda-calculus will be denoted by λH.

Types of system λH are the formulas of arithmetic over the simplified syntax, that is, formulas of the $(\forall, \rightarrow)$-fragment of second-order logic over the signature consisting of 0 and s. We assume all other symbols are appropriate abbreviations, in particular the symbol $=$ is used for Leibniz equality.

Terms of system λH are defined by the axioms and type assignment rules in Figure 12.1. Alpha-conversion is of course assumed.

$$\Gamma, x:\varphi \vdash x:\varphi$$

$$\Gamma \vdash \mathsf{P}_3 : \forall ab(sa = sb \rightarrow a = b) \qquad \Gamma \vdash \mathsf{P}'_4 : \forall ab(sa = 0 \rightarrow b = 0)$$

$$\frac{\Gamma \vdash M : \varphi \rightarrow \psi, \quad N : \varphi}{\Gamma \vdash (MN) : \psi} \qquad \frac{\Gamma, x:\varphi \vdash M : \psi}{\Gamma \vdash (\lambda x{:}\varphi.M) : \varphi \rightarrow \psi}$$

$$\frac{\Gamma \vdash M : \forall a\varphi}{\Gamma \vdash (Mt) : \varphi[a := t]} \qquad \frac{\Gamma \vdash M : \varphi}{\Gamma \vdash (\lambda a M) : \forall a\varphi} \ (a \notin \mathrm{FV}(\Gamma))$$

$$\frac{\Gamma \vdash M : \forall R\varphi}{\Gamma \vdash (M(\lambda \vec{a}.\psi)) : \varphi[R := \lambda \vec{a}.\psi]} \qquad \frac{\Gamma \vdash M : \varphi}{\Gamma \vdash (\Lambda R M) : \forall R\varphi} \ (R \notin \mathrm{FV}(\Gamma))$$

FIGURE 12.1: SYSTEM λH

Inhabitation in system λH represents second-order provability from the two specific axioms $(P_3)_L$ and $(P'_4)_L$. For our needs, the following properties of system λH are the most important.

12.7.1. LEMMA.

(i) *If a closed formula φ is a theorem of HAs then $(\varphi_L)^{\mathsf{Int}}$ is an inhabited type of system λH.*

(ii) *If a type ψ^{Int} is inhabited in λH then ψ is (classically) true in the standard model of arithmetic.*

12.7. LAMBDA-CALCULUS

PROOF. Part (i) is an immediate consequence of Lemma 12.6.7. For part (ii), one checks that the axioms are true in $\mathcal{N}$ and that $\mathcal{N} \models \psi \leftrightarrow \psi^{\mathsf{Int}}$. □

The reduction rules of system λH are the following beta-rules (the proper definition of substitution being left to the reader):

$$\begin{array}{rcl}(\lambda x{:}\varphi.M)N & \to & M[x := N]; \\ (\lambda a\, M)t & \to & M[a := t]; \\ (\Lambda R\, M)(\lambda \vec{a}.\psi) & \to & M[R := \lambda \vec{a}.\psi].\end{array}$$

(the second one called *first-order* reduction) and one specific rule for P_3:

$$\mathsf{P}_3 tt P^{st=st} \Rightarrow \Lambda R \lambda x^{Rt}\, x.$$

In the rule above, both sides are expressions of type $t = t$, that is, of type $\forall R\,(Rt \to Rt)$. The term P is a proof of $st = st$, and the role of this rule is to eliminate a detour in a proof of $t = t$, by appealing directly to reflexivity. There is no reduction rule for P'_4 and there is no constant corresponding to induction, so there is no reduction rule for induction either.

As the reduction rules do not have critical pairs (see Remark 6.3.10), we may state the following.

12.7.2. PROPOSITION. *System λH has the* WCR *property.* □

As for system $\lambda\mathsf{P}_1$ in Chapter 8, we now show the SN and CR property with the help of a contracting map $\mathfrak{b}$. Observe that by Definition 12.3.3 we have $\mathfrak{b}(\mathsf{Int}(t)) = \omega$ and (since we agreed to use Leibniz equality) we also have $\mathfrak{b}(t = s) = \forall r(r \to r)$, for all terms t and s.

12.7.3. DEFINITION. The contracting map $\mathfrak{b}$ maps terms of system λH to terms of system **F**. Below, *true* abbreviates $\forall r(r \to r)$.

- $\mathfrak{b}(x) = x$;
- $\mathfrak{b}(\mathsf{P}_3) = \mathfrak{b}(\mathsf{P}'_4) = \lambda x{:}true.\, x$;
- $\mathfrak{b}(\lambda x{:}\varphi\, M) = \lambda x{:}\mathfrak{b}(\varphi).\, \mathfrak{b}(M)$;
- $\mathfrak{b}(\lambda a\, M) = \mathfrak{b}(M)$;
- $\mathfrak{b}(\Lambda R\, M) = \Lambda r.\, \mathfrak{b}(M)$, where r corresponds to R as in Convention 12.3.2;
- $\mathfrak{b}(MN) = \mathfrak{b}(M)\mathfrak{b}(N)$;
- $\mathfrak{b}(Mt) = \mathfrak{b}(M)$;
- $\mathfrak{b}(M(\lambda \vec{a}\, \psi)) = \mathfrak{b}(M)\mathfrak{b}(\psi).$

12.7.4. LEMMA. *The operation $\mathfrak{b}$ commutes with substitution, i.e. the following equations always hold:*

(i) $\mathfrak{b}(\varphi[a := t]) = \mathfrak{b}(\varphi)$;

(ii) $\mathfrak{b}(\varphi[R := \lambda\vec{a}.\psi]) = \mathfrak{b}(\varphi)[r := \mathfrak{b}(\psi)]$;

(iii) $\mathfrak{b}(M[a := t]) = \mathfrak{b}(M)$;

(iv) $\mathfrak{b}(M[y := N]) = \mathfrak{b}(M)[y := \mathfrak{b}(N)]$;

(v) $\mathfrak{b}(M[R := \lambda\vec{a}.\psi]) = \mathfrak{b}(M)[r := \mathfrak{b}(\psi)]$.

PROOF. Similar to the proof of Lemma 8.7.7. □

The proofs of the following lemmas are routine and left to the reader.

12.7.5. LEMMA. *If $\Gamma \vdash M : \varphi$ in λH, then $\mathfrak{b}(\Gamma) \vdash \mathfrak{b}(M) : \mathfrak{b}(\varphi)$ in system **F**.*

12.7.6. LEMMA. *If $M \to N$ in system λH, then $\mathfrak{b}(M) \twoheadrightarrow \mathfrak{b}(N)$. In addition, $\mathfrak{b}(M) \twoheadrightarrow^+ \mathfrak{b}(N)$ holds in all cases except when the reduction step $M \to N$ is an application of first-order beta-reduction.*

12.7.7. THEOREM. *System λH has the strong normalization property.*

PROOF. Similar to the proof of Theorem 8.7.10. □

12.7.8. COROLLARY. *System λH has the Church-Rosser property.*

PROOF. From SN and Newman's lemma. □

By Theorem 12.7.7, if a formula φ (in the simplified syntax) is provable in HAs then there is a normal proof term of type φ^{Int} in λH. Our next objective is to describe how this normal proof should look like, depending on φ.

12.7.9. DEFINITION. The *target variable* of a formula (in the simplified syntax) is the rightmost relation variable occurring in it. That is, R is the target of $R\vec{t}$, and the target of $\forall a \varphi$, $\forall R \varphi$, and $\psi \to \varphi$ is the target of φ. We say that an environment Γ is *easy* if the target of every $\sigma \in \text{rg}(\Gamma)$ is free in σ.

Observe that if Γ is easy and an application beginning with x has type τ in Γ then the target of τ is the same as the target of $\Gamma(x)$.

12.7.10. LEMMA.

(i) *If M is a term in normal form then M is either an abstraction or an application $c\vec{e}$, where c is a constant or a variable, and $\vec{e}$ is a (possibly empty) sequence of terms and species.*

12.7. LAMBDA-CALCULUS

(ii) *A normal application $c\vec{e}$ beginning with a constant, and having no free individual variables, may only have one of the following types in an easy environment:*

- $\forall a \forall b (sa = sb \rightarrow a = b);$
- $\forall b (s(\underline{n}) = sb \rightarrow \underline{n} = b);$
- $s(\underline{n}) = s(\underline{m}) \rightarrow \underline{n} = \underline{m};$
- $\forall ab (sa = 0 \rightarrow b = 0);$
- $\forall b (s(\underline{n}) = 0 \rightarrow b = 0);$
- $s(\underline{n}) = 0 \rightarrow \underline{m} = 0.$

(iii) *A normal form without free individual variables, which has type $\underline{n} = \underline{m}$ in an easy environment must be of shape $\Lambda R \lambda x^{R\underline{n}}.x$. In particular, if $\underline{n} = \underline{m}$ is inhabited then $n = m$.*

(iv) *Let $\Gamma, x : \forall a(\theta \rightarrow \zeta \rightarrow S) \vdash M : \mathsf{Int}(\underline{n})$, where Γ is easy, S is a nullary relation variable, and M is a normal form without free individual variables. Let $\mathcal{O} = \lambda x^{\omega} \Lambda r \lambda y^{r \rightarrow r} \lambda z^r.xryz$. Then[1] $\mathcal{O}(\flat(M)) = \mathbf{c}_n$.*

PROOF. Part (i) is shown by induction with respect to M. If M is neither a constant nor an abstraction, then it must be an application $M_1 e$ and we can apply the induction hypothesis to M_1. But now M_1 cannot be an abstraction, because $M_1 e$ would be a redex.

Parts (ii) and (iii) are shown by parallel induction with respect to the size of terms. For part (ii) observe that in every application of the form $\mathsf{P}_3 \underline{n}\, \underline{m} P$, the term P has type $s\underline{n} = s\underline{m}$. By the induction hypothesis, part (iii), we have $n = m$, and thus our term is a redex—a contradiction. Similarly, an application $\mathsf{P}'_4 \underline{n}\, \underline{m} P$ would contain a closed term P of type $s\underline{n} = 0$ and this is again impossible by the induction hypothesis (iii).

For part (iii) observe that, by part (ii), a normal form of type $\underline{n} = \underline{m}$ cannot be an application beginning with a constant. It cannot be an application beginning with a variable, because the target of $\underline{n} = \underline{m}$ is not free.

For part (iv), observe that $\Gamma, x : \forall a(\theta \rightarrow \zeta \rightarrow S) \vdash M : \mathsf{Int}(\underline{n})$ implies, by parts (i) and (ii), that $M = \Lambda R \lambda y^{\forall a(Ra \rightarrow R(sa))} N$ where N has type $R0 \rightarrow R\underline{n}$ in the environment $\Gamma' = \Gamma \cup \{x : \forall a(\theta \rightarrow \zeta \rightarrow S), y : \forall a(Ra \rightarrow R(sa))\}$. We can also assume that $R \notin \mathsf{FV}(\Gamma)$. Then either $n = 1$ and $N = y0$, or $N = \lambda z^{R0} N'$ and we have $\Gamma', z : R0 \vdash N' : R\underline{n}$. In the latter case, by induction with respect to the size of N', it follows that $\flat(N) = y^n(z)$. Indeed, by parts (i) and (ii), either $N' = z$ (in which case $n = 0$), or $N' = y\underline{m}N''$ with N'' of type $R(\underline{m})$ (and then $n = m + 1$).

We conclude that $\flat(M)$ is either $\mathbf{c}_n$ or $\Lambda r \lambda y^{r \rightarrow r} y$ (and $n = 1$). In each case $\mathcal{O}(\flat(M)) = \mathbf{c}_n$. □

---
[1] Recall that not every closed term of type ω is of the form $\mathbf{c}_n$. The annoying exception is $\Lambda r \lambda y^{r \rightarrow r} y$, and this is what $\mathcal{O}$ takes care of.

We are now ready to state and prove the main theorem of this chapter.

12.7.11. THEOREM (Girard). *All functions provably total in* PAS *are definable in* **F**.

PROOF. If f is provably total in PAS, then by Theorem 12.5.6 it is also provably total in HAs. Without loss of generality (Exercise 12.15) we may assume that f is unary. By Lemma 12.7.1 there exists a closed term M of λH of the following type (where $t_f(\underline{n}, b) = 0$ is an abbreviation):

$$\forall a(\mathsf{Int}(a) \to \forall R(\forall b(\mathsf{Int}(b) \to (t_f(a,b)=0)^{\mathsf{Int}} \to R) \to R)).$$

Let $n \in \mathbb{N}$ and let N_n be a closed term of type $\mathsf{Int}(\underline{n})$, such that $\mathfrak{b}(N_n) = \mathbf{c}_n$. The application $M\underline{n}N_n$ is a closed term of type

$$\forall R(\forall b(\mathsf{Int}(b) \to (t_f(\underline{n},b)=0)^{\mathsf{Int}} \to R) \to R).$$

By Lemma 12.7.10(i),(ii), this term reduces to

$$\Lambda R \lambda x^{\forall b(\mathsf{Int}(b) \to (t_f(\underline{n},b)=0)^{\mathsf{Int}} \to R)}. x\underline{m}NN',$$

for some m, where

(a) $x : \forall b(\mathsf{Int}(b) \to (t_f(\underline{n},b)=0)^{\mathsf{Int}} \to R) \vdash N : \mathsf{Int}(\underline{m})$;

(b) $x : \forall b(\mathsf{Int}(b) \to (t_f(\underline{n},b)=0)^{\mathsf{Int}} \to R) \vdash N' : (t_f(\underline{n},\underline{m})=0)^{\mathsf{Int}}$.

From (b) it follows that

$$\vdash N'[R := (P \to P)][x := \lambda b \lambda y^{\mathsf{Int}(b)} \lambda z^{(t_f(\underline{n},b)=0)^{\mathsf{Int}}} \lambda v^P.v] : (t_f(\underline{n},\underline{m})=0)^{\mathsf{Int}},$$

where P is any nullary relation variable. That is, the type $(t_f(\underline{n},\underline{m})=0)^{\mathsf{Int}}$ is inhabited in λH, and thus $\mathcal{N} \models t_f(\underline{n},\underline{m}) = 0$, by Lemma 12.7.1(ii). It follows that $f(n) = \ell(m)$, where ℓ is the left converse to the pairing function.

Let $\sigma = \mathfrak{b}((t_f(\underline{n},b)=0)^{\mathsf{Int}})$. Then f is definable in system **F** by the term

$$F = \lambda n^\omega. L(\mathcal{O}(\mathfrak{b}(M)n\omega(\lambda m^\omega \lambda y^\sigma. m))),$$

where L is a term representing ℓ. Indeed, by inspecting the reduction of $M\underline{n}N_n$, and applying Lemma 12.7.6, we obtain that the term $\mathfrak{b}(M)\mathbf{c}_n$ reduces to $\Lambda r \lambda x^{\omega \to \sigma \to r}. x \mathfrak{b}(N) \mathfrak{b}(N')$. It follows that $\mathfrak{b}(M)\mathbf{c}_n\omega(\lambda m^\omega \lambda y^\sigma. m)$ reduces to $\mathfrak{b}(N)$. From (a) and Lemma 12.7.10(iv) we have $\mathcal{O}(\mathfrak{b}(N)) = \mathbf{c}_m$. Thus, the application $F\mathbf{c}_n$ reduces to $L(\mathbf{c}_m)$, where m is such that $f(n) = \ell(m)$. □

12.7.12. COROLLARY. *The class of functions definable in* **F** *coincides with the class of provably recursive functions of second-order Peano Arithmetic.*

PROOF. Immediate from Proposition 12.4.3 and Theorem 12.7.11. □

12.7.13. COROLLARY. *Strong normalization for* **F** *is not provable in* PAS.

PROOF. Similar to Exercise 10.13. □

12.8. Notes

In the 1928 book [231] of Hilbert and Ackermann, second-order predicate logic, called the "erweitere Funktionenkalkül" (the *extended* function calculus), is given as much attention as the first-order "engere Funktionenkalkül." In [231] the two systems were precisely defined, probably for the first time, and the distinction between them was clearly made. Later, first-order logic turned out to receive more attention, because of the completeness property, lacking for the second-order systems. It took years to discover that second-order logic is complete with respect to Henkin semantics [222]. In a Henkin model, relation variables range over a subclass of all relations, but it is required that interpretation of all formulas is possible within that subclass. (The definition of a full Kripke model in Chapter 11 follows the spirit of a Henkin model, while principal Kripke models of Remark 11.1.14 correspond to Tarskian ones.) Semantics for second-order intuitionistic predicate logic can also be defined in a similar style [404].

The fundamental progress in second-order proof theory—from Takeuti's conjecture (see Section 11.8) to Girard's discoveries—happened between the 50's and the 70's. Research on second-order and higher-order intuitionistic arithmetic resulted in Troelstra's seminal book [482].

The main result of this chapter, Theorem 12.7.11, occurred already in Girard's 1970 paper [179] and was soon extended to higher orders. The original proof used a variant of Gödel's *Dialectica* interpretation. Later proofs of this result are typically more or less based on the contracting map idea. Examples are Girard's proof [189] and the proof in [489] (both using the additional constant of type $\bot$ to avoid the problem with Peano's fourth axiom) or Krivine's proof [285] using a variant of modified realizability. See also [502] for another realizability proof. Our proof of Theorem 12.7.11 is a digested version of Leivant's [304].

Textbook presentations of classical second-order logic are e.g. in [53, 113, 337]. For intuitionistic higher-order logic, arithmetic, and the proof-theoretical issues see [285, 307, 471, 482, 488, 489]. Extensions of system **F** by generalized inductive types were introduced by Mendler [338] and the construction in Remark 12.3.4 is due to Geuvers and Spławski, see [452]. For data types in system **F** and polymorphic program extraction see also [288, 379, 382].

12.9. Exercises

12.1. The language of second-order logic can be extended by function variables of arbitrary arity (which may occur in terms quite like the signature function symbols) and quantification over such variables. Define the syntax and semantics of second-order classical logic with function variables. Show that this extension does not increase the expressive power: for every sentence with (bound) functional variables there exists a semantically equivalent sentence without functional variables.

12.2. Write a second-order sentence φ, which is classically true in a model $\mathcal{A}$ if and only if the domain of $\mathcal{A}$ is a finite set. Compare this to Exercise 8.17.

12.3. Write a second-order sentence φ, which is classically true in a model $\mathcal{A}$ if and only if $\mathcal{A}$ is isomorphic to the standard model of arithmetic.

12.4. Show that the set of tautologies of classical second-order logic is not recursively enumerable.

12.5. Show that there is a formula $P(x)$ of second-order classical logic such that $\mathcal{N} \models P(\underline{n})$ iff n is the Gödel number of a sentence $\zeta \in \text{Th}(\mathcal{N})$ (cf. Theorem 9.3.1).

12.6. Give an example of a function provably recursive in PAS but not in PA. *Hint:* Consider the function g of Exercise 10.12.

12.7. Let $a =_L b$ denote the formula $\forall R\,(Ra \to Rb)$, i.e. the Leibniz equality. Prove that axioms (E_1) to (E_4) of Definition 12.4.1 are provable in intuitionistic second-order logic if all occurrences of $=$ are replaced by $=_L$.

12.8. Let μ, in, and It be as in Remark 12.3.4, and assume that there is a combinator *Lift* of type $\forall p \forall q ((p \to q) \to \tau(p) \to \tau(q))$, i.e. $\tau(p)$ is *monotone* in p. Assuming *Lift* begins with four lambdas, define a term out : $\mu \to \tau(\mu)$, such that :

$$\text{out} \circ \text{in} =_\beta \textit{Lift}\,\mu\mu(\text{in} \circ \text{out})$$

12.9. Define a representation of lists in system **F**, based on Example 12.2.2(vi). Can you generalize it to an arbitrary type τ?

12.10. Show that all formulas of the form $\text{Int}(\underline{n})$ are provable in intuitionistic second-order logic. Then for every term t with $\text{FV}(t) = \{a_1, \ldots, a_k\}$ show that $\text{Int}(a_1), \ldots, \text{Int}(a_k) \vdash \text{Int}(t)$.

12.11. Show the correctness of Kolmogorov's translation for HAS and HAs (Proposition 12.5.2).

12.12. Why not simplify the proof of Theorem 12.5.6 by using an analogue of Lemma 9.5.4?

12.13. Show that the formula $\forall ab(a = b \leftrightarrow \forall R\,(Ra \to Rb))$ is provable in HAS, but not in PAs.

12.14. Give a definition of the three substitutions $M[x := N]$, $M[a := t]$, and $M[R := \lambda \vec{a}\,\psi]$, occurring in system λH. Show that substitution preserves types (for instance $\Gamma, x : \tau \vdash M : \sigma$ and $\Gamma \vdash N : \sigma$ implies $\Gamma \vdash M[x := N] : \sigma$). Then prove subject reduction.

12.15. Assume that all unary functions provably total in PAS are definable in system **F**. Prove that all functions provably total in PAS are definable in **F**.

Chapter 13

Dependent types

We have seen a limited use of dependent types in the lambda-calculi λP_1 (Section 8.7) and λH (Section 12.7). These were "special-purpose" calculi, designed as proof notation for the traditional *one-sorted* syntax of first-order and second-order logic. Dependent type constructors in λP_1 and λH correspond to first-order relations, and can only be applied to individual arguments. These arguments are algebraic terms—the only *object terms* (as opposed to *proof terms*). The object and proof terms are of course citizens of two distinct worlds. With only one sort of individuals, and no object terms representing functions, no type annotations are needed for object expressions.

However, if we want to model the use of dependent types in a programming language, there is no reason to restrict ourselves to the one-sorted case, and we need a more general framework. We want to apply type constructors to arguments of various types, including function types. This means that object expressions must be typed in a similar way as the proof terms are. In fact, it becomes convenient to no longer distinguish between these two categories of terms, and we may treat all terms uniformly. We accept the point of view that an object (closed term) of *type* τ can be seen as a proof of the *proposition* τ, even if τ is an atomic type.[1] In fact, a particular implementation of an atomic type may actually put this idea to life in a very exact sense: In the previous chapters we implemented integers as proofs of the formula $(p \to p) \to p \to p$.

Therefore, the *system* λP *of dependent types*, discussed in this chapter, makes no difference between object and proof terms. In particular, there is a uniform function abstraction mechanism, and quantification over functions is permitted.

---

[1] The book the reader is now holding in her hands is a proof of the proposition *book*.

13.1. The language of λP

As in system λP_1, dependent types are created with the help of type constructors applied to terms. These constructors are quite different than the constructors of system $\mathbf{F}_\omega$ of Section 11.7 (which are applied to types). But we still need to classify constructors according to their domains, so we use *kinds* again. For instance, we say that a unary array constructor **array**, like that of Example 8.7.1, is of kind **int** $\Rightarrow *$ (which we write as **array** : **int** $\Rightarrow *$), where $*$ is the kind of all types. Constructors may be of various arities, so we also have kinds of the form $\tau_1 \Rightarrow \cdots \Rightarrow \tau_n \Rightarrow *$.

This is however not enough. A type constructor can also be applied to an object of a dependent type. For instance, imagine that an array A : **array**(n) contains information about n documents to be processed, and that **store**(A) is the storage type needed for the processing of such documents. Here, **store** is a constructor of kind **array**$(n) \Rightarrow *$. But what if we want **store** to be parameterized by n, so that we have **store**(n) : **array**$(n) \Rightarrow *$? The kind of **store** can no longer be expressed with the function space operator $\Rightarrow$, and we need a notion of a constructor product, so that

$$\text{\bf store} : \Pi n{:}\text{\bf int}.\,\text{\bf array}(n) \Rightarrow *.$$

As we know (cf. Section 8.7), product is a generalization of function space. The operator $\Rightarrow$ can also be seen as a special case of a product; for instance the product $\Pi x{:}\tau.*$ is abbreviated by $\tau \Rightarrow *$.

Similarly, we now have universal types of the form $\forall x{:}\tau.\sigma$, and a function type $\tau \to \sigma$ can be defined as a special case of $\forall x{:}\tau.\sigma$, for x not free in σ. Thus, formally, our syntax does not have to include the symbol $\to$. Via the formulas-as-types correspondence, the calculus of dependent types corresponds to a certain logic. It follows from the above that implication in this logic is a special case of universal quantification.

The syntax of system λP is more complicated than any of the calculi we have discussed so far. As in $\mathbf{F}_\omega$, we have three sorts of expressions: object expressions (ranged over by M, N, etc.), constructors (ranged over by τ, φ, etc.) and kinds (ranged over by κ, κ', etc.). This time however, we cannot give an a priori definition of what a constructor, a type, or a kind is. An expression may be a well-formed type or not, depending on a given environment. For instance, if α is a constructor variable of kind $\tau \Rightarrow *$, then it depends on the type of N whether an application αN is a well-formed type or not. Also environments in λP can no longer be arbitrary sets of assumptions, as it was normally assumed in previous chapters. In order to declare a variable of type αx, one has to know that the application is legal, i.e. that the type of x fits the kind of α. Thus, environments in λP are defined as sequences, rather than sets, of assumptions, and not every sequence of declarations can be regarded as a *valid* environment. Being valid or not depends on derivability of certain judgements.

13.1. THE LANGUAGE OF λP

It follows that the definition of syntax must become part of the type assignment system. For instance, the judgement $\Gamma \vdash \tau : *$ asserts that τ is a type with respect to the environment Γ, and it also implicitly asserts that Γ itself is a valid environment.

13.1.1. DEFINITION.

(i) *Raw expressions* (raw environments Γ, raw kinds κ, raw constructors ϕ and raw terms M) are defined by the following grammar:

$$\Gamma ::= \varnothing \mid \Gamma, (x : \phi) \mid \Gamma, (\alpha : \kappa);$$
$$\kappa ::= * \mid (\Pi x{:}\phi\ \kappa);$$
$$\phi ::= \alpha \mid (\forall x{:}\phi\ \phi) \mid (\phi M) \mid (\lambda x{:}\phi\ \phi);$$
$$M ::= x \mid (MM) \mid (\lambda x{:}\phi\ M).$$

(ii) $\mathrm{dom}(\Gamma) = \{x \mid (x : \tau) \in \Gamma, \text{ for some } \tau\} \cup \{\alpha \mid (\alpha : \kappa) \in \Gamma, \text{ for some } \kappa\}$.

(iii) The free variables $\mathrm{FV}(E)$ in a raw expression E are defined as follows:

$$\mathrm{FV}(*) = \varnothing,\ \mathrm{FV}(\alpha) = \{\alpha\},\ \mathrm{FV}(x) = \{x\},$$
$$\mathrm{FV}(EE') = \mathrm{FV}(E) \cup \mathrm{FV}(E'),$$
$$\mathrm{FV}(\nabla x{:}\phi\ E) = \mathrm{FV}(\phi) \cup (\mathrm{FV}(E) - \{x\}),$$

for $\nabla \in \{\lambda, \Pi, \forall\}$. That is, the binding operators are lambda-abstraction, quantification and product. Alpha-conversion is assumed as always.

(iv) The definition of $E[x := M]$ is by the usual induction, for instance we have $(\nabla x{:}\phi\ E)[x := M] = (\nabla x{:}\phi[x := M].E[x := M])$. We write $\Gamma[x := M]$ for the raw environment obtained from Γ by applying the substitution to the right-hand sides of all declarations in Γ.

(v) If x is not free in κ then we write $\tau \Rightarrow \kappa$ instead of $(\Pi x{:}\tau\ \kappa)$. Of course $\tau \Rightarrow \cdots \Rightarrow \tau \Rightarrow \kappa$ with n occurrences of τ is abbreviated as $\tau^n \Rightarrow \kappa$. And if x is not free in σ then $\tau \to \sigma$ is used instead of $(\forall x{:}\tau\ \sigma)$.

13.1.2. CONVENTION. We use our ordinary conventions to avoid too many parentheses. In particular, we can write $\lambda x{:}\tau.\ MN$ or $\lambda x^\tau.\ MN$ instead of $(\lambda x{:}\tau\ (MN))$, and similar conventions apply to $\forall$ and Π as well. The metavariables Γ, κ always denote raw environments and raw kinds, respectively. Similarly, M, N will be used for raw terms and ϕ, τ, σ for raw constructors.

13.1.3. DEFINITION. Beta-reduction on raw expressions is defined as the compatible closure (cf. Section 1.3) of the following rules.

$$(\lambda x{:}\tau.\ M)N \to_\beta M[x := N];$$
$$(\lambda x{:}\tau.\ \phi)N \to_\beta \phi[x := N].$$

The notation $=_\beta$ and $\twoheadrightarrow_\beta$ is used accordingly.

13.1.4. LEMMA. *If $E \twoheadrightarrow_\beta E'$ and $N \twoheadrightarrow_\beta N'$ then $E[x := N] \twoheadrightarrow_\beta E'[x := N']$.*

PROOF. Left to the reader. □

13.1.5. LEMMA. *Beta-reduction on raw terms has the CR property.*

PROOF. An easy adaptation of the proof of Theorem 1.4.5.

13.2. Type assignment

13.2.1. DEFINITION. The rules of λP are given in Figure 13.1. There are three different forms of judgements in our system:

- Kind formation judgements of the form $\Gamma \vdash \kappa : \square$, read as "$\kappa$ is a kind in the environment Γ" (the symbol $\square$ itself is not a raw expression).

- Kinding judgements of the form $\Gamma \vdash \varphi : \kappa$, read as "$\varphi$ is a constructor of kind κ in the environment Γ."

- Typing judgements of the form $\Gamma \vdash M : \tau$, read as "M is a term of type τ in the environment Γ."

If $\Gamma \vdash \tau : *$ then we say that τ is a *type* in the environment Γ. A raw environment Γ occurring in a derivable judgement is called a *valid environment*. In this case Γ declares each variable only once (straightforward induction), so if $(x : \tau)$ or $(\alpha : \kappa)$ is in Γ then we write $\Gamma(x) = \tau$ or $\Gamma(\alpha) = \kappa$, respectively.

In addition to typing and kinding rules, there are six weakening rules in the system, but all obey the same pattern: an additional assumption does not hurt, as long as it is well-formed. We need explicit weakening (rather than relaxed axioms) because of the sequential structure of environments we must respect. There are also conversion rules, implementing the idea that beta-equal types and kinds should be identified.

Note that there is no restriction "$x \notin \mathrm{FV}(\Gamma)$" in the rules for lambda-abstraction. This is because $\Gamma, x : \tau$ is not a valid environment, if $x \in \mathrm{FV}(\Gamma)$. For a similar reason there is no need for an additional assumption $\Gamma \vdash \tau : *$ in the second kind formation rule. This is summarized by:

13.2.2. LEMMA. *If $\Gamma, x : \tau, \Delta$ (resp. $\Gamma, \alpha : \kappa, \Delta$) is a valid environment then $x \notin \mathrm{dom}(\Gamma)$ and $\Gamma \vdash \tau : *$ (resp. $\alpha \notin \mathrm{dom}(\Gamma)$ and $\Gamma \vdash \kappa : \square$), in particular Γ is valid. In addition the derivation of the latter judgement consists of fewer steps than the former one.*

PROOF. Easy induction. □

13.2. Type assignment

Kind formation rules

$$\vdash * : \square \qquad \dfrac{\Gamma, x{:}\tau \vdash \kappa : \square}{\Gamma \vdash \Pi x{:}\tau\, \kappa : \square}$$

Kinding rules

$$\dfrac{\Gamma \vdash \kappa : \square}{\Gamma, \alpha{:}\kappa \vdash \alpha : \kappa}\;(\alpha \notin \mathrm{dom}(\Gamma)) \qquad \dfrac{\Gamma, x{:}\tau \vdash \sigma : *}{\Gamma \vdash (\forall x{:}\tau\, \sigma) : *}$$

$$\dfrac{\Gamma \vdash \varphi : (\Pi x{:}\tau\, \kappa) \quad \Gamma \vdash M : \tau}{\Gamma \vdash (\varphi M) : \kappa[x := M]} \qquad \dfrac{\Gamma, x{:}\tau \vdash \varphi : \kappa}{\Gamma \vdash (\lambda x{:}\tau\, \varphi) : (\Pi x{:}\tau\, \kappa)}$$

Typing rules

$$\dfrac{\Gamma \vdash \tau : *}{\Gamma, x{:}\tau \vdash x : \tau}\;(x \notin \mathrm{dom}(\Gamma))$$

$$\dfrac{\Gamma \vdash M : (\forall x{:}\tau\, \sigma) \quad \Gamma \vdash N : \tau}{\Gamma \vdash (MN) : \sigma[x := N]} \qquad \dfrac{\Gamma, x{:}\tau \vdash M : \sigma}{\Gamma \vdash (\lambda x{:}\tau\, M) : (\forall x{:}\tau\, \sigma)}$$

Weakening rules ($x \notin \mathrm{dom}(\Gamma)$ at the left, $\alpha \notin \mathrm{dom}(\Gamma)$ at the right.)

$$\dfrac{\Gamma \vdash \tau : * \quad \Gamma \vdash \kappa : \square}{\Gamma, x{:}\tau \vdash \kappa : \square} \qquad \dfrac{\Gamma \vdash \kappa : \square \quad \Gamma \vdash \kappa' : \square}{\Gamma, \alpha{:}\kappa \vdash \kappa' : \square}$$

$$\dfrac{\Gamma \vdash \tau : * \quad \Gamma \vdash \varphi : \kappa}{\Gamma, x{:}\tau \vdash \varphi : \kappa} \qquad \dfrac{\Gamma \vdash \kappa : \square \quad \Gamma \vdash \varphi : \kappa'}{\Gamma, \alpha{:}\kappa \vdash \varphi : \kappa'}$$

$$\dfrac{\Gamma \vdash \tau : * \quad \Gamma \vdash M : \sigma}{\Gamma, x{:}\tau \vdash M : \sigma} \qquad \dfrac{\Gamma \vdash \kappa : \square \quad \Gamma \vdash M : \sigma}{\Gamma, \alpha{:}\kappa \vdash M : \sigma}$$

Conversion rules

$$\dfrac{\Gamma \vdash \varphi : \kappa \quad \Gamma \vdash \kappa' : \square}{\Gamma \vdash \varphi : \kappa'}\;(\kappa =_\beta \kappa') \qquad \dfrac{\Gamma \vdash M : \sigma \quad \Gamma \vdash \sigma' : *}{\Gamma \vdash M : \sigma'}\;(\sigma =_\beta \sigma')$$

FIGURE 13.1: SYSTEM λP

13.2.3. REMARK. Type checking in λP is not a trivial task, even if our language is Church-style. This is because of the conversion rules. A verification whether two types or kinds are equal may be as difficult as checking equality of simply typed lambda terms [453], and this decision problem is non-elementary (Theorem 3.7.1).

13.2.4. EXAMPLE. Let Γ be a valid environment such that $\Gamma \vdash \tau : *$ and containing the declarations $\alpha : \tau \Rightarrow *$ and $\beta : \tau \Rightarrow *$. Then

(i) The raw expression $\lambda x^{\forall z:\tau.\, \alpha z \to \beta z} \lambda y^{\forall z:\tau.\, \alpha z} \lambda z^\tau.\, xz(yz)$ is a term of type $(\forall x:\tau.\, \alpha x \to \beta x) \to (\forall x:\tau.\, \alpha x) \to \forall x:\tau.\, \beta x$ in Γ.

(ii) The term $\lambda f^{\tau \to \tau} \lambda y^{\forall x:\tau.\, \alpha x \to \alpha(fx)} \lambda x^\tau \lambda z^{\alpha x}.\, y(fx)(yxz)$ can be assigned type $\forall f:\tau \to \tau.\, (\forall x:\tau.\, \alpha x \to \alpha(fx)) \to \forall x:\tau.\, \alpha x \to \alpha(f(fx))$ in Γ.

(iii) If $\varphi = \forall f:\tau \to \tau \forall x:\tau.\, \alpha x \to \alpha(fx)$ and $\psi = \forall y:\tau.\, \alpha y \to \forall x:\tau.\, \alpha x$ then $\Gamma \vdash \lambda z^\varphi \lambda y^\tau \lambda u^{\alpha y} \lambda x^\tau.\, z(\lambda w^\tau\, x) y u : \varphi \to \psi$.

The following technical lemmas state some basic properties of our system. For readability we formulate these properties in the form of admissible rules.

13.2.5. LEMMA. *The substitution rules below are admissible in λP, that is, if premises are derivable judgements then the conclusion is derivable too.*

$$\frac{\Gamma, x:\tau, \Delta \vdash M : \sigma \quad \Gamma \vdash N : \tau}{\Gamma, \Delta[x := N] \vdash M[x := N] : \sigma[x := N]}$$

$$\frac{\Gamma, x:\tau, \Delta \vdash \varphi : \kappa \quad \Gamma \vdash N : \tau}{\Gamma, \Delta[x := N] \vdash \varphi[x := N] : \kappa[x := N]}$$

$$\frac{\Gamma, x:\tau, \Delta \vdash \kappa : \square \quad \Gamma \vdash N : \tau}{\Gamma, \Delta[x := N] \vdash \kappa[x := N] : \square}$$

PROOF. Induction with respect to the derivation of the left premise. As an example, consider a derivation ending with a constructor application

$$\frac{\Gamma, x:\tau, \Delta \vdash \varphi : (\Pi y{:}\sigma\, \kappa) \quad \Gamma \vdash Q : \sigma}{\Gamma, x:\tau, \Delta \vdash \varphi Q : \kappa[y := Q]}$$

where $y \notin \mathrm{FV}(N)$, and suppose $\Gamma \vdash N : \tau$. By the induction hypothesis:

$$\Gamma, \Delta[x := N] \vdash \varphi[x := N] : \Pi y{:}\sigma[x := N].\, \kappa[x := N],$$

$$\Gamma, \Delta[x := N] \vdash Q[x := N] : \sigma[x := N].$$

From this we derive

$$\Gamma, \Delta[x := N] \vdash \varphi[x := N] Q[x := N] : \kappa[x := N][y := Q[x := N]],$$

13.2. TYPE ASSIGNMENT   331

and it remains to note that $\kappa[x := N][y := Q[x := N]] = \kappa[x := N][y := Q]$, as y is not free in N. The other cases are also easy. Lemma 13.1.4 is needed when the proof ends with a conversion. □

13.2.6. LEMMA. *The following rules are admissible when* $y \notin \mathrm{dom}(\Gamma, \Delta)$.

$$\frac{\Gamma, x : \tau, \Delta \ \vdash \ \kappa : \Box}{\Gamma, y : \tau, \Delta[x := y] \ \vdash \ \kappa[x := y] : \Box}$$

$$\frac{\Gamma, x : \tau, \Delta \ \vdash \ \varphi : \kappa}{\Gamma, y : \tau, \Delta[x := y] \ \vdash \ \varphi[x := y] : \kappa[x := y]}$$

$$\frac{\Gamma, x : \tau, \Delta \ \vdash \ M : \tau}{\Gamma, y : \tau, \Delta[x := y] \ \vdash \ M[x := y] : \tau[x := y]}$$

PROOF. By induction with respect to derivations one proves that the conclusion can be derived in at most as many steps as the premise. □

13.2.7. LEMMA. *The following generalized weakening rules are admissible, provided* $x, \alpha \notin \mathrm{dom}(\Gamma) \cup \mathrm{dom}(\Delta)$.

$$\frac{\Gamma \vdash \tau : * \quad \Gamma, \Delta \vdash \kappa : \Box}{\Gamma, x : \tau, \Delta \vdash \kappa : \Box} \qquad \frac{\Gamma \vdash \kappa : \Box \quad \Gamma, \Delta \vdash \kappa' : \Box}{\Gamma, \alpha : \kappa, \Delta \vdash \kappa' : \Box}$$

$$\frac{\Gamma \vdash \tau : * \quad \Gamma, \Delta \vdash \varphi : \kappa}{\Gamma, x : \tau, \Delta \vdash \varphi : \kappa} \qquad \frac{\Gamma \vdash \kappa : \Box \quad \Gamma, \Delta \vdash \varphi : \kappa'}{\Gamma, \alpha : \kappa, \Delta \vdash \varphi : \kappa'}$$

$$\frac{\Gamma \vdash \tau : * \quad \Gamma, \Delta \vdash M : \sigma}{\Gamma, x : \tau, \Delta \vdash M : \sigma} \qquad \frac{\Gamma \vdash \kappa : \Box \quad \Gamma, \Delta \vdash M : \sigma}{\Gamma, \alpha : \kappa, \Delta \vdash M : \sigma}$$

PROOF. This is a routine induction with respect to the right premise. If Δ is empty, we apply weakening, otherwise we use the induction hypothesis. In some cases Lemma 13.2.6 is used to rename a variable. □

13.2.8. LEMMA. *The following rules are admissible whenever* $\tau =_\beta \tau'$ (*in the left column*) *and* $\kappa =_\beta \kappa'$ (*in the right column*).

$$\frac{\Gamma, x : \tau, \Delta \vdash \kappa'' : \Box \quad \Gamma \vdash \tau' : *}{\Gamma, x : \tau', \Delta \vdash \kappa'' : \Box} \qquad \frac{\Gamma, \alpha : \kappa, \Delta \vdash \kappa'' : \Box \quad \Gamma \vdash \kappa' : \Box}{\Gamma, \alpha : \kappa', \Delta \vdash \kappa'' : \Box}$$

$$\frac{\Gamma, x : \tau, \Delta \vdash \varphi : \kappa'' \quad \Gamma \vdash \tau' : *}{\Gamma, x : \tau', \Delta \vdash \varphi : \kappa''} \qquad \frac{\Gamma, \alpha : \kappa, \Delta \vdash \varphi : \kappa'' \quad \Gamma \vdash \kappa' : \Box}{\Gamma, \alpha : \kappa', \Delta \vdash \varphi : \kappa''}$$

$$\frac{\Gamma, x : \tau, \Delta \vdash M : \sigma \quad \Gamma \vdash \tau' : *}{\Gamma, x : \tau', \Delta \vdash M : \sigma} \qquad \frac{\Gamma, \alpha : \kappa, \Delta \vdash M : \sigma \quad \Gamma \vdash \kappa' : \Box}{\Gamma, \alpha : \kappa', \Delta \vdash M : \sigma}$$

PROOF. Induction with respect to the derivation of the left premise. As an example consider the case when Δ is empty and the derivation ends with

$$\frac{\Gamma \vdash \kappa : \Box}{\Gamma, \alpha : \kappa \vdash \alpha : \kappa}$$

where $\kappa =_\beta \kappa'$, and we also have $\Gamma \vdash \kappa' : \Box$. To derive $\Gamma, \alpha : \kappa' \vdash \alpha : \kappa$ we proceed as follows.

$$\frac{\dfrac{\Gamma \vdash \kappa' : \Box}{\Gamma, \alpha : \kappa' \vdash \alpha : \kappa'} \quad \dfrac{\Gamma \vdash \kappa : \Box \quad \Gamma \vdash \kappa' : \Box}{\Gamma, \alpha : \kappa' \vdash \kappa : \Box}}{\Gamma, \alpha : \kappa' \vdash \alpha : \kappa} \ (\kappa =_\beta \kappa')$$

The proofs of the remaining cases are either similar or are immediate consequences of the induction hypothesis. □

13.2.9. LEMMA (Generation Lemma).

(i) *If* $\Gamma \vdash x : \tau$ *then* $\Gamma(x) =_\beta \tau$;

(ii) *If* $\Gamma \vdash \alpha : \kappa$ *then* $\Gamma(\alpha) =_\beta \kappa$;

(iii) *If* $\Gamma \vdash (\Pi x{:}\tau.\,\kappa) : \Box$ *and* $x \notin \text{dom}(\Gamma)$ *then* $\Gamma, x{:}\tau \vdash \kappa : \Box$.

(iv) *If* $\Gamma \vdash (\lambda x{:}\tau.\,\varphi) : \kappa$ *and* $x \notin \text{dom}(\Gamma)$ *then* $\kappa =_\beta (\Pi x{:}\tau.\,\kappa')$, *for some* κ' *such that* $\Gamma, x{:}\tau \vdash \varphi : \kappa'$.

(v) *If* $\Gamma \vdash \varphi M : \kappa$ *then* $\Gamma \vdash \varphi : (\Pi x{:}\tau.\,\kappa')$, *for some* τ *and* κ' *such that* $\Gamma \vdash M : \tau$ *and* $\kappa =_\beta \kappa'[x := M]$.

(vi) *If* $\Gamma \vdash (\forall x{:}\tau.\,\sigma) : \kappa$ *then* $\kappa = *$ *and if* $x \notin \text{dom}(\Gamma)$ *then* $\Gamma, x{:}\tau \vdash \sigma : *$.

(vii) *If* $\Gamma \vdash MN : \sigma$ *then* $\Gamma \vdash M : (\forall x{:}\tau.\,\sigma')$, *for some* τ *and* σ' *such that* $\Gamma \vdash N : \tau$ *and* $\sigma =_\beta \sigma'[x := N]$.

(viii) *If* $\Gamma \vdash (\lambda x{:}\tau.\,M) : \sigma$ *and* $x \notin \text{dom}(\Gamma)$ *then* $\sigma =_\beta (\forall x{:}\tau.\,\sigma')$, *for some* σ' *such that* $\Gamma, x{:}\tau \vdash M : \sigma'$.

PROOF. Similar to the proof of Lemma 3.1.5. For part (vi), note that all reductions are inside constructors and terms, and thus $\kappa =_\beta *$ implies $\kappa = *$. For parts (iii), (iv), (vi) and (viii) one uses Lemmas 13.2.6 and 13.2.7. □

13.2.10. LEMMA. *If* $\Gamma \vdash \varphi : \kappa$ *then* $\Gamma \vdash \kappa : \Box$ *and if* $\Gamma \vdash M : \tau$ *then* $\Gamma \vdash \tau : *$.

PROOF. Simultaneous induction using Lemmas 13.2.5 and 13.2.9. □

13.3. STRONG NORMALIZATION

13.2.11. THEOREM (Subject reduction).

(i) If $\Gamma \vdash M : \sigma$ and $M \to_\beta M'$ then $\Gamma \vdash M' : \sigma$.
(ii) If $\Gamma \vdash \varphi : \kappa$ and $\varphi \to_\beta \varphi'$ then $\Gamma \vdash \varphi' : \kappa$.
(iii) If $\Gamma \vdash \kappa : \square$ and $\kappa \to_\beta \kappa'$ then $\Gamma \vdash \kappa' : \square$.

PROOF. The proof of all the three parts is by simultaneous induction with respect to the definition of $\to_\beta$. Consider the case of $\Gamma \vdash (\lambda x{:}\tau. M)N : \sigma$ and a reduction step of the form $(\lambda x{:}\tau. M)N \to_\beta M[x := N]$. Without loss of generality we can assume that $x \notin \mathrm{FV}(\Gamma)$. From Lemma 13.2.9(vii) and (viii) it follows that $\Gamma \vdash \lambda x{:}\tau. M : \forall x{:}\rho. \sigma'$, for some ρ, σ' such that $\sigma =_\beta \sigma'[x := N]$ and $\Gamma \vdash N : \rho$. In addition, $\forall x{:}\rho. \sigma' =_\beta \forall x{:}\tau. \sigma''$, for some type σ'' with $\Gamma, x{:}\tau \vdash M : \sigma''$. From Lemma 13.1.5 it follows that $\rho =_\beta \tau$ and $\sigma' =_\beta \sigma''$. By Lemma 13.2.10, we have $\Gamma \vdash \rho : *$ and $\Gamma \vdash \forall x{:}\rho. \sigma' : *$, and thus also $\Gamma, x{:}\rho \vdash \sigma' : *$. Therefore, $\Gamma, x{:}\rho \vdash M : \sigma'$, by Lemma 13.2.8 and conversion, so we can apply Lemma 13.2.5 (the first rule) to obtain $\Gamma \vdash M[x := N] : \sigma'[x := N]$. Using conversion again, we conclude with the desired $\Gamma \vdash M[x := N] : \sigma$.

The case of a redex of the form $(\lambda x{:}\tau. \varphi)N \to_\beta \varphi[x := N]$ is similar. Of the other cases, consider for example a reduction of the form $\varphi M \to_\beta \varphi M'$, where $M \to_\beta M'$ and $\Gamma \vdash \varphi M : \kappa$. By part (v) of Lemma 13.2.9, we have $\Gamma \vdash \varphi : (\Pi x{:}\tau. \kappa')$ and $\Gamma \vdash M : \tau$. In addition, $\kappa =_\beta \kappa'[x' := M']$. By the induction hypothesis, we obtain $\Gamma \vdash M' : \tau$, and thus $\Gamma \vdash \varphi M : \kappa'[x := M']$. We use Lemma 13.1.4 and conversion to conclude that $\Gamma \vdash \varphi M : \kappa$. □

13.3. Strong normalization

We show the strong normalization of system λP using a translation from λP to the simply typed lambda-calculus. For first-order logic (Theorem 8.7.10) we applied the contracting map $\mathfrak{b}$. For λP this will not work; our translation must be more sophisticated. But we begin with a contracting map.

13.3.1. DEFINITION. We first associate a propositional variable p_α to every constructor variable α, and we fix a type variable 0. The contraction map $\mathfrak{b}$ assigns a simple type to every (raw) kind and constructor, and it also maps raw terms of λP to raw terms of $\lambda_\to$.

- $\mathfrak{b}(*) = 0$;
- $\mathfrak{b}(\Pi x{:}\tau. \kappa) = \mathfrak{b}(\tau) \to \mathfrak{b}(\kappa)$;
- $\mathfrak{b}(\alpha) = p_\alpha$;
- $\mathfrak{b}(\forall x{:}\tau. \sigma) = \mathfrak{b}(\tau) \to \mathfrak{b}(\sigma)$;
- $\mathfrak{b}(\lambda x{:}\tau. \varphi) = \mathfrak{b}(\varphi)$;

- $\mathfrak{b}(\varphi M) = \mathfrak{b}(\varphi)$.

- $\mathfrak{b}(x) = x$;

- $\mathfrak{b}(MN) = \mathfrak{b}(M)\,\mathfrak{b}(N)$;

- $\mathfrak{b}(\lambda x{:}\tau.\,M) = \lambda x{:}\mathfrak{b}(\tau).\,\mathfrak{b}(M)$.

For any Γ, define $\mathfrak{b}(\Gamma) = \{(x : \mathfrak{b}(\tau)) \mid (x : \tau) \in \Gamma\} \cup \{(x_\alpha : \mathfrak{b}(\kappa)) \mid (\alpha : \kappa) \in \Gamma\}$, where x_α are fresh object variables.

The map $\mathfrak{b}$ on terms and constructors is like our good old contraction map. But $\mathfrak{b}$ also turns kinds into types, and thus flattens the three-level structure of $\lambda\mathrm{P}$ into the two-level structure of $\lambda_\to$. The next two lemmas are shown by routine induction.

13.3.2. LEMMA.

(i) *The contraction map satisfies the identities* $\mathfrak{b}(\varphi[x := M]) = \mathfrak{b}(\varphi)$ *and* $\mathfrak{b}(\kappa[x := M]) = \mathfrak{b}(\kappa)$.

(ii) *If* $E_1 =_\beta E_2$, *for raw constructors or kinds* E_1, E_2, *then* $\mathfrak{b}(E_1) = \mathfrak{b}(E_2)$.

13.3.3. LEMMA. *If* $\Gamma \vdash M : \tau$ *then* $\mathfrak{b}(\Gamma) \vdash \mathfrak{b}(M) : \mathfrak{b}(\tau)$ *in* $\lambda_\to$.

Lemma 13.3.3 is used in Section 13.4, but it is not sufficient for the proof of strong normalization (Exercise 13.11). We need another translation.

13.3.4. DEFINITION. Abusing the formalism a little, we assume that the following are different variables, not occurring in any $\lambda\mathrm{P}$-expression.

- x_α, for every constructor variable α;

- y_ρ, for every simple type ρ.

We assign raw terms of $\lambda_\to$ to raw terms and constructors of $\lambda\mathrm{P}$.

- $\alpha^\sharp = x_\alpha$,

- $(\varphi M)^\sharp = \varphi^\sharp M^\sharp$;

- $(\lambda x{:}\tau.\,\varphi)^\sharp = (\lambda z^0 \lambda x^{\mathfrak{b}(\tau)}.\,\varphi^\sharp)\tau^\sharp$;

- $(\forall x{:}\tau.\,\sigma)^\sharp = y_{\mathfrak{b}(\tau)} \tau^\sharp (\lambda x^{\mathfrak{b}(\tau)}.\,\sigma^\sharp)$;

- $x^\sharp = x$;

- $(MN)^\sharp = M^\sharp N^\sharp$;

- $(\lambda x{:}\tau.\,M)^\sharp = (\lambda z^0 \lambda x^{\mathfrak{b}(\tau)}.\,M^\sharp)\tau^\sharp$.

13.4. Dependent types à la Curry

Let $\mathfrak{b}^+(\Gamma) = \mathfrak{b}(\Gamma) \cup \{(y_\rho : 0 \to (\rho \to 0) \to 0) \mid \rho \text{ is a simple type}\}$.

13.3.5. LEMMA.

(i) *If* $\Gamma \vdash \varphi : \kappa$ *then* $\mathfrak{b}^+(\Gamma) \vdash \varphi^\sharp : \mathfrak{b}(\kappa)$ *in* $\lambda_\to$.

(ii) *If* $\Gamma \vdash M : \tau$ *then* $\mathfrak{b}^+(\Gamma) \vdash M^\sharp : \mathfrak{b}(\tau)$ *in* $\lambda_\to$.

(iii) *If* $M_1 \to_\beta M_2$ *then* $M_1^\sharp \twoheadrightarrow_\beta^+ M_2^\sharp$.

(iv) *If* $\varphi_1 \to_\beta \varphi_2$ *then* $\varphi_1^\sharp \twoheadrightarrow_\beta^+ \varphi_2^\sharp$.

PROOF. First of all we need to know that the translation commutes with substitution. That is, we must show (cf. Lemma 8.7.7) that:
$$(\varphi[x := M])^\sharp = \varphi^\sharp[x := M^\sharp];$$
$$(N[x := M])^\sharp = N^\sharp[x := M^\sharp].$$

Now, using a simultaneous induction, one proves parts (i) and (ii). The proof is fairly routine, with the help of the equations above and Lemma 13.3.2.

Parts (iii) and (iv) are shown by simultaneous induction with respect to the definition of reduction. □

13.3.6. THEOREM. *The system* $\lambda\mathrm{P}$ *has the strong normalization property, that is, if* $\Gamma \vdash M : \tau$ *then every reduction of* M *terminates.*

PROOF. By part (iii) of Lemma 13.3.5, an infinite reduction starting from M would be translated to an infinite reduction in $\lambda_\to$. □

An analogous result also holds for constructors and kinds (Exercise 13.13).

13.3.7. THEOREM. *The system* $\lambda\mathrm{P}$ *has the Church-Rosser property.*

PROOF. Show WCR and use Newman's lemma. □

13.4. Dependent types à la Curry

13.4.1. DEFINITION. We define a type erasing mapping $|\cdot|$ from raw terms of $\lambda\mathrm{P}$ to pure lambda-terms, as usual:

- $|x| = x$;
- $|\lambda x{:}\tau.M| = \lambda x\,|M|$;
- $|MN| = |M||N|$.

For a pure lambda-term N, we write $\Gamma \vdash_P N : \tau$ iff $N = |M|$, for some Church-style term M with $\Gamma \vdash M : \tau$. We say that a pure lambda-term N is *typable* iff $\Gamma \vdash_P N : \tau$ holds for some Γ, τ.

13.4.2. PROPOSITION. *The Curry-style variant of λP has the subject reduction property, that is if $\Gamma \vdash_P N : \tau$ and $N \twoheadrightarrow_\beta N'$ then also $\Gamma \vdash_P N' : \tau$.*

PROOF. The proof is an easy consequence of Theorem 13.2.11 and Exercise 13.9, and is left to the reader. □

13.4.3. PROPOSITION. *A pure lambda-term is typable in λP iff it is simply typable. Thus, the typability problem for λP is decidable in polynomial time.*

PROOF. Given a pure lambda-term N with $\Gamma \vdash_P N : \tau$, let M be such that $|M| = N$ and $\Gamma \vdash M : \tau$ in λP. By Lemma 13.3.3, we have $\mathfrak{b}(\Gamma) \vdash \mathfrak{b}(M) : \mathfrak{b}(\tau)$, in Church-style $\lambda_\rightarrow$. But $|\mathfrak{b}(M)| = |M| = N$, so $\mathfrak{b}(\Gamma) \vdash N : \mathfrak{b}(\tau)$ à la Curry.

It follows that all pure terms typable in λP are typable in simple types. By Proposition 3.4.5 the converse is obvious as $\lambda_\rightarrow$ is a fragment of λP. □

Here comes the surprise: it is not that easy with type checking.

13.4.4. THEOREM (Dowek). *Type checking in λP à la Curry is undecidable.*

PROOF. We reduce the Post Correspondence Problem of Section A.4 to type checking. Let $(w_1, v_1), \ldots, (w_k, v_k)$ be an instance of the Post Correspondence Problem over the alphabet $\{a, b\}$, and let Γ be the environment:

$0 : *, \ x : 0, \ y : 0, \ a : 0 \to 0, \ b : 0 \to 0, \ \alpha : 0 \Rightarrow *, \ F : \forall x{:}0.\, \alpha x \to 0.$

Using the variables in Γ we encode any word $w = e_1 \ldots e_n$ over $\{a, b\}$ as the pure λ-term $\overline{w} = \lambda x.e_1(e_2(\ldots(e_n x)\ldots))$. Define:

$$M = \mathbf{K}x(\lambda f \lambda g \lambda h.\, gABCD),$$

where the components of M are as follows:

$$\begin{aligned} A &= h(f\overline{w}_1 \ldots \overline{w}_k); \\ B &= h(f\overline{v}_1 \ldots \overline{v}_k); \\ C &= Fx(f\mathbf{I} \ldots \mathbf{I}); \\ D &= Fy(f(\mathbf{K}y) \ldots (\mathbf{K}y)), \end{aligned}$$

and each occurrence of f has k arguments. We claim that $\Gamma \vdash_P M : 0$ iff

$$w_{i_1} w_{i_2} \ldots w_{i_m} = v_{i_1} v_{i_2} \ldots v_{i_m},$$

for some $m > 0$ and some $i_1, \ldots, i_m$. Suppose first that the equation holds. If $R = \overline{w}_{i_1}(\overline{w}_{i_2}(\cdots(\overline{w}_{i_m}x)\cdots))$ and $R' = \overline{v}_{i_1}(\overline{v}_{i_2}(\cdots(\overline{v}_{i_m}x)\cdots))$ then $R =_\beta R'$, so $\alpha(R) =_\beta \alpha(R')$, and for $\sigma = \forall x_1^{0 \to 0} \cdots \forall x_k^{0 \to 0}.\, \alpha(x_{i_1}(x_{i_2}(\ldots(x_{i_m}(x))\ldots)))$,

$$\begin{aligned} \Gamma, f : \sigma &\vdash f\overline{w}_1 \ldots \overline{w}_k : \alpha(R); \\ \Gamma, f : \sigma &\vdash f\overline{v}_1 \ldots \overline{v}_k : \alpha(R); \\ \Gamma, f : \sigma &\vdash f\mathbf{I} \ldots \mathbf{I} : \alpha x; \\ \Gamma, f : \sigma &\vdash f(\mathbf{K}y) \ldots (\mathbf{K}y) : \alpha y. \end{aligned}$$

It is now readily seen that M is obtained by type erasure from a well-typed term, in which the abstractions $\ldots \lambda f \lambda g \lambda h \ldots$ are decorated as follows:

$$\ldots \lambda f^\sigma \lambda g^{0\to 0\to 0\to 0} \lambda h^{\alpha(R)\to 0} \ldots$$

For the opposite direction we show that essentially no other typing of M is possible. Assume that M is an erasure of a term where f has type σ. Since the application $f\overline{w}_1 \ldots \overline{w}_k$ and the terms $Fx(f\mathbf{I}\ldots \mathbf{I})$ and $Fy(f(\mathbf{K}y)\ldots(\mathbf{K}y))$ are well-typed, we must have $\sigma =_\beta \forall x_1^{0\to 0} \cdots \forall x_k^{0\to 0}.\alpha U$, for some term U, with $U[x_1,\ldots,x_k := \mathbf{I},\ldots,\mathbf{I}] =_\beta x$ and $U[x_1,\ldots,x_k := \mathbf{K}y,\ldots,\mathbf{K}y] =_\beta y$. The term U has type 0 in the environment Γ, $x_1 : 0 \to 0, \ldots, x_k : 0 \to 0$, so the normal form of U must be of shape $x_{i_1}(x_{i_2}(\ldots(x_{i_m}x)\ldots))$. Here, $m \neq 0$, as otherwise $U[x_1,\ldots,x_k := \mathbf{K}y,\ldots,\mathbf{K}y] = x$. Now $f\overline{w}_1 \ldots \overline{w}_k$ and $f\overline{v}_1 \ldots \overline{v}_k$ are of types $\alpha U[x_1,\ldots x_k := \overline{w}_1,\ldots,\overline{w}_k]$ and $\alpha U[x_1,\ldots x_k := \overline{v}_1,\ldots,\overline{v}_k]$, respectively. Because of the applications $h(f\overline{w}_1 \ldots \overline{w}_k)$ and $h(f\overline{v}_1 \ldots \overline{v}_k)$, these types must be β-equal. It follows that $w_{i_1}w_{i_2}\ldots w_{i_m} = v_{i_1}v_{i_2}\ldots v_{i_m}$. □

13.5. Correspondence with first-order logic

In order to compare the calculus λP of dependent types and the system λP$_1$ of Chapter 8, one must either add the missing logical connectives to λP, or restrict attention to the universally-implicational fragment of λP$_1$. For simplicity, we choose the latter option. In this section, we thus consider formulas involving only $\to$ and $\forall$.

The system λP extends first-order logic in a variety of ways. In particular, the presence of quantification over functions of all finite types means that λP can express many languages that are commonly classified as higher-order (see [214]). We would however prefer not to use the expression "higher-order logic" here, as there is no quantification over types (propositions) nor constructors, unlike the "true" higher-order logic (cf. Chapters 11 and 12). A more adequate statement is perhaps that λP gives a many-sorted first-order representation of higher-order logic.

13.5.1. REMARK. There are features of λP that go beyond the standard understanding of first-order. The identification of proofs and objects has the following important consequences:

- Formulas can refer to properties of proofs. (For instance the formula $\forall x\!:\!(\forall y^0.Py).Qx$ reads "*Every proof of $\forall y^0.Py$ satisfies Q*.")

- We can no longer assume that domains of quantification are non-empty. For instance, the formula $(\forall x\!:\!\tau.\sigma) \to \sigma$, where $x \notin \mathrm{FV}(\sigma)$, does not have to be valid (inhabited), although its first-order analogue is a special case of axiom (Q1) in Section 8.3. This becomes quite apparent if τ and σ are propositional variables, say p,q, and our formula becomes the clearly unwanted $(p \to q) \to q$. See also Exercise 13.15(i).

The system λP does not distinguish between propositions and domains nor between proofs and objects. This may or may not be convenient with respect to various applications. For instance, it may actually be desirable that a lambda-calculus representing a logical system satisfies one of the following additional assumptions:

- Extensionality for propositions: If φ and ψ are equivalent propositions, i.e. if $\varphi \leftrightarrow \psi$ is provable, then φ and ψ are identified.

- Proof irrelevance: If x and y are proofs of the same proposition, then x and y are identified.

It should be clear, that the existence of two maps $f : A \to B$ and $g : B \to A$ is not a reasonable cause to identify two *sets* A and B—this would apply to any two non-empty domains. Also proof irrelevance would have a catastrophic consequence if every object is a proof: all non-empty domains become singletons. Therefore, various systems of dependent types distinguish between propositions and object domains (cf. Section 14.5).

CONSERVATIVITY. If we single out a domain for individuals, then a small fragment of λP suffices to represent the $(\forall, \to)$-fragment of first-order logic.

13.5.2. DEFINITION. Let Σ be a first-order signature. Then Γ_Σ is the (valid) environment consisting of the following declarations:

- A declaration $(0 : *)$ of the basic domain;

- For every k-ary relation symbol α in Σ, a declaration $\alpha : 0^k \Rightarrow *$;

- For every k-ary function symbol f in Σ, a declaration $f : 0^k \to 0$.

If X is a sequence of declarations of the form $(a : 0)$, then we write Γ_Σ^X for Γ_Σ, X. It should be obvious that Γ_Σ^X is always a valid environment.

Up to slight syntactic detail (notation $\forall a$ instead of $\forall a{:}0$, and $\psi \to \vartheta$ instead of $\forall x{:}\psi.\,\vartheta$) each first-order formula φ over Σ is a raw constructor. It is now easy to see that $\Gamma_\Sigma^X \vdash \varphi : *$ holds, provided $\mathrm{FV}(\varphi) \subseteq \mathrm{dom}(X)$.

13.5.3. DEFINITION. A *first-order environment* is one of the form Γ_Σ^X, Γ, where each type declared in Γ is a first-order formula.

13.5.4. PROPOSITION (Conservativity). *Let Γ_Σ^X, Γ be a first-order environment, and assume that $\Gamma_\Sigma^X, \Gamma \vdash M : \varphi$, where φ is a first-order formula. Then $\mathrm{rg}(\Gamma) \vdash \varphi$ in the $(\forall, \to)$-fragment of intuitionistic first-order logic.*

PROOF. We proceed by induction with respect to M. By Theorem 13.3.6, we can assume that M is in normal form. Thus M is either an abstraction or an application $x\vec{P}$, where x is a variable, and $\vec{P}$ is a (possibly empty) sequence of terms. In the former case, the variable x must be declared in the environment Γ_Σ^X, Γ. Because the type of M is a first-order formula, the variable x cannot be a function symbol in Σ, and its type must be a formula too. Summarizing, only the following cases are possible. In (ii) and (v), the term N is not an abstraction.

(i) M is a variable declared in Γ.

(ii) $M = NQ$, where $\Gamma_\Sigma^X, \Gamma \vdash N : \psi \to \varphi$ and $\Gamma_\Sigma^X, \Gamma \vdash Q : \psi$, and $\psi \neq 0$.

(iii) $M = \lambda x{:}\psi.N$, where $\Gamma_\Sigma^X, \Gamma, x{:}\psi \vdash N : \vartheta$ and $\varphi = \psi \to \vartheta$, and $\psi \neq 0$.

(iv) $M = \lambda a{:}0.N$, where $\Gamma_\Sigma^X, \Gamma, a{:}0 \vdash N : \psi$ and $\varphi = \forall a{:}0\, \psi$.

(v) $M = Nt$, where $\Gamma_\Sigma^X, \Gamma \vdash N : \forall a{:}0\, \psi$ and $\Gamma_\Sigma^X, \Gamma \vdash t : 0$ and $\varphi = \psi[x := t]$.

Case (i) is obvious. Cases (ii) and (iii) follow immediately from the induction hypothesis. In case (iv) one has to observe that $\Gamma_\Sigma^X, \Gamma, a{:}0 \vdash N : \psi$ implies $\Gamma_\Sigma^Y, \Gamma \vdash N : \psi$ for $Y = X, a : 0$. Finally, case (v) needs the observation that a normal form of type 0 in Γ_Σ^X, Γ must be an algebraic term. □

13.5.5. REMARK. Proposition 13.5.4 states the conservativity of λP over first-order logic understood as in our Chapter 8. This should not be confused with much stronger results (like that of [174]) concerning many-sorted logics with higher-order functions and function abstraction.

As we know, first-order intuitionistic logic is undecidable. And we have just observed that λP is conservative over first-order logic.

13.5.6. THEOREM (Bezem, Springintveld). *Inhabitation in λP is undecidable.*

PROOF. Immediate from Corollary 8.8.3 and Proposition 13.5.4. □

13.6. Notes

Dependent types are almost as old as the whole idea of *propositions-as-types*. As pointed out in [241], the list of symbols in [107] reveals that the concept must have been known to Curry in the 1950's. The operator $G\alpha\beta$, corresponding in our notation to $\forall x{:}\alpha.\beta(x)$, occurs in the second volume [108], and has later been studied by Seldin, and discussed in depth in [241, Chapter 16]. Also in de Bruijn's language Automath, dependent types are extensively used. De Bruijn's first paper [58] contains for instance an explicit definition of implication in terms of the universal quantifier. At approximately the same time Martin-Löf's first works on type theory appeared, cf. [365]. Various other systems aimed at constructing and verifying formal proofs, inspired by de Bruijn and Martin-Löf, emerged soon. Edinburgh LCF [197] and Constable's systems PL and PRL [79, 80] also involved dependent type machinery.

A theory of dependently typed calculi was developed in mid 1980's by various authors, including Meyer and Reinhold [340], Harper, Honsell and Plotkin [214] and Coquand and Huet [90]. The Edinburgh Logical Framework (LF) of [214] is very close to λP; for instance Theorem 13.3.6 and the translation $\natural$ of Section 13.3 are from [214]. Our definition of λP follows Barendregt's [32].

Currently dependent types are a standard feature of many logical frameworks and proof-assistants including e.g. Nuprl or Coq [81, 82, 33]. The use of dependent types in programming is also an emerging subject [19, 331, 509].

Theorem 13.4.4 is from [131], and Theorem 13.5.6 from [46]. The reader has perhaps noticed that the proof of the former uses very little of the language of λP that is not expressible in the "first-order" system λP_1 of Chapter 8. However, the type checking problem was investigated and shown decidable in [429] for an appropriate Curry-style (or "domain-free", cf. [37]) version of system λP_1.

13.7. Exercises

13.1. Assume that $\text{string}(n)$ is the type of binary strings of length n. For every such string w we form a record type $\text{record}(n)(w)$ with an integer field for every 1 in w and a boolean field for every 0 in w. For example, $\text{record}(5)(01101)$ will be $\text{int} \wedge \text{bool} \wedge \text{bool} \wedge \text{int} \wedge \text{bool}$. What kinds are string and record of?

13.2. Let Γ be a valid environment such that $\Gamma \vdash \tau : *$ and $\Gamma \vdash \sigma : *$, and assume that Γ contains the declarations: $p : *$, $\alpha : \tau \Rightarrow *$, $\delta : \sigma \Rightarrow *$, $f : \tau \to \sigma$. Find a term of type $\forall x^\tau (\alpha x \to \delta(fx)) \to \forall x^\tau (\alpha x \to \forall y^\sigma (\delta y \to p) \to p)$ in Γ.

13.3. For $\tau = q \to p$ and $\sigma = (\tau \to p) \to p$, find a term which has type

$$\forall f^{\tau \to \sigma} \forall g^{\sigma \to \tau}(\forall x^\tau \forall y^q(\alpha(g(fx)y) \to \alpha(xy)) \to r) \to r$$

in the environment $\{p : *, q : *, r : *, \alpha : \tau \Rightarrow *\}$.

13.4. (From [401].) Are the following derivable if $\Gamma = \{\alpha : *, P : \alpha \Rightarrow *, x : \alpha\}$?

(i) $\Gamma \vdash (\lambda x{:}Px.\, x) : \forall x{:}Px.\, Px$;

(ii) $\Gamma \vdash (\lambda x{:}Px.\, x) : \forall y{:}Px.\, Px$.

13.5. Does the uniqueness of types property (Proposition 3.4.1) hold for Church-style λP?

13.6. Show that if $\Gamma \vdash E_1 : E_2$ then $\text{FV}(E_1) \cup \text{FV}(E_2) \cup \text{FV}(\Gamma) \subseteq \text{dom}(\Gamma)$.

13.7. Show that an environment Γ is valid iff $\Gamma \vdash * : \square$.

13.8. Suppose we delete the last sentence in Lemma 13.2.2. Will the proof of Lemma 13.3.5 still work?

13.9. Let M be a raw term and let $|M| = M_1 \to_\beta N_1$. Show that $M \to_\beta N$ for some N with $|N| = N_1$.

13.10. Prove Lemma 13.3.3. Also prove that $\Gamma \vdash M : \tau$ and $M \to_\beta M'$ implies $\flat(M) \to_\beta \flat(M')$ or $\flat(M) = \flat(M')$, and that the latter case only happens when the reduction is performed within a type contained in M.

13.11. Find a mistake in the following attempt to prove strong normalization for λP by induction with respect to length of terms: *By Exercise 13.10, an infinite*

13.7. EXERCISES

reduction of a term M must, from some point on, consist exclusively of reduction steps performed within types. But all terms occurring in types are proper subterms of M, so one can apply the induction hypothesis.

13.12. Show an example of λP terms M and M' of the same type such that $\mathfrak{b}(M) = \mathfrak{b}(M')$, but $M \neq_\beta M'$.

13.13. Show that if $\Gamma \vdash \phi : \kappa$ then both ϕ and κ are strongly normalizable.

13.14. Prove strong normalization for Curry-style λP.

13.15. Let Γ consists of the declarations: $p : *, q : *, \alpha : p \Rightarrow *, \gamma : (p \to q) \Rightarrow *$. Are the following types inhabited in Γ?

(i) $(\forall x^p.\alpha x \to q) \to (\forall x^p.\alpha x) \to q$;

(ii) $\forall x^{p \to q}(\gamma x \to \gamma(\lambda y^p.xy))$.

13.16. Let $\mathbf{bool} = p \to p \to p$ and let **true** and **false** be as in Chapter 1. Is the type $\beta(\mathbf{true}) \to \beta(\mathbf{false}) \to \forall x^{\mathbf{bool}} \beta x$ inhabited in $\{p : *, \beta : \mathbf{bool} \Rightarrow *\}$?

13.17. Let $\omega = (p \to p) \to (p \to p)$, and let **0** and **succ** be as in Chapter 1. Is the induction scheme $\beta \mathbf{0} \to \forall x^\omega (\beta x \to \beta(\mathbf{succ}\ x)) \to \forall x^\omega \beta x$ inhabited in the environment $\{p : *, \beta : \omega \Rightarrow *\}$?

Chapter 14

Pure type systems and the λ-cube

The simply typed λ-calculus $\lambda_\to$, the system λP of dependent types, the polymorphic λ-calculus λ2, and other typed λ-calculi differ in various aspects but have a lot in common. Each one involves one or more forms of abstraction and application. The abstraction always lives in a certain product domain (sometimes disguised as a function type) and the application returns an element of a component of the product.

Many properties of these systems are similar and are established with recurring proof techniques. This calls for a certain form of unification of definitions and for the development of a general theory.

In this last chapter we aim at such a unification by introducing the notion of a *pure type system* (abbreviated PTS). We will see how this unification simplifies the presentation of the calculi already considered, and how other useful concepts emerge in a natural way from the PTS approach.

A particularly interesting PTS, which encompasses most of what we have discussed so far, is called the Calculus of Constructions (λC). This system is presented in Section 14.3 in the form of *Barendregt's cube*.

14.1. System λP revisited

The reader must have become a little upset by the numerous occurrences in Chapter 13 of similar definitions, rules and lemmas, repeating the same patterns in various syntactic forms. We have chosen this presentation of system λP in order to make it easier to understand, by highlighting the syntactical hierarchy of terms, types and kinds. The price we paid was for instance the need to consider as many as six weakening rules where a more general syntax would make it possible to state only one. Once however we become familiar with the system, we may look for such a general syntax and a more concise presentation. In the following definition we unify the syntax for terms, constructors and kinds, and we abandon the distinction between $\Rightarrow$ and $\to$ and between Π and $\forall$.

14.1.1. DEFINITION (Compact λP).

(i) The set $\mathcal{S}$ of *sorts* is defined by $\mathcal{S} = \{*, \square\}$. We use s, s', s_1, s_2, etc. to range over sorts.

(ii) *Raw expressions* A and *raw environments* Γ are defined by:
$$A ::= x \mid s \mid (A\,A) \mid (\lambda x{:}A\,A) \mid (\Pi x{:}A\,A);$$
$$\Gamma ::= \varnothing \mid \Gamma, (x : A).$$

(iii) The operators Π and λ are assumed to bind variables. The definition of the set $\mathrm{FV}(A)$ of *free variables* of A and of the substitution $A[x := B]$ is routine. We write $A \to B$ for $(\Pi x{:}A\,B)$ when $x \notin \mathrm{FV}(B)$. As usual, we identify expressions that only differ in their bound variables, and we apply our ordinary notational conventions.

(iv) The relation $\to_\beta$ on raw expressions is the compatible closure of
$$(\lambda x{:}A\,B)C \to_\beta B[x := C].$$
The notation $\twoheadrightarrow_\beta$ and $=_\beta$ is used as before.

(v) The rules in Figure 14.1 derive judgements of the form $\Gamma \vdash A : B$, where Γ is a raw environment and A and B are raw expressions. In order to distinguish these rules from the system λP defined in Chapter 13, the system of Figure 14.1 is called *compact* λP.

It should be intuitively obvious[1] that, up to syntactic details, the two formulations of system λP are equivalent. Indeed, most of the rules in Figure 14.1 are simply generic patterns for the corresponding rules in Figure 13.1. There are only three differences. The first two are the additional premises in the compact rules for product and abstraction. These are necessary, because A in rule (Prod) should not be of sort $\square$, and the product in (Abs) ought to be well-formed. This problem did not occur in Chapter 13, because the syntax of types and kinds was different. The last difference is in rule (Conv), where it is only implicit that B and B' are of the same sort.

14.2. Pure type systems

The rules of compact λP turn out to be surprisingly universal. Up to some syntactic sugar, nearly all Church-style typed systems discussed in this book can be obtained by minor modifications of these rules. In fact, one has to modify just one rule: the product rule.

For instance, suppose we restrict rule (Prod) to $s = *$. One easily shows that if $\Gamma \vdash (\Pi x{:}A.\,B) : *$ is derivable in such a system then $x \notin \mathrm{FV}(B)$

---
[1] A formal proof requires a few technicalities, see Exercise 14.4.

14.2. Pure type systems

(Ax)	$\emptyset \vdash * : \square$
(Var)	$\dfrac{\Gamma \vdash A : s}{\Gamma, x:A \vdash x : A}$ $\quad (x \notin \mathrm{dom}(\Gamma))$
(Prod)	$\dfrac{\Gamma \vdash A : * \quad \Gamma, x:A \vdash B : s}{\Gamma \vdash (\Pi x{:}A.\,B) : s}$
(Abs)	$\dfrac{\Gamma, x:A \vdash B : C \quad \Gamma \vdash (\Pi x{:}A.\,C) : s}{\Gamma \vdash (\lambda x{:}A.\,B) : (\Pi x{:}A.\,C)}$
(App)	$\dfrac{\Gamma \vdash A : (\Pi x{:}B.\,C) \quad \Gamma \vdash D : B}{\Gamma \vdash (AD) : C[x := D]}$
(Weak)	$\dfrac{\Gamma \vdash A : B \quad \Gamma \vdash C : s}{\Gamma, x:C \vdash A : B}$ $\quad (x \notin \mathrm{dom}(\Gamma))$
(Conv)	$\dfrac{\Gamma \vdash A : B \quad \Gamma \vdash B' : s}{\Gamma \vdash A : B'}$ $\quad (B =_\beta B')$

FIGURE 14.1: COMPACT λP

and thus all product types are implications. This fragment of λP is thus essentially the same as the Church-style simply typed lambda-calculus.

As a next example, consider the following variant of rule (Prod):

$$\dfrac{\Gamma \vdash A : s \quad \Gamma, x:A \vdash B : *}{\Gamma \vdash (\Pi x{:}A.\,B) : *}$$

This rule makes it possible to construct function types (when $s = *$) and polymorphic types $\Pi p{:}{*}.\,B$ (when $s = \square$). Up to minor syntactic differences (like $\Pi p{:}{*}.\,B$ written instead of $\forall p\, B$) this is Girard's system **F** of Chapter 11.

In general one can consider an arbitrary set of sorts, additional axioms and various product rules. The following definition covers all such systems.

14.2.1. DEFINITION.

(i) A *pure type system* (*PTS*) is determined by a triple $(\mathcal{S}, \mathcal{A}, \mathcal{R})$ where

(1) $\mathcal{S}$ is a set of *sorts*;

(2) $\mathcal{A} \subseteq \mathcal{S} \times \mathcal{S}$ is a set of *axioms*;

(3) $\mathcal{R} \subseteq \mathcal{S} \times \mathcal{S} \times \mathcal{S}$ is a set of *rules*.

(ii) Raw expressions and raw environments for a PTS are defined exactly as in Definition 14.1.1. The same holds for the notions of free variables, substitution and reduction. In addition to the standard notational conventions, we also define $A \to B$ as $(\Pi x{:}A\ B)$ when $x \notin \mathrm{FV}(B)$.

(iii) Type assignment rules for $(\mathcal{S}, \mathcal{A}, \mathcal{R})$ are found in Figure 14.2.

If $s_2 = s_3$ then a rule (s_1, s_2, s_3) is abbreviated (s_1, s_2). An axiom (s_1, s_2) is often written as $s_1 : s_2$. Rules in $\mathcal{R}$ represent possible product constructions, or *dependencies*. For instance, in system **F**, we have two forms of dependency, that is, two forms of abstraction: $\lambda x{:}\tau.\,M$ (where the object expression M *depends* on an object argument x) and $\lambda p{:}{*}.\,M$ (where M depends on a type argument p). These two dependencies are represented by the rules $(*, *, *)$ and $(\Box, *, *)$, abbreviated to $(*, *)$ and $(\Box, *)$.

We are not going to further study pure type systems; we only mention that subject reduction holds for arbitrary PTS.

14.2.2. THEOREM. *If $\Gamma \vdash A : B$ and $A \twoheadrightarrow A'$ then also $\Gamma \vdash A' : B$.*

PROOF. See Exercises 14.1–14.3 for an outline and [32, 175] for details. □

14.3. The Calculus of Constructions

We have dealt so far with two sorts: the sort $*$ of types and the sort $\Box$ of kinds, and we have the axiom $* : \Box$. There are four possible rules of the form (s_1, s_2), where $s_1, s_2 \in \{*, \Box\}$. Assuming that rule $(*, *)$ should occur in any reasonable PTS, we can make eight combinations of other rules and define eight different systems, one for each such combination. These eight systems are presented in Figure 14.3 as *Barendregt's cube*.

The lower left corner of the cube is the simply typed lambda-calculus. The only rule available in that system is

$(*,*)$ *Objects may depend on objects.*

The edges of the cube extend into the three dimensions. Each dimension represents one additional dependency. The edge oriented to the right points to the system $\lambda \mathrm{P}$, obtained by adding to $\lambda_\to$ the rule:

$(*, \Box)$ *Types may depend on objects.*

The upward edge pointing to $\lambda 2$ (i.e. to system **F**) adds polymorphism, expressed by the rule

$(\Box, *)$ *Objects may depend on types.*

14.3. THE CALCULUS OF CONSTRUCTIONS

$$
\begin{array}{ll}
(\text{Ax}) & \varnothing \vdash s_1 : s_2, \quad \text{when } (s_1, s_2) \in \mathcal{A} \\[1em]
(\text{Var}) & \dfrac{\Gamma \vdash A : s}{\Gamma, x : A \vdash x : A} \quad (x \notin \mathrm{dom}(\Gamma)) \\[1em]
(\text{Prod}) & \dfrac{\Gamma \vdash A : s_1 \quad \Gamma, x : A \vdash B : s_2}{\Gamma \vdash (\Pi x : A.\, B) : s_3} \quad ((s_1, s_2, s_3) \in \mathcal{R}) \\[1em]
(\text{Abs}) & \dfrac{\Gamma, x : A \vdash B : C \quad \Gamma \vdash (\Pi x : A.\, C) : s}{\Gamma \vdash (\lambda x : A.\, B) : (\Pi x : A.\, C)} \\[1em]
(\text{App}) & \dfrac{\Gamma \vdash A : (\Pi x : B.\, C) \quad \Gamma \vdash D : B}{\Gamma \vdash (AD) : C[x := D]} \\[1em]
(\text{Weak}) & \dfrac{\Gamma \vdash A : B \quad \Gamma \vdash C : s}{\Gamma, x : C \vdash A : B} \quad (x \notin \mathrm{dom}(\Gamma)) \\[1em]
(\text{Conv}) & \dfrac{\Gamma \vdash A : B \quad \Gamma \vdash B' : s}{\Gamma \vdash A : B'} \quad (B =_\beta B')
\end{array}
$$

FIGURE 14.2: A PURE TYPE SYSTEM

Finally, the third dimension corresponds to adding the rule:

$(\square, \square)$ *Types may depend on types.*

This edge points to a system denoted by $\lambda\underline{\omega}$, a fragment of system $\mathbf{F}_\omega$, where it is possible to form type constructors like $\lambda p:*.\, p \to p$ of kind $* \to *$, but polymorphism is not permitted.

System $\mathbf{F}_\omega$ occurs in the cube in the back-upper-left corner, under its alternative name $\lambda\omega$. This system uses the two dependencies $(\square, *)$ and $(\square, \square)$ in adddition to the basic $(*, *)$. Systems obtained by other combinations of axioms occupy the positions determined by adding the appropriate vectors.

The back-upper-right corner is the most distant from our departure point and combines all dimensions. This system, called the *Calculus of Constructions* (also denoted by CC), is the most powerful in the cube.

As every PTS, the Calculus of Constructions has the subject reduction property. In addition, we have the following fact.

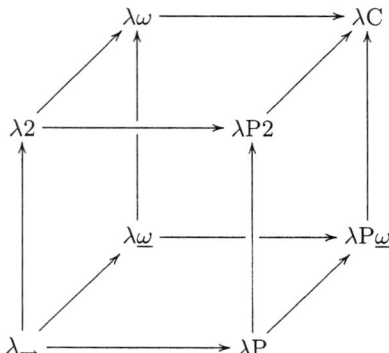

FIGURE 14.3: BARENDREGT'S CUBE

14.3.1. PROPOSITION.

(i) *If $\Gamma \vdash A : B$ and $\Gamma \vdash A : B'$ then $B =_\beta B'$.*

(ii) *If $\Gamma \vdash A : B$ then exactly one of the following holds: either $\Gamma \vdash B : *$ or $\Gamma \vdash B : \square$ or $B = \square$.*

PROOF. Follows by induction from an appropriate generation lemma (Exercise 14.1). Details are again in [32, 175]. □

Although the syntax of λC is "homogeneous" (there is no a priori distinction between terms, constructors and kinds), with Proposition 14.3.1 we can classify expressions into these three levels (relative to a given environment).

14.3.2. DEFINITION. If $\Gamma \vdash A : \square$ (resp. $\Gamma \vdash A : *$) then we say that A is a *kind* (resp. *a type*) in the environment Γ. If $\Gamma \vdash A : B$ and B is a type (resp. kind) in Γ then A is a *term of type B* (resp. *constructor of kind B*) in Γ. If a declaration $x : A$ is in Γ and A is a type (resp. constructor) in Γ, then x is an *object variable* (resp. *constructor variable*) of type (resp. kind) A.

In what follows we use the notation $\tau, \sigma, \ldots$ for types, $\kappa, \ldots$ for kinds, $M, N, \ldots$ for terms, and $\varphi, \psi, \ldots$ for constructors.

14.4. Strong normalization

We derive strong normalization for the Calculus of Constructions from strong normalization of Curry-style system $\mathbf{F}_\omega$ (Theorem 11.7.5). This is achieved by means of a translation similar to the one we used in Chapter 13. In fact, we had not one translation there but two, and now we are going to have three, because the situation is even more complicated. Indeed, terms and

14.4. Strong normalization

types of system λP do not contain kinds, and thus a reduction-preserving translation of kinds was not necessary in Section 13.3. This time, we have to cope with all the three levels of the syntax: kinds, constructors and object terms, and we define the following translations:

- A contracting map $\flat$, which maps kinds of λC into kinds of $\mathbf{F}_\omega$;
- A translation $\mathfrak{c}$, from kinds and constructors to constructors;
- And a translation $\sharp$, which maps everything into pure lambda-terms.

Of course, the above classification into kinds, constructors, and terms is only applicable in a fixed environment. Therefore, the translations are defined relative to an environment, and strictly speaking, we should e.g. use the notation $\mathfrak{c}_\Gamma$ rather than just $\mathfrak{c}$.[2]

14.4.1. DEFINITION. First we define a contracting map on kinds.

- $\flat(*) = *$;
- $\flat(\Pi\alpha{:}\kappa_1.\kappa_2) = \flat(\kappa_1) \Rightarrow \flat(\kappa_2)$, when κ_1 and κ_2 are kinds.
- $\flat(\Pi x{:}\tau.\kappa) = \flat(\kappa)$, when τ is a type and κ is a kind.

It will be also useful to assume that $\flat(\square) = *$.

14.4.2. DEFINITION. The following translation $\mathfrak{c}$ assigns types of $\mathbf{F}_\omega$ to kinds of λC and constructors of $\mathbf{F}_\omega$ to constructors of λC. Below, 0 denotes a fixed fresh type variable, and $\alpha^{\flat(\kappa)}$ is a constructor variable of $\mathbf{F}_\omega$ of kind $\flat(\kappa)$.

- $\mathfrak{c}(*) = 0$;
- $\mathfrak{c}(\alpha) = \alpha^{\flat(\kappa)}$, if α is a constructor variable of kind κ;
- $\mathfrak{c}(\Pi\alpha{:}\kappa.\,A) = \forall\alpha^{\flat(\kappa)}.\,\mathfrak{c}(\kappa) \to \mathfrak{c}(A)$;
- $\mathfrak{c}(\Pi x{:}\tau.\,A) = \mathfrak{c}(\tau) \to \mathfrak{c}(A)$;
- $\mathfrak{c}(\lambda\alpha{:}\kappa.\,\varphi) = \lambda\alpha^{\flat(\kappa)}.\,\mathfrak{c}(\varphi)$;
- $\mathfrak{c}(\lambda x{:}\tau.\,\varphi) = \mathfrak{c}(\varphi)$;
- $\mathfrak{c}(\varphi\psi) = \mathfrak{c}(\varphi)\mathfrak{c}(\psi)$;
- $\mathfrak{c}(\varphi M) = \mathfrak{c}(\varphi)$.

---

[2]For instance, we should have $\mathfrak{c}_\Gamma(\Pi\alpha{:}\kappa.\,A) = \forall\alpha^{\flat(\kappa)}.\,\mathfrak{c}_\Gamma(\kappa) \to \mathfrak{c}_{\Gamma,\alpha:\kappa}(A)$ in the third clause of Def. 14.4.2.

We also take $\mathfrak{c}(\square) = 0$, and we write $\mathfrak{c}(\Gamma)$ for the $\mathbf{F}_\omega$ environment consisting of the declarations $x_* : 0$, $y : \forall pq(p \to q \to 0)$, and the union

$$\{(x : \mathfrak{c}(\tau)) \mid (x : \tau) \in \Gamma\} \cup \{(x_\alpha : \mathfrak{c}(\kappa)) \mid (\alpha : \kappa) \in \Gamma\}.$$

14.4.3. LEMMA. *Let* $\Gamma \vdash A : B$ *in* $\lambda \mathrm{C}$, *and either* $B = \square$ *or* $\Gamma \vdash B : \square$. *Then* $\mathfrak{c}(A)$ *is a constructor of* $\mathbf{F}_\omega$ *of kind* $\mathfrak{b}(B)$. *In particular, if* $B = \square$ *or* $B = *$ *then* $\mathfrak{c}(A)$ *is a type. In addition, the free constructor variables in* $\mathfrak{c}(A)$ *are all of the form* $\alpha^{\mathfrak{b}(\kappa)}$, *where* $\alpha \in \mathrm{FV}(A)$ *is of kind* κ.

PROOF. One first proves the following properties of substitution:

$$\begin{aligned} \mathfrak{b}(\kappa[\alpha := \psi]) &= \mathfrak{b}(\kappa); \\ \mathfrak{b}(\kappa[x := M]) &= \mathfrak{b}(\kappa). \end{aligned}$$

Then apply routine induction with respect to derivations. For the conversion rule, note that $\kappa_1 =_\beta \kappa_2$ implies $\mathfrak{b}(\kappa_1) = \mathfrak{b}(\kappa_2)$. □

14.4.4. DEFINITION. For any expression A of $\lambda \mathrm{C}$, we now define a pure lambda-term $A^\sharp$. Assume that a new object variable x_α is associated to every constructor variable α and let x_* and y be two more fresh object variables.

- $*^\sharp = x_*$;
- $x^\sharp = x$;
- $\alpha^\sharp = x_\alpha$;
- $(\Pi \alpha{:}\kappa.\, A)^\sharp = y\kappa^\sharp(\lambda x_\alpha.\, A^\sharp)$;
- $(\Pi x{:}\tau.\, A)^\sharp = y\tau^\sharp(\lambda x.\, A^\sharp)$;
- $(\lambda \alpha{:}\kappa.\, A)^\sharp = (\lambda z \lambda x_\alpha.\, A^\sharp)\kappa^\sharp$, where $z \notin \mathrm{FV}(A^\sharp)$;
- $(\lambda x{:}\tau.\, A)^\sharp = (\lambda z \lambda x.\, A^\sharp)\tau^\sharp$, where $z \notin \mathrm{FV}(A^\sharp)$;
- $(A\psi)^\sharp = A^\sharp \psi^\sharp$;
- $(AM)^\sharp = A^\sharp M^\sharp$.

14.4.5. LEMMA. *If* $\Gamma \vdash A : B$ *in* $\lambda \mathrm{C}$ *then* $\mathfrak{c}(\Gamma) \vdash A^\sharp : \mathfrak{c}(B)$, *in* $\mathbf{F}_\omega$.

PROOF. For the proof we need two properties of substitution (where A is either a kind or a type in Γ). We leave them to the reader.

$$\begin{aligned} \mathfrak{c}(A[\alpha := \psi]) &= \mathfrak{c}(A)[\alpha := \mathfrak{c}(\psi)]; \\ \mathfrak{c}(A[x := M]) &= \mathfrak{c}(A). \end{aligned}$$

It follows that $B_1 =_\beta B_2$ implies $\mathfrak{c}(B_1) =_\beta \mathfrak{c}(B_2)$. The proof of the lemma is by induction, and we demonstrate the case of constructor abstraction.
  Let $\Gamma \vdash \lambda\alpha{:}\kappa.B \,:\, \Pi\alpha{:}\kappa.C$ be obtained from $\Gamma, \alpha{:}\kappa \vdash B \,:\, C$ and $\Gamma \vdash \kappa \,:\, \square$. From the induction hypothesis we have $\mathfrak{c}(\Gamma), x_\alpha : \mathfrak{c}(\kappa) \vdash B^\sharp : \mathfrak{c}(C)$, so $\mathfrak{c}(\Gamma) \vdash \lambda x_\alpha B^\sharp : \mathfrak{c}(\kappa) \to \mathfrak{c}(C)$ and $\mathfrak{c}(\Gamma) \vdash (\lambda z \lambda x_\alpha B^\sharp) : 0 \to \mathfrak{c}(\kappa) \to \mathfrak{c}(C)$. We also know that $\mathfrak{c}(\Gamma) \vdash \kappa^\sharp : 0$, and thus $(\lambda z \lambda x_\alpha B^\sharp)\kappa^\sharp$ is of type $\mathfrak{c}(\kappa) \to \mathfrak{c}(C)$. By Lemma 14.4.3, the variable $\alpha^{\mathfrak{b}(\kappa)}$ is not free in $\mathfrak{c}(\Gamma)$, because α was not free in Γ, and we have that $\mathfrak{c}(\Gamma) \vdash (\lambda z \lambda x_\alpha B^\sharp)\kappa^\sharp : \forall \alpha^{\mathfrak{b}(\kappa)}(\mathfrak{c}(\kappa) \to \mathfrak{c}(C))$. □

14.4.6. LEMMA. *If $A_1 \to_\beta A_2$ then $A_1^\sharp \to_\beta^+ A_2^\sharp$.*

PROOF. The proof is again a straightforward induction using the equations:
$$\begin{aligned}(A[\alpha := \psi])^\sharp &= A^\sharp[x_\alpha := \psi^\sharp]; \\ (A[x := M])^\sharp &= A^\sharp[x := M^\sharp].\end{aligned}$$
□

14.4.7. THEOREM. *The Calculus of Constructions has the strong normalization property.*

PROOF. Immediate from Lemmas 14.4.5–14.4.6 and Theorem 11.7.5. □

14.5. Beyond the cube

The notion of a PTS is very general and a variety of systems can be presented as PTS. Here are a few examples.

14.5.1. EXAMPLE. The system $\mathrm{N}^\omega \mathrm{JP}$ of Berardi [42] has four sorts:
$$\textbf{Set}, \textbf{Prop}, \square^s \text{ and } \square^p$$
and two axioms:
$$\textbf{Set} : \square^s \text{ and } \textbf{Prop} : \square^p.$$
This system represents a higher-order logic where propositions and domains (sets) are treated as members of different sorts **Prop** and **Set**. There are five rules. Three of them:
$$(\textbf{Prop}, \textbf{Prop}),\ (\textbf{Set}, \textbf{Prop}),\ (\square^p, \textbf{Prop})$$
make it possible to form implications and to quantify over elements of an arbitrary domain, as well as over propositions (and other type constructors). The other two rules
$$(\textbf{Set}, \square^p) \text{ and } (\square^p, \square^p)$$
can be used to define constructors of such kinds as e.g. $\Pi x{:}A.\textbf{Prop}$ or $\textbf{Prop} \to \cdots \to \textbf{Prop}$.

System $\lambda\text{PRED}_\omega$ (see [32]) has in addition a function sort **Fun** and two more rules

(**Set**, **Set**, **Fun**) and (**Set**, **Fun**, **Fun**),

representing the ability to create function domains $\Pi x : A. B$ and $\Pi x : A. F$, where $A, B : $ **Set** and $F : $ **Fun**.

14.5.2. EXAMPLE. Our next example is a subsystem ECC_Π of the Extended Calculus of Constructions (ECC) of Luo [317] (The full ECC is strictly speaking not a PTS, for instance because of the sum operator Σ.) The system ECC_Π has infinitely many sorts and infinitely many axioms:

$$\textbf{Prop} : \textbf{Type}_0 : \textbf{Type}_1 : \textbf{Type}_2 : \ldots$$

Rules of ECC_Π are:

(**Prop**, **Prop**), (**Type**$_i$, **Prop**) and (**Type**$_i$, **Type**$_j$, **Type**$_{\max(i,j)}$).

for arbitrary i, j. The infinite hierarchy of sorts is a way to avoid the problems caused by the axiom **Type** : **Type** (see below).

14.6. Girard's paradox

As we can see from the examples above, it is quite easy to define some very strong calculi in the form of PTS. However, it is also easy to obtain calculi that are *too strong*, namely inconsistent. This happened in the past to many formal systems. Typically, such systems are "too permissive", so that various kinds of "paradoxical" objects can be constructed. A famous example is naive set theory and the notion of the set of all sets leading to Russell's paradox.

Russell's paradox can be turned into a theorem of axiomatic set theory stating that no set A can contain its own power set, i.e. that $\mathcal{P}(A) \subseteq A$ never holds. (Otherwise, define $\Delta = \{a \in A \mid a \notin a\}$ and show that $\Delta \in \Delta$ is equivalent to $\Delta \notin \Delta$.) Cantor's theorem draws a contradiction from a weaker assumption: that there exists a surjection $f : A \to \mathcal{P}(A)$. This time we define the "paradoxical" set as $\Delta = \{a \in A \mid a \notin f(a)\}$. Now we choose a such that $f(a) = \Delta$, and we ask if $a \in \Delta$.

It is natural to suspect that some PTS axioms and rules might cause such a paradoxical behaviour too. An obvious candidate is the axiom $* : *$, expressed by the slogan "*Type is a type*" strongly reminiscent of the naive idea of the set of all sets. This system is indeed inconsistent, but it turns out that the problem also occurs in seemingly innocent cases.

14.6.1. DEFINITION. *System* **U** is a PTS with sorts $\{*, \square, \triangle\}$, axioms $* : \square$ and $\square : \triangle$, and rules $\{(*, *), (\square, *), (\square, \square), (\triangle, *), (\triangle, \square)\}$. The fragment of system **U** without rule $(\triangle, *)$ is denoted by $\textbf{U}^-$. We also define $\textbf{F}^+_\omega$ as $\textbf{F}_\omega$ extended with the third sort $\triangle$ and the axiom $\square : \triangle$ (and no added rules).

14.6. GIRARD'S PARADOX

The purpose of the extension $\mathbf{F}_\omega^+$ is to make it possible to declare kind variables $k : \square$. In system $\mathbf{U}^-$ one can define "polymorphic kinds" and assign them to "generic constructors" in a similar way as polymorphic types are assigned to terms of system $\mathbf{F}$. In this respect, system $\mathbf{U}^-$ can be seen as a natural extension of $\mathbf{F}_\omega$ and $\mathbf{F}_\omega^+$. There seems to be nothing wrong for instance with a generic constructor like this one (where k is a kind variable):

$$\lambda k^\square \lambda \alpha^{k \to k} \lambda \beta^k . \alpha(\alpha \beta) : \Pi k : \square ((k \to k) \to k \to k).$$

In system $\mathbf{U}$ one can in addition define types with quantified kind variables. But the polymorphic kinds available in $\mathbf{U}^-$ are enough to cause inconsistency, and we are now going to derive it. More precisely, we show that in system $\mathbf{U}^-$ there exists a closed term of type $\bot = \forall p : * . p$. It follows that every type is inhabited, i.e. every proposition is provable (Theorem 14.6.5).

The idea is to express a paradox of set theory within the system, by constructing a paradoxical set and then deriving a contradiction. Sets are interpreted as kinds (not types). This makes it possible to define

$$\mathcal{P}(\kappa) = \kappa \to *,$$

so that both a "set" κ and its "powerset" $\mathcal{P}(\kappa)$ live in the same sort.

14.6.2. CONVENTION. Although, formally, the universal quantifier $\forall$ does not occur in the PTS syntax, we use it below to enhance readability. The symbols $\bot$, $\neg$, $\wedge$, $\exists$ can also be used, because the logical connectives are definable in a similar way as in system $\mathbf{F}$ (Section 11.3). The term syntax (pairs, projections, **let**, etc.) is used accordingly. We also often write $\Gamma \vdash \varphi$ (meaning that φ is provable from Γ) when there exists an M with $\Gamma \vdash M : \varphi$.

14.6.3. EXAMPLE. This example shows how Cantor's theorem can be formalized in $\mathbf{F}_\omega^+$. The environment Γ represents a hypothetical situation where a power set $\mathcal{P}(k)$ is (almost) a retract of k: there are functions $el : \mathcal{P}(k) \to k$ and $set : k \to \mathcal{P}(k)$ such that $set \circ el$ is (almost) identity on $\mathcal{P}(k)$.

$$\Gamma = \{k : \square,\ el : \mathcal{P}(k) \to k,\ set : k \to \mathcal{P}(k),\ V : \forall X^{\mathcal{P}(k)} \forall \alpha^k (X\alpha \leftrightarrow set(elX)\alpha)\}$$

The assumptions in Γ do not even postulate an actual retraction but only the logical equivalence of $X\alpha$ and $set(elX)\alpha$. This is perfectly enough to derive a contradiction from Γ with the help of the following diagonal argument.

Define $\Delta = el(\lambda \alpha^k . \neg set \alpha \alpha)$, and let $M = V(\lambda \alpha^k . \neg set \alpha \alpha) \Delta$. Then $\Gamma \vdash M : \neg set \Delta \Delta \leftrightarrow set \Delta \Delta$. Apply Exercise 4.12 to obtain $\Gamma \vdash \bot$.

Suppose we can "implement" the environment Γ of Example 14.6.3 by providing a kind κ and constructors set, el and V of appropriate kinds. That would imply inconsistency. However, the requirement that $set \circ el$ is the identity seems to be difficult to fulfill. In a sense, it amounts to constructing a fixpoint of the operator $k \mapsto \mathcal{P}(k)$ and we have good reasons to suspect that such a construction may be impossible [452]. However, we have already seen in Section 12.3 that inductive types, which behave to some extent as least fixpoints, are definable with the help of type polymorphism. In system $\mathbf{U}^-$ we can use kind polymorphism to obtain an "inductive kind" κ representing the fixpoint of the map $k \mapsto \mathcal{P}(k)$. This kind represents a slightly weaker but still paradoxical behaviour. One can show that there are mappings $el : \mathcal{P}(\kappa) \to \kappa$ and $set : \kappa \to \mathcal{P}(\kappa)$ such that, informally,

$$set(el(X)) = \{el(set(x)) \mid x \in X\}, \qquad (*)$$

for all $X \subseteq \kappa$. In other words, we have $set \circ el = El \circ Set$, where $El(X)$ (respectively $Set(X)$) is the image of X under el (respectively set). This cannot happen in set theory, and the reader may wish to verify that (Exercise 14.12) before proceeding to the formalization given by the next lemma.

14.6.4. LEMMA. *Let $=^k$ denote the Leibniz equality on k, i.e. $\alpha =^k \beta$ stands for $\forall \gamma^{k \to *}. \gamma \alpha \to \gamma \beta$. Define:*

$$\Gamma = \{k : \Box,\ el : \mathcal{P}(k) \to k,\ set : k \to \mathcal{P}(k),$$
$$V : \forall X^{\mathcal{P}(k)} \forall \alpha^k (set(elX)\alpha \leftrightarrow \exists \beta^k (X\beta \wedge \alpha =^k el(set\beta)))\}.$$

Then $\Gamma \vdash \bot$ in $\mathbf{F}_\omega^+$.

PROOF. The type assigned to V by the environment Γ is of course meant to represent the assumption $(*)$. The idea of the proof is to formalize in $\mathbf{F}_\omega^+$ the set-theoretical argument used in Exercise 14.12. We begin the formalization by expressing the notion of an equivalence relation. Define

$$Eqv = \lambda R^{k \to k \to *} \forall \alpha^k \beta^k \gamma^k ((R\alpha\alpha \wedge (R\alpha\beta \to R\beta\alpha)) \wedge (R\alpha\beta \to R\beta\gamma \to R\alpha\gamma)).$$

Of course we have $\Gamma \vdash Eqv : (k \to k \to *) \to *$. In what follows, we use the following notation:

- We abbreviate $el(set\,\alpha)$ by α';

- We write $\alpha \in \beta$ for $\exists \gamma^k (\alpha \approx \gamma \wedge set\beta\gamma)$ and $\alpha \notin \beta$ for $\neg(\alpha \in \beta)$;

- And $\alpha \approx \beta$ stands for $\forall R^{k \to k \to *}(Eqv\,R \to \forall \gamma^k R\gamma\gamma' \to R\alpha\beta)$.

Observe that $\alpha \approx \beta$ and $\beta \in \alpha$ are types in the environment $\Gamma, \alpha : k, \beta : k$, and that $\Gamma \vdash \forall \alpha^k (\alpha \approx \alpha')$. We prove in order the following claims:

(i) $\Gamma, \alpha : k, \beta : k \vdash set\,\alpha\beta \to set\,\alpha'\beta'$;

(ii) $\Gamma \vdash Eqv(\lambda \alpha^k \beta^k.\alpha \approx \beta)$;

(iii) $\Gamma, \alpha : k, \beta : k \vdash \alpha \approx \beta \to \forall \delta^k (\alpha \in \delta \leftrightarrow \beta \in \delta)$;

14.6. GIRARD'S PARADOX   355

(iv) $\Gamma, \alpha:k, \beta:k \vdash \beta \,\epsilon\, \alpha \leftrightarrow \beta \,\epsilon\, \alpha'$;

(v) $\Gamma, \alpha:k, \beta:k \vdash \alpha \approx \beta \to \forall \gamma^k (\gamma \,\epsilon\, \alpha \leftrightarrow \gamma \,\epsilon\, \beta)$;

(vi) $\Gamma, \beta:k \vdash \beta \,\epsilon\, \Delta \leftrightarrow \beta \notin \beta$, where $\Delta = el(\lambda \alpha^k . \alpha \notin \alpha)$.

From (vi) it follows that $\Gamma \vdash \Delta \,\epsilon\, \Delta \leftrightarrow \Delta \notin \Delta$ and this implies $\Gamma \vdash \bot$. First we show (i). Observe that

$$\Gamma, \alpha:k, \beta:k \vdash V(set\,\alpha)\beta' : set\,\alpha'\beta' \leftrightarrow \exists \gamma^k (set\,\alpha\gamma \wedge \beta' =^k \gamma'),$$

and remember that $\leftrightarrow$ abbreviates a conjunction of two implications. Let us assume that I is a proof of $\beta' =^k \beta'$. Then we derive

$$\Gamma, \alpha:k, \beta:k, u:set\,\alpha\beta \vdash \pi_2(V(set\,\alpha)\beta')[\beta, \langle u, I \rangle]_{\exists \gamma (set\,\alpha\gamma \wedge \beta' =^k \gamma')} : set\,\alpha'\beta'.$$

To show (ii) one has to define three terms A_1, A_2, A_3 such that

- $\Gamma, \alpha:k, \beta:k, \gamma:k \vdash A_1 : \alpha \approx \alpha$;
- $\Gamma, \alpha:k, \beta:k, \gamma:k \vdash A_2 : \alpha \approx \beta \to \beta \approx \alpha$;
- $\Gamma, \alpha:k, \beta:k, \gamma:k \vdash A_3 : \alpha \approx \beta \to \beta \approx \gamma \to \alpha \approx \gamma$.

As an example, we show a definition of A_2, leaving A_1 and A_3 to the reader.

$$A_2 = \lambda x^{\alpha \approx \beta} \lambda R^{k \to k \to *} \lambda y^{EqvR} \lambda z^{\forall \gamma^k . R\gamma\gamma'} . \pi_2(\pi_1(y\alpha\beta\gamma))(xRyz).$$

Now consider (iii) and define (remember that $\alpha \,\epsilon\, \delta$ is an existential type):

$$B = \mathbf{let}\ [\gamma, z:(\beta \approx \gamma \wedge set\,\delta\gamma)] = y\ \mathbf{in}\ [\gamma, \langle A_3 x(\pi_1(z)), \pi_2(z) \rangle]_{\alpha \,\epsilon\, \delta},$$

where A_3 is as in the proof of (ii). We have

$$\Gamma, \alpha:k, \beta:k, x:(\alpha \approx \beta), \delta:k, y:(\beta \,\epsilon\, \delta) \vdash B : (\alpha \,\epsilon\, \delta).$$

We leave to the reader to complete the proof of (iii), and we proceed to (iv). First observe that (i) implies

$$\Gamma, \alpha:k, \beta:k, \gamma:k \vdash (\beta \approx \gamma \wedge set\,\alpha\gamma) \to (\beta \approx \gamma' \wedge set\,\alpha'\gamma'),$$

and thus

$$\Gamma \vdash \exists \gamma^k (\beta \approx \gamma \wedge set\,\alpha\gamma) \to \exists \gamma^k (\beta \approx \gamma \wedge set\,\alpha'\gamma).$$

To complete the proof of (iv) we also need the converse. For this, observe that $set\,\alpha'\gamma = set\,(el(set\,\alpha))\gamma$ is equivalent to $\exists \delta^k (set\,\alpha\delta \wedge \gamma =^k \delta')$. Thus the assumption $\beta \approx \gamma \wedge set\,\alpha'\gamma$ implies $\exists \delta^k (\beta \approx \delta \wedge set\,\alpha\delta)$ (because $\beta \approx \gamma =^k \delta'$ implies $\beta \approx \delta' \approx \delta$).

We can now show (v). Take $R = \lambda \alpha^k \beta^k \forall \gamma^k (\gamma \,\epsilon\, \alpha \leftrightarrow \gamma \,\epsilon\, \beta)$. From (iv) we know that $\Gamma, \alpha:k \vdash R\alpha\alpha'$, so it suffices to construct an inhabitant of $Eqv(R)$, and apply the definition of $\approx$. For this, one has to prove that R is a reflexive, symmetric and transitive relation. Details are left to the reader.

The "only if" part of (vi) amounts to proving that the assumptions $\beta \approx \gamma$, and $set \Delta \gamma$, contradict $\beta \in \beta$. But $set \Delta \gamma$ implies $\exists \delta^k (\gamma \approx \delta' \wedge \delta \not\in \delta)$, so it suffices to infer $\bot$ from the set $\{\beta \approx \gamma, \beta \in \beta, \gamma \approx \delta', \delta \not\in \delta\}$. This is an immediate consequence of (iii) and (v), because $\beta \approx \gamma \approx \delta'$ implies $\beta \approx \delta$.

For the "if" part, note that $\beta \not\in \beta$ implies $\exists \gamma^k (\gamma \not\in \gamma \wedge \beta' =^k \gamma')$, which is equivalent to $set \Delta \beta'$. Since $\beta \approx \beta'$, we conclude with $\beta \in \Delta$. □

14.6.5. THEOREM (Girard's paradox). *Systems* **U** *and* **U**$^-$ *are inconsistent.*

PROOF. From Lemma 14.6.4 we know that it suffices to define a kind κ and terms $el : \mathcal{P}(\kappa) \to \kappa$ and $set : \kappa \to \mathcal{P}(\kappa)$, so that the type

$$\forall X^{\mathcal{P}(\kappa)} \forall \alpha^\kappa (set(elX)\alpha \leftrightarrow \exists \beta^\kappa (X\beta \wedge \alpha =^\kappa el(set\beta)))$$

is inhabited. The reader who worked out Exercise 12.8 should now guess that our definition of κ follows the pattern of an inductive type:

$$\kappa = \forall k^\square (\forall \iota^\square ((\iota \to k) \to \mathcal{P}(\iota) \to k) \to k).$$

The functions el and set are defined thus:

$$el = \lambda X^{\mathcal{P}(\kappa)} \lambda k^\square \lambda y^{\forall \iota : \square . (\iota \to k) \to \mathcal{P}(\iota) \to k}. y\kappa(\lambda \beta^\kappa. \beta ky) X;$$
$$set = \lambda \beta^\kappa. \beta(\mathcal{P}(\kappa))\psi,$$

where ψ is a constructor of kind $\forall \iota^\square ((\iota \to \mathcal{P}(\kappa)) \to \mathcal{P}(\iota) \to \mathcal{P}(\kappa))$, such that, informally speaking, $\psi \iota f X$ is the image of $X \subseteq \iota$ under $el \circ f$:

$$\psi = \lambda \iota^\square \lambda f^{\iota \to \mathcal{P}(\kappa)} \lambda X^{\mathcal{P}(\iota)} \lambda \alpha^\kappa. \exists \beta^\iota (X\beta \wedge \alpha =^\kappa el(f\beta)).$$

We now have (omitting type annotations for clarity):

$set(elX)\alpha =_\beta (\lambda \beta. \beta(\mathcal{P}(\kappa))\psi)(elX)\alpha =_\beta elX(\mathcal{P}(\kappa))\psi\alpha$
$=_\beta \psi\kappa(\lambda\beta. \beta(\mathcal{P}(\kappa))\psi)X\alpha = \psi\kappa\, set X\alpha =_\beta \exists \beta(X\beta \wedge \alpha =^\kappa el(set\beta))$,

so that we have in fact shown more than needed: beta-equality rather than logical equivalence. □

14.6.6. COROLLARY ("Type" is not a type). *Let $\lambda *$ be the PTS with only one sort $*$ and with the axiom $* : *$ and rule $(*, *)$. Then $\lambda *$ is inconsistent.*

PROOF. Take a term M such that $\vdash M : \bot$ in system **U**$^-$, and replace all occurrences of $\square$ and $\triangle$ in M by $*$. The resulting term has type $\bot$ in $\lambda *$. □

Normalization is often used to prove consistency of various systems. Equivalently, we can say that inconsistency unavoidably causes divergence:

14.6.7. PROPOSITION. *If a PTS inhabits $\Pi p:*p$ then it is not normalizing.*

PROOF. There is no closed normal form of type $\Pi p:*p$ (i.e. of type $\bot$). □

14.7. Notes

The Calculus of Constructions was proposed in 1985 by T. Coquand and studied and implemented in CAML by Coquand and Huet [84, 90]. The name obviously bears historical connotations and alludes not only to the BHK interpretation of intuitionistic logic and Howard's seminal [247] but also refers to the long standing effort towards exhibiting the computational contents of proofs. Various "calculi of constructions" or "theories of constructions" by Kreisel (1962), Goodman (1968), Scott (1968) and others, appeared earlier in the literature, but it seems that now the name is finally won by system λC, because of its generality and flexibility. As mentioned in [84], the inspiration for λC came from the work of Martin-Löf, Girard, and de Bruijn.

The eight basic extensions of the system $\lambda_\to$ were all essentially known to Curry [241, p. 233]. But the presentation of λC in the form of a cube is (of course) due to Barendregt. It was conceived in the late 80's and presented in [27, 32]. The notion of a PTS (initially called "generalized type system") seems to be independently invented by Berardi and Terlouw, see [27, 32, 42, 475]. The strong normalization proof in Section 14.4 is from [175]. Other SN proofs can be found in [32, 88, 173, 336, 475].

It is a natural idea to generalize the notion of a pure type system to Curry-style. A result similar to Proposition 13.4.3, due to Paulin [386], states that all pure λ-terms typable in the Calculus of Constructions are typable already in system $\mathbf{F}_\omega$ (see Exercise 14.6). Curry-style systems corresponding to all the calculi in Barendregt's cube are discussed in [23]. It turns out that the combination of polymorphism and dependent types makes the situation less transparent than in the monomorphic case. While every judgement $\Gamma \vdash M : \tau$ of λC erases to a correct type assignment judgement $|\Gamma| \vdash |M| : |\tau|$ (where $|\ |$ denotes an appropriate type-erasing map), the converse is not true: There are derivable Curry-style judgements, not corresponding to any typed λC judgement (Exercise 14.7).

This incompatibility between typing à la Church and à la Curry is not so surprising if we realize the following. A Church-style variant of a typed calculus may have several forms of abstraction $\lambda x : A. B$, always with a "domain" A attached (sometimes the domain is implicit, like $*$ in case of the capital Λ of system $\mathbf{F}$). In contrast, in the pure λ-terms used in Curry style only one form of abstraction can occur (with no domain attached). There is, however, a middle path, in which we have the same forms of abstraction as in the Church-style variant, but where we do not attach any entity to the abstractions. Such *domain-free* systems were studied by Barthe and Sørensen [37].

The inconsistency of system $\mathbf{U}$ was first proved by Girard in [180]. The consequence of Girard's result was that an initial version of Martin-Löf's theory of types turned out to be inconsistent, because it involved the notion of "strong sum", powerful enough to interpret the "type is a type" paradigm. A discussion of Girard's paradox can be found in Coquand's paper [83], and Barendregt's chapter [32].

Meyer and Reinhold conjectured in [340] that the axiom $* : *$ should make it possible to define a fixed-point combinator. The conjecture seems to be quite justified intuitively. In fact, already in Prawitz's old book [403, p. 95] one finds an example of a non-normalizing proof corresponding to Russell's paradox. But the problem raised in [340] remains open. Howe [248] used Girard's paradox to obtain *looping combinators*: a sequence of terms Y_n such that $Y_n F$ reduces to $F(Y_{n+1} F)$.

A generic construction of looping combinators, applicable to various inconsistent systems, is given in [89]. Most of these terms were in fact quite complex: Howe's example of [248] (formalized in Nuprl) does not fit in the paper, and the Automath example [32] takes 3 full pages in a font so tiny that it is barely readable. Shorter examples and simpler proofs can be found in the paper [251] by Hurkens. Our proof of Theorem 14.6.5 is based on that work.

Systems of dependent types based on the Calculus of Constructions, its modifications and extensions, find their applications in computer-assisted reasoning. One important example is system Coq [44]. The survey articles [33, 35] are recommended for further reading on the subject.

THE BARENDREGT-GEUVERS-KLOP CONJECTURE. We conclude by stating a major open problem in the field. As we have seen, all the systems of the λ-cube are strongly normalizing, but certain PTS, like $\lambda*$ and $\mathbf{U}$, are not. But these systems are not even "weakly" normalizing: There are expressions without normal forms. It is conjectured [36, 176] that it is always the case.

14.7.1. CONJECTURE (Barendregt, Geuvers, Klop). *If all terms of a PTS have normal forms then the system has the strong normalization property.*

14.8. Exercises

14.1. Formulate and prove a generation lemma for PTS, similar to Lemma 13.2.9. *Hint:* This requires proving a number of lemmas in the spirit of Section 13.2.

14.2. Show that if $\Gamma \vdash A : B$ in a PTS then $\Gamma \vdash B : s$ or B is a sort. If A is an abstraction or an application then $\Gamma \vdash B : s$.

14.3. Prove subject reduction for arbitrary PTS. *Hint:* Use Exercises 14.1 and 14.2.

14.4. For a raw expression E of λP, by $\overline{E}$ we denote an expression of compact λP obtained from E by replacing all $\forall$ by Π, and ignoring the difference between object and constructor variables. Define $\overline{\Gamma}$ accordingly. Prove that

  (i) If $\Gamma \vdash E : F$ in λP then $\overline{\Gamma} \vdash \overline{E} : \overline{F}$ in compact λP.

  (ii) If $\overline{\Gamma} \vdash A : B$ in compact λP, then there are Δ, E and F, with $\overline{\Delta} = \Gamma$, $\overline{E} = A$, $\overline{F} = B$, and such that $\Delta \vdash E : F$ in λP.

14.5. There is an apparent difference between the definition of system $\mathbf{F}$ given in Chapter 11 and the definition of system $\mathbf{F}$ as a PTS with rules $(*, *)$ and $(\square, *)$. The rule $(\square, *)$ makes it possible to define types of the form $\forall \alpha{:}\kappa.\sigma$, for an arbitrary kind κ. Explain why the difference is not essential.

14.6. The contracting map $\flat$ and the type-erasure map $|\;|$ from Chapter 13 naturally extend to raw terms of λC. Therefore one can adjust Definition 13.4.1 and consider pure λ-terms *typable* in λC. Prove an analogue of Proposition 13.4.3: If a term is typable in λC then it is typable in $\mathbf{F}_\omega$.

14.7. (After [23].) Define a Curry-style λC with pure λ-terms occurring in types (instead of terms of Church-style λC). Show that Proposition 11.4.2(i) does not generalize to λC. That is, show a derivable Curry-style judgement that cannot be obtained by erasing types in a judgement of λC. *Hint:* First do Exercise 11.18.

14.8. Define $\exists x^p \varphi$ as $\forall q^* (\forall x^p (\varphi \to q) \to q)$, where q is not free in φ. Find a term M such that $p : *,\; R : p \to * \vdash M : \exists x^p (Rx) \to p$ holds in λC.

14.9. (Geuvers) Show that $p : *,\; R : p \to *,\; q : * \vdash \exists x^p (Rx) \to \forall x^p q \to q$ in λC.

14.8. EXERCISES

14.10. Let $W(n)$ denote the following induction scheme [502]:

$$\forall p^* \forall X^{p \to *} \forall s^{p \to p} [\forall x^p (Xx \to X(sx)) \to \forall x^p (Xx \to X(nsx))].$$

Prove that $W(\mathbf{c}_n)$ is inhabited in $\lambda \mathrm{C}$, for every Church numeral $\mathbf{c}_n$.

14.11. (Geuvers) Find a term M such that in the Calculus of Constructions

$$A : *, \; a : A, \; p : *, \; P : * \to * \vdash M : P(\forall x^A \, p) \to \exists r^* \, P(r \to p),$$

where $\exists r^* \varphi$ abbreviates $\forall q^* (\forall r^* (\varphi \to q) \to q)$. Then replace $A : *$ by $A : \mathbf{Set}$ and replace all other occurrences of $*$ in the judgement above by $\mathbf{Prop}$. Can you derive the resulting judgement in $\lambda \mathrm{PRED}_\omega$?

14.12. Let A be a set, and let $el : \mathcal{P}(A) \to A$ and $set : A \to \mathcal{P}(A)$ be any functions. Show that there is a subset X of A such that $set(el(X)) \neq \{el(set(x)) \mid x \in X\}$.

14.13. For the "inductive kind"

$$\kappa = \forall k^\square (\forall \iota^\square ((\iota \to k) \to \mathcal{P}(\iota) \to k) \to k),$$

used in the proof of Theorem 14.6.5, define the operators in and It similar to those of Section 12.3. Then define el and set using these operators.

14.14. Which typed systems discussed in this book are *not* PTS's?

14.15. (Gajda) We have noted in Chapters 3 and 11 that type checking problems are usually "harder" than typability problems, because typability easily reduces to type checking: To find out if M is typable, ask if $x : p \vdash \mathbf{K}x(\lambda \vec{y} M) : p$. Does this reduction work for every PTS?

Appendix A

Mathematical background

This final chapter collects various definitions and facts used in the book. Part of the material is included here to help the reader recall some less commonly used notions, other issues are mentioned just to fix notation and terminology. Of course, not everything could be included, and some choice had to be made. For instance, we omitted definitions related to Turing Machines, a part of standard curriculum, because we only refer to Turing machines on a fairly high level of abstraction. On the other hand, we included Kleene's definition of recursive functions, as we believe it may be less familar to many computer science students. In general, most of the contents of this chapter is probably well known to the reader. We recommend to consult it only in case of doubt, and this is why we placed it at the end.

A.1. Set theory

Our notation is quite standard. For instance we write $\cup, \cap, -, \varnothing$ for set-theoretic union, intersection, difference and empty set, respectively. The sum (resp. intersection) of a family $\mathcal{A}$ of sets is written as $\bigcup \mathcal{A}$ (resp. $\bigcap \mathcal{A}$), e.g. $\bigcup \{X, Y\} = X \cup Y$. The symbols $\in$ and $\subseteq$ denote set membership and set inclusion. The powerset of A (the set of all subsets of A) is denoted by $\mathcal{P}(A)$, and we write $\{a \in A \mid W(a)\}$ for the set of all elements of A satisfying $W(a)$. Also $\{a, b, c\}$ is the set of (at most three) elements a, b and c. Ordered pairs are written as $\langle a, b \rangle$ and the Cartesian product of A and B is $A \times B$. The Cartesian power $A \times \cdots \times A$ with n coordinates is abbreviated as A^n.

We often write $\vec{a}$ to denote a sequence of the form $a_1, \ldots, a_n$. The vector notation and the symbol $\in$ are used quite liberally, for instance we may write $a \in \vec{a}$ when a is one of $a_1, \ldots, a_n$, or $\vec{a} \in A$ when all $a_1, \ldots, a_n$ are in A. At various occasions we also find it convenient to confuse a sequence $a_1, \ldots, a_n$ with the set $\{a_1, \ldots, a_n\}$.

We have the following notation for the commonly used sets of numbers: $\mathbb{N}$ is the set of natural numbers, including zero, $\mathbb{Q}$ is the set of rationals and $\mathbb{R}$ is the set of reals. We write (a, b) for an open segment of $\mathbb{R}$, i.e. $(a, b) = \{x \in \mathbb{R} \mid a < x < b\}$. The set of all words (strings) over an alphabet A is denoted by A^*.

RELATIONS AND FUNCTIONS. Relations are identified with subsets of the appropriate Cartesian products. For instance, a binary relation r on A is a subset of A^2.

Such a relation is:
- *reflexive* iff $\langle a, a \rangle \in r$, for all $a \in A$;
- *symmetric* iff $\langle a, b \rangle \in r$ implies $\langle b, a \rangle \in r$;
- *anti-symmetric* iff $\langle a, b \rangle, \langle b, a \rangle \in r$ implies $a = b$;
- *transitive* iff $\langle a, b \rangle, \langle b, c \rangle \in r$ implies $\langle a, c \rangle \in r$.

The *transitive closure* of a relation r on A is the smallest (with respect to inclusion) transitive relation containing r. Similar terminology applies for other properties of relations. An *equivalence relation* is one which is reflexive, transitive and symmetric. An *equivalence class* of an equivalence relation r in A, determined by $a \in A$, is the set $[a]_r = \{b \in A \mid \langle a, b \rangle \in r\}$. Any element of an equivalence class can be called a *representative* of the class. The set of all equivalence classes is called the *quotient set* and denoted by $A/_r$.

A function $f : A \to B$ is defined as a relation $f \subseteq A \times B$, such that for all $a \in A$ there is exactly one $b \in B$ with $\langle a, b \rangle \in f$. The set A is then called the *domain* of f and denoted by $\text{dom} f$. The *range* of f is the set $\text{rg} f = \{b \in B \mid \exists a \in A (f(a) = b)\}$. If $f : A \to B$ and $C \subseteq A$ then the *image* of C under f is the set $f(C) = \{f(a) \mid a \in C\}$. Iterated composition of a function $f : A \to A$ with itself is denoted by f^i. A *partial function from A to B* is an arbitrary function $f : A' \to B$, where $A' \subseteq A$. If $A' = A$ we may say that f is *total*.

We often write $f(a \mapsto b)$ for a function defined as follows:

$$f(a \mapsto b)(x) = \begin{cases} b, & \text{if } x = a; \\ f(x), & \text{otherwise.} \end{cases}$$

The notion of a finite Cartesian product is generalized as follows. Given a family of sets A_t for $t \in T$ we define $\Pi_{t \in T} A_t$ as the set of all functions f such that $\text{dom} f = T$ and $f(t) \in A_t$ for all $t \in T$. If all A_t are equal to some A then $\Pi_{t \in T} A_t$ becomes the set of all functions from T to A, often denoted by A^T. Observe that $A^\varnothing$ (which is identified with A^0) is a one-element set. Thus a nullary function $f : A^0 \to B$ is naturally identified with an element of B (a constant).

As we said, relations are typically understood as sets of pairs. But it is sometimes convenient to identify a relation r with a function ranging over $\{0, 1\}$ so that we may write $r(\vec{a}) = 1$ iff $\vec{a} \in r$. Observe that this interpretation of relations suggests the following identification: a nullary relation on an arbitrary set is simply a (classical) truth value.

MULTISETS. A fundamental property of sets is *extensionality*: If sets A and B have the same elements then they are equal (A and B are names of the same object). But sometimes it is convenient to consider collections where a member can occur many times. For this purpose one introduces the notion of a *multiset*, which is like a set with repetitions of elements. Formally, a multiset of elements of a set Ψ is defined as a function $\Gamma : \Psi \to \mathbb{N}$. We say that $\psi \in \Psi$ is an *element* of Γ, when $\Gamma(\psi) > 0$. If $\Gamma(\psi) = n$ then we say that ψ occurs n times in Γ. The notation $\varnothing$, $\{\psi, \psi\}$, etc. applies to multisets with the obvious meaning. The union of two multisets Γ' and Γ'' is defined by the equation

$$(\Gamma' \sqcup \Gamma'')(\psi) = \Gamma'(\psi) + \Gamma''(\psi).$$

More about multisets can be found e.g. in Chapter 2 of [21].

ORDERED SETS. A binary relation $\leq$ on A is a *partial order* iff it is transitive, reflexive and anti-symmetric. The pair $\langle A, \leq \rangle$ is then called a *partially ordered set*.

A.1. SET THEORY

This terminology is used quite liberally, e.g. one can say that A itself is a partially ordered set or a partial order. A partial order is called *total* or *linear* if each two elements a, b are *comparable*, i.e. either $a \leq b$ or $b \leq a$ always holds.

Observe that each family of sets $\mathcal{F}$ is partially ordered by inclusion. We adopt the convention that whenever we talk about a family of sets, and we mention any order-related notions (e.g. lower and upper bounds defined below) we always mean the partial order given by set inclusion.

An element $a \in A$ is *maximal* (resp. *minimal*) in A iff $a \leq b$ (resp. $a \geq b$) implies $a = b$, for all $b \in A$. Observe that the top (resp. bottom) element is always maximal (resp. minimal), but not conversely.

If A is a partial order then a subset $B \subseteq A$ is *upward-closed* (resp. *downward-closed*) iff $a \geq b \in B$ (resp. $a \leq b \in B$) implies $a \in B$. An element a of a partially ordered set A is an *upper* (resp. *lower*) *bound* of $B \subseteq A$ iff $a \geq b$ (resp. $a \leq b$) for all $b \in B$. We say that a is the *supremum* of B (and we write $a = \sup_A B$), when it is the *least upper bound* of B, i.e.

- a is an upper bound of B;
- $a \leq b$, for any other upper bound b of B.

Observe that there is at most one element satisfying these conditions. Dually, we define the *infimum* of a subset B (written $\inf_A B$) as the *greatest lower bound* of B. If A is known from the context, we may omit the subscript $_A$ in the notation $\sup_A B$ and $\inf_A B$. But in some cases the subscript is important. Observe for instance that if $C \subseteq B \subseteq A$ then $\sup_B C \geq \sup_A C$ (assuming both sides exist) but not necessarily $\sup_B C = \sup_A C$.

The reader can easily verify that $\sup_A \varnothing$ (resp. $\inf_A \varnothing$), if it exists, must be the bottom (resp. top) element of A. Note also that if $\mathcal{A} \subseteq \mathcal{P}(X)$ then $\sup_{\mathcal{P}(X)} \mathcal{A} = \bigcup \mathcal{A}$ and $\inf_{\mathcal{P}(X)} \mathcal{A} = \bigcap \mathcal{A}$.

The following result is often useful, but the proof of it (based on the axiom of choice, see e.g. [212, 290]) is highly non-constructive.

A.1.1. LEMMA (Kuratowski, Zorn). *Let A be partially ordered. If every chain (linearly ordered subset) has an upper bound in A then A has a maximal element.* □

If A and B are partially ordered then a function $f : A \to B$ is *monotone* iff $a_1 \leq a_2$ implies $f(a_1) \leq f(a_2)$. The following result is sometimes called *Knaster-Tarski fixed point theorem*:

A.1.2. THEOREM. *Let A be a complete lattice (all subsets have suprema and infima) and let $f : A \to A$ be monotone. Define $a_0 = \inf\{a \in A \mid f(a) \leq a\}$. Then a_0 is the least fixed point of f. That is, $f(a_0) = a_0$ and $a_0 \leq b$, whenever $f(b) = b$.*

PROOF. Let $B = \{a \in A \mid f(a) \leq a\}$. We have $f(a_0) \leq f(b) \leq b$, for $b \in B$ and thus $f(a_0) \leq a_0$, i.e. a_0 itself is in B. But then $f(a_0) \in B$, because of monotonicity, so that also $a_0 \leq f(a_0)$. □

A.1.3. REMARK. Define $a_1 = \inf\{a \in A \mid \forall b \in A (b \leq a \to f(b) \leq a)\}$, under the assumptions of Theorem A.1.2. Then $a_1 = a_0$, i.e. we have an alternative definition of the least fixed point. Note however that monotonicity is essential. For instance, let $f : \mathcal{P}(\mathbb{N}) \to \mathcal{P}(\mathbb{N})$ be such that $f(A) = A \cup \{\min(\mathbb{N} - A)\}$ for finite A and $f(A) = A - \{\min A\}$ for infinite A. Then $a_0 = \varnothing$, $a_1 = \mathbb{N}$, and f has no fixed point. But observe that $f(a_1) \leq a_1$ holds for arbitrary f.

INDUCTION. A partially ordered set A is *well-founded* iff every non-empty subset of A has a minimal element. Observe that if A and B are well-founded then the *lexicographic* order on $A \times B$, given by $(a,b) \leq (a',b')$ iff $a < a'$, or $a = a'$ and $b \leq b'$, is also well-founded. If A is a well-founded set then properties of elements of A can be proved by induction with respect to the ordering.

A.1.4. LEMMA (Principle of induction). *Let $B \subseteq A$, where A is a well-founded set. If $\forall a \in A(\forall b \in A(b < a \rightarrow b \in B) \rightarrow a \in B)$ then $B = A$.*

PROOF. Otherwise the set $B' = \{a \in A \mid a \notin B\}$ has a minimal element a' satisfying $\forall b \in A(b < a' \rightarrow b \in B)$. □

Many proofs by "structural induction" are based on the principle above. To show a statement of the form $\forall x \in X \ (W_x)$ one assigns an element $f(x)$ of a well-founded set A to every $x \in X$. Then the problem is reduced to showing that $B = A$ where $B = \{a \in A \mid \forall x \in X(f(x) = a \rightarrow W_x)\}$.

In particular it may happen that a set is well-founded because it has been defined by induction. For instance, the transitive and reflexive closure r^* of a relation r is the least set such that $r \subseteq r^*$ and for $\langle a,b \rangle, \langle b,c \rangle \in r^*$ also $\langle a,c \rangle \in r^*$. This induces a stratification of r^* as the union of an infinite sequence of sets: $r_0 = r$, $r_{n+1} = r_n \cup \{\langle a,c \rangle \mid \langle a,b \rangle, \langle b,c \rangle \in r_n\}$, and we have a well-founded order on r^*:

$$\langle a,b \rangle \leq \langle c,d \rangle \quad \text{iff} \quad \forall n(\langle c,d \rangle \in r_n \rightarrow \langle a,b \rangle \in r_n).$$

That is, $\langle a,b \rangle \leq \langle c,d \rangle$ holds when $\langle a,b \rangle$ falls into the union r^* at an earlier (or the same) stage than $\langle c,d \rangle$.

The above justifies a proof method called *induction with respect to the definition of r^**. To show that a property $P(a,b)$ holds for each pair $\langle a,b \rangle \in r^*$ one first proves $P(a,b)$ for all $\langle a,b \rangle \in r$ and then makes the induction step by showing that $P(a,b)$ and $P(b,c)$ implies $P(a,c)$. Similar machinery occurs whenever one talks about induction with respect to various inductive definitions.

TOPOLOGICAL SPACES. A *topological space* is a pair $\mathcal{T} = \langle T, \mathcal{O} \rangle$, consisting of a set T together with a family $\mathcal{O} \subseteq \mathcal{P}(T)$ of sets, called *open sets* of T, which is required to satisfy the following conditions:

- $\emptyset, T \in \mathcal{O}$;

- $A, B \in \mathcal{O}$ implies $A \cap B \in \mathcal{O}$;

- $\mathcal{R} \subseteq \mathcal{O}$ implies $\bigcup \mathcal{R} \in \mathcal{O}$.

That is, arbitrary unions and finite intersections of open sets must be open. The *interior* of a set $A \subseteq T$, denoted by $\text{Int}(A)$ is the largest open set contained in A (i.e. the union of all open subsets of A). The complement of an open set is called a *closed set*. The *closure* of A is the smallest closed set containing A. For $\mathcal{T} = \langle T, \mathcal{O} \rangle$ we use the notation $\mathcal{O} = \mathcal{O}(\mathcal{T})$, and we often make no distinction between $\mathcal{T}$ and T.

Many topological spaces can be defined using a measure of *distance* between points. Such spaces are called *metric spaces*. Examples of metric spaces are the set $\mathbb{Q}$ of rationals, the set $\mathbb{R}$ of reals, as well as all Euclidean spaces $\mathbb{R}^k$. In all these cases the topology is defined according to a common pattern. Let $\varrho(a,b)$ denote the distance between points a and b. A set A is *open* iff for every $a \in A$ there is an $r > 0$ with $\{b \mid \varrho(a,b) < r\} \subseteq A$. In the space of real numbers, the distance between x and y is $|x-y|$. We say that $a \in \mathbb{R}$ is an *accumulation point* of $A \subseteq \mathbb{R}$ when a is a limit of a sequence x_n of elements of A, and all x_n are different from a.

A.2. Algebra and unification

A *signature* is a family of function and relation symbols, each with a fixed arity (possibly zero). Nullary symbols are called *constants*. If not stated otherwise, signatures are assumed to be finite.

Assume a fixed infinite set $\{a_0, a_1, \ldots\}$ of *individual variables*. An *algebraic term*[1] over a signature Σ, or just *term*, is either an *individual variable*, or an expression of the form $(ft_1 \ldots t_n)$, where f is an n-ary function symbol, and $t_1, \ldots, t_n$ are terms. We usually omit outermost parentheses when writing terms. The symbol $FV(t)$ denotes the set of all individual variables occurring in a term t. A term is *closed* iff $FV(t) = \varnothing$.

The formal definition of an algebraic term involves a prefix application of function symbols. Of course, there is a tradition to write some binary function symbols in the infix style, and we normally do it this way. Our most beloved signature is the one that has the (infix) arrow as the only symbol. It is not difficult to see that algebraic terms over this signature can be identified with simple types, or with implicational formulas if you prefer.

In this section we consider only *algebraic signatures*, i.e., signatures without relation symbols. An *algebra* over such a signature $\Sigma = \{f_1, \ldots, f_n\}$, where the arity of each f_i is r_i, is a set A together with functions $f_i^A : A^{r_i} \to A$, i.e. a system $\mathcal{A} = \langle A, f_1^A, \ldots, f_n^A \rangle$. (We often forget about the notational distinction between $\mathcal{A}$ and its domain A.) A subset B of A is a *subalgebra* of A iff B is closed under all operations. That is, whenever $\vec{b} \in B$ then also $f_i^A(\vec{b}) \in B$, for all $i = 1, \ldots, n$ (assuming the appropriate length of $\vec{b}$). If $\langle A, f_1^A, \ldots, f_n^A \rangle$ and $\langle B, f_1^B, \ldots, f_n^B \rangle$ are two algebras of the same signature then a function $h : A \to B$ satisfying the condition $h(f_i^A(\vec{a})) = f_i^B(h(\vec{a}))$ is a *homomorphism*. If, in addition, h is a bijection, and the converse function is also a homomorphism then h is called an *isomorphism* and the two algebras are *isomorphic* to each other.

UNIFICATION. A *substitution* is a function from variables to terms which is identity almost everywhere. Such a function S is extended to a function from terms to terms by $S(ft_1 \ldots t_n) = fS(t_1) \cdots S(t_n)$ (i.e. $S(f) = f$, when f is a constant). In case $S(a) = s$ and $S(b) = b$, for all other variables b, one may write $t[a := s]$ for $S(t)$.

If P and R are substitutions then $P \circ R$ is defined by $(P \circ R)(x) = P(R(x))$. A substitution S is an *instance* of another substitution R (written $R \leq S$) iff $S = P \circ R$, for some substitution P.

An *equation* is a pair of terms, written "$t = u$". A *system of equations* is a finite set of equations. Variables occurring in a system of equations are called *unknowns*. A substitution S is a solution of an equation "$t = u$" iff $S(t) = S(u)$ (meaning that $S(t)$ and $S(u)$ is the same term). It is a solution of a system E of equations iff it is a solution of all equations in E. A solution R of a system E is *principal* iff the following equivalence holds for all substitutions S:

$$S \text{ is a solution of } E \quad \text{iff} \quad R \leq S.$$

Thus, for instance, the equation $f(gab)a = fc(fbb)$ has a principal solution S, such that $S(a) = fbb$, $S(b) = b$ and $S(c) = g(fbb)b$ (and many other solutions too), while the equation $f(gab)h = fc(fbb)$, where h is a constant, has no solution. This

---

[1] Do not confuse algebraic terms with lambda-terms, and do not confuse individual variables with propositional or lambda-variables.

is because no substitution can turn fbb into h, or vice versa. Another example with no solution is $f(gab)a = fa(fbb)$, but this time the reason is different: if S were a solution then $S(a)$ would be a proper subterm of itself.

The problem of determining whether a given system of equations has a solution is called the *unification problem*. It is not difficult to see that there is no loss of generality in considering single equations rather than systems of equations and just one binary function symbol rather than arbitrary signatures (Exercise 3.7). The unification algorithm of Robinson [418] can be found in various textbooks, for instance [21, 349, 474]. We have:

A.2.1. LEMMA. *If a system of equations has a solution then it has a principal one, of size at most exponential in the size of the input.*

A.3. Partial recursive functions

A.3.1. DEFINITION. Let $g : \mathbb{N}^k \to \mathbb{N}$ and $h_1, \ldots, h_k : \mathbb{N}^m \to \mathbb{N}$ be partial functions. The *composition*

$$f(n_1, \ldots, n_m) = g(h_1(n_1, \ldots, n_m), \ldots, h_k(n_1, \ldots, n_m))$$

is defined iff:

- $h_i(n_1, \ldots, n_m)$ is defined, for all $i = 1, \ldots, k$;
- $g(a_1, \ldots, a_k)$ is defined, where $a_i = h_i(n_1, \ldots, n_m)$, for $i = 1, \ldots, k$.

A.3.2. DEFINITION. Let $g : \mathbb{N}^m \to \mathbb{N}$ and $h : \mathbb{N}^{m+2} \to \mathbb{N}$ be partial functions. We say that a partial function $f : \mathbb{N}^{m+1} \to \mathbb{N}$ is defined from g and h by *primitive recursion* iff for all $n, n_1, \ldots, n_m \in \mathbb{N}$,

$$\begin{aligned} f(0, n_1, \ldots, n_m) &= g(n_1, \ldots, n_m); \\ f(n+1, n_1, \ldots, n_m) &= h(f(n, n_1, \ldots, n_m), n, n_1, \ldots, n_m). \end{aligned}$$

The equations above are understood so that the left-hand-side is defined if and only if the right-hand-side is defined.

A.3.3. DEFINITION. If $g : \mathbb{N}^{m+1} \to \mathbb{N}$ is a partial function, then the notation $\mu n[g(n_1, \ldots, n_m, n) = 0]$ stands for the least number n such that

- $g(n_1, \ldots, n_m, n) = 0$.
- $g(n_1, \ldots, n_m, n')$ is defined and not equal to 0, for all $n' \in \{0, \ldots, n-1\}$.

If there is no such n then $\mu n[g(n_1, \ldots, n_m, n) = 0]$ is undefined. The partial function $f : \mathbb{N}^m \to \mathbb{N}$ given by

$$f(n_1, \ldots, n_m) = \mu n[g(n_1, \ldots, n_m, n) = 0]$$

is said to be defined from g by *minimization*.

The following introduces an important subclass of recursive functions.

A.3.4. DEFINITION. The class of *primitive recursive functions* is the smallest class of total numeric functions containing the *initial functions*

(i) *projections:* $p_i^m(n_1, \ldots, n_m) = n_i$ for all $1 \leq i \leq m$;
(ii) *successor:* $s(n) = n + 1$;
(iii) *zero:* $z(n) = 0$.

A.3. PARTIAL RECURSIVE FUNCTIONS

and closed under composition and primitive recursion. A predicate (set) $P \subseteq \mathbb{N}^k$ is *primitive recursive* iff its characteristic function is primitive recursive.

Most computable functions of everyday use are in fact primitive recursive. (An exception is for instance the *Ackermann function* of Exercise 10.5.) In particular, the following bijective *pairing function*

$$p(m,n) = \frac{(m+n)(m+n+1)}{2} + m,$$

is primitive recursive, as well as the left and right converse functions ℓ and r, satisfying $n = p(\ell(n), r(n))$ for all n. There are more sophisticated examples of primitive recursive functions and predicates. One is as follows. Consider a fixed enumeration M_k of all Turing Machines that are capable of processing an input $\vec{n} \in \mathbb{N}$ and returning an output $m \in \mathbb{N}$. (If a partial function f is computed this way by a Turing machine, we call it *Turing computable* and we say that k is a *Gödel number* of f.) Define *Kleene's T predicate*:

$T(k, \vec{n}, m)$ iff M_k halts on input $\vec{n}$ in $r(m)$ steps with output $\ell(m)$.

The proof of the following can be found in [109, 269, 337].

A.3.5. LEMMA. *The predicate $T(k, \vec{n}, m)$ is primitive recursive.*

A.3.6. DEFINITION.

(i) The class of *partial recursive functions* is the smallest class of partial functions containing the initial functions and closed under composition, primitive recursion and minimization.
(ii) A set $A \subseteq \mathbb{N}^k$ is *recursively enumerable* iff it is the domain of some partial recursive function.
(iii) A *recursive function* is a partial recursive function that is total.
(iv) A set $A \subseteq \mathbb{N}^k$ is *recursive* iff its characteristic function is recursive.

It is well-known that (partial) recursive functions coincide with (partial) Turing-computable functions, see e.g. [337].

When we talk about a recursive or partially recursive *function* we often mean a particular algorithm (Turing machine) computing this function. The following notation follows this convention.

A.3.7. NOTATION. Let f be a partial recursive function computed by the machine M_k. Then the characteristic function of the binary predicate $T(k, \cdot, \cdot)$ is denoted by t_f, that is,

$t_f(\vec{n}, m) = 0$ iff M_k halts on input $\vec{n}$ in $r(m)$ steps with output $\ell(m)$.

Clearly, t_f is primitive reccursive, and we have the following result:

A.3.8. THEOREM (Kleene's normal form). *Every partial recursive function f can be written as*

$$f(\vec{n}) = \ell(\mu y[t_f(\vec{n}, y) = 0]),$$

where t_f and ℓ are primitive recursive functions.

A.3.9. COROLLARY. *A set $B \subseteq \mathbb{N}$ is recursively enumerable if and only if there is a primitive recursive set $A \subseteq \mathbb{N}^2$ such that*

$n \in B$ iff   *there exists m with $(n, m) \in A$.*

A.4. Decision problems

Of course not every recursively enumerable set is recursive. For instance, the set $\{(k,n) \in \mathbb{N} \mid M_k \text{ halts on input } n\}$ is not, and we usually express this fact by saying that "the halting problem for Turing machines is undecidable". Formally, a *decision problem* is defined as an arbitrary set of numbers, or more generally (and often more conveniently) an arbitrary set of words over a fixed alphabet. Observe that a numerical input is always represented as a word, and a word can easily be encoded as a number. A problem P is *decidable* iff the set P is recursive. Otherwise it is *undecidable*. In practice, the notion of a "decision problem" is used in a more relaxed way. For instance, each of the following questions is a decision problem:

Does a given Turing machine halt on a given input?

Is a given first-order formula a classical tautology?

Here, the input data is not an arbitrary number or an arbitrary word, but a certain finite object (a machine and an input, a formula). This is not a problem, because we can encode Turing machines as words or numbers. A formula is a word itself, so we do not even need to encode it. But not every word is a formula, and a more adequate formulation of the second problem is

Is a given word a well-formed formula that is a classical tautology?

It does not really matter which question we actually ask, because there is a simple algorithm to decide whether a given word is a formula. But consider this one:

Is a given first-order classical tautology an intuitionistic tautology?

The question whether an input word is a correct instance of this problem is no longer trivial. In fact, it is undecidable. There is no effective encoding of first-order classical tautologies as all natural numbers. Thus the "decision problem" above must be classified as ill-formed according to our definition.

A.4.1. EXAMPLE. Consider the following "decision problem": Given a Turing machine which halts on input zero, is the result of the computation also equal to zero? There is a natural algorithm to solve it. Just run the machine and check the output. This algorithm is guaranteed to terminate because... the difficulty is hidden in the definition of legal inputs. To see that the "termination property" of our algorithm is of no actual value, observe the following: *For every computable function f, every solution of the problem above requires more than $f(n)$ steps on some inputs of size n.* In other words, our "decidable" problem has no time complexity!

Indeed, otherwise consider a language L which is computable, but requires time greater than $2^{f(2n)}$, and let L be accepted by a machine M_L. For a given word w, let M^w be a machine which takes an integer input, runs M_L on w, and returns 0 if $w \in L$ and otherwise returns 1. Observe that M^w halts for every input, in particular for zero. To decide if $w \in L$ we now construct M^w (which takes time equal to the length of w plus a constant) and then we ask if the output of M^w is 0. By our assumption, this can be done in time at most proportional to $f(2n)$, contradicting the definition of L.

To show that a particular decision problem P is undecidable, we usually find another problem, which is already known to be undecidable, and we *reduce* it to P.

A.4. DECISION PROBLEMS

A.4.2. DEFINITION. A problem P_1 is *reducible* to P_2, written $P_1 \leq P_2$, iff there is a computable function f, such that

$$w \in P_1 \quad \text{iff} \quad f(w) \in P_2,$$

for all inputs w. Clearly, if $P_1 \leq P_2$ and P_1 is undecidable then so is P_2.

POST CORRESPONDENCE PROBLEM. *Post Correspondence Problem* (PCP) is the following decision problem. Given pairs of words $(w_1, v_1), \ldots, (w_n, v_n)$, find a non-empty sequence of indices $i_1, \ldots, i_k$, such that

$$w_{i_1} w_{i_2} \ldots w_{i_k} = v_{i_1} v_{i_2} \ldots v_{i_k}.$$

The sequence $i_1, i_2, \ldots, i_k$ is then called a *solution* of the problem. Also the word $w_{i_1} w_{i_2} \ldots w_{i_k}$ is often called a solution. An instance $\{(w_i, v_i) \mid i = 1, \ldots, n\}$ of Post Correspondence Problem is often displayed as:

w_1	w_2	$\ldots$	w_n
v_1	v_2	$\ldots$	v_n

For instance, 1213 is a solution of

a^2	b^2	ab^2
$a^2 b$	ba	b

because $a^2 \cdot b^2 \cdot a^2 \cdot ab^2 = a^2 b^2 a^3 b^2 = a^2 b \cdot ba \cdot a^2 b \cdot b$, and the system

$a^2 b$	a
a^2	ba^2

has no solution. It is known that PCP is undecidable, see [245, 256].

TWO-COUNTER AUTOMATA. A *two-counter automaton* $\mathcal{A} = \langle Q, q_0, q_f, I \rangle$ consists of a finite set of *states* Q, the distinguished initial state q_0 and final state q_f, and a set of instructions, each of one of the following forms:

(u$_i$) $(q : c_i := c_i + 1;\ \text{goto}\ p)$;

(d$_i$) $(q : c_i := c_i - 1;\ \text{goto}\ p)$;

(z$_i$) $(q : \text{if}\ c_i = 0\ \text{then goto}\ p\ \text{else goto}\ r)$;

where $i = 1, 2$ and $q, p, r \in Q$. A *configuration* of $\mathcal{A}$ is a triple $\langle q, m, n \rangle$ consisting of a state $q \in Q$ and numbers $m, n \in \mathbb{N}$, understood as current values of two counters c_1 and c_2. There is one *initial configuration*, namely $\langle q_0, 0, 0 \rangle$, and configurations $\langle q_f, m, n \rangle$ are called *final*.

The meaning of the instructions should be obvious. For configurations $\mathcal{C}_1, \mathcal{C}_2$, we write $\mathcal{C}_1 \to_{\mathcal{A}} \mathcal{C}_2$ when $\mathcal{C}_2$ is obtained from $\mathcal{C}_1$ in one step. (We assume that no (d$_i$) step is possible when $c_i = 0$, i.e. that the machine gets stuck when trying to subtract 1 from 0.) The notation $\twoheadrightarrow_{\mathcal{A}}$ stands for the transitive and reflexive closure of $\to_{\mathcal{A}}$, and if $\mathcal{C}_1 \twoheadrightarrow_{\mathcal{A}} \mathcal{C}_2$ then $\mathcal{C}_2$ is *reachable* from $\mathcal{C}_1$. The *halting problem* for two-counter automata is to decide if a given automaton *halts*, that is, if a final configuration is reachable from the initial one. This problem is well-known to be undecidable, see e.g. [245, 276].

A.5. Hard and complete

This section recalls a few basic notions from complexity theory. The reader is referred to standard textbooks, like [245, 256], for a more comprehensive discussion.

A.5.1. DEFINITION. The notation LOGSPACE and PTIME refers to the classes of languages (decision problems) solvable by deterministic Turing Machines in logarithmic space[2] and polynomial time, respectively. The class PSPACE consists of problems solvable by Turing Machines (deterministic or not) in polynomial space, and EXPTIME refers to deterministic time $2^{p(n)}$, where $p(n)$ can be any polynomial. We have the following inclusions: LOGSPACE $\subseteq$ PTIME $\subseteq$ PSPACE $\subseteq$ EXPTIME.

A.5.2. DEFINITION. We say that a language L_1 is *reducible* to a language L_2 *in logarithmic space* (or LOGSPACE-*reducible*) when there is a LOGSPACE-computable function f, such that

$$w \in L_1 \quad \text{iff} \quad f(w) \in L_2,$$

for all inputs w. Two languages are LOGSPACE-*equivalent* iff there are LOGSPACE-reductions each way.

That is, to decide if $w \in L_1$ one can ask if $f(w) \in L_2$, and the cost of the translation is only shipping and handling. (Observe that $\log n$ space is exactly what is needed to store the position of the machine head on an input of length n.) Note that a logarithmic space reduction is also a polynomial time reduction.

A.5.3. DEFINITION. We say that a language L is *hard* for a complexity class $\mathcal{C}$, iff every language $L' \in \mathcal{C}$ is reducible to L in logarithmic space. If we have $L \in \mathcal{C}$ in addition, then we say that L is *complete* in the class $\mathcal{C}$, or simply $\mathcal{C}$-*complete*.

UNIFICATION. As mentioned in Section A.2, the unification problem is decidable, and the solution can be of exponential size. In fact, Robinson's algorithm can be optimized to work in polynomial time, provided we only need to check whether a solution exists, and we do not need to *write it down* explicitly, cf. Exercise 3.6. Also the verification whether a *given* solution is correct can be done in polynomial time. The following is from Dwork *et al* [138].

A.5.4. THEOREM. *The unification problem is* PTIME-*complete*.

QUANTIFIED BOOLEAN FORMULAS. A standard example of a problem which is PSPACE-complete is the *classical* validity problem for *second-order propositional formulas* (see Definition 11.1.1), called also *quantified Boolean formulas* (QBF) in this context. A value of a second-order propositional formula under a binary valuation v is defined in a similar way as for ordinary propositional formulas, with the following clauses in addition:

$$[\![\forall p\, \alpha]\!]_v = \max\{[\![\alpha]\!]_{v(p \mapsto 1)}, [\![\alpha]\!]_{v(p \mapsto 0)}\};$$
$$[\![\exists p\, \alpha]\!]_v = \min\{[\![\alpha]\!]_{v(p \mapsto 1)}, [\![\alpha]\!]_{v(p \mapsto 0)}\}.$$

A second-order classical propositional *tautology* is a formula true under all valuations. The *QBF problem* is as follows:

> Is a given second-order propositional formula a tautology?

In fact it suffices to only consider *closed* formulas, where each variable is bound by a quantifier. In this case the *validity* problem above is the same as the satisfiability

---

[2] One counts only the work tapes, not the input or output tapes.

A.5. HARD AND COMPLETE

problem for QBF. A closed formula is either a tautology, or it is not satisfiable. Proofs of the following fact can be found in [245, 256].

A.5.5. THEOREM. *The QBF problem is* PSPACE-*complete.*

FURTHER READING. For more background information we recommend:

- Set theory, Boolean algebras: Davey and Priestley [119], Halmos [212, 213], Kuratowski and Mostowski [290], Monk [353].

- Turing machines, theory of computation and complexity: Boolos, Burgess and Jeffrey [53], Cutland [109], Hopcroft, Motwani and Ullman [245], Jones [256], Kozen [276], Mendelson [337].

- Classical logic: Adamowicz and Zbierski [6], Bell and Machover [39], Huth and Ryan [252], Mendelson [337], Monk [352], van Dalen [113].

- Rewriting, unification (and *all that*): Baader and Nipkow [21], Terese [474].

- Programming languages: Mitchell [349, 350], Pierce [393].

Appendix B

Solutions and hints to selected exercises

Solutions to Chapter 1

1.2. The *capture-avoiding* substitution can be defined as follows:
$$x[x := N] = N;$$
$$y[x := N] = y, \text{ if } y \neq x;$$
$$(PQ)[x := N] = P[x := N]Q[x := N];$$
$$(\lambda x P)[x := N] = \lambda x P$$
$$(\lambda y P)[x := N] = \lambda y.P[x := N], \text{ if } y \neq x, \text{ and } y \notin \text{FV}(N);$$
$$(\lambda y P)[x := N] = \lambda z.P[y := z][x := N], \text{ if } y \neq x \text{ and } y \in \text{FV}(N).$$

In the last clause, the variable z is "fresh", i.e. $z \notin \text{FV}(P) \cup \text{FV}(N) \cup \{x, y\}$. To ensure that the result of substitution is unique, one takes the least such z with respect to a fixed enumeration of all variables.

It is not difficult to see that $M[x := N]$ is always of the form $M_1[x := N]$, where $M =_\alpha M_1$ and M_1 is such that $M_1[x := N]$ is defined according to Definition 1.2.4. The remaining part of the exercise is thus an immediate consequence of Lemma 1.2.11. A (non-trivial) direct proof can be found in [241]. □

1.3. We recast the function as a relation, namely the least $r \subseteq \Lambda \times \Lambda$ such that:
- $r(x_i, N_i)$ for all i;
- $r(y, y)$ for all y not among $\vec{x}$;
- if $r(P, P')$ and $r(Q, Q')$ then $r(PQ, P'Q')$;
- if $r(P, P')$ and for all i we have $y \neq x_i$ and $y \notin \text{FV}(N_i)$ then $r(\lambda y P, \lambda y P')$.

We need to prove that for every P there is exactly one P' with $r(P, P')$. To show existence we proceed by induction on the size of M. The interesting case is when M is an abstraction. By Lemma 1.2.20(i) we can assume M is of form $\lambda y P$ where, for all i, it holds that $y \neq x_i$ and $y \notin \text{FV}(N_i)$. By the induction hypothesis there is P' with $r(P, P')$ and then also $r(\lambda y P, \lambda y P')$.

For uniqueness, first observe that $r(M, M')$ and $u \in \text{FV}(M')$ implies either $u \in \text{FV}(M)$ or $u \in \text{FV}(N_i)$, for some i. Then we establish the following by induction on the size of M:

Assume that $u, v \neq x_i$ and $u, v \notin \text{FV}(N_i)$, for all i.
If $r(M, M')$ then $r(M[v := u], M'[v := u])$.

We consider the case where M is an abstraction. Then $M = \lambda y P$, and $M' = \lambda y P'$, where y, P, P' are as stated in the last clause of the definition of r. Let z be fresh. By the induction hypothesis (twice) $r(P[y := z][v := u], P'[y := z][v := u])$. Then also $r(\lambda z.P[y := z][v := u], \lambda z.P'[y := z][v := u])$, and it is easy to see that $M[v := u] = \lambda z.P[y := z][v := u]$ and similarly with M'.

Using the property above, we show by induction on M that $r(M, M'_1)$ and $r(M, M'_2)$ implies $M'_1 = M'_2$. If M is an abstraction then $M = \lambda v P_1$ and $M'_1 = \lambda v P'_1$. Similarly, we have $M = \lambda w P_2$ and $M'_2 = \lambda w P'_2$. If $v = w$ then we apply the induction hypothesis. Otherwise we have $P_1[v := w] = P_2$, so by the preceding property, $r(P_2, P'_1[v := w])$. By the induction hypothesis, $P'_2 = P'_1[v := w]$, and thus $M'_1 = \lambda v P'_1 = \lambda w.P'_1[v := w] = \lambda w P'_2 = M'_2$. (Note that $w \notin \mathrm{FV}(P'_1)$. Indeed, otherwise $w \in \mathrm{FV}(P_1)$ or $w \in \mathrm{FV}(N_i)$ for some i. The second possibility is excluded by the definition of $r(M, M_2)$. □

1.4. By Lemma 1.2.20(iv),(v), we have $Q = N$ and $M = P[x := y]$. Therefore $M[y := N] = P[x := y][y := N] = P[x := N]$, by Lemma 1.2.20(iii). □

1.5. For the "only if" part we use Remark 1.3.7. If $M = xM_1 \ldots M_n$, where $n > 0$, then $M = (z\, M_2 \ldots M_n)[z := x\, M_1]$. The other cases are immediate. For "if" one has to prove that for any normal forms N, M, the term $N[x := yM]$ is also a normal form. This goes by induction on N using Remark 1.3.7. □

1.7. (i) true; (ii) false; (iii) false. □

1.9. (i) true; (ii) false; (iii) true; (iv) false. □

1.10. Let $M = (\lambda x_1.(\lambda x_2 z)\mathbf{I})(\mathbf{III})$ and $N = (\lambda x z)\mathbf{I}$. Then $M \overset{i}{\twoheadrightarrow}_\beta N$ (where non-parallel reduction is required on $\mathbf{III}$, and $M \Rightarrow_\beta N$, (where reduction of the head redex is required). It is clearly not the case that $M \overset{i}{\Rightarrow}_\beta N$. □

1.15. Suppose that M is a fixed-point combinator in a normal form, and take a fresh variable x. We have $x(Mx) =_\beta Mx$ and thus the normal form of Mx must begin with x. It follows that M must be an abstraction of the form $\lambda z.zN$. Thus $x(xN[z := x]) =_\beta x((\lambda z.zN)x) =_\beta (\lambda z.zN)x =_\beta xN[z := x]$. This would be an equality between two distinct normal forms. □

1.18. Take $P_1 = \lambda k.k(\lambda xy.y\mathbf{I}x)$ and $P_2 = \lambda k.k(\lambda xy.x\mathbf{I}y)$. A three-argument projection can be defined as $P_2^3 = \lambda k.k(\lambda xyz.x\mathbf{I}z\mathbf{I}y)$. □

1.20. The proof is by induction with respect to the number of steps in the reduction $M[z := \lambda x N] \twoheadrightarrow_\beta \lambda u Q$. The base case of zero steps is immediate, so let us consider the first step $M[z := \lambda x N] \to_\beta M_1$.

First suppose that the variable z occurs only once in M. We have assumed that $\lambda x N$ is normal, and thus one of the following cases must hold:

CASE 1: The reduction takes place within M, i.e. we have $M \to_\beta M'$ for some M' with $M'[z := \lambda x N] = M_1$. Then we can apply the induction hypothesis to M_1.

CASE 2: The operator part of the redex is $\lambda x N$ substituted for the only occurrence of z in M. This means that z is applied in M to an argument, and it must remain applied after any beta-reductions. But $M \twoheadrightarrow_\beta z$, a contradiction.

If M has more than one occurrence of z, then still only one of these occurrences is "responsible" for the normal form. More precisely, one easily shows that there exists a term M' with only one occurence of z and with $\mathrm{FV}(M') = \mathrm{FV}(M) \cup \{z'\}$, where z' is a fresh variable, satisfying:

$$M'[z' := z] = M \quad \text{and} \quad M' \twoheadrightarrow_\beta z.$$

APPENDIX B. SOLUTIONS AND HINTS   375

Take $M'' = M'[z' := \lambda x N]$. Clearly, we have $M'' \twoheadrightarrow_\beta z[z' := \lambda x N] = z$ and $M''[z := \lambda x N] = M[z := \lambda x N]$. In addition, z occurs only once in M'' so we are back in the previous case. □

1.21. We must prove that the term $F c_{n_1} \ldots c_{n_m}$, in the proof of Theorem 1.7.5, is strongly normalizing, whenever $f(n_1, \ldots, n_m)$ is defined. Consider the test
$$T = G y c_{n_1} \ldots c_{n_m} (\lambda v. \text{false}) \text{true}.$$
Observe that T is strongly normalizing, because by our assumption, $T[y := d_n]$ is strongly normalizing for all n.

We first show that if $T[y := d_n]$ reduces to an abstraction P, then $P = \text{true}$ or $P = \text{false}$, depending on n. Indeed, suppose that $G d_n c_{n_1} \ldots c_{n_m}$ reduces to c_0, and let z be a fresh variable. Then $M = G d_n c_{n_1} \ldots c_{n_m} (\lambda v. \text{false}) z$ reduces to z and we can apply Exercise 1.20. If $G d_n c_{n_1} \ldots c_{n_m}$ reduces to a non-zero numeral then we apply a similar argument.

It follows that true is the only abstraction to which T itself (with y free) can be reduced. To see this, first assume that $T[y := d_{n'}] \twoheadrightarrow \text{false}$ and $T[y := d_{n''}] \twoheadrightarrow \text{true}$ for some n' and n''. In this case T cannot be reduced to an abstraction at all. Otherwise, substituting $d_{n'}$ and $d_{n''}$ for y would turn this abstraction respectively into false and true. We know that the minimum is defined, so the only other possibility is that $T[y := d_n]$ reduces to true for all n. (This may happen when the minimum equals zero.) Then T can be reduced to an abstraction, namely to true.

As the next step, observe that the term $W' = W[x_1, \ldots, x_m := c_{n_1}, \ldots, c_{n_m}]$ is strongly normalizing. Indeed, W' can only reduce to:
- $\lambda y. T'(\lambda w L')(\lambda w. w R' w)$;
- $\lambda y. (\lambda x w. L')(\lambda w. w R' w)$;
- $\lambda y. (\lambda w L')$,

where T', L' and R' are reducts of T, Ly and succ y respectively.

Let k be the smallest number such that $g(k, n_1, \ldots, n_m) = 0$, and let $n \leq k$. Take any W_1, W_2 such that $W' \twoheadrightarrow_\beta W_1$ and $W' \twoheadrightarrow_\beta W_2$. We claim that $W_1 d_n W_2$ is strongly normalizing, and this of course suffices to complete our proof. First suppose that $n = k$, i.e. our test evaluates to true. By the consideration above, any sequence of reductions of $W_1 d_n W_2$ must eventually lead to $\text{true}(\lambda w. L')(\lambda w. w R' w) W_2'$, where L', R', and W_2' are reducts of $L d_n$, d_{n+1}, and W_2 respectively. It is easy to see that all further reductions must terminate.

For $n < k$, we proceed by induction backward. The test evaluates to false and $W_1 d_n W_2$ must reduce to $\text{false}(\lambda w. L')(\lambda w. w R' w) W_2'$, and eventually to $W_2' R'' W_2''$, where $W_2 \twoheadrightarrow W_2'$, $W_2 \twoheadrightarrow W_2''$ and $d_{n+1} \twoheadrightarrow R''$. But $W_2' d_{n+1} W_2'' \twoheadrightarrow W_2' R'' W_2''$, and $W_2' d_{n+1} W_2''$ is strongly normalizable by the induction hypothesis. □

Solutions to Chapter 2

2.1. If $e + \pi$ and $e\pi$ were both algebraic then the coefficients of the polynomial $x^2 - x(e + \pi) + e\pi = (x - e)(x - \pi)$ would be algebraic and thus the roots e and π would have to be algebraic too [296]. But which of $e + \pi$ and $e\pi$ is not algebraic? □

2.2. The most confusing case is probably (ix) $\neg\neg(p \vee \neg p)$. Suppose we have a construction of $\neg(p \vee \neg p)$ and we want to give a construction of $\bot$. To achieve this, we first show a construction of $\neg p$. For a given construction of p, we first turn it into a construction of $p \vee \neg p$, and pass it to our construction of $\neg(p \vee \neg p)$ to obtain a construction of $\bot$.

So we have a construction of $\neg p$. Then we make a construction of $p \vee \neg p$ and obtain a construction of $\bot$, just like above. (Now try a construction for the double negation of Peirce's law.) □

2.3. We show the closure under weakening. The proof is by induction with respect to the size of the proof of $\Gamma \vdash \varphi$. We proceed by cases depending on the last rule used in this proof. Recall that the notation Γ, ψ stands for the union $\Gamma \cup \{\psi\}$, whether or not $\psi \in \Gamma$. That's why e.g. Case 3 below works.

CASE 1: The proof consists only of a single application of an axiom scheme, that is φ is an element of Γ. Then $\Gamma, \psi \vdash \varphi$ is also an axiom.

CASE 2: The proof ends with an application of ($\wedge$I). That is, φ has the form $\varphi_1 \wedge \varphi_2$ and we have proven $\Gamma \vdash \varphi_1$ and $\Gamma \vdash \varphi_2$. The proofs of these judgements are contained in the proof of $\Gamma \vdash \varphi$, so we can apply the induction hypothesis to obtain $\Gamma, \psi \vdash \varphi_1$ and $\Gamma, \psi \vdash \varphi_2$. By an application of ($\wedge$I) we can derive $\Gamma, \psi \vdash \varphi_1 \wedge \varphi_2$.

CASE 3: The proof ends with an application of ($\vee$E). That is, we have $\Gamma \vdash \vartheta_1 \vee \vartheta_2$, for some formulas ϑ_1, ϑ_2, such that $\Gamma, \vartheta_1 \vdash \varphi$ and $\Gamma, \vartheta_2 \vdash \varphi$. These proofs are all shorter, thus we can apply the induction hypothesis to obtain $\Gamma, \psi \vdash \vartheta_1 \vee \vartheta_2$ and $\Gamma, \vartheta_1, \psi \vdash \varphi$ and $\Gamma, \vartheta_2, \psi \vdash \varphi$. It remains to use rule ($\vee$E).

Other cases are similar. The part of the lemma concerning substitutions is proved by an analogous induction. □

2.4. Proofs of (ii) and (iii) are given in Example 2.2.3. Hints for the other formulas:
  (i) Begin with the axiom $\bot \vdash \bot$. Apply ($\bot$E), to derive $\bot \vdash p$ and ($\to$I) to derive $\vdash \bot \to p$.
  (iv) Begin with $p, p \to \bot \vdash p$ and $p, p \to \bot \vdash p \to \bot$. Apply ($\to$E) to get $p, p \to \bot \vdash \bot$, then twice ($\to$I) to get $\vdash p \to (p \to \bot) \to \bot$.
  (v) First show that $\neg\neg\neg p, \neg p, p \vdash \bot$ (unfold the $\neg$). Thus, $\neg\neg\neg p, p \vdash \neg p \to \bot$, i.e., $\neg\neg\neg p, p \vdash \neg\neg p$. But $\neg\neg\neg p = \neg\neg p \to \bot$ and one can derive $\neg\neg\neg p, p \vdash \bot$. It remains to use ($\to$I).
  (vi) First show that $p \to q, q \to \bot, p \vdash \bot$. Then apply ($\to$I) three times.
  (vii) Remember that $\leftrightarrow$ abbreviates a conjunction, so ($\wedge$I) will be the last rule. For the "$\to$" part begin with $\neg(p \vee q), p \vdash p \vee q$. Then derive $\neg(p \vee q), p \vdash \bot$ using (Ax) and ($\to$E), and apply ($\to$I) to obtain $\neg(p \vee q) \vdash \neg p$. In a similar way derive $\neg(p \vee q) \vdash \neg q$. Now apply ($\wedge$I). For the converse, observe that $\neg p \wedge \neg q, p \vdash \bot$, because both p and $\neg p$ can be derived. Similarly, $\neg p \wedge \neg q, q \vdash \bot$ and by ($\vee$E) one gets $\neg p \wedge \neg q, p \vee q \vdash \bot$. The conclusion is obtained with the help of ($\to$I).
  (viii) Again ($\wedge$I) is the last rule. One implication uses ($\wedge$E), the other one uses ($\wedge$I).
  (ix) First derive $(p \vee \neg p) \to \bot, p \vdash \bot$, using rules ($\vee$I) and ($\to$E). By ($\to$I) obtain $(p \vee \neg p) \to \bot \vdash \neg p$. Use ($\vee$I) and ($\to$E) again to get $(p \vee \neg p) \to \bot \vdash \bot$.
  (x) Begin with $\neg p, \neg\neg p \vdash \bot$ and use ($\bot$E) to obtain $\neg p, \neg\neg p \vdash p$. Now apply ($\vee$E) to the latter and $p, \neg\neg p \vdash p$. □

2.5. Every provable formula is a classical tautology. (This is a special case of the soundness theorem, which can be proved by a straightforward induction.) Since $\bot$ is not valid, it cannot be provable. □

2.6. The relation $\leq$ is reflexive, because $a \sqcup a = a$. It is transitive because $a \sqcup b = b$ and $b \sqcup c = c$ imply $a \sqcup c = a \sqcup (b \sqcup c) = (a \sqcup b) \sqcup c = b \sqcup c = c$. And it is anti-symmetric, because $a \sqcup b = b$ and $b \sqcup a = a$ imply $a = b \sqcup a = a \sqcup b = b$.

To show that $\sqcup$ is the upper bound, first note that $a \sqcup (a \sqcup b) = (a \sqcup a) \sqcup b = a \sqcup b$. Then suppose that $a, b \leq c$. We have $c = a \sqcup c = a \sqcup (b \sqcup c) = (a \sqcup b) \sqcup c$, that is,

APPENDIX B. SOLUTIONS AND HINTS   377

$a \sqcup b \leq c$. It remains to check that $\sqcap$ is the lower bound. We begin with a useful observation: The conditions $a \sqcap b = a$ and $a \sqcup b = b$ are equivalent. Indeed, $a \sqcap b = a$ implies $b = b \sqcup (a \sqcap b) = b \sqcup a = a \sqcup b$, and similarly the other way around. Then we proceed as for $\sqcup$. □

2.8. First note that the reverse inequality holds in every lattice, as well as this one: $(a \sqcap b) \sqcup c \leq (a \sqcup c) \sqcap (b \sqcup c)$. Now observe that $(a \sqcap b) \sqcup c = (a \sqcap b) \sqcup (a \sqcap c) \sqcup c \geq (a \sqcap (b \sqcup c)) \sqcup c = (a \sqcap (b \sqcup c)) \sqcup (c \sqcap (b \sqcup c)) \geq (a \sqcup c) \sqcap (b \sqcup c)$, where the assumption is used twice. □

2.9. Let $d = (a \sqcap c) \sqcup (b \sqcap c)$. Then $a \sqcap c, b \sqcap c \leq d$, so $a, b \leq c \Rightarrow d$. It follows that $a \sqcup b \leq c \Rightarrow d$, and thus $(a \sqcup b) \sqcap c \leq d$. Now apply Exercise 2.8. □

2.10. All the graphs represent lattices. Only (b) is not a Heyting algebra, because it is not distributive, for example $(2 \sqcup 3) \sqcap 4 = 4$ and $(2 \sqcap 4) \sqcup (3 \sqcap 4) = 2$. Also note that $4 \Rightarrow 2$ does not exist.

But none of our examples is a Boolean algebra. To verify this, there is a choice: Either show that certain elements do not have complements, or prove that the cardinality of a finite Boolean algebra must be of the form 2^n. □

2.12. Suppose $\Gamma \vdash \varphi$. Then $\Gamma_0 \vdash \varphi$, for some finite $\Gamma_0 \subseteq \Gamma$. As in the proof of Theorem 2.4.7, we have $[\![\Gamma_0]\!]_v \leq [\![\varphi]\!]_v$. Then $[\![\varphi]\!]_v \in F$, because all components of the intersection $[\![\Gamma_0]\!]_v$ are in F. □

2.15. To show that the map $a \mapsto Z_a$ is injective, take $a \neq b$. Then $a \not\leq b$ or $b \not\leq a$. Suppose the latter. Then the set $b\!\uparrow = \{c \in B \mid b \leq c\}$ is a proper filter, and $a \notin b\!\uparrow$. By Lemma 2.5.7, there exists a prime filter F such that $b \in F$ but $a \notin F$. It follows that $Z_a \neq Z_b$. We now verify that our embedding preserves the Boolean algebra operations. The inclusion from right to left in $Z_{a \sqcup b} = Z_a \cup Z_b$ is obvious, the other follows from primality. The equality $Z_{a \sqcap b} = Z_a \cap Z_b$ is immediate. We also have $Z_{-a} = -Z_a$. Indeed, since $a \sqcup -a = 1$ and 1 belongs to every filter, we must always have either a or $-a$ in a prime filter. To complete the proof that $a \mapsto Z_a$ is an isomorphic embedding (in particular that $\{Z_a \mid a \in \mathcal{B}\}$ is a Boolean algebra), it remains to note that $Z_0 = \emptyset$ and $Z_1 = \mathcal{Z}$. □

2.16. The only property which is not immediate is that $\Rightarrow$ is a relative pseudocomplement, i.e. that $A \cap C \subseteq B$ is equivalent to $C \subseteq \text{Int}(-A \cup B)$. First note that the condition $A \cap C \subseteq B$ is equivalent to $C \subseteq -A \cup B$, for all A, B and C. For the left-to-right implication observe that $X \subseteq Y$ implies $X \subseteq \text{Int}(Y)$, whenever X is an open set. The converse follows from $\text{Int}(-A \cup B) \subseteq -A \cup B$. □

2.17. The construction in the proof of Lemma 2.5.8 turns a finite Heyting algebra into a finite Kripke model, so one can use Theorem 2.4.12. But a direct proof yields a better size bound.

Suppose that $\not\Vdash \varphi$, i.e. $\mathcal{C}, c_0 \not\Vdash \varphi$ for some $\mathcal{C} = \langle C, \leq, \Vdash \rangle$ and some $c_0 \in C$. Let Θ be the set of all subformulas of φ. Write $c \sim c'$ if the conditions $c \Vdash \alpha$ and $c' \Vdash \alpha$ are equivalent for all $\alpha \in \Theta$. Note that the equivalence relation $\sim$ has a finite number of classes (exponential in the size of φ).

Write $[c_1]_\sim \vartriangleleft [c_2]_\sim$ when there are $c'_1 \sim c_1$ and $c'_2 \sim c_2$ with $c'_1 \leq c'_2$, and let $\leq$ be the transitive closure of the relation $\vartriangleleft$. Then $\leq$ is a partial order in $C/\!\sim$, and the following monotonicity property is useful to prove anti-symmetry: If $[c]_\sim \leq [c']_\sim$ and $c \Vdash \alpha$, where $\alpha \in \Theta$, then also $c' \Vdash \alpha$. Define a new model $\mathcal{C}/\!\sim = \langle C/\!\sim, \leq, \Vdash \rangle$ where $[c]_\sim \Vdash p$ holds iff $p \in \Theta$ and $c \Vdash p$. One proves by induction the equivalence

$$\mathcal{C}/\!\sim, [c]_\sim \Vdash \alpha \quad \text{iff} \quad \mathcal{C}, c \Vdash \alpha, \tag{$*$}$$

for all $\alpha \in \Theta$. For instance, assume that $c \Vdash \alpha$ for $\alpha = \psi \to \vartheta$. If $[c]_\sim \leq [c']_\sim$ then, by our monotonicity property, also $c' \Vdash \psi \to \vartheta$. If $[c']_\sim \Vdash \psi$ then also $c' \Vdash \psi$, by the induction hypothesis, and thus $c' \Vdash \vartheta$. Apply the induction hypothesis again to get $[c']_\sim \Vdash \vartheta$, and conclude $[c]_\sim \Vdash \psi \to \vartheta$. From (*) we obtain $[c_0]_\sim \not\Vdash \varphi$. □

2.18. First we consider counterexamples with open subsets of the real line. In what follows we use the convention that $v(p) = P$, $v(q) = Q$, etc., and we write $\sim A$ for $\mathrm{Int}(-A)$. Formulas (i) and (ii) are considered in Example 2.4.9.

(iii) Take $P = \mathbb{R} - \{0\}$. Then $\sim P$ is empty and $\sim\sim P$ is the full set $\mathbb{R}$. Thus $\sim\sim P \Rightarrow P \neq \mathbb{R}$.
(iv) Let $P = \mathbb{R} - \{0\}$ and $Q = \mathbb{R} - \{0,1\}$. Then $\sim P = \sim Q = \varnothing$ and thus the value of the left-hand side is $\mathbb{R}$. But $P \not\subseteq Q$, so the right-hand side is not $\mathbb{R}$.
(v) Take $P = (-\infty, 0)$ and $Q = (0, \infty)$.
(vi) Take $P = (0, \infty)$ and $Q = \mathbb{R} - \{0\}$.
(vii) Let $Q = R = (0, \infty)$ and let $P = \mathbb{R} - \{0\}$. Then $(P \Rightarrow Q) \cap (Q \Rightarrow P)$ is equal to Q and R. Thus the value of $(p \leftrightarrow q) \leftrightarrow r$ is $\mathbb{R}$. But the value of the right-hand side is $(P \Rightarrow \mathbb{R}) \cap (\mathbb{R} \Rightarrow P) = P \neq \mathbb{R}$.
(viii) Define P and Q as the same open segment.
(ix) Same as (v).
(x) That can't be simpler. Take any open segment.

A Kripke counter-model for (ii) was given in Example 2.5.4, and it also works in cases (v), (ix), and (x). In the remaining cases one can take:

(i) Two states $c < c'$, where $c \not\Vdash p$, $c' \Vdash p$, and $c, c' \not\Vdash q$. The same for (iii).
(iv) Two states $c < c'$, where $c, c' \Vdash p$ and $c' \Vdash q$ and $c \not\Vdash q$.
(vi) Two states $c < c'$, where $c' \Vdash p, q$ and c forces nothing. The same for (viii)
(vii) Three states $c < c' < c''$, where $c' \Vdash p$, $c'' \Vdash p, q, r$ and c forces nothing. □

2.19. Use n rays beginning in $(0,0)$ to divide the space $\mathbb{R}^2$ into n disjoint angles and take their interiors as values of p_i. The point $(0,0)$ does not belong to the interpretation of our formula. Or take a Kripke model with all sets $\{c \geq c_0 \mid c \Vdash p_i\}$ different. Then c_0 does not force our formula. □

2.20. Only (i) and (iii). A hint for the latter is to begin with $p \to \neg q, q \vdash \neg p$. □

2.21. Valid formulas are (i), (ii), (iv), (viii) and (ix). Here are a few hints:

(ii) Let $\Gamma = \{((p \to q) \to p) \to p) \to q, (p \to q) \to p\}$. First show that $\Gamma, p \vdash q$. Then derive $\Gamma \vdash p$ and then $((p \to q) \to p) \to q \vdash ((p \to q) \to p) \to p$.
(viii) First show that $(p \vee \neg p) \to \neg q, q \vdash \neg p$.
(ix) Use Exercise 2.20(i). □

2.22. The valid equivalences are (iii) and (iv). The right to left part of (iii) can be obtained as follows. First show $p \leftrightarrow (p \to q) \vdash p \to q$. Then derive $p \leftrightarrow (p \to q) \vdash p$ and finally $p \leftrightarrow (p \to q) \vdash q$. □

2.23. Only (iii) is not (think of positive and negative reals). Take (iv) as an example. Let $\Gamma = \{\neg\neg\varphi \to \neg\neg\psi, \neg(\varphi \to \psi)\}$. To derive $\Gamma \vdash \bot$, first note that $\Gamma, \varphi \to \psi \vdash \bot$. Now, using $\neg\varphi \vdash \varphi \to \psi$ and $\psi \vdash \varphi \to \psi$, derive $\Gamma, \neg\varphi \vdash \bot$ and $\Gamma, \psi \vdash \bot$, that is $\Gamma \vdash \neg\neg\varphi$ and $\Gamma \vdash \neg\psi$. □

2.26. To show that $\vee$ is not definable from $\wedge, \to,$ and $\bot$, consider the Kripke model in Example 2.5.4. For a formula φ built solely from the latter three connectives, one shows (by induction with respect to φ) that the set $\{c \mid c \Vdash \varphi\}$ never consists of two elements. Thus no such formula can be equivalent to $p \vee q$.

APPENDIX B. SOLUTIONS AND HINTS   379

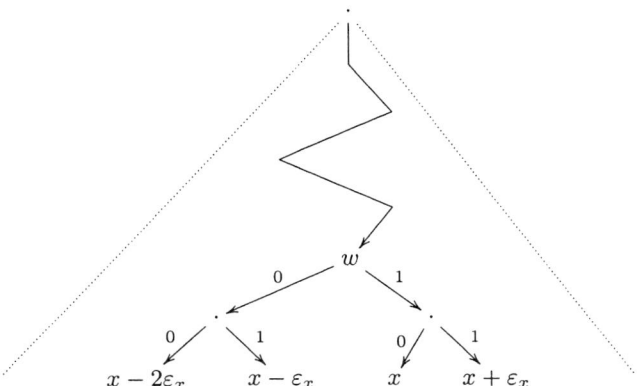

FIGURE B.1: EXERCISE 2.29

Now consider algebra (d) of Exercise 2.10. The subset $\{0, 2, 3, 1\}$ is closed under the operations $\sqcup$ and $\Rightarrow$. But it is not closed under $\sqcap$, because $2 \sqcap 3 = 4$. We conclude that conjunction is not definable from the other connectives.

For implication, consider a valuation v in $\mathcal{O}(\mathbb{R})$ such that $v(p) = \mathbb{R} - \{0\}$ and $v(q) = \mathbb{R} - \{0,1\}$. Then $[\![p \to q]\!]_v = \mathbb{R} - \{1\}$, but $[\![\varphi]\!]_v \neq \mathbb{R} - \{1\}$ for all formulas φ built from p and q with the help of $\vee, \wedge, \bot$ (and even $\neg$).

Finally, for $\bot$ take a valuation which assigns $\mathbb{R}$ to every variable. □

2.27. Let $K(\varphi, \psi, \vartheta)$ stand for $((\varphi \vee \psi) \wedge \neg \vartheta) \vee (\neg \varphi \wedge (\psi \leftrightarrow \vartheta))$. Then
  $\neg p$ is equivalent to $K(p, p, p)$;
  $\bot$ is equivalent to $\neg K(p, \neg p, \neg p)$;
  $p \vee q$ is equivalent to $K(K(p, \neg q, \top), K(p, q, \bot), \top)$;
  $p \leftrightarrow q$ is equivalent to $K(K(p, q, \top), K(\bot, q, p), \top)$.
For conjunction and implication use Exercise 2.22. □

2.29. A number $x = 0.b_1 b_2 \ldots \in (0, 1)$ is conveniently represented as the infinite path $b_1 b_2 \ldots$ in the full binary tree of all finite words over 0 and 1. Numbers with finite binary unfoldings can also be identified with nodes of the tree. Note that numbers represented by any two "neighbour" nodes at depth d are at a distance 2^{-d}. Another useful observation is that if $\ell(b_1 b_2 \ldots b_n) = c$ then $\pi(x) \geq c$, for all numbers $x = 0.b_1 b_2 \ldots b_n \ldots$, except perhaps when x is exactly $0.b_1 b_2 \ldots b_n$ and $b_n = 1$ in addition (Figure B.1).

We prove condition (i) in the "finite" case when $x = 0.b_1 b_2 \ldots b_n 1000 \ldots$ and $\pi(x) = \ell(b_1 b_2 \ldots b_n) = c$, leaving to the reader the similar proof in the "infinite" case. Define $\varepsilon_x = 2^{-(n+2)}$ and observe that $|x - y| < \varepsilon_x$ implies that the binary notation for y begins with $0.b_1 b_2 \ldots b_n 01 \ldots$ or $0.b_1 b_2 \ldots b_n 1 \ldots$ In particular the path representing y begins with $w = b_1 b_2 \ldots b_n$, which corresponds to $x - 2\varepsilon_x$. Since all labels below $w = b_1 b_2 \ldots b_n$ are greater than or equal c, and the number $x - 2\varepsilon_x = 0.b_1 b_2 \ldots b_n 00 \ldots$ is too far from x, it must be the case that $\pi(y) \geq c$.

For part (ii), observe that if $\pi(x) = \ell(w) = c' \leq c$ then the label c occurs in the tree infinitely many times below w at an arbitrarily close distance to x. □

2.30. Let $A_\varphi = \pi^{-1}(\{c \in C \mid c \Vdash \varphi\})$. The only non-trivial case in the proof of $[\![\varphi]\!]_v = A_\varphi$ is when $\varphi = \psi \to \vartheta$. From the induction hypothesis applied to ψ and ϑ we have $[\![\varphi]\!]_v = A_\psi \Rightarrow A_\vartheta$, so we need to prove that $A_\psi \Rightarrow A_\vartheta = A_{\psi \to \vartheta}$.

To show "⊆", we assume that $x \in A_\psi \Rightarrow A_\vartheta$ and we prove that $\pi(x) \Vdash \psi \to \vartheta$. Suppose $\pi(x) \leq c' \Vdash \psi$. Since $A_\psi \Rightarrow A_\vartheta$ is open, its image under π is upward-closed by Exercise 2.29, i.e. there is $y \in A_\psi \Rightarrow A_\vartheta$ with $\pi(y) = c'$. Thus $y \in A_\psi$ and $y \in A_\psi \Rightarrow A_\vartheta$, so $y \in A_\vartheta$, that is, $c' \Vdash \vartheta$, as needed. For the other inclusion it is enough to notice that $A_{\psi \to \vartheta} \cap A_\psi \subseteq A_\vartheta$. □

2.31. From the previous exercise one easily derives a completeness theorem for the algebra $\mathcal{O}((0,1))$, and thus also for its isomorphic image $\mathcal{O}(\mathbb{R})$. The construction in Exercises 2.29 and 2.30 can also be applied to rational numbers. A generalization to products is easy. □

2.34. Assume that φ is a classical tautology (the other direction is obvious), but $\neg\neg\varphi$ is not valid intuitionistically. This means that there is a Kripke model with a state c, such that $c \Vdash \neg\varphi$. Without loss of generality, we can assume that c determines all propositional variables in φ. Indeed, suppose that c does not determine a variable p. Then there is a $c' \geq c$ with $c' \Vdash p$, and we can take c' instead. From Exercise 2.32 we obtain that $c \Vdash \varphi$, a contradiction. □

Solutions to Chapter 3

3.2. No. A typable example is $\lambda x.\mathbf{K}x(xy) \to \mathbf{I}$. □

3.3. Reduce $(\lambda x.xx)\mathbf{I}$ and it becomes typable. □

3.4. One possibility is $\lambda x.(\lambda y.xyy)\mathbf{I}$ reducing to $\lambda x.x\mathbf{II}$. The first term is typable but it cannot be assigned type $((p \to p) \to (q \to q) \to r) \to r$. □

3.5. Define $t_p = x_p$, and $t_{\sigma \to \tau} = \lambda x.\mathbf{K}t_\tau(\lambda zy.z(yx)(yt_\sigma))$. □

3.6. It is a direct consequence of Lemma A.2.1, that the length of the principal type is at most exponential in the length of the term. To see that this is also a lower bound, define $M_n = \lambda y x_0 \ldots x_n.y(x_n(x_n x_{n-1})) \ldots (x_1(x_1 x_0))$. The length of M_n is linear in n and a type assigned to M_n is at least of length exponential in n. (Despite this, typability is in PTIME because so is unification, cf. Theorem A.5.4.) □

3.7. The trick is to represent n-ary operators as complex expressions built from the single binary symbol. For this, one can use non-unifiable terms as different flags. For instance, suppose that the signature consists of two function symbols f and g, where f is unary and g is binary. We encode terms over this signature as terms in the signature consisting of one binary symbol $\to$ written in infix notation. Our flags are y and $y \to y$ for a fixed variable y, not occurring in the unification problem to be encoded. Define $\overline{x} = x$ if x is a variable, and $\overline{f(t)} = y \to \overline{t}$ and $\overline{g(t,u)} = (y \to y) \to (\overline{t} \to \overline{u})$. Now the equation $t = u$ has a solution iff its translation $\overline{t} = \overline{u}$ has a solution. □

3.8. Replace a system of equations $\{t_i = u_i \mid i = 1, \ldots, m\}$ by the single equation $t_1 \to t_2 \to \cdots \to t_m = u_1 \to u_2 \to \cdots \to u_m$, where $\to$ is binary. □

3.9. By Exercises 3.7 and 3.8, it suffices to consider a single unification equation "$\tau = \sigma$" over the signature consisting exclusively of the binary function symbol $\to$. Let x, y be new variables and let t_τ and t_σ be as in Exercise 3.5. Our equation has a solution if and only if $\lambda xy.x(yt_\tau)(yt_\sigma)$ is typable. □

3.11. From the proof of Theorem 3.2.7 and from Exercise 3.9 we know that typability (i) and type checking (iv) are LOGSPACE-equivalent to unification. Here are the remaining reductions:
(i) $\leq$ (ii): To reduce $? \vdash M : ?$ to $\vdash M : ?$, observe that a term M with free variables $x_1, \ldots, x_n$ is typable if and only if $\lambda x_1 \ldots x_n . M$ is typable.
(iii) $\leq$ (iv): To decide $\Gamma \vdash M : ?$ ask if $\Gamma, x : p \vdash \mathbf{K} x M : p$.
(iv) $\leq$ (v): To decide $\Gamma \vdash M : \tau$ ask if $\vdash \lambda x_1 \ldots x_n . M : \tau_1 \to \cdots \to \tau_n \to \tau$, where $FV(M) = \{x_1, \ldots, x_n\}$ and $\Gamma(x_i) = \tau_i$ for $i = 1, \ldots, n$.
(vi) $\leq$ (iii): To decide $? \vdash M : \tau$ ask if $x : \tau \vdash \lambda \vec{x} y z . y(zx)(zM) : \sigma$, for any σ, where $\vec{x}$ are all free variables of M.
Problem (ii) is a special case of (i), and (v) is a special case of (vi), so these reductions are trivial. □

3.12. This only guarantees that the type of M is an instance of τ. □

3.13. Take the term $(\lambda v . \mathbf{K} v (\lambda x_{p_1} \ldots x_{p_n} y z . y(zv)(zt_\tau)))N$, where N is any inhabitant of τ, and t_τ is as in Exercise 3.5. □

3.16. Consider $A = \{a, b, c, d\}$ and the relation $\to$ given by the graph below:

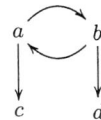

□

3.18. Reduce $|M_1|$ and $|M_2|$ to the (same) normal form, and apply Proposition 3.4.5 and Exercise 3.17. To see that the claim does not hold for =, consider for instance the terms $\mathbf{K}x(\lambda y : p . y)$ and $\mathbf{K}x(\lambda y : q . y)$. □

3.19. Let $M_0 \twoheadrightarrow_\beta M_1$ and $M_0 \twoheadrightarrow_\beta M_2$. Then $|M_0| \twoheadrightarrow_\beta |M_1|$ and $|M_0| \twoheadrightarrow_\beta |M_2|$. By the Church-Rosser property for untyped lambda-terms, there exists a (typable) term P with $|M_1| \twoheadrightarrow_\beta P$ and $|M_2| \twoheadrightarrow_\beta P$. By Proposition 3.4.5, there is M_3 such that $|M_3| = P$ and $M_1 \twoheadrightarrow_\beta M_3$. And there is M_4 such that $|M_4| = P$ and $M_2 \twoheadrightarrow_\beta M_4$. If M_3 and M_4 are the same then we are done, but it does not have to be the case. Using Theorem 3.5.1 we can however assume that P is in normal form, and apply Exercise 3.17. Observe that we had to appeal to normalization to complete this, seemingly obvious, argument. See also the discussion in [349, pp. 269, 559]. □

3.21. A term of depth d has less than 2^d redexes of maximal degree. It takes less than 2^d reduction steps to eliminate all of them, if we always choose the rightmost one, as in the proof of Theorem 3.5.1. It remains to see that the resulting term is of depth at most 2^d. This can be done by induction with respect to d. The only nontrivial case is when M is an application, say $M = NP$. With our reduction strategy we first reduce M to NP_1 and then to $N_1 P_1$ where N_1 and P_1 are of degree less than δ, and of depth at most 2^{d-1} (so that $N_1 P_1$ is of depth $2^{d-1} + 1 \leq 2^d$. If $N_1 P_1$ is not a redex, or if it is a redex of degree smaller than δ, then we are done. If $N_1 P_1$ is a redex of degree δ then the depth of the result is at most $d(N_1) + d(P_1) \leq 2^d$. □

3.23. The answer is again "a tower of 2's of height proportional to n". Exercise 3.20 gives the lower bound. The upper bound obtained from Exercises 3.21–3.22 depends on the size of types. By Exercise 3.6, these are at most exponential in n, so our tower grows by only one floor, and is still of height $\mathcal{O}(n)$. □

3.24. Addition and multiplication can be defined as in Example 1.7.3, and we can take $\lambda mnkfx . m(\lambda y . nfx)(kfx)$ for the conditional. □

3.25. The proof is by induction with respect to the length of M, which can be assumed normal. Our induction hypothesis is required to hold for arbitrary r.

If M is a variable x_i then $P(m,n) = 0$. Otherwise, M is an application, and it is not hard to see that one of the following must hold.[1]

$$M = fM_1 \quad \text{or} \quad M = aNM_1 \quad \text{or} \quad M = bNM_1,$$

where $\Gamma \vdash N : p \to p$ and $\Gamma \vdash M_1 : p$. Let i and $P_1(m,n)$ be obtained from the induction hypothesis for M_1. If $M = fM_1$ then $P(m,n) = P_1(m,n) + 1$, so suppose that $M = aNM_1$, the case of $M = bNM_1$ being similar. Without loss of generality, we can assume that N is an abstraction $N = \lambda x_{r+1} M_2$, where $\Gamma, x_{r+1} : p \vdash M_2 : p$. Indeed, otherwise one can replace N by $\lambda x_{r+1}.Nx_{r+1}$, and $M[a := \mathbf{c}_m, b := \mathbf{c}_n]$ remains the same (up to beta-conversion). By the induction hypothesis we have $M_2[a := \mathbf{c}_m, b := \mathbf{c}_n] =_\beta f^{P_2(m,n)} x_j$, for some polynomial P_2 and some $j \leq r+1$. It follows that $M[a := \mathbf{c}_m, b := \mathbf{c}_n] =_\beta \mathbf{c}_m(\lambda x_{r+1}.f^{P_2(m,n)} x_j)(f^{P_1(m,n)} x_i)$. Now, if $j \leq r$ then the latter is β-equal to $f^{P_2(m,n)} x_j$. If $j = r+1$ we obtain $f^{P(m,n)} x_i$, where $P(m,n) = m \cdot P_2(m,n) + P_1(m,n)$. □

3.26. Suppose that a function g is definable by a term G in normal form. Assume for simplicity that g has two arguments. Without loss of generality, we can write $G = \lambda a^{\mathbf{int}} \lambda b^{\mathbf{int}} \lambda f^{p \to p} \lambda x_1^p . M$. By Exercise 3.25, there is a polynomial $P(m,n)$ such that $g(m,n) = P(m,n)$, for all non-zero arguments. Using a similar technique, one proves that there are unary polynomials Q and R such that $g(n,0) = Q(n)$ and $g(0,n) = R(n)$ holds for all n. Let $g(0,0) = k$. It follows that

$$g(m,n) = \text{ if } m = 0 \text{ then if } n = 0 \text{ then } k \text{ else } R(n)$$
$$\text{else if } n = 0 \text{ then } Q(m) \text{ else } P(m,n).$$
□

Solutions to Chapter 4

4.1. (i) No. (ii) Yes. □

4.5. If a type has infinitely many inhabitants, these are arbitrarily long. Therefore, the algorithm of Lemma 4.2.3 (where we restrict attention to environments with no repeated types) can generate infinitely many inhabitants, provided of course that we allow for arbitrarily long runs.

Each run can be seen as a tree, labeled with questions of the form $\Gamma \vdash ? : \tau$, where branching nodes correspond to applications. If there are only finitely many runs, the answer is negative. Otherwise, there is a successful run where the same question is repeated on some branch of the tree, and thus there are infinitely many inhabitants. In addition, if such runs exist, we can find one of a bounded depth. See [237] for details. □

4.6. The algorithm works in PSPACE so it can only make an exponential number of steps. For the lower bound define

$$\tau_n = (p_n \to p_n \to p_{n-1}) \to (p_{n-1} \to p_{n-1} \to p_{n-2}) \to \cdots \to (p_1 \to p_1 \to p_0) \to p_n \to p_0.$$ □

4.7. In a sense, nothing will change. It is easy to prove that the sets of provable theorems are the same. (Prove that if φ can be deduced from assumptions in Γ then it can be deduced from a subset of Γ under the complete discharge convention.)

From the point of view of reductions the situation becomes much worse. We can see it when we attempt to represent such proofs by lambda-terms, and we notice

---

[1] Note that a normal non-abstraction typable in Γ can only be of type p or $p \to p$.

APPENDIX B. SOLUTIONS AND HINTS 383

for instance that in $\lambda x{:}p.zy$ the variable y cannot be of type p. But the judgement $z : p \to q, y : p \vdash (\lambda u{:}q\lambda x{:}p.u)(zy)$ should be legal and we see that subject reduction is violated. Put it differently, the proof represented by the latter term reduces to $\lambda x{:}p.zx$ rather than to $\lambda x{:}p.zy$.

Under this peculiar regime, the strong normalization property is lost, as shown by Statman and Leivant [301]. Let p and q be two different type variables. To simplify notation we now assume that lower-case and upper-case characters are used as variables of type p and q, respectively. Consider the application MN, where

$$M = \lambda a.(\lambda b.A)((\lambda B.a)((\lambda c.C)a));$$
$$N = (\lambda D.d)((\lambda e.E)((\lambda F.f)G)).$$

By reducing the outermost redex, the variables E and G become captured by λB and must be renamed to B. Thus MN reduces in two steps to $(\lambda b.A)(M'N')$, where

$$M' = \lambda B.(\lambda D.d)((\lambda e.B)((\lambda F.f)B));$$
$$N' = (\lambda c.C)((\lambda D.d)((\lambda e.E)f).$$

Observe that M' and N' are "mirror images" of M and N, respectively. To get one from the other, we swap types p and q and rename variables (pressing the Shift key as needed). By repeating similar reduction steps we can keep our computation running forever. □

4.9. Assume all τ_i's are inhabited. We prove indirectly that then φ is not inhabited. Assume $\vdash N_i : \tau_i$ and $\vdash M : \varphi$. Then $\vdash M N_1 \cdots N_n : p$, which is a contradiction because type variables are not inhabited.

The proof that φ is inhabited if there is a τ_i that is not inhabited is by induction on the size of φ. Since p is the only type variable in φ, type τ_i has the form $\sigma_1 \to \cdots \to \sigma_m \to p$. If any of the σ_j's were not inhabited then by the induction hypothesis τ_i would be inhabited, but we know it is not. Thus there are N_j's with $\vdash N_j : \sigma_j$ and $\vdash \lambda x_1 \cdots x_n. x_i N_1 \ldots N_m : \varphi$. □

4.10. Prove by induction that every implicational formula built from only one propositional variable p is intuitionistically equivalent to $\top$ or to p. For a classical tautology the second case is impossible.[2] □

4.11. First reduce the general case to the case when τ is a type variable p. For the special case, to show $\vdash \tau_1 \to \cdots \to \tau_n \to p$, consider a substitution S such that $S(q) = \tau_1 \to \cdots \to \tau_n \to q$, for all variables q. For all types ρ, we now show that $\tau_1, \ldots, \tau_n \vdash S(\rho) \leftrightarrow \rho$. In particular, $\tau_1, \ldots, \tau_n \vdash \tau_i \to S(\tau_i)$, and "unfolding" $S(\tau_i)$ yields $\vdash S(\tau_i)$. It follows that $\vdash S(p)$. □

4.12. The term $\lambda x^{p \to \neg p} y^{\neg p \to p}. x(y(\lambda z^p.xzz))(y(\lambda z^p.xzz))$ has the desired type. □

4.15. For a type τ over $\to$ and $\wedge$, one defines a finite sequence $[\tau]$ of simple types: $[p] = p$, $[\tau \wedge \sigma] = [\tau][\sigma]$, and $[\tau \to \sigma] = \tau_1 \to \cdots \to \tau_m \to \sigma_1, \ldots, \tau_1 \to \cdots \to \tau_m \to \sigma_n$, where $[\sigma] = \sigma_1, \ldots, \sigma_n$, and $[\tau] = \tau_1, \ldots, \tau_m$. For an environment Γ, define $[\Gamma]$ as follows: If $\Gamma(x) = \tau$ and $[\tau] = \tau_1, \ldots, \tau_n$ then $[\Gamma](x_i) = \tau_i$, where x_i are fresh variables associated with x.

Now, if $\Gamma \vdash M : \tau$ in the calculus with products, then one can define a sequence $[M]$ of terms, such that $[\Gamma] \vdash [M] : [\tau]$, where $[M] : [\tau]$ is understood componentwise. In addition, if $M \to N$ and $[M] = M_1, \ldots, M_n$ and $[N] = N_1, \ldots, N_n$ then $M_i \to_\beta N_i$ for at least one i among $1, \ldots, n$ (and $M_i = N_i$ otherwise). Thus, if M has an infinite reduction sequence, the same must hold for at least one term M_i. □

---

[2] An alternative solution is to use the previous exercise.

4.16. Let $\mathbf{b} = (p \to p) \vee (p \to p)$, where p is a fixed type variable. One can "define" the product type $\tau \wedge \sigma$ as $\mathbf{b} \to (\tau \vee \sigma)$. A pair $\langle M^\tau, N^\sigma \rangle$ of type $\tau \wedge \sigma$ is then represented as $\lambda x^{\mathbf{b}}.\mathbf{case}\ x\ \mathbf{of}\ [y]\,\mathbf{in}_1(M)\ \mathbf{or}\ [z]\,\mathbf{in}_2(N)$. The projections can be defined so that $\pi_1 = \lambda f^{\tau \wedge \sigma}.\mathbf{case}\ f(\mathbf{in}_1(\lambda u^p u))\ \mathbf{of}\ [x]x\ \mathbf{or}\ [y]v$ and similarly $\pi_2 = \lambda f^{\tau \wedge \sigma}.\mathbf{case}\ f(\mathbf{in}_2(\lambda u^p u))\ \mathbf{of}\ [x]u\ \mathbf{or}\ [y]y$, where $v : \tau$ and $u : \sigma$ are some fixed variables. Clearly, we have $\pi_1(\langle M, N \rangle) \twoheadrightarrow^+ M$ and $\pi_2(\langle M, N \rangle) \twoheadrightarrow^+ N$. In this way, an infinite reduction sequence involving product types could be translated into an infinite reduction sequence without product types. □

4.17. The notion of η-reduction may be understood as follows. An elimination followed by an introduction of the same connective should be ignored. We can write the following eta rule for $\wedge$:
$$\langle \pi_1(M), \pi_2(M) \rangle \to_\eta M.$$
This idea does not generalize to disjunction. The problem is that the conclusion of the elimination rule is not necessarily identical to a premise of any of the introduction rules. Thus, in an elimination-introduction pair, one eliminates and introduces possibly different things.

However, the eta rule for both implication and conjunction can also be read as a postulate of extensionality: Every function is an abstraction. Every object of a product type is actually a pair. This hints at the following eta rule for disjunction:
$$\mathbf{case}\ M\ \mathbf{of}\ [x]\mathbf{in}_1(x)\ \mathbf{or}\ [y]\mathbf{in}_2(y) \to M,$$
or even a more general one:
$$\mathbf{case}\ M\ \mathbf{of}\ [x]F(\mathbf{in}_1(x))\ \mathbf{or}\ [y]F(\mathbf{in}_2(y)) \to FM. \quad \square$$

4.19. Let $M = \langle \pi_1(F\mathbf{K}), \pi_2(F\mathbf{K}) \rangle$, where $F = \lambda x.\langle x, \lambda y.x \rangle$. Then F can be assigned all types of the form $\tau \to (\tau \wedge (\sigma \to \tau))$, and since $\mathbf{K}$ has both the types $q \to q \to q$ and $q \to (q \to q) \to q$, we get $M : (q \to q \to q) \wedge (p \to q \to (q \to q) \to q)$. But $M \to_\eta F\mathbf{K}$, and $F\mathbf{K}$ cannot be assigned this type. □

4.20. The example term in the solution above is an untyped η-redex that is *not* an erasure of a typed η-redex. The Church-style version of that term uses two different $\mathbf{K}$'s. □

Solutions to Chapter 5

5.1. Hint: You may want to use axiom (A2) in the following forms:
- $(p \to q \to r) \to (p \to q) \to p \to r$;
- $((q \to r) \to (p \to q \to r) \to (p \to q) \to p \to r) \to$
  $((q \to r) \to p \to q \to r) \to (q \to r) \to (p \to q) \to p \to r$;
- $((q \to r) \to (p \to q) \to p \to r) \to$
  $((q \to r) \to p \to q) \to (q \to r) \to p \to r$;
- $((p \to q) \to ((q \to r) \to p \to q) \to (q \to r) \to p \to r) \to$
  $((p \to q) \to (q \to r) \to p \to q) \to (p \to q) \to (q \to r) \to p \to r$. □

5.6. One possibility is $\mathsf{S}(\mathsf{S}(\mathsf{KS})\mathsf{K})\mathsf{I}$. □

5.10. Definition 5.3.2 and Lemma 5.3.3 remain the same. Lemma 5.3.4 becomes:

If $M[x := N]\vec{P} \in [\![\tau]\!]$ and $N \in \mathrm{SN}$ then $(\lambda x\, M)N\vec{P} \in [\![\tau]\!]$.

But in the last step we deal with free variables and substitution, and Lemma 5.3.5 requires a stronger induction hypothesis. We must show that if M is of type τ and M' is obtained by a (type-preserving) substitution of computable terms for the free variables of M, then M' is computable (cf. Lemma 10.3.7). □

5.14. The term S(K(SII))(K(SII)) is in normal form. But just apply it... □

5.15. It suffices to define **K** and **S** using terms in A. By Exercise 4.2 all we need is to prove their types. □

5.16. It should be easy to see that all types that can be assigned to **K**, **I** and **S**° must be respectively of the form:
- $\tau \to \sigma \to \tau$;
- $\tau \to \tau$;
- $(\tau \to \tau) \to (\tau \to \tau \to \tau) \to (\tau \to \tau) \to \tau \to \tau$.

We can take all instances of the formulas above as our Hilbert style axioms. But we can easily simplify the system, replacing the last axiom by:
- $(\tau \to \tau \to \tau) \to (\tau \to \tau) \to \tau \to \tau$.

A Hilbert style proof will now correspond to a combinatory term built from the combinators of Exercise 5.13, and without loss of generality, we can deal only with terms in normal forms.

Suppose we have a proof of $(p \to p \to q) \to p \to q$, where p and q are propositional variables. Let M be the corresponding combinatory term in the language of Exercise 5.13, and assume that M is in a normal form. Clearly, M is neither $\mathbf{K}_{\sigma,\tau}$, nor $\mathbf{S}_{\tau,\tau,\tau}$, nor $\mathbf{I}_\tau$. If $M = \mathbf{K}_{\sigma,\tau} N$, then $\sigma = p \to q$ and N proves $p \to q$, which is not a valid tautology. Also $M = \mathbf{S}_{\tau,\tau,\tau} PQ$ or $M = \mathbf{S}_{\tau,\tau,\tau} N$ is impossible, because types of these expressions do not match $(p \to p \to q) \to p \to q$. □

5.17. One can define I as CKK. Indeed, CKK$x \twoheadrightarrow_w$ KxK $\twoheadrightarrow_w x$. □

5.18. Not true. A counterexample is $(p \to p \to p) \to p \to p$. □

5.20. Here is the multiset variant of rule (MP):

$$(\text{MP*}) \quad \frac{\Gamma' \vdash \varphi \quad \Gamma'' \vdash \varphi \to \psi}{\Gamma' \sqcup \Gamma'' \vdash \psi}$$

Let Γ be a multiset. We write $\Gamma \vdash_A \varphi$, if the judgement can be derived from the axioms of the **BCK**-logic with the help of (MP*) and (Id). And we write $\Gamma \vdash_L \varphi$, when it is derived from the **BCI** axioms using (MP*) and

$$(\text{Id-R}) \quad \varphi \vdash \varphi.$$

If Γ is an environment then the *multi-range* of Γ is the multiset $\text{mrg}(\Gamma)$ defined by $\text{mrg}(\varnothing) = \varnothing$ and $\text{mrg}(\Gamma, x : \varphi) = \text{mrg}(\Gamma) \sqcup \{\varphi\}$. One can now show that the deduction theorem holds for the two systems, and that
- If $\Gamma \vdash M : \varphi$, and M is an affine term, then $\text{mrg}(\Gamma) \vdash_A \varphi$.
- If $\Gamma \vdash_A \varphi$, then there exists an affine term M, and an environment Δ, such that $\text{mrg}(\Delta) = \Gamma$ and $\Delta \vdash M : \varphi$.
- If $\Gamma \vdash M : \varphi$, where M is linear and $\text{FV}(M) = \text{dom}(\Gamma)$, then $\text{mrg}(\Gamma) \vdash_L \varphi$.
- If $\Gamma \vdash_L \varphi$, then there exists a linear term M, and an environment Δ, such that $\text{mrg}(\Delta) = \Gamma$, $\text{dom}(\Delta) = \text{FV}(M)$, and $\Delta \vdash M : \varphi$. □

5.22. First show that if (Γ, τ) is a principal pair of an affine term then each type variable occurs in (Γ, τ) at most twice. (By Exercise 5.21, the principal pair of an affine term is the same as the principal pair of its normal form.) Now assume that each type variable occurs in (Γ, τ) at most twice and prove that there is at most one Church-style affine term M in long normal form such that $\Gamma \vdash M : \tau$. The proof is by induction with respect to the total number of arrows in (Γ, τ). Assume first that $\tau = \tau_1 \to \tau_2$. Then M cannot be of the form $xM_1 \ldots M_n$, as it would not be in long normal form (x has too few arguments). Thus, $M = \lambda y.N$, with long normal N, and we apply the induction hypothesis to the pair $(\Gamma \cup \{y : \tau_1\}, \tau_2)$.

The remaining case is when τ is a type variable p. There is at most one variable x declared in Γ to be of type $\sigma_1 \to \cdots \to \sigma_n \to p$, since p occurs at most twice. Thus $M = xN_1...N_p$. Apply the induction hypothesis to pairs (Γ', σ_j), where Γ' is Γ without the declaration $(x : \sigma_1 \to \cdots \to \sigma_n \to p)$. It is shown in [242], that affine terms with the same principal pair must actually be β-equal. □

5.23. Assume that a user-friendly term M is not strongly normalizing. Beginning with $M_0 = M$, we define an infinite sequence of user-friendly terms M_n which are not strongly normalizing. If M_n is of the form $\lambda x M'$ then we take $M_{n+1} = M'$. If M_n has the form $xP_1 \ldots P_n$, or the form $(\lambda x P_1)P_2P_3 \ldots P_n$, then we look for the least i such that P_i is not strongly normalizing, and define $M_{n+1} = P_i$ if that is possible. Otherwise, $M_n = (\lambda x P_1)P_2P_3 \ldots P_n$, and we take $M_{n+1} = P_1[x := P_2]P_3 \ldots P_n$ (note that a user-friendly term reduces to a user-friendly term).

Call an abstraction $\lambda x M$ *duplicating*, if x has more than one free occurrence in M. Let $d(M)$ be the number of duplicating abstractions in M and let $|M|$ be the length of M. We obtain a contradiction, by proving that the lexicographically ordered measure $\langle d(M_n), |M_n| \rangle$ is strictly decreasing in n.

If M_{n+1} is a proper subterm of M_n then of course $d(M_{n+1}) \le d(M_n)$ and $|M_{n+1}| < |M_n|$, so assume that M_{n+1} is obtained by head-reducing M_n. If the abstraction $\lambda x P_1$ is non-duplicating, then $d(M_{n+1}) \le d(M_n)$ and $|M_{n+1}| < |M_n|$. Otherwise, the reduction step removes one duplicating abstraction, and no new duplicating abstractions are created. Indeed, M_n is user-friendly (and thus abstractions inside P_2 are non-duplicating) and there is no external abstraction that could be turned into a duplicating one. □

Solutions to Chapter 6

6.3. Note that the deduction theorem holds for both systems. With that tool, part (i) is easy. In part (ii) define $\bot = \neg(p \to p)$. From $(A5_\neg)$ in the form $(\neg\bot \to \neg\varphi) \to (\neg\bot \to \varphi) \to \bot$ we obtain $\neg\varphi, \varphi \vdash \bot$. It follows that $\neg\varphi \vdash \varphi \to \bot$ and thus also $(\varphi \to \bot) \to \bot \vdash \neg\varphi \to \bot$. On the other hand, if $\psi = p \to p$ then axiom $(A5_\neg)$ yields $\neg\varphi \to \bot \vdash \varphi$ and we finally have $(\varphi \to \bot) \to \bot \vdash \varphi$. □

6.5. One possibility for part (ii) is $a \Rightarrow b = 2$ if $a < b$, and $a \Rightarrow b = 0$, otherwise. For part (iii), a possible choice (found by a computer program) is $0 \Rightarrow a = a$, $2 \Rightarrow 1 = 2$ and $a \Rightarrow b = 0$ otherwise (note that $(2 \Rightarrow (2 \Rightarrow 1)) \Rightarrow ((2 \Rightarrow 2) \Rightarrow (2 \Rightarrow 1))$ is 2). □

6.10. Adding the rule would imply:
- $\lambda x^p.(\mu a^{\neg(p \to p)}.m)x \to \mu a^{\neg(p \to p)}.m$;
- $\lambda x^p.(\mu a^{\neg(p \to p)}.m)x \to \lambda x^p.\mu b^{\neg p}.m$.

The two reduced terms have no common reduct—this would require rule (ζ'). □

6.18. For (i) there is the following derivation:

$$\dfrac{\dfrac{\dfrac{\dfrac{\dfrac{\varphi, (\varphi \vee (\varphi \to \psi)) \to \psi \vdash (\varphi \vee (\varphi \to \psi)) \to \psi \quad \dfrac{}{\varphi, (\varphi \vee (\varphi \to \psi)) \to \psi \vdash \varphi} \qquad \varphi, (\varphi \vee (\varphi \to \psi)) \to \psi \vdash \varphi \vee (\varphi \to \psi)}{\varphi, (\varphi \vee (\varphi \to \psi)) \to \psi \vdash \psi}}{(\varphi \vee (\varphi \to \psi)) \to \psi \vdash \varphi \to \psi}}{(\varphi \vee (\varphi \to \psi)) \to \psi \vdash \varphi \vee (\varphi \to \psi)}}{\vdash \varphi \vee (\varphi \to \psi)}$$

where (P) is used in the last step. □

6.19. The next day the poor shepherd brings to the king's palace a huge machine. The machine has two openings. One is marked "Put the philosopher's stone here!" and on the other it reads "The gold will fall out from here". That will perfectly work as long as the king cannot put the philosopher's stone into the machine. But what if, somehow, the king comes into the possession of the philosopher's stone? Then the shepherd's brother, hidden inside the machine, will grab the stone, and hand it discretely to the shepherd. The shepherd now can say: "Oops, Your Majesty, I've been mistaken. Here is the philosopher's stone!" □

Solutions to Chapter 7

7.3. From left to right use (Cut). □

7.4. Using Exercise 7.3 it is easy to see that the new rule is derived in LJ. Conversely, to show that any sequent of LJ is derivable in the new system, show by simultaneous induction:
- if $\Gamma \vdash \varphi$ in LJ then $\Gamma \vdash \varphi$ in the new system.
- if $\Gamma \vdash$ in LJ then $\Gamma \vdash \bot$ in the new system. □

7.8. In the case of contraction. □

7.10. $\Gamma \vdash_{LK} \Delta$ entails $\Gamma, \Gamma' \vdash \Delta, \Delta'$ in the new system for any Γ', Δ'. □

7.11. Using contraction it is easy to see that the new rules are derivable in LK. Conversely, show that $\Gamma \vdash_{LK} \Delta$ implies $\Gamma' \vdash \Delta'$ in the new system, where Γ', Δ' arise from Γ, Δ, respectively, by taking each formula precisely once. □

7.12. (After [489].) Clearly, the new rules are derivable in LK and LJ, respectively. For the converse direction, consider the classical case. Show that if there is a derivation of $\Gamma \vdash \Delta$ in the modified system, then there is a derivation also of $\Gamma', \Gamma \vdash \Delta, \Delta'$ for arbitrary Γ', Δ', of at most the same depth. So weakening holds derived.

Call a rule of some system *invertible* if, whenever there is a derivation of some depth of the conclusion, there is also a derivation of each premise of at most the same depth. Show that all the logical rules of the modified system are invertible, and use this to show by induction on derivations that the contraction rules hold as derived rules (also here one must keep track of the depth).

With weakening and contraction now available, it is easy to show that derivability in LK implies derivability in the modified system.

For LJ, again show that weakening is derived, and show that all rules except (L'''→) are invertible, and show that the latter rule is "half invertible," that is, if there is a derivation of $\Gamma, \varphi \to \psi \vdash \sigma$, then there is a another one of at most the same depth of $\Gamma, \psi \vdash \sigma$. Using this, then prove derivability of contraction as in the classical case. □

7.13. Consider the system obtained from LJ by the following steps (see Figure B.2):
- Read all left-hand sides as sets.
- Replace (Ax) by $\Gamma, \varphi \vdash \varphi$.
- Replace (L$\bot$) by the rule in Exercise 7.4.
- Replace the left logical rules with rules à la Exercise 7.11.
- Omit all structural rules and (Cut).

Derivability in the modified system entails derivability in LJ and conversely. To decide whether $\vdash \varphi$ can be derived in the modified LJ, it suffices (as in the proof of Corollary 7.3.9) to look for derivations in which no judgement occurs twice in the

---

Axiom:

$$\Gamma, \varphi \vdash \varphi \quad (\text{Ax})$$

Logical Rules:

$$\frac{\Gamma \vdash \bot}{\Gamma \vdash \sigma} \; (\text{L}\bot)$$

$$\frac{\Gamma, \varphi \wedge \psi, \varphi \vdash \sigma}{\Gamma, \varphi \wedge \psi \vdash \sigma} \; (\text{L}\wedge) \frac{\Gamma, \varphi \wedge \psi, \psi \vdash \sigma}{\Gamma, \varphi \wedge \psi \vdash \sigma} \qquad \frac{\Gamma \vdash \varphi \quad \Gamma \vdash \psi}{\Gamma \vdash \varphi \wedge \psi} \; (\text{R}\wedge)$$

$$\frac{\Gamma, \varphi \vee \psi, \varphi \vdash \sigma \quad \Gamma, \varphi \vee \psi, \psi \vdash \sigma}{\Gamma, \varphi \vee \psi \vdash \sigma} \; (\text{L}\vee) \qquad \frac{\Gamma \vdash \varphi}{\Gamma \vdash \varphi \vee \psi} \; (\text{R}\vee) \frac{\Gamma \vdash \psi}{\Gamma \vdash \varphi \vee \psi}$$

$$\frac{\Gamma, \varphi \to \psi \vdash \varphi \quad \Gamma, \varphi \to \psi, \psi \vdash \sigma}{\Gamma, \varphi \to \psi \vdash \sigma} \; (\text{L}{\to}) \qquad \frac{\Gamma, \varphi \vdash \psi}{\Gamma \vdash \varphi \to \psi} \; (\text{R}{\to})$$

FIGURE B.2: EXERCISE 7.13

same branch. Let n be the length of φ. Each time we move from a conclusion to a premise, the left hand side of the sequent remains the same or has a subformula of φ added. There are n such subformulas, so the adding can happen at most n times. However each addition can be separated by a segment of rule applications where the left hand side remains the same, but at each step the right hand side changes. The length of each such segment is also at most n, so the total length of any branch is at most n^2 (cf. the argument used in the proof of Lemma 4.2.3). □

7.16. Since (L$\to$) cannot be used, $\bot$ works exactly like a variable. □

7.17. Suppose the second claim were true. Since $\vdash_{\text{LP}} \varphi, \bot \to \bot$ we would have $(\bot \to \bot) \to \bot \vdash_P \varphi$. We also have $\vdash_P \bot \to ((\bot \to \bot) \to \bot)$, so $\vdash_P \bot \to \varphi$. Here $\bot$ is just a propositional variable, so substitution yields $\vdash_P (q \to q) \to \varphi$ and then $\vdash_P \varphi$, i.e. the system P and then also C would be inconsistent. □

7.18. No, e.g. not $(\lambda x^{\tau \to \sigma}.x)yz$.

7.19. Part (ii): $\pi_1(\textbf{case } z \textbf{ of } [x]v \textbf{ or } [y]v)$. Part (iii): Show that a term M is in $c\beta$-normal form iff either
- $M = x$ is a variable, or
- $M = \lambda x P$, or
- $M = P[y := xQ]$, or
- $M = \langle P, Q \rangle$, or
- $M = P[y := \pi_1(x)]$, or
- $M = P[y := \pi_2(x)]$, or
- $M = \textbf{in}_1(P)$, or

- $M = \mathbf{in}_2(P)$, or
- $M = \mathbf{case}\ x\ \mathbf{of}\ [v]Q\ \mathbf{or}\ [w]R$,

where P, Q, and R are normal forms. Then proceed as in the case without conjunction and disjunction. □

7.20. If $\vdash_{\mathsf{LP}} \varphi$ then $\vdash_{\mathsf{LP}} \varphi[\bot := p]$, and then $\vdash_{\mathsf{LK}} \varphi[\bot := p]$, so there is a winning proponent strategy for $\varphi[\bot := p]$ in the original Lorenzen type game. Then there is a winning proponent strategy for φ in the modified type of game (where $\bot$ is treated exactly like a propositional variable). The converse steps also hold. □

Solutions to Chapter 8

8.1. Both arguments are correct, but beware of the replacement. The *catch* here is that $\neg p \to \neg q$, $p \to \neg q \vdash \neg q$ but $\neg p \to q$, $p \to q \nvdash q$. □

8.4. The definition of φ' is by induction with respect to the length of the quantifier prefix of φ. If φ is quantifier-free then $\varphi' = \varphi$. If $\varphi = \forall a\, \psi$ then $\varphi' = \psi'$. Finally, if $\varphi = \exists a\, \psi(a)$ then $\varphi' = (\psi(b))'$, where b is such that $\vdash \psi(b)$. □

8.5. A quantifier-free formula φ' is *adequate* for a prenex formula φ iff φ' is obtained from φ by removing quantifiers and by replacing some existentially quantified variables by other variables. It follows from Exercise 8.4 that $\vdash \varphi$ implies $\vdash \varphi'$ for some φ', adequate for φ. The converse also holds, and can be shown by induction with respect to φ. In addition, up to variable replacement, there is only a finite number of formulas adequate for a given prenex formula φ. To verify provability of φ, one checks provability of this finite collection of open formulas. (Provability of open formulas is decidable, just like provability in propositional logic). □

8.6. The classical case is an easy application of the classical tautologies from Example 8.2.1(ii,ix) and Example 8.2.2(ii,ix) and of the definition of $\varphi \to \psi$ as $\neg \varphi \vee \psi$. In the intuitionistic case one can of course argue as follows: If every intuitionistic first-order formula was equivalent to a prenex formula, then intuitionistic first-order logic would be decidable by the previous exercise. Note however that there is a little gap in this argument. We know that every formula φ has an equivalent one in a prenex normal form, but the decision procedure must be able to *effectively find* this prenex formula. For this, list all possible proofs until a proof of $\varphi \leftrightarrow \psi$ is found, for a prenex normal form ψ. □

8.7. Here are the natural deduction rules as found e.g. in [161, 405, 489]. In ($\forall$I) and ($\exists$E) we require $b \notin \mathrm{FV}(\forall a\, \varphi)$ rather than $b \notin \mathrm{FV}(\varphi)$, so that $b = a$ is possible.

$$(\forall \mathrm{I})\ \frac{\Gamma \vdash \varphi[a := b]}{\Gamma \vdash \forall a\, \varphi}\ (b \notin \mathrm{FV}(\Gamma, \forall a\, \varphi))$$

$$(\exists \mathrm{E})\ \frac{\Gamma \vdash \exists a\, \varphi \quad \Gamma, \varphi[a := b] \vdash \psi}{\Gamma \vdash \psi}\ (b \notin \mathrm{FV}(\Gamma, \psi, \exists a\, \varphi))$$ □

8.9. The proof is similar to that of Theorem 2.3.11. For a given $c \in X$ we transform a $\mathcal{P}(X)$-structure $\mathcal{A}$ into an ordinary model $\mathcal{A}^c$, by defining

$$(\mathbf{a}_1, \ldots, \mathbf{a}_n) \in r^{\mathcal{A}^c} \quad \text{iff} \quad c \in r^{\mathcal{A}}(\mathbf{a}_1, \ldots, \mathbf{a}_n).$$

Both our models have the same domain A. One proves by induction that:

$$\mathcal{A}^c, \varrho \models \varphi \quad \text{iff} \quad c \in \llbracket \varphi \rrbracket_\varrho^{\mathcal{A}},$$

for all φ and all valuations ϱ in $\mathcal{A}$. (We write $\llbracket \varphi \rrbracket_\varrho^{\mathcal{A}}$ for values taken in $\mathcal{A}$.) Note that in the left to right direction in the case of $\forall$, one uses the fact that lower bounds in

$\mathcal{P}(X)$ are actually intersections (if c belongs to all sets in a family then it belongs to the infimum of that family). Now if $\mathcal{A}, \varrho \not\models \varphi$ then there is $c \notin [\![\varphi]\!]_\varrho$, and thus $\mathcal{A}^c, \varrho \not\models \varphi$ as well. □

8.11. Let $\mathcal{H} = \mathcal{P}(\mathbb{R})$ and let A be the set of all half-lines of the form (r, ∞), for any $r \in \mathbb{R}$. Then $\bigcap A = \varnothing$, and by Lemmas 2.5.6 and 2.5.7, there is a prime filter F containing A. Let $\mathcal{G}$ be the algebra $\mathcal{P}(\mathcal{Z})$ of sets of prime filters in $\mathcal{H}$. By Exercise 2.15, we can consider $\mathcal{H}$ as a subalgebra of $\mathcal{G}$ under the embedding $a \mapsto Z_a$. Let $\mathcal{A}$ be the image of A under this embedding. Then $\inf_\mathcal{G} \mathcal{A}$ is the intersection of all sets Z_a for $a \in A$, that is the set $D = \{F \in \mathcal{Z} \mid A \subseteq F\}$. Since A is contained in at least one prime filter, we have $D \neq \varnothing$, i.e., the infimum of $\mathcal{A}$ is non-zero. □

8.12. First observe that $X \subseteq con(X)$, and dually, $X \subseteq X \Downarrow\Uparrow$. Then note that

$$\text{if } X \subseteq Y \text{ then } Y\Downarrow \subseteq X\Downarrow \text{ and } Y\Uparrow \subseteq X\Uparrow. \qquad (*)$$

In particular $X \subseteq Y$ implies $con(X) \subseteq con(Y)$, and since $X\Uparrow \subseteq X\Uparrow\Downarrow\Uparrow$, we also have $con(con(X)) \subseteq con(X)$, for all $X \subseteq \mathcal{H}$. This implies part (i).

It is easy to prove that an intersection of a family of cones is always a cone. Thus, Con is a complete lattice with $\inf_{Con} \mathcal{A} = \bigcap \mathcal{A}$ and $\sup_{Con} \mathcal{A} = con(\bigcup \mathcal{A}) = \bigcap \{Y \mid Y \in Con \text{ and } \bigcup \mathcal{A} \subseteq Y\}$, for $\mathcal{A} \subseteq Con$.

The inclusion from right to left in (ii) is a simple consequence of $X\Uparrow \subseteq (X \cdot Y)\Uparrow$ and of $(*)$. To prove the other inclusion, assume that $p \in con(X) \cap con(Y)$, and let $q \in (X \cdot Y)\Uparrow$. We need to show that $p \leq q$.

Let $y \in Y$. For all $x \in X$ we have $x \sqcap y \leq q$, and thus $x \leq y \Rightarrow q$, where $\Rightarrow$ is the relative pseudo-complement in $\mathcal{H}$. Hence $X \subseteq (y \Rightarrow q)\Downarrow$. It follows that $p \in con(X) \subseteq con((y \Rightarrow q)\Downarrow) = (y \Rightarrow q)\Downarrow$ so we have $p \leq y \Rightarrow q$. In other words, $y \leq p \Rightarrow q$, for all $y \in Y$, so $Y \subseteq (p \Rightarrow q)\Downarrow$. But $p \in con(Y) \subseteq con((p \Rightarrow q)\Downarrow) = (p \Rightarrow q)\Downarrow$, thus $p \leq p \Rightarrow q$, i.e. $p \leq q$, and this completes the proof of part (ii).

To show part (iii), first observe that $X \cdot Y = X \cap Y$ for downward-closed X, Y. It is also easy to check that for downward-closed Z, in particular for $Z \in Con$,

$$Z \subseteq X \Rrightarrow Y \quad \text{iff} \quad Z \cap X \subseteq Y, \qquad (**)$$

It remains to show that $X, Y \in Con$ implies $X \Rrightarrow Y \in Con$. By $(**)$ this amounts to proving $X \cap con(X \Rrightarrow Y) \subseteq Y$. Since $X \Rrightarrow Y$ is downward-closed, we have:

$$X \cap con(X \Rrightarrow Y) = con(X) \cap con(X \Rrightarrow Y) =$$
$$con(X \cdot (X \Rrightarrow Y)) = con(X \cap (X \Rrightarrow Y)) \subseteq con(Y) = Y. \qquad \square$$

8.13. (After A.G. Dragalin.) We show that the mapping $i : \mathcal{H} \to Con$ (see Exercise 8.12) given by $i(p) = p\Downarrow$ is such an embedding. It is easy to see that i is monotone. To prove that i preserves the existing lower bounds, assume that $a = \inf_\mathcal{H} X$. Then $\inf_{Con} i(X) = \bigcap i(X) = \{q \mid q \leq p, \text{ for all } p \in X\} = \{q \mid q \leq a\} = a\Downarrow$. Now assume $a = \sup_\mathcal{H} X$. Then $\sup_{Con} i(X) = con(X) = a\Downarrow$. Finally, we have

$$p\Downarrow \Rrightarrow q\Downarrow = \{x \mid p\Downarrow \cap x\Downarrow \subseteq q\Downarrow\} = \{x \mid (p \sqcap x)\Downarrow \subseteq q\Downarrow\} =$$
$$\{x \mid p \sqcap x \leq q\} = \{x \mid x \leq p \Rightarrow q\} = (p \Rightarrow q)\Downarrow,$$

so that i also preserves the relative pseudo-complement. □

8.14. Enumerate all variables as $a_0, a_1, a_2, \ldots$. In Γ and φ replace every a_i by a_{2i} to obtain Γ' and φ'. Show that $\Gamma \vdash \varphi$ iff $\Gamma' \vdash \varphi'$ and that $\Gamma \models \varphi$ iff $\Gamma' \models \varphi'$. □

8.15. (Tarski) First we define a sequence of elements $b_n \in A_n$ so that the sets $Z_n = \{a, -b_0 \sqcup a_0, \ldots, -b_{n-1} \sqcup a_{n-1}\}$ satisfy $\inf Z_n \neq 0$. As b_n take any $b \in A_n$ such that $z_n \sqcap (-b \sqcup a_n) \neq 0$, where $z_n = \inf Z_n$. Such b must exist, because otherwise $(z_n \sqcap -b) \sqcup (z_n \sqcap a_n) = 0$, for all $b \in A_n$, so that $z_n \leq a_n$ and $z_n \leq -a_n$, i.e. $z_n = 0$.

Now let Z be the union of all Z_n. By Lemmas 2.5.6 and 2.5.7 there is a prime filter F such that $Z \subseteq F$. This is the filter we are looking for. Indeed, if $A_n \subseteq F$ then $a_n = b_n \sqcap (-b_n \sqcup a_n) \in F$. □

8.16. Suppose that $\nvdash \varphi$. Without loss of generality we can assume that φ does not contain existential quantifiers. Consider the Lindenbaum algebra $\mathcal{L}$, as in the proof of Theorem 8.5.6. Let $a = [\neg\varphi]_\sim$ and let $A_n = \{[\psi_n[a := t]]_\sim \mid t \text{ is a term}\}$ where ψ_n is an enumeration of all formulas. Observe that $a \neq 0$ and that for $a_n = [\forall a\, \varphi]_\sim$ we have $a_n = \inf A_n$. Apply the Rasiowa-Sikorski lemma and let F be the appropriate filter. Since $\psi \vee \neg\psi$ is valid for all ψ, we have either $[\psi]_\sim \in F$ or $[\neg\psi]_\sim \in F$. Define a term model by letting $r^{\mathcal{A}}(t_1,\ldots,t_k) = 1$ when $[rt_1,\ldots,t_k]_\sim \in F$ and $r^{\mathcal{A}}(t_1,\ldots,t_k) = 0$ otherwise. Then prove by induction that $[\![\psi]\!]_\varrho = 1$ iff $[\psi]_\sim \in F$, for all ψ, where $\varrho(a) = a$ for all a. Since $[\neg\varphi]_\sim \in F$ we have $[\![\varphi]\!]_\varrho \neq 1$, i.e. $\nvDash \varphi$. □

8.17. If Γ is not satisfiable then $\Gamma \vDash \bot$, and thus $\Gamma \vdash \bot$ by completeness. The proof uses only finitely many assumptions, so there is a finite subset $\Gamma_0 \subseteq \Gamma$ with $\Gamma_0 \vdash \bot$. It follows that Γ_0 is not satisfiable.

Now suppose that Δ is satisfied in all finite models, and let Γ be obtained from Δ by adding all sentences of the form $\exists a_1 \ldots a_n.\, \zeta(a_1,\ldots,a_n)$, where $\zeta(a_1,\ldots,a_n)$ is the conjunction of all $a_i \neq a_j$ for $i,j \leq n$, $i \neq j$. All finite subsets of Γ are satisfied (in sufficiently large finite sets), but a model of the whole set Γ must be infinite. □

8.20. In all the counterexamples below, the domain of individuals is $A = \mathbb{N} - \{0\}$. The underlying Heyting algebra is $\mathcal{O}(\mathbb{Q})$ in part (viii) and $\mathcal{O}(\mathbb{R})$ otherwise. Compare (ii), (iv) and (iii) to Example 2.1.4(v), (ix) and Exercise 2.21(v) respectively.

- If we take $r^{\mathcal{A}}(n) = (-\infty, 0)$ for even n and $r^{\mathcal{A}}(n) = (0, \infty)$ for odd n, then formulas (ii), (iii) and (iv) are all of value $\mathbb{R} - \{0\}$.
- The case of (v) reduces to (ii) if we take $p^{\mathcal{A}} = \varnothing$.
- In (vi) put $r^{\mathcal{A}}(2k+1) = q^{\mathcal{A}}(2k) = (0, \infty)$, $r^{\mathcal{A}}(2k) = (-1, \infty)$ and define $q^{\mathcal{A}}(2k+1) = (1, \infty)$, for all k.
- For part (viii), let χ_n be a decreasing sequence of positive irrationals with limit zero, and let $r^{\mathcal{A}}(n) = \{x \in \mathbb{Q} \mid x < -\chi_n \text{ or } \chi_n < x\}$.
- The counterexample for (i) works also for (x). □

8.21. As an example we give an inhabitant of the left-to-right implication in part (ii):
$$\lambda x^{\neg\neg \exists a\, \neg\neg ra} \lambda y^{\neg \exists a\, ra}.\, x(\lambda z^{\exists a\, \neg\neg ra}.\, \textbf{let } [b, u : \neg\neg rb] = z \textbf{ in } u(\lambda v^{rb}.y[b, v]_{\exists a\, ra})).$$ □

8.22. First show the following lemma: If A is an open segment of $\mathbb{R}$ and B, C are disjoint open sets then $A \subseteq B \cup C$ implies $A \subseteq B$ or $A \subseteq C$. In order to show that our formula is valid in $\mathcal{O}(\mathbb{R})$-structures, it suffices to know that for an arbitrary family of open sets $\{P_t \mid t \in T\}$ the following inclusion holds:
$$\text{Int}(\bigcap_{t \in T}(P_t \cup \sim P_t)) \subseteq \bigcup_{t \in T} P_t \cup \text{Int}(\bigcap_{t \in T} \sim P_t).$$

Assume that a point **a** belongs to the left hand side. Since the set is open, there is an open segment A containing **a** and enclosed in $\bigcap_{t \in T}(P_t \cup \sim P_t)$, i.e., enclosed in every set $P_t \cup \sim P_t$. From our little lemma it follows that $A \subseteq P_t$ or $A \subseteq \sim P_t$ for each t. If it happens that $A \subseteq P_t$, for at least one t, then we are done. Otherwise $A \subseteq \bigcap_{t \in T}(\sim P_t)$. But A is open, so it is contained in the interior. □

8.24. We only give a few example solutions. A hint for the other parts: First construct counterexamples for appropriate propositional formulas. For instance, for part (ii) consider Example 2.1.4(v). In the solutions below we abuse the notation writing e.g. $c \Vdash \varphi(n)$ instead of $c, \varrho \Vdash \varphi(a)$, for $\varrho(a) = n$.

(i) Let $\mathcal{C} = \langle \mathbb{N}, \leq, \{\mathcal{A}_n \mid n \in \mathbb{N}\}\rangle$ where $\mathcal{A}_n = \langle \mathbb{N}, r_n \rangle$ and $r_n(j)$ holds iff $j < n$. Then $1 \not\Vdash \neg\neg\forall a(ra \vee \neg ra)$. Otherwise there would be a number n such that $n \Vdash \forall a(ra \vee \neg ra)$, in particular $n \Vdash r(n) \vee \neg r(n)$, which does not hold. The same example applies to (x).

(iv) The model consists of two states $c < d$, with $\mathcal{A}_c = \{1\}$ and $\mathcal{A}_d = \{1, 2\}$. Let $c, d \not\Vdash r(1)$ and $d \Vdash r(2)$. This works also for (vii) and (viii).

(ix) The model consists of two states $c < d$, with $\mathcal{A}_c = \{1\}$ and $\mathcal{A}_d = \{1, 2\}$. Let $c \Vdash r(1)$ and $c \not\Vdash p$, and let $d \Vdash p$ and $d \Vdash r(1)$, but $d \not\Vdash r(2)$.

8.25. Our first example shows that $\forall$ is not definable from $\exists$ and the other connectives. Let the domain of individuals be $\mathcal{A} = \mathbb{N}$, and the underlying Heyting algebra be $\mathcal{O}(\mathbb{R})$. Define $r^{\mathcal{A}}(n) = \mathbb{R} - \{n\}$. Then $[\![\forall a\, ra]\!] = \mathbb{R} - \mathbb{N}$. But if ψ has no occurrences of $\forall$, then the value of ψ is either empty or has a finite complement (easy induction). It follows that no formula written without $\forall$ is equivalent to $\forall a\, ra$.

To show that $\exists$ is not definable, assume that our signature consists of a single unary relation symbol r. Take a Kripke model with two states $c < d$. The structures associated to these states are $\mathcal{A}_c = \langle \{0\}, r^c \rangle$ and $\mathcal{A}_d = \langle \{0, 1\}, r^d \rangle$, where $r^c = \emptyset$ and $r^d = \{1\}$. Let $\varrho(a) = 0$, for all variables a. One proves by induction that $c, \varrho \Vdash \psi$ iff $d, \varrho \Vdash \psi$, for all formulas without $\exists$. It follows that no $\exists$-free sentence is equivalent to $\exists a\, ra$. □

8.26. The implication from right to left is always valid so we only consider the implication $\forall a(\psi \vee \varphi(a)) \to \psi \vee \forall a\, \varphi(a)$. Assume that $\mathcal{A}_c = \mathcal{A}$ for all c in our model. Let $c, \varrho \Vdash \forall a(\psi \vee \varphi(a))$, and suppose that $c, \varrho \not\Vdash \psi$. Thus for all $\mathbf{a} \in \mathcal{A}$ it must be that $c, \varrho(a \mapsto \mathbf{a}) \Vdash \varphi(a)$. By monotonicity, we have $c', \varrho(a \mapsto \mathbf{a}) \Vdash \varphi(a)$, for all $c' \geq c$. It follows that $c, \varrho \Vdash \forall a\, \varphi(a)$. □

8.28. First observe that $c'', \varrho \Vdash \varphi \vee \neg\varphi$ holds if c'' is a maximal state. Now suppose $c, \varrho \not\Vdash \neg\neg\forall a(\varphi \vee \neg\varphi)$ in a finite model. There is $c' \geq c$ with $c', \varrho \Vdash \neg\forall a(\varphi \vee \neg\varphi)$. Take a maximal state $c'' \geq c'$ and we obtain $c'', \varrho(a \mapsto \mathbf{a}) \not\Vdash \varphi \vee \neg\varphi$, for some $\mathbf{a}$. □

8.29. To derive (E3) take $c = d$ as φ in (E2). For (E4) take $b = c$ and use (E1). □

8.32. Recall that $\varphi \leftrightarrow \psi$ abbreviates $(\varphi \to \psi) \wedge (\psi \to \varphi)$. Note the similarity between (iii) and the combinator **S**. Terms for (v) and (vi) are given in Example 8.7.3 and for (i) in the proof of Proposition 8.7.4.

(ii) $\langle \lambda x^{\neg\exists a\, \varphi} \lambda a\, \lambda y^{\varphi}. x[a, y]_{\exists a\, \varphi}, \lambda x^{\forall a \neg \varphi} \lambda y^{\exists a\, \varphi}. \mathbf{let}\ [a, z] = y\ \mathbf{in}\ xaz \rangle$.

(iii) $\lambda x^{\forall a(\varphi \to \psi)} \lambda y^{\forall a\, \varphi} \lambda a. xa(ya)$.

(iv) $\lambda x^{\forall a(\varphi \to \psi)} \lambda y^{\exists a\, \varphi}. \mathbf{let}\ [a, z] = y\ \mathbf{in}\ [a, xaz]$.

(vii) $\lambda x^{\psi} \lambda a. x$.

(viii) $\lambda x^{\tau} \lambda a.\ \mathbf{case}\ \pi_1(x)a\ \mathbf{of}\ [y] y\ \mathbf{or}\ [z]\varepsilon_{\varphi}(\pi_2(x)(\lambda v^{\forall a\, \varphi}. z(va))$, where $\tau = \forall a(\varphi \vee \neg\varphi) \wedge \neg\neg\forall a\, \varphi$.

(ix) $\langle M, N \rangle$, where $N = \lambda x^{\psi \wedge \exists a\, \varphi(a)}. \mathbf{let}\ [a, y] = \pi_2(x)\ \mathbf{in}\ [a, \langle \pi_1(x), y \rangle]$, and $M = \lambda x^{\exists a(\psi \wedge \varphi)}. \langle \mathbf{let}\ [a, y] = x\ \mathbf{in}\ \pi_1(y), \mathbf{let}\ [a, y] = x\ \mathbf{in}\ \pi_2(y)] \rangle$.

(x) $\lambda x^{\neg\neg\forall a\, \varphi} \lambda a \lambda y^{\neg\varphi}. x(\lambda z^{\forall a\, \varphi}. y(za))$. □

8.34. In the proof of Theorem 8.8.2 we have used a possibly unbounded number of binary relation symbols $R_1, \ldots, R_n$ corresponding to the internal states of a given counter automaton. One can "encode" these predicates using fresh variables $c^1, \ldots, c^n$ and one ternary relation symbol P, so that $Pc^i ab$ can be used instead of $R_i ab$ in the construction. Another possibility is to choose a fixed two-counter automaton $\mathcal{A}$, and a halting problem of the form: *Given m, n, does $\mathcal{A}$ halt if started in the configuration $\langle q_0, m, n \rangle$?* For a suitable $\mathcal{A}$, this specific problem is undecidable too [245]. □

Solutions to Chapter 9

9.2. (i) We prove $\forall c\,((a+b)+c = a+(b+c))$ by induction and then generalize over a and b. The base step $(a+b)+0 = a+(b+0)$ follows from the axiom $\forall a\,(a+0 = a)$, with the help of (9.1) and transitivity and symmetry of equality. Then we must derive $(a + b) + sc = a + (b + sc)$ from $(a + b) + c = a + (b + c)$. Using (9.1), the axioms $\forall a \forall b\,(a = b \to sa = sb)$ and $\forall a \forall b\,(a + sb = s(a + b))$, transitivity and symmetry of equality, we can formalize the calculation $(a+b)+sc = s((a+b)+c) = s(a+(b+c)) = a + s(b+c) = a + (b + sc)$.

(ii) First prove by induction that $\forall b\,(0+b = b+0)$. This is the base of the main induction (with respect to a). The induction step is to derive $\forall b\,(sa + b = b + sa)$ from $\forall b\,(a + b = b + a)$. This also goes by induction (with respect to b). □

9.3. To prove (9.1) we proceed by induction with respect to b', and for the base step we show $\forall a \forall a'\,(a = a' \to b = 0 \to a + b = a' + 0)$. This is easy, but requires induction with respect to b. The induction step for b' also goes by induction on b. In a similar way one shows (9.2). To prove (9.3), generalize (9.1) and (9.2) to arbitrary terms and use Exercise 8.29. □

9.5. Assume the left-hand side and prove $\forall a \forall b(b < a \to \varphi(b))$ by induction. □

9.6. We proceed by induction with respect to $n + m$, using the induction scheme of Exercise 9.5. If $n = m = 1$ then $u = 1$ and $v = 0$. Otherwise, let e.g. $n < m$ and let $k = m - n$. By the induction hypothesis we have $u' \cdot n - v' \cdot k = 1$, for some u', v' and we can take $u = u' + v'$, $v = v'$. For $n > m$ the proof is similar. □

9.7. Peano Arithmetic is effectively axiomatizable and thus recursively enumerable. Therefore, if it were complete it would have to be recursive. (To check if a sentence φ is a theorem of PA, one would run two tests in parallel: one for φ and another for $\neg\varphi$. One of these would give a positive result in a finite number of steps.) On the other hand, the completeness of PA would imply $PA = Th(\mathcal{N})$, and we know that $Th(\mathcal{N})$ is undecidable.

The proof above is only seemingly simpler than that of Section 9.3. When we say that Peano Arithmetic is recursively enumerable, we actually refer to a coding machinery, essentially equivalent to Gödel's. □

9.8. Arithmetic is about numbers, not about truth. To assert that two plus two is four we must know how much is two plus two. We do not need to know if "2 plus 2 is 4" is true. The statements "2 plus 2 is 4" and "2 plus 2 is 4 is true" have quite different meaning, and should not be confused with each other. □

9.9. Let $A \subseteq \mathbb{N}$ be recursively enumerable. There exists a recursive set B such that $A = \{n \mid (n,m) \in B$, for some $m\}$, and a recursive function f of two arguments such that $B = f^{-1}(\{0\})$. If the function f is representable by a formula $\varphi(a,b,c)$ then A is weakly representable by $\exists b\,\varphi(a,b,0)$. □

9.10. Let $A \subseteq \mathbb{N}$ be recursively enumerable. From Exercise 9.9 we know that A is weakly representable by some formula φ. If PA was decidable one could decide membership in A by verifying whether $\varphi(\underline{n})$ is provable in PA. □

9.11. Yes. All you need is a soundness theorem restricted to one particular finite model $\mathcal{A}$ of your choice. Just fix a set, say $\{1,2,3\}$, define relations (say, all of them empty) and functions (say, all of them constantly 1), and proceed by induction to show that $\vdash \varphi$ implies $\mathcal{A} \models \varphi$ (in the classical sense). All this argument can be formalized in arithmetic. (Observe that the requirement that $\mathcal{A}$ is finite is essential. Consistency of PA is also a consequence of $\mathcal{N} \models PA$ but $\mathcal{N}$ is infinite.) □

9.12. Unfortunately yes. Consider this one: "a is not a Gödel number of a proof of **Con**", where **Con** is the formula expressing consistency of PA. □

9.15. Take $Z \vee \neg Z$ where $\text{PA} \nvdash Z$ and $\text{PA} \nvdash \neg Z$. □

9.16. The proof is by induction with respect to a. One first shows the base step:
$$\forall b (0 = b \vee \neg(0 = b)),$$
by induction with respect to b. The induction step uses the axiom $\forall a(sa = 0 \to \bot)$. Then from the induction hypothesis $\forall b\, (a = b \vee \neg(a = b))$ one derives
$$\forall b (sa = b \vee \neg(sa = b)),$$
again by induction with respect to b. □

9.17. From $\text{HA} \vdash \neg\neg \exists x\, \varphi(x)$ it follows that $\text{PA} \vdash \exists x\, \varphi(x)$, and thus $\mathcal{N} \models \varphi(\underline{k})$, for some k. But we have $\text{HA} \vdash \varphi(\underline{k}) \vee \neg \varphi(\underline{k})$ and thus, by the disjunction property (Proposition 9.6.6(ii)), either $\varphi(\underline{k})$ or $\neg\varphi(\underline{k})$ is a theorem of HA. In the first case we conclude that $\text{HA} \vdash \exists x\, \varphi(x)$, the second case is impossible. See [488, vol. II] for a proof that HA is closed under the general case of Markov's rule. □

9.19. The realizers are
- $e_1 = \Lambda a.\Lambda b.a$;
- $e_2 = \Lambda a.\Lambda b.\Lambda c.\{\{a\}(c)\}(\{b\}(c))$.

Do these look familiar? □

9.20. One shows $\text{PA} \vdash \varphi$ by a trivial application of the De Morgan's law and *tertium non datur*. On the other hand, $\neg \varphi$ is realizable because φ is not. Indeed, realizability of φ would imply decidability of the problem "*Given a number a, does M_a halt on input a?*". (Recall that $T(a,a,c)$ is an atomic formula). □

9.21. In HA, the formula $\forall a \exists b \forall c (T(a,a,b) \vee \neg T(a,a,c))$ (where $T(a,a,b)$ is a single equation) is equivalent to the formula in Exercise 9.20. □

9.22. The very definition of realizability. This is a binary relation between the realizer and (the number of) the formula realized. To formalize the consistency proof one has to express this binary relation as a formula of arithmetic. But our definition is by induction. See also the discussion following Theorem 10.3.8. □

9.23. The apparent "proof" is certainly an algorithm capable to verify each particular instance of Fermat's Last Theorem. But it does not obey "Kreisel's dictum:"

We recognize a proof (...) when we see it.

What we would like to accept as a proof should be a *proof term* that can be assigned type $\forall mnk\ell (m^n + m^k \neq m^\ell)$. That should be a static typing, independent of any concrete values m, n, k, ℓ. Our construction can only be verified dynamically. □

Solutions to Chapter 10

10.3. Let $M = \lambda fxy.\mathbf{R}_{\text{int}}(\lambda uv.fv)xy$. If F_k defines f_k then $F_{k+1} = \lambda x.MF_k xx$ defines f_{k+1}. □

10.4. If functions g, h are respectively defined by terms G and H and
$$\begin{aligned} f(0, n_1, \ldots, n_k) &= g(n_1, \ldots, n_k); \\ f(m+1, n_1, \ldots, n_k) &= h(m, n_1, \ldots, n_k, f(m, n_1, \ldots, n_k)), \end{aligned}$$
then the function f is defined by $F = \lambda x \vec{y}.\mathbf{R}_{\text{int}}(\lambda uv.Hu\vec{y}v)(G\vec{y})x$.

To show that all functions definable with the help of $\mathbf{R}_{\text{int}}$ are primitive recursive, one proceeds by induction with respect to the size of closed long normal forms of types $\text{int}^k \to \text{int}$, which can be assumed to be of shape $\lambda \vec{x}.M^{\text{int}}$. By analyzing the shapes of normal forms with all free variables among $\vec{x}$, we find out that M is either a variable or zero or it must be of the form sQ or $\mathbf{R}_{\text{int}}NPQ$, where $N : \text{int} \to \text{int} \to \text{int}$ and $P, Q : \text{int}$. Apply the induction hypothesis to $\lambda \vec{x} uv.N$, $\lambda \vec{x}.P$, and $\lambda \vec{x}.Q$, and use primitive recursion and composition. □

10.6. The definition is $F_\omega = \lambda x. \mathbf{R}_{\text{int} \to \text{int}}(\lambda z \varphi y. M \varphi yy) sxx$, where M is as in Exercise 10.3. Indeed, the operator $\lambda \varphi y. M \varphi yy$ applied to F_k returns F_{k+1}. □

10.8. Yes, it will come to an end, but not so soon. In a similar style one can define a transfinite sequence of definable functions f_α, for $\alpha < \epsilon_0$. Here, ϵ_0 is the least ordinal such that $\omega^{\epsilon_0} = \epsilon_0$, i.e., it is the limit of the sequence $\alpha_0 = 0$, $\alpha_{n+1} = \omega^{\alpha_n}$. See [434, 444] for more on this. □

10.9. The shortest solution is to apply Schwichtenberg's Theorem 3.7.4. Indeed, all extended polynomials are primitive recursive. An "educated" argument is to use the fact that strong normalization for the simply typed lambda calculus is provable within PA. Thus, every function definable in $\lambda_\to$ is provably total in PA. A direct solution is to "translate" definitions in $\lambda_\to$ using the following identity

$$\mathbf{R}_{\text{int}}(\lambda z f) x \overline{m} =_\beta f^m(x).$$

□

10.10. All terms of type $\text{int} \to \text{int}$ can be effectively enumerated, and each of these terms defines a total function (because of strong normalization). But total recursive functions cannot be effectively enumerated. □

10.12. Otherwise, this function would be definable in $\mathbf{T}$, by a certain term G. The following is a term of type $\text{int} \to \text{int}$, not equal to any of the F_n's:

$$\lambda x. \mathbf{R}_{\text{int}}(\lambda uv.su)0(s(Gxx)).$$

□

10.13. Let $\{F_n\}_{n \in \mathbb{N}}$ and g be as in Exercise 10.12. If we could prove strong normalization in PA, we could prove in PA the totality of the function g. □

10.16. It suffices to prove that every function strongly representable in HA is definable in $\mathbf{T}$. Suppose that f is strongly represented by a formula φ. Proceeding as in the proof of Theorem 10.4.9, we obtain a term M which ⊢-m-realizes the formula $\forall a \exists b \varphi(a,b)$. Thus, for every n, there is m such that $\pi_1(M\overline{n}) =_T \overline{m}$, and $\pi_2(M\overline{n})$ ⊢-m-realizes $\varphi(\underline{n}, \underline{m})$, in particular HA ⊢ $\varphi(\underline{n}, \underline{m})$. It follows that $\varphi(\underline{n}, \underline{m})$ holds in the standard model, and thus $m = f(n)$. We conclude that $\lambda x. \pi_1(Mx)$ defines f in system $\mathbf{T}$. □

Solutions to Chapter 11

11.1. In part (i) it is enough to show $\neg(p \vee \neg p) \vdash \bot$ and in part (ii) one should observe that $p \leftrightarrow (\neg q \vee q) \vdash \neg \neg p$. Part (iii) goes by induction with respect to φ. Hints for parts (iv) and (v) are encoded below as terms of system $\mathbf{F}$:
(iv) Let $\tau(\varrho) = (\varrho \to p) \to q$. Then $\lambda x^{\tau(\varphi)} \Lambda r \lambda y^{\tau(r) \to r}. y(\lambda z^{r \to p} x(\lambda u^\varphi z(ury)))$ has type $\tau(\varphi) \to \varphi$ and $\lambda x^\varphi \lambda y^{\varphi \to p}. xq(\lambda z^{\tau(q)}. z(\lambda u^q. yx))$ has type $\varphi \to \tau(\varphi)$.
(v) Let $\tau = ((s \to r) \to q) \to p$ and $\sigma = \forall t(((t \to r) \to q) \to (t \to s) \to p)$. Then $\lambda x \Lambda t \lambda yz.x(\lambda u.y(\lambda w.u(zw)))$ has type $\tau \to \sigma$, and $\lambda xz.xsz(\lambda vv)$ has type $\sigma \to \tau$ (types of bound variables omitted for brevity). □

11.2. If $t \notin Z$ is an accumulation point of T then an open segment $(t - \varepsilon, t + \varepsilon)$ is disjoint from Z. If $x, y \in T$ are close enough to t then $\frac{1}{2}(|x - Z| + |y - Z|) \geq |x - y|$.

Now assume that $z \in Z$ is not an accumulation point of T, i.e. $|z - T| \geq \varepsilon$, for some ε (note that $Z \cap A = \varnothing$). But z is in the closure of A, so there is $y \in A$

as close to z as we need. By the maximality of T, there is a number $x \in T$, with $|x - y| < \frac{1}{2}(|x - Z| + |y - Z|)$. Then $|z - x| \leq |z - y| + |y - x| < \varepsilon$, if $|z - y|$ is small enough. See [399] for details. □

11.3. Formula (i) is not valid even with respect to the classical zero-one semantics, and formula (ii) is not satisfied in the metric space $\{0\} \cup \{\frac{1}{n} \mid n \in \mathbb{N} - \{0\}\}$, if p is interpreted as the set $A = \{\frac{1}{n} \mid n \in \mathbb{N} - \{0\}\}$, because $A \subseteq \mathord{\sim} B \cup B$ for all B.

To see that formula (i) is valid in $\mathcal{O}(\mathbb{R})$, observe that $x \notin (-\infty, x) \cup (x, \infty)$, for every $x \in \mathbb{R}$. In part (ii) we recall the argument in [399]. Let A, Z, and T be as in Exercise 11.2 (use Lemma A.1.1 to see that such T must exist). Then $A \cap Z = \varnothing$, and if we take $B = A - T = A - (Z \cup T)$ then B is an open set with $\mathord{\sim} B \cap A = \varnothing$. Now define $C = A \Rightarrow (\mathord{\sim} B \cup B) = \text{Int}(-A \cup \mathord{\sim} B \cup B)$ and observe that the complement $-C = -\text{Int}(-A \cup B) = -\text{Int}(-T)$ is the closure of T and thus $Z \subseteq -C$. It follows that $\mathord{\sim\sim} A \subseteq -C \cup A$, and finally $\mathord{\sim\sim} A \subseteq (C \Rightarrow A)$. □

11.4. If v is a valuation in $\mathcal{O}(\mathbb{R})$ and $P \subseteq \mathbb{R}$ is an open segment (not necessarily proper) then by v_P we denote the valuation in $\mathcal{O}(P)$ defined by $v_P(p) = v(p) \cap P$. A segment P is *easy* for v iff for all p either $P \subseteq v(p)$ or $P \cap v(p) = \varnothing$. In this case, v induces a binary valuation ϱ such that $\varrho(p) = 1$ if $P \subseteq v(p)$ and $\varrho_P(p) = 0$ otherwise. For any φ, write $\overline{v}_P(\varphi) = P$ when $[\![\varphi]\!]_\varrho = 1$ and $\overline{v}_P(\varphi) = \varnothing$ otherwise. By induction with respect to φ one proves for all open segments P and S that:

If P is easy for v and $P \subseteq S$, then $\overline{v}_P(\varphi) = [\![\varphi]\!]_{v_P} = [\![\varphi]\!]_{v_S} \cap P$.

The non-trivial case is when $\varphi = \exists p\,\psi$. There is no problem if either $[\![\psi]\!]_{v_P(p \mapsto \varnothing)}$ or $[\![\psi]\!]_{v_P(p \mapsto P)}$ is equal to P, so let us assume that both are empty. Take any open set Y and let $w = v(p \mapsto Y)$. Suppose that a segment $Q \subseteq P$ is easy for w. From the induction hypothesis for ψ, applied twice with respect to Q, we have

$$[\![\psi]\!]_{w_P} \cap Q = [\![\psi]\!]_{w_Q} = \overline{w}_Q(\psi) = \overline{u}_Q(\psi) = [\![\psi]\!]_{u_P} = \varnothing,$$

where $u = v(p \mapsto P)$, when $Q \subseteq Y$ and $u = v(p \mapsto \varnothing)$ otherwise. Thus $[\![\psi]\!]_{w_P}$ is disjoint from all segments easy for w and contained in P. It follows that it is disjoint from $P \cap (Y \cup \mathord{\sim} Y)$, i.e. empty. We conclude that $[\![\varphi]\!]_{v_P} = \varnothing$.

Now let $\mathcal{O}(\mathbb{R}) \models \varphi$. If $\mathbb{R}$ is easy for v then $[\![\varphi]\!]_v = \mathbb{R}$, and thus $[\![\varphi]\!]_\varrho = 1$ for all binary valuations ρ. □

11.5. The first part is similar to Exercise 8.26. In the second part consider a Kripke model $\mathcal{C} = \langle C, \leq, \{D_c \mid c \in C\}\rangle$, where $C = \{1, 2, 3, 4\}$ and the ordering relation is such that $1 < 2 < 3, 4$ and $3, 4$ are incomparable. For $c \neq 1$, let the sets D_c contain all upward-closed subsets of C, and let $D_1 = \{\varnothing, \{3, 4\}, \{2, 3, 4\}, \{1, 2, 3, 4\}\}$.

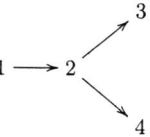

Then $1, v \Vdash \forall p(q \vee \neg p \vee \neg\neg p)$ but $1, v \nVdash q \vee \forall p(\neg p \vee \neg\neg p)$, for $v(q) = \{2, 3, 4\}$. Indeed, $1, v \nVdash q$ and $2, v' \nVdash \neg p \vee \neg\neg p$, for $v'(p) = \{3\}$.

To show that the model is full, we prove that for all φ and all valuations v ranging over D_1, the set $[\![\varphi]\!]_v = \{c \mid c, v \Vdash \varphi\}$ is in D_1. Here we can use the following lemma. For $c \in C$ let $\underline{c}$ be defined by: $\underline{1} = 1$, $\underline{2} = 2$, $\underline{3} = 4$, $\underline{4} = 3$, i.e. one swaps 3 and 4. If $A \subseteq C$, let $\underline{A} = \{\underline{c} \mid c \in A\}$ and for any valuation v, let $\underline{v}(c) = v(\underline{c})$. Then $c, \underline{v} \Vdash \varphi$ iff $\underline{c}, v \Vdash \varphi$, in particular if v ranges over D_1 then $[\![\varphi]\!]_v = \underline{[\![\varphi]\!]_v}$. □

APPENDIX B. SOLUTIONS AND HINTS   397

11.6. This exercise contains a little trap: The notion of a full model refers to all formulas, including those containing $\vee$, $\wedge$ and $\exists$. But one can also consider Kripke models for the language containing only $\to$ and $\forall$, satisfying the appropriate (weaker) notion of fullness. Then the semantics of the formula

$$\tau + \sigma := \forall p((\tau \to p) \to (\sigma \to p) \to p)$$

does not have to coincide with the expected semantics of $\tau \vee \sigma$. Indeed, consider a model of three states 0, 1 and 2, ordered so that $0 < 1, 2$ and $1, 2$ are incomparable, and with a constant domain $D = \{\varnothing, \{1\}, \{2\}, \{0, 1, 2\}\}$ for each state. This model is full with respect to $\to$ and $\forall$, and we have $0, v \Vdash q + r$ and $0, v \not\Vdash q \vee r$, for $v(q) = \{1\}$ and $v(r) = \{2\}$.

The case of $\wedge$ is different: if a model is full with respect to $\to$ and $\forall$, then $c \Vdash \forall p((\tau \to \sigma \to p) \to p)$ iff $c \Vdash \tau$ and $c \Vdash \sigma$.   □

11.8. (Sobolev) Yes. One can turn a full model $\mathcal{C} = \langle C, \leq, \{D_c \mid c \in C\}\rangle$ into an "equivalent" model $\mathcal{C}' = \langle C, \leq, \{D'_c \mid c \in C\}\rangle$, satisfying the stronger definition of fullness. Let $d\!\uparrow = \{d' \in C \mid d' \geq d\}$ and define $D'_c = \{y \cap d\!\uparrow \mid d \leq c \text{ and } y \in D_d\}$. Then $\mathcal{C}, c, v \Vdash \varphi$ is equivalent to $\mathcal{C}', c, v' \Vdash \varphi$, whenever $v(p) \cap c\!\uparrow = v'(p) \cap c\!\uparrow$, for all $p \in \mathrm{FV}(\varphi)$. In particular $\mathcal{C} \Vdash \varphi$ iff $\mathcal{C}' \Vdash \varphi$.   □

11.9. If we could define a formula $\psi(p)$ equivalent to $\exists q((p \to \neg q \vee q) \to p)$ using only propositional connectives, then we would obtain $\mathcal{O}(\mathbb{R}) \models \neg\neg p \to \psi(p)$ from Exercise 11.3(ii). By Theorem 2.4.11, the formula $\neg\neg p \to \psi(p)$ would be intuitionistically valid. And then so would $\neg\neg p \to \exists q((p \to \neg q \vee q) \to p)$. See [281] for more examples.   □

11.10. Suppose that $\forall p(p \vee \neg p)$ is equivalent to a $\forall$-free formula ϑ. We can assume that ϑ is closed (otherwise add some $\exists$'s at the beginning). By Exercises 11.3(i) and 11.4, the negation $\neg\vartheta$ is a tautology. But $\neg\forall p(p \vee \neg p)$ is not.   □

11.12. The reader might already have noticed that there is a common pattern in the definitions of Section 11.3. We follow this pattern. A word over $\{a, b\}$ is either the empty word or it is obtained by adding a or b at the beginning. Thus, all words are generated with the help of two unary constructors and one constant (a nullary constructors). If we define $\mathbf{word} = \forall p((p \to p) \to (p \to p) \to p \to p)$, then the three constructors are represented by the three arguments. Unlabeled binary trees are built from a nullary constructor (leaf) and one binary constructor, so the appropriate type is $\mathbf{tree} = \forall p((p \to p \to p) \to p \to p)$. Trees labeled by integers require a unary leaf constructor, and a ternary one for internal nodes, and this time we define $\omega\text{-}\mathbf{tree} = \forall p((\omega \to p \to p \to p) \to (\omega \to p) \to p)$.   □

11.15. In the environment $\{y : \bot,\ z : \forall r(p \to r)\}$ one can derive $zyz : q$. Now assume $\{y : \bot,\ x : \forall pq(\forall r(p \to r) \to q)\}$ and derive $xx : \bot$. Thus our term has type $\forall pq(\forall r(p \to r) \to q) \to \bot$ in the environment $\{y : \bot\}$.   □

11.16. Think of types as finite binary trees with leaves labeled by type variables and internal nodes corresponding to arrows. Some of the nodes are also labeled by quantifiers. Here is a a tree view of $\forall p(p \to q \to p)$:

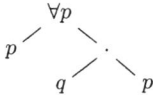

Suppose that yy is typable in an environment containing the declaration $(y : \tau)$. The type τ must begin with one or more universal quantifiers, and one of these quantifiers must bind a type variable occurring at the very end of the leftmost path of the type. (Otherwise self-application is impossible.) Thus, the type of $\lambda y.yy$ (the same as the type of x) must have the form $\forall \vec{q}(\forall \vec{p}\tau \to \sigma)$ with some $p \in \vec{p}$ at the end of the leftmost path. Repeating a similar argument for x, we find out that two different quantifiers attempt to bind the same variable, as shown below.

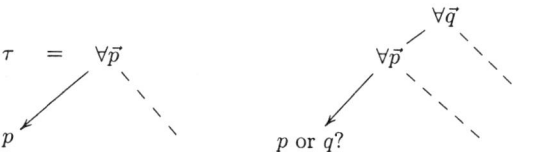

11.17. Yes. For instance the term discussed in Exercise 11.15 becomes untypable. An even simpler example is $\lambda y.y(\lambda x.xx)y$. This "predicative" instantiation can be generalized to form an infinite (even transfinite) hierarchy, see [114, 306, 457]. □

11.18. Not always. Let $M^0 = u(\lambda f.\mathbf{K}x(fy))$ and $M^1 = u(\lambda f.\mathbf{K}x(fz))$. In the environment $\Gamma = \{x:p,\ y:r,\ z:s,\ u:\forall r(((r\to r)\to p)\to q)\}$, terms M^0 and M^1 are of type q. Without loss of generality we can assume that $N^0 = ur(\lambda f^{r\to r}.L^0 x(fy))$ and $N^1 = us(\lambda f^{s\to s}.L^0 x(fz))$, for some L^0 and L^1 such that $|L^0| = |L^1| = \mathbf{K}$, and then $N^0 \neq_\beta N^1$. □

11.19. We have the following correct type assignment:
$$x : p \to \forall q(q \to q) \vdash \lambda y.xy : p \to q \to q,$$
and the eta-reduction $\lambda y.xy \to_\eta x$. However,
$$x : p \to \forall q(q \to q) \nvdash x : p \to q \to q.$$
That is, eta-reductions do *not* preserve types of Curry-style terms, although in the Church-style system **F** one has type-preserving eta-reductions:
$$\lambda x{:}\tau.Mx \to_\eta M,\ \text{for } x \notin \mathrm{FV}(M);$$
$$\Lambda p.Mp \to_\eta M,\ \text{for } p \notin \mathrm{FV}(M).$$
To see why this happens, observe that the Church-style term corresponding to $\lambda y.xy$ in our example is $\lambda y{:}p.xyq$ and is *not* an η-redex. Thus, the reason why Curry-style system **F** is not closed under η-reductions is that there are Curry-style η-redexes that do not correspond to any Church-style redex. □

11.20. An example known to the authors is $N = \lambda a.(\lambda xyzu.a(xy)(zy)(zu))XYZU$, where $X = \lambda v.v(\lambda xy.yay)$, $Y = \lambda x.\mathbf{K}(xx)(xaa)$, $Z = \lambda u.u(\lambda xy.a)$, $U = \lambda x.xxx$. The term M is obtained by replacing the second y in N by $V = \lambda w.y(\lambda v_1 v_2.wv_1 v_2)$. With $z : \forall r((\sigma \to r) \to r)$ and $x : \forall r((\tau \to r) \to r)$, where $\sigma = \forall p_1 p_2(p_1 \to p_2 \to q)$ and $\tau = \forall p(p \to \bot \to q)$, one can assign a type to M, because z can be applied to V. To verify that N is untypable, consider the type of Y as a tree. Starting from the root, follow the path of alternating left and right steps, until a leaf (a type variable) is reached. This variable must be bound by two different quantifiers, and we obtain a contradiction as in Solution 11.16. *Warning:* The proof requires a tedious analysis of a number of cases. □

11.21. Observe that $\tau \sqsubseteq \sigma$ implies the existence of a *coercion* $\lambda x\, c : \tau \to \sigma$ such that $c \twoheadrightarrow_\eta x$. If M has type τ then $c[x := M]$ has type σ and η-reduces to M. This gives one direction of the equivalence. For the other direction, one first shows that the system with $\sqsubseteq$ can equivalently be defined by the axiom $\Gamma, x:\tau \vdash x : \tau$ and the following rules:

APPENDIX B. SOLUTIONS AND HINTS   399

$$\frac{\Gamma \vdash M : \forall \vec{p}(\sigma \to \tau) \quad \Gamma \vdash N : \forall \vec{p}\sigma}{\Gamma \vdash MN : \forall \vec{p}\tau} \qquad \frac{\Gamma, x : \sigma \vdash M : \tau}{\Gamma \vdash \lambda x M : \forall \vec{p}(\sigma \to \tau)} \qquad \frac{\Gamma \vdash M : \tau}{\Gamma \vdash M : \sigma}\,(\tau \sqsubseteq \sigma)$$

In the second rule it is assumed that $\vec{p} \notin \mathrm{FV}(\Gamma)$. This type assignment is preserved by η-reductions: If $\Gamma \vdash \lambda x.Mx : \rho$ and $x \notin \mathrm{FV}(M)$ then we have $\Gamma \vdash M : \forall \vec{p}(\zeta \to \tau_0)$ and $\Gamma, x : \sigma \vdash x : \forall \vec{p}\zeta$, where $\sigma \sqsubseteq \forall \vec{p}\zeta$, $\forall \vec{p}\tau_0 \sqsubseteq \tau$, $\forall \vec{q}(\sigma \to \tau) \sqsubseteq \rho$, and $\vec{p}, \vec{q} \notin \mathrm{FV}(\Gamma)$.
It follows that $\forall \vec{p}(\zeta \to \tau_0) \sqsubseteq \sigma \to \tau$ and hence $\forall \vec{q}\vec{p}(\zeta \to \tau_0) \sqsubseteq \forall \vec{q}(\sigma \to \tau) \sqsubseteq \rho$.
Since $\vec{q} \notin \mathrm{FV}(\Gamma)$, we have $\Gamma \vdash M : \forall \vec{q}\vec{p}(\zeta \to \tau_0)$, and finally $\Gamma \vdash M : \rho$. □

11.22. Unfortunately, the subject reduction property does not hold any more. The following example is based on an idea used in [26] for *union types* (which cause similar problems). Consider a context Γ consisting of the declarations:

$$x : \forall p((p \to p) \to (p \to p) \to q), \quad z : r, \quad y : r \to \exists p(p \to p).$$

Then $\Gamma \vdash_\exists x(\mathbf{I}yz)(\mathbf{I}yz) : q$, but $\Gamma \nvdash_\exists x(\mathbf{I}yz)(yz) : q$. Indeed, to derive the latter judgement, one would have to use rule ($\exists$E) twice. In the first application of ($\exists$E) the side-condition $p \notin \mathrm{FV}(\Gamma, \psi)$ would have to be violated:

$$\frac{\Gamma \vdash \mathbf{I}yz : \exists p(p \to p), \quad \Gamma, u : p \to p \vdash xu : (p \to p) \to q}{\Gamma \vdash x(\mathbf{I}yz) : (p \to p) \to q} \qquad \qquad \square$$

11.23. We give a solution to the first part (after [495]). The type assigned to **K** must have the form $\sigma = \forall \vec{p}(\tau \to \forall \vec{q}(\mu \to \tau'))$, where τ' is an instance of τ. Observe that **22K** reduces to $\mathbf{2}(\lambda x.\mathbf{K}(\mathbf{K}x))$, and types of the two copies of **K** must be different. To make their composition well-typed, polymorphism is necessary, and one of the variables $\vec{p}$, say p, must occur at the end of the rightmost path in τ. The same p must remain at the end of the rightmost path in τ', at the same depth.

Since **22K** reduces to $\mathbf{2}(\mathbf{2K})$, the type of the second **2** in **22K** must have the form $\forall \vec{r}(\sigma_0 \to \varrho)$, where σ can be obtained from σ_0 by instantiating $\vec{r}$. Now, σ_0 is the type of f in $f(fx)$ and ϱ is the type of $\lambda x.f(fx)$. Type σ_0 begins with $\forall \vec{p}$ and has occurrences of p at the same places as σ does. In particular, there is an occurrence of p at the end of the rightmost path of the left subtree of σ_0, at some depth n, and an occurrence of p at depth $n+1$ at the rightmost path of the right subtree of σ_0. No matter what the type of x is, the asymmetry of σ_0 is doubled in ϱ, and thus the rightmost path in ϱ must be of length at least $n+3$. It follows that **2** cannot be composed with itself, as the positions of p in $\forall \vec{r}(\sigma_0 \to \varrho)$ cannot be changed by just instantiating $\vec{r}$. □

11.25. One can normalize the terms and compare the obtained normal forms. But the problem does not classify as "decidable" because it is not a decision problem at all (cf. Example A.4.1). □

11.26. Suppose that C is typable in such a way that the two copies of $F = \lambda fg.gf$ are of types $\sigma \to \sigma$ and $\tau \to \tau$, respectively. The type assigned to g in the first copy of F must be a polymorphic type of the form $\forall \vec{p}\varrho$ with a variable $p \in \vec{p}$ occurring at the end of the leftmost path. This p also occurs at the end of the leftmost path in σ, bound by a $\forall p$ occurring at the left child of the root of σ. Using the same argument for τ, we obtain a similar conflict as in Solution 11.16, when trying to apply the expression $xF(\lambda z\,1)$ of type σ to the argument $yF(\lambda z\,0)$ of type τ.

It follows that $\nvdash C : \omega \to \omega \to \omega$, i.e. the particular algorithm determined by C is not type-correct, cf. [284]. But the function c is of course definable, and the term C is typable because it is a normal form. □

11.27. An infinite sequence of Church-style beta-reductions $M_i \to_\beta M_{i+1}$ erases to an infinite sequence of Curry-style terms $|M_i|$, where at each step we either have

$|M_i| \to_\beta |M_{i+1}|$ or $|M_i| = |M_{i+1}|$. The latter case must hold for almost all i, and is only possible when almost all steps $M_i \to_\beta M_{i+1}$ are caused by type reductions of the form $(\Lambda p.M)\tau \to_\beta M[p := \tau]$. But each of these steps decreases the number of Λ's in our term, so this process must also terminate. □

11.28. Every primitive recursive function is definable in system **F** by a closed term F of type $\omega^m \to \omega$. Then $Fc_{n_1} \ldots c_{n_m}(\lambda v.\text{false})\text{true}$ has type **bool** and is strongly normalizing by Theorem 11.5.7. □

11.30. One can reduce the problem to strong normalization in the simply typed lambda-calculus. Kinds translate to types built from a single base type $*$, and constructors translate to Church-style terms. The connectives $\to$ and $\forall$ are represented by fixed variables (think of $\forall \alpha^\kappa \varphi$ as of an application of $\forall$ to $\lambda \alpha^\kappa \varphi$.) □

Solutions to chapter 12

12.2. *Hint:* A set A is finite iff every partial order on A has a maximal element. □

12.3. Take the conjunction of $\forall a\, \text{Int}(a)$, definitions of addition and multiplication, and the axioms $\forall a \forall b\, (sa = sb \to a = b)$ and $\forall a(sa = 0 \to \bot)$. □

12.4. Let φ be as in Exercise 12.3. For all sentences ψ in the language of second-order arithmetic, show that $\mathcal{N} \models \psi$ if and only if $\varphi \to \psi$ is a tautology. Now let $\psi(a)$ represent a non-r.e. set, and consider sentences of the form $\varphi \to \psi(\underline{n})$. □

12.5. Formalize the definition of $\mathcal{N} \models \varphi$. □

12.6. Strong normalization of system **T**, provable in PAS, implies totality of g. □

12.7. The only doubtful case is the symmetry axiom (E_2). In order to derive the formula $\forall R\, (Rb \to Ra)$ from $\forall R\, (Ra \to Rb)$ one instantiates R by $\lambda c.\, Rc \to Ra$. □

12.8. Take out $= \text{It}(\tau(\mu))(\Lambda p \lambda f^{p \to \tau(\mu)}.\, \textit{Lift}\, p\mu(\text{in} \circ f))$. □

12.9. This construction works for monotone $\tau(p)$. Let *Lift* be as in Exercise 12.8. For $\mu' = \forall q\, ((\tau(q) \to q) \to q)$, define an iterator $\text{It}' = \Lambda q \lambda y^{\tau(q) \to q} \lambda z^{\mu'}.\, zqy$ and an embedding $\text{in}' = \lambda x^{\tau(\mu')} \Lambda q \lambda y^{\tau(q) \to q}.\, y(\textit{Lift}\, \mu' q(\lambda z^{\mu'}\, zqy)x)$. Then $\text{in}' : \tau(\mu') \to \mu'$ (note the analogy between in' and the argument for $f(a_0) \leq a_0$ in the proof of Theorem A.1.2), and $\text{It}' : \forall q\, ((\tau(q) \to q) \to \mu' \to q)$. The induction principle for μ' has the form $\text{It}'\sigma Z(\text{in}'M) =_\beta Z(\textit{Lift}\, \mu'\sigma(\text{It}'\sigma Z)M)$. See [2, 329, 338, 452] for further information on inductive and recursive types. □

12.11. This is mostly routine but one has to show that translations of axioms of PAS (respectively PAs) are provable in HAS (respectively HAs). The induction axiom can be treated similarly as in the first-order case. Consider for instance the axiom P'_4 of PAs. The translation of this axiom is equivalent to $\forall ab\, (\neg\neg sa = 0 \to \neg\neg b = 0)$, and follows from P'_4, because $\varphi \to \psi$ implies $\neg\neg\varphi \to \neg\neg\psi$. □

12.12. In HAs we can't prove $\neg\neg R\bar{a} \to R\bar{a}$, where R is a relation variable. □

12.13. The proof of the implication $\forall R\, (Ra \to Rb) \to a = b$ follows by instantiating R by $\lambda c\, (a = c)$. The proof of the other implication goes by induction, say with respect to b. The base step $a = 0 \to \forall R\, (Ra \to R0)$ is then shown by induction with respect to a. The base case of this internal induction is the trivial implication $0 = 0 \to \forall R\, (R0 \to R0)$. The induction step $sa = 0 \to \forall R\, (R(sa) \to R0)$ follows by an application of P_4. The main induction step $a = sb \to \forall R\, (Ra \to R(sb))$ is also shown by induction with respect to a. Now the base step follows from P_4. For the induction step, we assume $sa = sb$ and we need to derive $\forall R\, (R(sa) \to R(sb))$.

APPENDIX B. SOLUTIONS AND HINTS   401

From P_3 we have $a = b$ and we can use the (main) induction hypothesis to obtain $\forall R\,(Ra \to Rb)$. It remains to instantiate R with $\lambda c.R(sc)$.

To show the unprovability of our formula in PAs, consider a model with the domain $\mathbb{N}$, where $=$ is interpreted as the full relation $\mathbb{N} \times \mathbb{N}$. This model satisfies the axioms of PAs, but does not satisfy $\forall ab(a = b \leftrightarrow \forall R\,(Ra \to Rb))$. □

12.15. Let p be a pairing function. For a binary f_2 consider a unary function f_1 such that $f_1(p(m,n)) = f_2(m,n)$. If F_1 defines f_1 and P defines p then $\lambda xy.F_1(pxy)$ defines f_2. Functions of arbitrary arity are treated similarly. □

Solutions to Chapter 13

13.1. We should have **string** : **int** $\Rightarrow *$ and **record** : $(\Pi n\!:\!\mathbf{int}.\mathbf{string}(n) \Rightarrow *)$. □

13.2. $\lambda z^{\forall x^\tau (\alpha x \to \delta(fx))} \lambda x^\tau \lambda u^{\alpha x} \lambda v^{\forall y^\sigma (\delta y \to p)}. v(fx)(zxu)$. □

13.3. Let $there = \lambda x^\tau \lambda y^{\tau \to p}. yx$ and $back = \lambda x^\sigma \lambda y^q. x(\lambda z^{q \to p}. zy)$. The inhabitant is $\lambda X^{\forall f \cdots}. X\,there\,back(\lambda x^\tau \lambda y^q \lambda z^{\alpha(xy)}. z)$ because $\alpha(xy) =_\beta \alpha(back(there\,x)y)$. □

13.4. The second judgement is equivalent to $\Gamma \vdash (\lambda y\!:\!Px.y) : Px \to Px$ and is derivable. The first one is not, because $\forall x\!:\!Px.Px$ is the same as $\forall y\!:\!Px.Py$ and P cannot be applied to a variable of type Px. □

13.5. Only up to β-equality. □

13.8. Not the one we know. For instance, when proving part (ii) in the case of an abstraction $\lambda x\!:\!\sigma.M$, one must apply induction to $\Gamma \vdash \sigma : *$ to derive $\Gamma \vdash \sigma^\sharp : \Box$. □

13.9. Prove that $|M[x := N]| = |M|[x := |N|]$ holds for arbitrary raw terms M and N. Then proceed by induction with respect to the definition of $\to_\beta$. □

13.11. Consider the term $(\lambda x^\tau \lambda y^{\sigma(xN)}.P)Q$, which reduces to $\lambda y^{\sigma(QN)}.P[x := Q]$. The induction hypothesis does not apply to QN. □

13.12. Let $\Gamma = \{p : *,\ \alpha : p \Rightarrow *,\ y : p,\ z : p\}$. Then $(\lambda v^{\alpha y \to \alpha y}\,y)(\lambda x^{\alpha y}\,x)$ and $(\lambda v^{\alpha z \to \alpha z}\,y)(\lambda x^{\alpha z}\,x)$ are of type p in Γ, and $\flat$ maps both to $(\lambda v^{\alpha \to \alpha}\,y)(\lambda x^\alpha\,x)$. □

13.13. First show that every constructor occurs in a term and apply Theorem 13.3.6. For kinds, use induction ($\Pi x\!:\!\tau\,\kappa$ can only be reduced inside τ or κ). □

13.14. Apply Proposition 13.4.3. □

13.15. No. There are no normal forms of such types. □

Solutions to Chapter 14

14.4. In part (i) use Lemmas 13.2.2 and 13.2.10. Part (ii) requires subject reduction for compact λP (Exercise 14.3). □

14.5. Because $\Gamma \vdash \kappa : \Box$ entails $\kappa = *$. □

14.7. Let $M^0 = u(\lambda f.\mathbf{K}x(fy))$ and $M^1 = u(\lambda f.\mathbf{K}x(fz))$ and let

$$\Gamma = \{p\!:\!*,\ q\!:\!*,\ r\!:\!*,\ s\!:\!*,\ x\!:\!p,\ y\!:\!p,\ z\!:\!s,\ u\!:\!\forall r^*(r \to r) \to q\}.$$

Then $\Gamma,\ \alpha\!:\!q \to *,\ a\!:\!\alpha M^0 \to p,\ b\!:\!\alpha M^1 \vdash ab : p$, but we do not have β-equal Church style terms of type p that would erase to M^0 and M^1. (The argument used in Exercise 11.18 still works.) □

14.8. The term $M = \lambda y^{\exists x^p(Rx)}. yp(\lambda x^p \lambda u^{Rx}\,x)$ has the required type. Observe that M applied to $[t^p, N^{Rt}]_{\exists x^p(Rx)}$ (defined as $\lambda q^* \lambda y^{\forall x^p (\varphi \to q) \to q}. ytN$) yields a witness t of the existential statement $\exists x^p(Px)$ as the result. □

14.9. One possible solution is to apply the previous exercise. Another one is the term $\lambda y^{\exists x^p(Rx)} \lambda z^{\forall x^p q}. yq(\lambda x^p \lambda u^{Rx}. zx)$. □

14.12. Suppose otherwise, and define $\approx$ as the least equivalence relation on A such that $a \approx el(set(a))$, for all $a \in A$. For $a, b \in A$, write $a \, \epsilon \, b$ when $a \approx c$ holds for some $c \in set(b)$. Show that $\approx$ is a congruence with respect to ϵ, i.e. that $a \, \epsilon \, b$ implies $a' \, \epsilon \, b'$ for $a \approx a'$ and $b \approx b'$. Then define $\Delta = el(\{a \in A \mid \neg(a \, \epsilon \, a)\})$ and prove that $\Delta \, \epsilon \, \Delta$ is equivalent to $\neg(\Delta \, \epsilon \, \Delta)$. □

14.13. The constructor $Lift = \lambda \iota^\square \lambda k^\square \lambda f^{\iota \to k} \lambda X^{\mathcal{P}(\iota)} \lambda \alpha^k. \exists \beta^\iota (\alpha =^k f\beta \wedge X\beta)$ is of kind $\forall \iota^\square k^\square ((\iota \to k) \to \mathcal{P}(\iota) \to \mathcal{P}(k))$ and can be used to translate the solution of Exercise 12.8. Informally, $Lift \, \iota k f$ maps a "subset" X of ι into its image $f(X)$. □

14.14. Curry-style systems, systems with $\wedge$ and $\vee$, the $\lambda\mu$-calculus, etc. □

14.15. Not for system λP. Indeed, the single declaration $x : p$ does not make a valid environment. What is worse, to type an expression one has to declare all type variables used in the typing. And we do not *a priori* know how many type variables are needed. Paradoxically, λP is just the very system in which typability is decidable and type-checking is not. □

Bibliography

[1] M. Abadi, L. Cardelli, P.-L. Curien, and J.-J. Lévy. Explicit substitutions. *Journal of Functional Programming*, 1(4):375–416, 1991.

[2] A. Abel and T. Altenkirch. A predicative strong normalisation proof for a λ-calculus with interleaving inductive types. In T. Coquand, P. Dybjer, B. Nordström, and J. Smith, editors, *Types for Proofs and Programs, International Workshop TYPES'99*, volume 1956 of *Lecture Notes in Computer Science*, pages 21–40. Springer-Verlag, 2000.

[3] H. Abelson and G.J. Sussman with J. Sussman. *The Structure and Interpretation of Computer Programs*. MIT Press, second edition, 1996.

[4] S. Abramsky. Computational interpretations of linear logic. *Theoretical Computer Science*, 111(1-2):3–57, 1993.

[5] S. Abramsky and R. Jagadeesan. Games and full completeness for multiplicative linear logic. *Journal of Symbolic Logic*, 59(2):543–574, 1994.

[6] Z. Adamowicz and P. Zbierski. *Logic of Mathematics. A Modern Course of Classical Logic*. John Wiley & Sons, 1997.

[7] Y. Akama. A lambda-to-CL translation for strong normalization. In de Groote and Hindley [208], pages 1–10.

[8] R.M. Amadio and P.-L. Curien. *Domains and Lambda-Calculi*, volume 46 of *Cambridge Tracts in Theoretical Computer Science*. Cambridge University Press, 1998.

[9] A.R. Anderson and N.A. Belnap. *Entailment. The Logic of Relevance and Necessity*, volume I. Princeton University Press, 1975.

[10] A.R. Anderson, N.A. Belnap, and J.M. Dunn. *Entailment. The Logic of Relevance and Necessity*, volume II. Princeton University Press, 1992.

[11] A. Appel. *Compiling with Continuations*. Cambridge University Press, 1992.

[12] K.R. Apt. Ten years of Hoare's logic: A survey—Part I. *ACM Transactions on Programming Languages and Systems*, 3(4):431–483, 1981.

[13] Z.M. Ariola and H. Herbelin. Minimal classical logic and control operators. In J.C.M. Baeten, J.K. Lenstra, J. Parrow, and G.J. Woeginger, editors, *Automata, Languages, and Programming*, volume 2719 of *Lecture Notes in Computer Science*, pages 871–885. Springer-Verlag, 2003.

[14] A. Arnold and D. Niwiński. *Rudiments of μ-Calculus*, volume 146 of *Studies in Logic and the Foundations of Mathematics*. North-Holland, 2001.

[15] S.N. Artemov. Explicit provability and constructive semantics. *Bulletin of Symbolic Logic*, 7(1):1–36, 2001.

[16] S.N. Artemov and L.D. Beklemishev. Provability logic. In D. Gabbay and F. Guenthner, editors, *Handbook of Philosophical Logic*, volume 13, pages 229–403. Kluwer, second edition, 2004.

[17] A. Asperti and S. Guerrini. *Optimal Implementation of Functional Programming Languages*, volume 45 of *Cambridge Tracts in Theoretical Computer Science*. Cambridge University Press, 1998.

[18] A. Asperti and G. Longo. *Categories, Types and Structures. An Introduction to Category Theory for the Working Computer Scientist*. The MIT Press, 1991.

[19] D. Aspinall and M. Hofmann. Dependent types. In B.C. Pierce, editor, *Advanced Topics in Types and Programming Languages*, chapter 2, pages 45–86. The MIT Press, 2005.

[20] J. Avigad and S. Feferman. Gödel's functional ("Dialectica") interpretation. In Buss [63], chapter V, pages 337–405.

[21] F. Baader and T. Nipkow. *Term Rewriting and All That*. Cambridge University Press, 1998.

[22] R. Backhouse, P. Chisholm, G. Malcolm, and E. Saaman. Do-it-yourself type theory. *Formal Aspects of Computing*, 1(1):19–84, 1989.

[23] S. van Bakel, L. Liquori, S. Ronchi Della Rocca, and P. Urzyczyn. Comparing cubes of typed and type assignment systems. *Annals of Pure and Applied Logic*, 86(3):267–303, 1997.

[24] F. Barbanera and S. Berardi. Extracting constructive content from classical proofs via control-like reductions. In Bezem and Groote [45], pages 45–59.

[25] F. Barbanera and S. Berardi. A symmetric lambda-calculus for "classical" program extraction. *Information and Computation*, 125(2):103–117, 1996.

[26] F. Barbanera, M. Dezani-Ciancaglini, and U. de'Liguoro. Intersection and union types: syntax and semantics. *Information and Computation*, 119:202–230, 1995.

[27] H. Barendregt. Introduction to generalized type systems. *Journal of Functional Programming*, 1(2):125–154, 1991.

[28] H. Barendregt. The impact of the lambda calculus in logic and computer science. *Bulletin of Symbolic Logic*, 3(2):181–215, 1997.

[29] H. Barendregt and S. Ghilezan. Lambda terms for natural deduction, sequent calculus, and cut elimination. *Journal of Functional Programming*, 10(1):121–134, 2000.

[30] H. Barendregt and T. Nipkow, editors. *Types for Proofs and Programs, International Workshop TYPES'93*, volume 806 of *Lecture Notes in Computer Science*. Springer-Verlag, 1994.

[31] H.P. Barendregt. *The Lambda Calculus. Its Syntax and Semantics*. North-Holland, second, revised edition, 1984.

[32] H.P. Barendregt. Lambda calculi with types. In S. Abramsky, D.M. Gabbay, and T.S.E. Maibaum, editors, *Handbook of Logic in Computer Science*, volume II, pages 117–309. Oxford University Press, 1992.

[33] H.P. Barendregt and H. Geuvers. Proof-assistants using dependent type systems. In J.A. Robinson and A. Voronkov, editors, *Handbook of Automated Reasoning*, volume 2, pages 1149–1238. Elsevier, 2001.

[34] E.M. Barth and E.C.W. Krabbe. *From Axiom to Dialogue. A Philosophical Study of Logics and Argumentation.* Walter de Gruyter, 1982.
[35] G. Barthe and T. Coquand. An introduction to dependent type theory. In G. Barthe, P. Dybjer, L. Pinto, and J. Saraiva, editors, *Applied Semantics, International Summer School, APPSEM 2000*, volume 2395 of *Lecture Notes in Computer Science*, pages 1–41. Springer, 2002.
[36] G. Barthe, J. Hatcliff, and M.H. Sørensen. Weak normalization implies strong normalization in a class of non-dependent pure type systems. *Theoretical Computer Science*, 269(1-2):317–361, 2001.
[37] G. Barthe and M.H. Sørensen. Domain-free pure type systems. *Journal of Functional Programming*, 10(5):417–452, 2000.
[38] M. Beeson. *Foundations of Constructive Mathematics.* Springer-Verlag, 1985.
[39] J.L. Bell and M. Machover. *A Course in Mathematical Logic.* North-Holland, 1977.
[40] J. van Benthem. *Language in Action. Categories, Lambdas and Dynamic Logic*, volume 130 of *Studies in Logic and the Foundations of Mathematics.* North-Holland, 1991.
[41] J. van Benthem. *Logic in Games.* Universiteit van Amsterdam, 1999–2005.
[42] S. Berardi. *Type Dependence and Constructive Mathematics.* PhD thesis, Università di Torino, 1990.
[43] U. Berger, W. Buchholz, and H. Schwichtenberg. Refined program extraction from classical proofs. *Annals of Pure and Applied Logic*, 114(1-3):3–25, 2002.
[44] Y. Bertot and P. Castéran. *Interactive Theorem Proving and Program Development. Coq'Art: The Calculus of Inductive Constructions.* Texts in Theoretical Computer Science. An EATCS Series. Springer-Verlag, 2004.
[45] M. Bezem and J.F. Groote, editors. *Typed Lambda Calculi and Applications*, volume 664 of *Lecture Notes in Computer Science.* Springer-Verlag, 1993.
[46] M. Bezem and J. Springintveld. A simple proof of the undecidability of inhabitation in λP. *Journal of Functional Programming*, 6(5):757–761, 1996.
[47] G. Bezhanishvili and M. Gehrke. Completeness of S4 with respect to the real line: revisited. *Annals of Pure and Applied Logic*, 131(1-3):287–301, 2005.
[48] E.A. Bishop and D.S. Bridges. *Constructive Analysis.* Springer-Verlag, 1985.
[49] A.R. Blass. Is game semantics necessary? In E. Börger, Y. Gurevich, and K. Meinke, editors, *Computer Science Logic. 7th Workshop, CSL'93*, volume 832 of *Lecture Notes in Computer Science*, pages 66–77. Springer-Verlag, 1994.
[50] R. Bloo and H. Geuvers. Explicit substitution: On the edge of strong normalization. *Theoretical Computer Science*, 211(1-2):375–395, 1999.
[51] R. Bloo and K. Rose. Preservation of strong normalisation in named lambda calculi with explicit substitution and garbage collection. In *CSN'95— Computer Science in the Netherlands*, pages 62–72, 1995.
[52] C. Böhm and A. Berarducci. Automatic synthesis of types λ-programs on term algebras. *Theoretical Computer Science*, 39:135–154, 1985.
[53] G.S. Boolos, J.P. Burgess, and R.C. Jeffrey. *Computability and Logic.* Cambridge University Press, fourth edition, 2002.

[54] V. Breazu Tannen, D. Kesner, and L. Puel. A typed pattern calculus. In *Logic in Computer Science*, pages 262–274. IEEE Computer Society Press, 1993.

[55] S. Broda and L. Damas. On principal types of combinators. *Theoretical Computer Science*, 247(1-2):277–290, 2000.

[56] L.E.J. Brouwer. *Collected Works: Philosophy and Foundations of Mathematics* (A. Heyting, editor), volume 1. North-Holland, 1975.

[57] N.G. de Bruijn. On the roles of types in mathematics. In de Groote [206], pages 27–54.

[58] N.G. de Bruijn. The mathematical language AUTOMATH, its usage, and some of its extensions. In Laudet et al. [299], pages 29–61. Reprinted in Nederpelt et al [360], pages 73–100.

[59] N.G. de Bruijn. Lambda calculus notation with nameless dummies, a tool for automatic formula manipulation. *Indagationes Mathematicae*, 34:381–392, 1972. Reprinted in Nederpelt et al [360], pages 375–388.

[60] N.G. de Bruijn. A survey of the project AUTOMATH. In Seldin and Hindley [441], pages 579–606. Reprinted in Nederpelt et al [360], pages 141–161.

[61] W. Buchholz. Ein ausgezeichnetes Modell für die intuitionistische Typenlogik. *Archiv für mathematische Logik und Grundlagenforschung*, 17:55–60, 1975.

[62] M.W. Bunder. Propositional and predicate calculuses based on illative combinatory logic. *Notre Dame Journal of Formal Logic*, 15(1):25–32, 1974.

[63] S.R. Buss, editor. *Handbook of Proof Theory*, volume 137 of *Studies in Logic and the Foundations of Mathematics*. North-Holland, Amsterdam, 1998.

[64] W. Buszkowski. The logic of types. In J. Srzednicki, editor, *Initiatives in Logic*, pages 180–206. Martinus Nijhoff (Kluwer), 1987.

[65] V. Capretta and S. Valentini. A general method to prove the normalization theorem for first and second order typed λ-calculi. *Mathematical Structures in Computer Science*, 9(6):719–739, 1999.

[66] L. Cardelli and P. Wegner. On understanding types, data abstraction, and polymorphism. *ACM Computing Surveys*, 17(4):471–522, 1985.

[67] S. Cerrito and D. Kesner. Pattern matching as cut elimination. In *Logic in Computer Science*, pages 98–108. IEEE Computer Society Press, 1999.

[68] A. Chagrov and M. Zakharyaschev. *Modal Logic*, volume 35 of *Oxford Logic Guides*. Clarendon Press, Oxford, 1997.

[69] A. Church. A set of postulates for the foundation of logic. *Annals of Mathematics*, 33(2):346–366, 1932.

[70] A. Church. A set of postulates for the foundation of logic. (Second paper.) *Annals of Mathematics*, 34(4):839–864, 1933.

[71] A. Church. A note on the Entscheidungsproblem. *Journal of Symbolic Logic*, 1(1):40–41, 1936.

[72] A. Church. An unsolvable problem of elementary number theory. *American Journal of Mathematics*, 58(2):345–363, 1936.

[73] A. Church. A formulation of the simple theory of types. *Journal of Symbolic Logic*, 5(2):56–68, 1940.

[74] A. Church. *The Calculi of Lambda-Conversion.* Princeton University Press, 1941.

[75] A. Church. *Introduction to Mathematical Logic, Volume I.* Princeton University Press, 1956. Reprinted by Dover, 1982.

[76] A. Church and J.B. Rosser. Some properties of conversion. *Transactions of the American Mathematical Society*, 39(3):472–482, 1936.

[77] C. Consel and O. Danvy. For a better support of static data flow. In J. Hughes, editor, *Conference on Functional Programming and Computer Architecture*, volume 523 of *Lecture Notes in Computer Science*, pages 495–519. Springer-Verlag, 1991.

[78] R. Constable and C. Murthy. Finding computational contents in classical proofs. In G. Huet and G. Plotkin, editors, *Logical Frameworks*, pages 341–362. Cambridge University Press, 1991.

[79] R.L. Constable. Constructive mathematics and automatic program writers. In *Proceddings of the IFIP Congress*, pages 229–233, Ljubljana, 1971.

[80] R.L. Constable. Programs as proofs: A synopsis. *Information Processing Letters*, 16(3):105–112, 1983.

[81] R.L. Constable. The structure of Nuprl's type theory. In H. Schwichtenberg, editor, *Logic of Computation*, volume 157 of *NATO Advanced Study Institute, Series F: Computer and Systems Sciences*, pages 123–156. Springer-Verlag, 1997.

[82] R.L. Constable. Types in logic, mathematics and programming. In Buss [63], chapter X, pages 683–786.

[83] T. Coquand. An analysis of Girard's paradox. In *Logic in Computer Science*, pages 227–236. IEEE Computer Society Press, 1986.

[84] T. Coquand. Metamathematical investigations of a calculus of constructions. In Odifreddi [367], pages 91–122.

[85] T. Coquand. On the analogy between propositions and types. In Huet [250], chapter 17, pages 399–417.

[86] T. Coquand. A semantics of evidence of classical arithmetic. *Journal of Symbolic Logic*, 60(1):325–337, 1995.

[87] T. Coquand and P. Dybjer. Intuitionistic model constructions and normalization proofs. *Mathematical Structures in Computer Science*, 7(1):75–94, 1997.

[88] T. Coquand and J. Gallier. A proof of strong normalization for the theory of constructions using a Kripke-like interpretation. In G. Huet, editor, *Proceedings of the First Annual Workshop on Logical Frameworks*, 1990.

[89] T. Coquand and H. Herbelin. A-translation and looping combinators in pure type systems. *Journal of Functional Programming*, 4(1):77–88, 1994.

[90] T. Coquand and G. Huet. The calculus of constructions. *Information and Computation*, 76(2-3):95–120, 1988.

[91] T. Crolard. A confluent lambda-calculus with a catch/throw mechanism. *Journal of Functional Programming*, 9(6):625–647, 1999.

[92] J.N. Crossley. Reminiscences of logicians. In J.N. Crossley, editor, *Algebra and Logic*, volume 450 of *Lecture Notes in Mathematics*, pages 1–62. Springer-Verlag, 1975.

[93] J.N. Crossley and M.A.E. Dummett, editors. *Formal Systems and Recursive Functions, Proceedings of the Eighth Logic Colloquium, Oxford, July 1963*, Studies in Logic and the Foundations of Mathematics. North-Holland, 1965.

[94] P.-L. Curien. Categorical combinators. *Information and Control*, 69(1-3):188–254, 1986.

[95] P.-L. Curien. Introduction to linear logic and ludics, Part I. *Advances in Mathematics (China)*, 34(1-3):513–544, 2005.

[96] P.-L. Curien and H. Herbelin. The duality of computation. In *International Conference on Functional Programming*, pages 233–243. ACM Press, 2000.

[97] H.B. Curry. An analysis of logical substitution. *American Journal of Mathematics*, LI(3):363–384, 1929.

[98] H.B. Curry. Grundlagen der kombinatorischen Logik. Teil I. *American Journal of Mathematics*, LII(3):509–536, 1930.

[99] H.B. Curry. Grundlagen der kombinatorischen Logik. Teil II. *American Journal of Mathematics*, LII(4):789–834, 1930.

[100] H.B. Curry. Functionality in combinatory logic. *Proceedings of the National Academy of Science USA*, 20(11):584–590, 1934.

[101] H.B. Curry. Some properties of equality and implication in combinatory logic. *Annals of Mathematics*, 35(2):849–860, 1934.

[102] H.B. Curry. The combinatory foundations of mathematical logic. *Journal of Symbolic Logic*, 7(2):49–64, 1942.

[103] H.B. Curry. The inconsistency of certain formal logics. *Journal of Symbolic Logic*, 7(3):115–117, 1942.

[104] H.B. Curry. The inconsistency of the full theory of combinatory functionality. *Journal of Symbolic Logic*, 20(1):91, 1955.

[105] H.B. Curry. Consistency of the theory of functionality. *Journal of Symbolic Logic*, 21(1):110, 1956.

[106] H.B. Curry. *Foundations of Mathematical Logic*. McGraw-Hill, 1963. Second edition by Dover, 1977.

[107] H.B. Curry and R. Feys. *Combinatory Logic, volume I*. Studies in Logic and the Foundations of Mathematics. North-Holland, 1958.

[108] H.B. Curry, J.R. Hindley, and J.P. Seldin. *Combinatory Logic, volume II*, volume 65 of *Studies in Logic and the Foundations of Mathematics*. North-Holland, 1972.

[109] N. Cutland. *An Introduction to Recursive Function Theory*. Cambridge University Press, 1980.

[110] D. van Dalen. Intuitionistic logic. In Gabbay and Guenthner [155], chapter 4, pages 225–339.

[111] D. van Dalen. L.E.J. Brouwer's intuitionism: A revolution in two installments. In *Logic in Computer Science*, pages 228–241. IEEE Computer Society Press, 1998.

[112] D. van Dalen. *Mystic, Geometer, and Intuitionist: The Life of L.E.J. Brouwer. Volume 1: The Dawning Revolution*. Clarendon Press, Oxford, second edition, 2002.

[113] D. van Dalen. *Logic and Structure.* Universitext. Springer-Verlag, fourth edition, 2004.

[114] N. Danner and D. Leivant. Stratified polymorphism and primitive recursion. *Mathematical Structures in Computer Science*, 9(4):507–522, 1999.

[115] V. Danos, J.-B. Joinet, and H. Schellinx. Sequent calculi for second order logic. In Girard et al. [188], pages 211–224.

[116] V. Danos, J.-B. Joinet, and H. Schellinx. A new deconstructive logic: Linear logic. *Journal of Symbolic Logic*, 62(3):755–807, 1997.

[117] V. Danos and J.-L. Krivine. Disjunctive tautologies and synchronisation schemes. In P.G. Clote and H. Schwichtenberg, editors, *Computer Science Logic, 14th Annual Conference of the EACSL*, volume 1862 of *Lecture Notes in Computer Science*, pages 292–301. Springer-Verlag, 2000.

[118] V. Danos and L. Regnier. Proof-nets and the Hilbert space. In Girard et al. [188], pages 307–328.

[119] B.A. Davey and H.A. Priestley. *Introduction to Lattices and Order.* Cambridge University Press, 2nd edition, 2002.

[120] R. David. Normalization without reducibility. *Annals of Pure and Applied Logic*, 107(1-3):121–130, 2001.

[121] R. David and G. Mounier. An intuitionistic lambda calculus with exceptions. *Journal of Functional Programming*, 15(1):33–52, 2005.

[122] R. David and K. Nour. A short proof of the strong normalization of classical natural deduction with disjunction. *Journal of Symbolic Logic*, 68(4):1277–1288, 2003.

[123] R. Davies and F. Pfenning. A modal analysis of staged computation. *Journal of the ACM*, 48(3):555–604, 2001.

[124] M. Davis. *The Undecidable.* New York, Raven Press, 1960. Reprinted with corrections by Dover, 2004.

[125] A. Degtyarev and A. Voronkov. Decidability problems for the prenex fragment of intuitionistic logic. In *Logic in Computer Science*, pages 503–512. IEEE Computer Society Press, 1996.

[126] W. Dekkers. Inhabitation of types in the simply typed λ-calculus. *Information and Computation*, 119(1):14–17, 1995.

[127] W. Dekkers, M. Bunder, and H. Barendregt. Completeness of the propositions-as-types interpretation of intuitionistic logic into illative combinatory logic. *Journal of Symbolic Logic*, 63(3):869–890, 1998.

[128] M. Dezani-Ciancaglini, S. Ghilezan, and B. Venneri. The "relevance" of intersection and union types. *Notre Dame Journal of Formal Logic*, 38(2):246–269, 1997.

[129] M. Dezani-Ciancaglini and G. Plotkin, editors. *Typed Lambda Calculi and Applications*, volume 902 of *Lecture Notes in Computer Science*. Springer-Verlag, 1995.

[130] K. Došen and P. Schroeder-Heister, editors. *Substructural Logics.* Oxford University Press, 1993.

[131] G. Dowek. The undecidability of typability in the lambda-pi-calculus. In Bezem and Groote [45], pages 139–145.

[132] A.G. Dragalin. The computability of primitive recursive terms of finite type, and primitive recursive realization. In A.O Slisenko, editor, *Studies in Constructive mathematics and mathematical logic II*, volume 8 of *Seminars in Mathematics, V.A. Steklov Mathematical Institute*, pages 13–18. 1970.

[133] A.G. Dragalin. New forms of realizability and Markov's rule. *Soviet Mathematics Doklady*, 21:461–464, 1980.

[134] A.G. Dragalin. *Mathematical Intuitionism. Introduction to Proof Theory*, volume 67 of *Translations of Mathematical Monographs*. AMS, 1988.

[135] B.F. Duba, R. Harper, and D. MacQueen. Typing first-class continuations in ML. In *Symposium on Principles of Programming Languages*. ACM Press, 1991.

[136] M. Dummett. *Elements of Intuitionism*, volume 39 of *Oxford Logic Guides*. Clarendon Press, second edition, 2000.

[137] J.M. Dunn. Relevance logic and entailment. In Gabbay and Guenthner [155], chapter 3, pages 117–224.

[138] C. Dwork, P.C. Kanellakis, and J.C. Mitchell. On the sequential nature of unification. *Journal of Logic Programming*, 1(1):35–50, 1984.

[139] P. Dybjer. Inductive families. *Formal Aspects of Computing*, 6(4):440–465, 1994.

[140] P. Dybjer. A general formulation of simultaneous inductive-recursive definitions in type theory. *Journal of Symbolic Logic*, 65(2):525–549, 2000.

[141] R. Dyckhoff. Contraction-free sequent calculi for intuitionistic logic. *Journal of Symbolic Logic*, 57(3):795–807, 1992.

[142] R. Dyckhoff and L. Pinto. Permutability of proofs in intuitionistic sequent calculi. *Theoretical Computer Science*, 212(1-2):141–155, 1999.

[143] R. Dyckhoff and C. Urban. Strong normalization of Herbelin's explicit substitution calculi with substitution propagation. *Journal of Logic and Computation*, 13(5):689–706, 2003.

[144] T. Ehrhard, J.-Y. Girard, P. Ruet, and P. Scott, editors. *Linear Logic in Computer Science*, volume 316 of *London Mathematical Society Lecture Note Series*. Cambridge University Press, 2004.

[145] M. Felleisen, D. Friedman, E. Kohlbecker, and B. Duba. A syntactic theory of sequential control. *Theoretical Computer Science*, 52(3):205–237, 1987.

[146] M. Felleisen and R. Hieb. The revised report on the syntactic theories of sequential control and state. *Theoretical Computer Science*, 103:235–271, 1992.

[147] W. Felscher. Dialogues as a foundation for intuitionistic logic. In Gabbay and Guenthner [155], chapter 5, pages 341–372.

[148] J.E. Fenstad, editor. *Proc. of the Second Scandinavian Logic Symposium*, volume 63 of *Studies in Logic and the Foundations of Mathematics*. North-Holland, 1971.

[149] A.J. Field and P.G. Harrison. *Functional Programming*. Addison-Wesley, 1988.

[150] A. Filinski. Declarative continuations: An investigation of duality in programming language semantics. In D.H. Pitt, D.E. Ryderhead, P. Dybjer,

A.M. Pitts, and A. Poigné, editors, *Category Theory and Computer Science*, volume 389 of *Lecture Notes in Computer Science*, pages 224–249. Springer-Verlag, 1989.

[151] M. Fitting. *Intuitionistic Logic, Model Theory and Forcing*. Studies in Logic and the Foundations of Mathematics. North-Holland, 1969.

[152] M. Fitting. *Proof Methods for Modal and Intuitionistic Logics*, volume 169 of *Synthese Library*. D. Reidel, 1983.

[153] S. Fortune, D. Leivant, and M. O'Donnell. The expresssiveness of simple and second-order type structures. *Journal of the ACM*, 30(1):151–185, 1983.

[154] H. Friedman. Classically and intuitionistically provably recursive functions. In G.H. Müller and D.S. Scott, editors, *Higher Set Theory*, volume 669 of *Lecture Notes in Mathematics*, pages 21–28. Springer-Verlag, 1978.

[155] D. Gabbay and F. Guenthner, editors. *Handbook of Philosophical Logic*, volume III. D. Reidel, 1986.

[156] D.M. Gabbay. On 2nd order intuitionistic propositional calculus with full comprehension. *Archiv für mathematische Logik und Grundlagenforschung*, 16:177–186, 1974.

[157] D.M. Gabbay. *Semantical Investigations in Heyting's Intuitionistic Logic*, volume 148 of *Synthese Library*. D. Reidel, 1981.

[158] D.M. Gabbay and R.J.G.B. de Queiroz. Extending the Curry-Howard interpretation to linear, relevant and other resource logics. *Journal of Symbolic Logic*, 57(4):1319–1366, 1992.

[159] D.M Gabbay and J. Woods, editors. *The Handbook of the History of Logic*. Elsevier, 2006. To appear.

[160] J. Gallier. On the correspondence between proofs and λ-terms. In de Groote [206], pages 55–138.

[161] J. Gallier. Constructive logics, part I: A tutorial on proof systems and typed λ-calculi. *Theoretical Computer Science*, 110(2):249–339, 1993.

[162] J.H. Gallier. *Logic for Computer Science. Foundations of Automatic Theorem Proving*. John Wiley & Sons, 1987.

[163] J.H. Gallier. On Girard's "candidats de reductibilité". In Odifreddi [367], pages 123–203.

[164] R. Gandy. The confluence of ideas in 1936. In R. Herken, editor, *The Universal Turing Machine. A Half-Century Survey*, pages 55–112. Oxford University Press, 1988.

[165] R.O. Gandy. The simple theory of types. In R.O. Gandy and J.M.E. Hyland, editors, *Logic Colloquium 76*, volume 87 of *Studies in Logic and the Foundations of Mathematics*, pages 173–181. North-Holland, 1977.

[166] R.O. Gandy. An early proof of normalization by A.M. Turing. In Seldin and Hindley [441], pages 453–455.

[167] R.O. Gandy. Proofs of strong normalization. In Seldin and Hindley [441], pages 457–477.

[168] A.O. Gelfond. *Transcendental and Algebraic Numbers*. Dover, New York, 1960.

[169] G. Gentzen. Über das Verhältnis zwischen intuitionistischer und klassischer Aritmetik. Unpublished galley proof for *Mathematische Annalen*, 1933. English translation: On the relation between intuitionist and classical arithmetic, in Szabo [465], pages 53–67.

[170] G. Gentzen. Die Widerspruchsfreiheit der reinen Zahlentheorie. *Mathematische Annalen*, 112:493–565, 1935. English translation: The consistency of elementary number theory, in Szabo [465], pages 132–200.

[171] G. Gentzen. Untersuchungen über das logische Schliessen. *Mathematische Zeitschrift*, 39:176–210, 405–431, 1935. English translation: Investigations into logical deduction, in Szabo [465], pages 68–131.

[172] H. Geuvers. Conservativity between logics and typed lambda-calculi. In Barendregt and Nipkow [30], pages 79–107.

[173] H. Geuvers. A short and flexible proof of strong normalization for the calculus of constructions. In *Types for Proofs and Programs, International Workshop TYPES'94*, volume 996 of *Lecture Notes in Computer Science*, pages 14–38. Springer-Verlag, 1995.

[174] H. Geuvers and E. Barendsen. Some logical and syntactical observations concerning the first-order dependent type system λP. *Mathematical Structures in Computer Science*, 9(4):335–359, 1999.

[175] H. Geuvers and M.-J. Nederhof. A modular proof of strong normalization for the calculus of constructions. *Journal of Functional Programming*, 1(2):155–189, 1991.

[176] J.H. Geuvers. *Logics and Type Systems*. PhD thesis, University of Nijmegen, 1993.

[177] P. Giannini and S. Ronchi Della Rocca. Characterization of typings in polymorphic type discipline. In *Logic in Computer Science*, pages 61–70. IEEE Computer Society Press, 1988.

[178] J.-Y. Girard. Linear logic: A survey. In de Groote [206], pages 193–255.

[179] J.-Y. Girard. Une extension du système de fonctionelles recursives de Gödel et son application aux fondements de l'analyse. In Fenstad [148], pages 63–92.

[180] J.-Y. Girard. Interprétation fonctionelle et élimination des coupures dans l'arithmétique d'ordre supérieur. Thèse d'État, Université Paris 7, 1972.

[181] J.-Y. Girard. Three-valued logic and cut-elimination: the actual meaning of Takeuti's conjecture. *Dissertationes Math. (Rozprawy Mat.)*, 136:1–49, 1976.

[182] J.-Y. Girard. The system F of variable types, fifteen years later. *Theoretical Computer Science*, 45:159–192, 1986. Reprinted in Huet [250], pages 87–126.

[183] J.-Y. Girard. Linear logic. *Theoretical Computer Science*, 50(1):1–101, 1987.

[184] J.-Y. Girard. *Proof Theory and Logical Complexity*, volume 1. Bibliopolis, 1987.

[185] J.-Y. Girard. A new constructive logic: classical logic. *Mathematical Structures in Computer Science*, 1(3):255–296, 1991.

[186] J.-Y. Girard. Locus Solum: From the rules of logic to the logic of rules. *Mathematical Structures in Computer Science*, 11(3):301–506, 2001.

[187] J.-Y. Girard. From foundations to ludics. *Bulletin of Symbolic Logic*, 9(2):131–168, 2003.

[188] J.-Y. Girard, Y. Lafont, and L. Regnier, editors. *Advances in Linear Logic*, volume 222 of *London Mathematical Society Lecture Note Series*. Cambridge University Press, 1995.

[189] J.-Y. Girard, Y. Lafont, and P. Taylor. *Proofs and Types*, volume 7 of *Cambridge Tracts in Theoretical Computer Science*. Cambridge University Press, 1989.

[190] V.I. Glivenko. Sur quelques points de la logique de M. Brouwer. *Bulletin de l'Académie Royale de Belgique, Classe des Sciences, ser. 5*, 15:183–188, 1929. English translation: On some points of the logic of Mr. Brouwer, in Mancosu [320], pages 301–305.

[191] C.A. Goad. Proofs as descriptions of computation. In W. Bibel and R. Kowalski, editors, *5th Conference on Automated Deduction*, volume 87 of *Lecture Notes in Computer Science*, pages 39–52. Springer-Verlag, 1980.

[192] K. Gödel. In what sense is intuitionistic logic constructive? In Gödel [195, vol. III], pages 189–200.

[193] K. Gödel. Zur intuitionistichen Aritmetik und Zahlentheorie. *Ergebnisse eines mathematischen Kolloquiums*, 4:34–38, 1933. English translation : On intuitionistic arithmetic and number theory, in Davis [124], pages 75–81. Also in Gödel [195, vol. I], pp 286–295.

[194] K. Gödel. Über eine bisher noch nicht benüntze Erweiterung des finiten Standpunktes. *Dialectica*, 12:280–287, 1958. English translation: On a hitherto unutilized extension of the finitary standpoint, in Gödel [195, vol. II], pages 241–251. Also in *J. Philos. Logic*, 9:133–142, 1980.

[195] K. Gödel. *Collected Works*, volume I-V. Oxford University Press, 1986, 1990, 1995, 2003, 2003. Edited by S. Feferman *et al*.

[196] L. Gordeev. On cut elimination in the presence of Peirce rule. *Archiv für mathematische Logik und Grundlagenforschung*, 26:147–164, 1987.

[197] M. Gordon. From LCF to HOL: A short history. In G. Plotkin, C. Stirling, and M. Tofte, editors, *Proof, Language and Interaction. Essays in Honour of Robin Milner*, pages 169–185. The MIT Press, 2000.

[198] S. Görnemann. A logic stronger than intuitionism. *Journal of Symbolic Logic*, 36(2):249–261, 1971.

[199] I. Gørtz, S. Reuss, and M.H. Sørensen. Strong normalization from weak normalization by translation into the lambda-I-calculus. *Higher-Order and Symbolic Computation*, 16(3):253–285, 2003.

[200] P. Graham. *ANSI Common Lisp*. Prentice-Hall, 1995.

[201] T.G. Griffin. A formulae-as-types notion of control. In *Symposium on Principles of Programming Languages*, pages 47–58. ACM Press, 1990.

[202] P. de Groote. The conservation theorem revisited. In Bezem and Groote [45], pages 163–178.

[203] P. de Groote. A CPS-translation of the $\lambda\mu$-calculus. In S. Tison, editor, *Colloquium on Trees in Algebra and Programming*, volume 787 of *Lecture Notes in Computer Science*, pages 85–99. Springer-Verlag, 1994.

[204] P. de Groote. On the relation between the $\lambda\mu$-calculus and the syntactic theory of sequential control. In F. Pfenning, editor, *Logic Programmming and*

Automated Reasoning, volume 822 of *Lecture Notes in Artificial Intelligence*, pages 31–43. Springer-Verlag, 1994.

[205] P. de Groote. A simple calculus of exception handling. In Dezani-Ciancaglini and Plotkin [129], pages 201–215.

[206] P. de Groote, editor. *The Curry-Howard Isomorphism*, volume 8 of *Cahiers du centre de logique*. Universite catholique de Louvain, 1995.

[207] P. de Groote. Strong normalization of classical natural deduction with disjunction. In S. Abramsky, editor, *Typed Lambda Calculi and Applications*, volume 2044 of *Lecture Notes in Computer Science*, pages 182–196. Springer-Verlag, 2001.

[208] P. de Groote and J.R. Hindley, editors. *Typed Lambda Calculi and Applications*, volume 1210 of *Lecture Notes in Computer Science*. Springer-Verlag, 1997.

[209] A. Grzegorczyk. A philosophically plausible formal interpretation of intuitionistic logic. *Indagationes Mathematicae*, 26:596–601, 1964.

[210] A. Grzegorczyk. Recursive objects in all finite types. *Fundamenta Mathematicae*, LIV:73–93, 1964.

[211] C.A. Gunter. *Semantics of Programming Languages. Structures and Techniques*. The MIT Press, 1992.

[212] P.R. Halmos. *Naive Set Theory*. Van Nostrand, 1960. Reprinted by Springer-Verlag in 1998.

[213] P.R. Halmos. *Lectures on Boolean Algebras*. Van Nostrand, 1963. Reprinted by Springer-Verlag in 1974.

[214] R. Harper, F. Honsell, and G. Plotkin. A framework for defining logics. *Journal of the ACM*, 40(1):143–184, 1993.

[215] R. Harper and M. Lillibridge. Explicit polymorphism and CPS conversion. In *Symposium on Principles of Programming Languages*, pages 206–219. ACM Press, 1993.

[216] R. Harper and M. Lillibridge. Polymorphic type assignment and CPS conversion. *LISP and Symbolic Computation*, 6(3-4):361–380, 1993.

[217] S. Hayashi and H. Nakano. *PX: A Computational Logic*. The MIT Press, 1988.

[218] J. van Heijenoort, editor. *From Frege to Gödel. A Source Book in Mathematical Logic, 1879–1931*. Harvard University Press, 1967.

[219] J. Heller. *Catch-22*. Simon and Schuster, New York, 1961.

[220] L. Henkin. The completeness of the first-order functional calculus. *Journal of Symbolic Logic*, 14(3):159–166, 1949.

[221] L. Henkin. An algebraic characterization of quantifiers. *Fundamenta Mathematicae*, 37:63–74, 1950.

[222] L. Henkin. Completeness in the theory of types. *Journal of Symbolic Logic*, 15(2):81–91, 1950.

[223] H. Herbelin. Games and weak-head reduction for classical PCF. In de Groote and Hindley [208], pages 214–230.

[224] H. Herbelin. A λ-calculus structure isomorphic to Gentzen-style sequent calculus structure. In L. Pacholski and J. Tiuryn, editors, *Computer Science Logic. 8th Workshop, CSL '94*, volume 933 of *Lecture Notes in Computer Science*, pages 61–75. Springer-Verlag, 1995.

[225] H. Herbelin. *Séquents qu'on Calcule. De l'interprétation du calcul des séquents comme calcul de λ-termes et comme calcul de stratégies gagnantes*. PhD thesis, Université Paris 7, 1995.

[226] O. Hermant. Semantic cut elimination in the intuitionistic sequent calculus. In P. Urzyczyn, editor, *Typed Lambda Calculi and Applications*, volume 3461 of *Lecture Notes in Computer Science*, pages 221–233. Springer-Verlag, 2005.

[227] A. Heyting. Die formalen Regeln der intuitionistischen Logik. *Sitzungsberichte der Preussischen Akademie der Wissenschaften. Physikalisch-mathematische Klasse, Jahrgang 1930*, pages 42–56. English translation: The formal rules of intuitionistic logic, in Mancosu [320], pages 311-327.

[228] A. Heyting. Die intuitionistische Grundlegung der Mathematik. *Erkenntniss*, 2:106–115. English translation: The intuitionist foundations of mathematics, in P. Benacerraf and H. Putnam, editors, *Philosophy of Mathematics. Selected Readings*, Cambridge University Press, second edition, 1983, pages 52–61.

[229] A. Heyting. *Mathematische Grundlagenforschung. Intuitionismus. Beweisteorie*. Verlag von Julius Springer, 1934.

[230] A. Heyting. *Intuitionism. An Introduction*. Studies in Logic and the Foundations of Mathematics. North-Holland, 1956. Third edition 1971.

[231] D. Hilbert and W. Ackermann. *Grundzüge der theoretischen Logik*. Verlag von Julius Springer, 1928. English translation: *Principles of Mathematical Logic*, AMS, 1958.

[232] D. Hilbert and P. Bernays. *Grundlagen der Mathematik*, volume I. Springer-Verlag, second edition, 1968.

[233] J.R. Hindley. M.H. Newman's typability algorithm for lambda-calculus. *Journal of Logic and Computation*. To appear.

[234] J.R. Hindley. The principal type scheme of an object in combinatory logic. *Transactions of the American Mathematical Society*, 146:29–60, 1969.

[235] J.R. Hindley. Combinatory reductions and lambda reductions compared. *Zeitschrift für mathematische Logik und Grundlagen der Mathematik*, 23:169–180, 1977.

[236] J.R. Hindley. BCK-combinators and linear λ-terms have types. *Theoretical Computer Science*, 64(1):97–105, 1989.

[237] J.R. Hindley. *Basic Simple Type Theory*, volume 42 of *Cambridge Tracts in Theoretical Computer Science*. Cambridge University Press, 1997.

[238] J.R. Hindley and F. Cardone. History of λ-calculus and combinatory logic. In Gabbay and Woods [159]. To appear.

[239] J.R. Hindley, B. Lercher, and J.P. Seldin. *Introduction to Combinatory Logic*. Cambridge University Press, 1972.

[240] J.R. Hindley and D. Meredith. Principal type-schemes and condensed detachment. *Journal of Symbolic Logic*, 51(1):90–105, 1990.

[241] J.R. Hindley and J.P. Seldin. *Introduction to Combinators and λ-calculus*. Cambridge University Press, 1986. Second edition to appear in 2006.

[242] S. Hirokawa. Principal types of BCK-lambda terms. *Theoretical Computer Science*, 107(2):253–276, 1993.

[243] S. Hirokawa, Y. Komori, and I. Takeuti. A reduction rule for the Peirce formula. *Studia Logica*, 56(3):419–426, 1996.

[244] M. Hofmann and T. Streicher. Continuation models are universal for $\lambda\mu$-calculus. In *Logic in Computer Science*, pages 387–397. IEEE Computer Society Press, 1997.

[245] J.E. Hopcroft, R. Motwani, and J.D. Ullman. *Introduction to Automata Theory, Languages, and Computation*. Addison-Wesley, second edition, 2001.

[246] W. Howard. Assignment of ordinals to terms for primitive recursive arithmetic. In Kino et al. [266], pages 443–458.

[247] W. Howard. The formulae-as-types notion of construction. In Seldin and Hindley [441], pages 479–490. Reprinted in de Groote [206], pages 15–26..

[248] D.J. Howe. The computational behaviour of Girard's paradox. In *Logic in Computer Science*, pages 205–214. IEEE Computer Society Press, 1987.

[249] G. Huet. Confluent reductions: Abstract properties and applications to term rewriting systems. *Journal of the ACM*, 27(4):797–821, 1980.

[250] G. Huet, editor. *Logical Foundations of Functional Programming*. Addison-Wesley, 1990.

[251] A.J.C. Hurkens. A simplification of Girard's paradox. In Dezani-Ciancaglini and Plotkin [129], pages 266–278.

[252] M. Huth and M. Ryan. *Logic in Computer Science: Modelling and Reasoning about Systems*. Cambridge University Press, second edition, 2004.

[253] D. Jarden. A simple proof that a power of an irrational number to an irrational exponent may be rational. *Scripta Mathematica*, 19:229, 1953.

[254] S. Jaśkowski. On the rules of suppositions in formal logic. *Studia Logica*, 1, 1934. Reprinted in McCall [332], pages 232–258.

[255] F. Joachimski and R. Matthes. Short proofs of normalization for the simply-typed λ-calculus, permutative conversions and Gödel's T. *Archive for Mathematical Logic*, 42(1):59–87, 2003.

[256] N.D. Jones. *Computability and Complexity from a Programming Perspective*. MIT Press, 1997.

[257] F. Kamareddine, T. Laan, and R. Nederpelt. Types in logic and mathematics before 1940. *Bulletin of Symbolic Logic*, 8(2):185–245, 2002.

[258] F. Kamareddine and A. Ríos. A λ-calculus à la de Bruijn with explicit substitutions. In M. Hermenegildo and S.D. Swierstra, editors, *Programming Languages: Implementations, Logics and Programs*, volume 982 of *Lecture Notes in Computer Science*, pages 45–62. Springer-Verlag, 1995.

[259] P.C. Kanellakis, H.G. Mairson, and J.C. Mitchell. Unification and ML type reconstruction. In J.-L. Lassez and G. Plotkin, editors, *Computational Logic. Essays in Honor of Alan Robinson*, chapter 13, pages 444–478. The MIT Press, 1991.

[260] S. Kanger. *Provability in Logic*, volume 1 of *Acta Universitatis Stockholmiensis*. Almqvist & Wiksell, Stockholm, 1957.

[261] M. Karr. "Delayability" in proofs of strong normalizability in the typed lambda calculus. In H. Ehrig, C. Floyd, M. Nivat, and J. Thatcher, editors, *Mathematical Foundations of Software Development, vol. 1: CAAP'85*, volume 185 of *Lecture Notes in Computer Science*, pages 208–222. Springer-Verlag, 1985.

[262] H. Kennedy. *Peano. Life and Works of Giuseppe Peano*. D. Reidel, 1980.

[263] A.J. Kfoury, J. Tiuryn, and P. Urzyczyn. An analysis of ML-typability. *Journal of the ACM*, 41(2):368–398, 1994.

[264] A.J. Kfoury and J. Wells. New notions of reduction and non-semantic proofs of strong β-normalization in typed λ-calculi. In *Logic in Computer Science*, pages 311–321. IEEE Computer Society Press, 1995.

[265] Z. Khasidashvili. The longest perpetual reductions in orthogonal expression reduction systems. In A. Nerode and Yu. V. Matiyasevich, editors, *Symposium on Logical Foundations of Computer Science*, volume 813 of *Lecture Notes in Computer Science*, pages 191–203. Springer-Verlag, 1994.

[266] A. Kino, J. Myhill, and R.E. Vesley, editors. *Intuitionism and Proof Theory*. Studies in Logic and the Foundations of Mathematics. North-Holland, 1970.

[267] S.C. Kleene. λ-definability and recursiveness. *Duke Mathematical Journal*, 2:340–353, 1936.

[268] S.C. Kleene. On the interpretation of intuitionistic number theory. *Journal of Symbolic Logic*, 10(4):109–124, 1945.

[269] S.C. Kleene. *Introduction to Metamathematics*. Van Nostrand, 1952.

[270] S.C. Kleene. Origins of recursive function theory. *Annals of the History of Computing*, 3(1):52–67, 1981.

[271] S.C. Kleene and J.B. Rosser. The inconsistency of certain formal logics. *Annals of Mathematics*, 36(3):630–636, 1935.

[272] J.W. Klop. *Combinatory Reduction Systems*, volume 127 of *Mathematical Centre Tracts*. Mathematisch Centrum, Amsterdam, 1980.

[273] A.N. Kolmogorov. (Колмогоров, А.Н.) О принципе tertium non datur. *Математический Сборник*, 32:646–667, 1925. English translation: On the principle of excluded middle, in Heijenoort [218], pages 414–437.

[274] A.N. Kolmogorov. Zur Deutung der intuitionistischen Logik. *Mathematische Zeitschrift*, 35(1):58–65, 1932. English translation: On the interpretation of intuitionistic logic, in Mancosu [320], pages 328–334.

[275] Y. Komori and H. Ono. Logics without the contraction rule. *Journal of Symbolic Logic*, 50(1):169–201, 1985.

[276] D. Kozen. *Automata and Computability*. Undergraduate Texts in Computer Science. Springer-Verlag, New York, 1997.

[277] G. Kreisel. Interpretation of analysis by means of functionals of finite type. In A. Heyting, editor, *Constructivity in Mathematics*, Studies in Logic and the Foundations of Mathematics, pages 101–128. North-Holland, 1959.

[278] G. Kreisel. Foundations of intuitonistic logic. In E. Nagel, P. Suppes, and A. Tarski, editors, *Logic, Methodology and Philosophy of Science: Proceedings of the 1960 International Congress*, pages 198–210. Stanford University Press, 1962.

[279] G. Kreisel. On weak completeness of intuitionistic predicate logic. *Journal of Symbolic Logic*, 27(2):139–158, 1962.

[280] G. Kreisel. Some uses of proof theory for finding computer programs. In *Colloque International de Logique, Clermont-Ferrand, 18-25 julliet 1975*, pages 123–134. CNRS, 1977.

[281] G. Kreisel. Monadic operators defined by means of propositional quantification in intuitionistic logic. *Reports on Mathematical Logic*, 12:9–15, 1981.

[282] P. Kremer. On the complexity of propositional quantification in intuitionistic logic. *Journal of Symbolic Logic*, 62(2):529–544, 1997.

[283] S.A. Kripke. Semantical analysis of intuitionistic logic I. In Crossley and Dummett [93], pages 92–129.

[284] J.-L. Krivine. Un algorithme non typable dans le système F. *Comptes Rendus Acad. Sc. Paris, Série I*, 304(5):123–126, 1987.

[285] J.-L. Krivine. *Lambda-Calculus, Types and Models*. Ellis Horwood Series in Computers and their Applications. Masson and Ellis Horwood, 1993.

[286] J.-L. Krivine. Classical logic, storage operators, and second-order λ-calculus. *Annals of Pure and Applied Logic*, 68(1):53–78, 1994.

[287] J.-L. Krivine. Typed lambda-calculus in classical Zermelo-Fraenkel set theory. *Archiv für mathematische Logik und Grundlagenforschung*, 40(3):189–205, 1994.

[288] J.-L. Krivine and M. Parigot. Programming with proofs. *J. Inf. Process. Cybern. EIK*, 26(3):149–167, 1990.

[289] K. Kuiper, editor. *Merriam Webster's Encyclopedia of Literature*. Merriam-Webster Inc. Publishers, Springfield, MA, 1995.

[290] K. Kuratowski and A. Mostowski. *Set Theory with an Introduction to Descriptive Set Theory*, volume 86 of *Studies in Logic and the Foundations of Mathematics*. North-Holland, 1976.

[291] A.V. Kuznetsov. Analogs of the "Sheffer stroke" in constructive logic. *Soviet Mathematics Doklady*, 6:70–74, 1965.

[292] R.E. Ladner. The computational complexity of provability in systems of modal propositional logic. *SIAM Journal of Computing*, 6(3):467–480, 1977.

[293] J. Lambek. The mathematics of sentence structure. *American Mathematical Monthly*, 65(3):154–170, 1958.

[294] J. Lambek and P.J. Scott. *Introduction to Higher-Order Categorical Logic*. Cambridge Studies in Advanced Mathematics. Cambridge University Press, 1986.

[295] P.J. Landin. A correspondence between ALGOL 60 and Church's lambda-notation. *Communications of the ACM*, 8(2-3):89–101,158–165, 1965.

[296] S. Lang. *Algebra*, volume 211 of *Graduate Texts in Mathematics*. Springer-Verlag, revised third edition, 2002.

[297] H. Läuchli. Intuitionistic propositional calculus and definably non-empty terms. *Journal of Symbolic Logic*, 30(2):263, 1965.

[298] H. Läuchli. An abstract notion of realizability for which intuitionistic predicate calculus is complete. In Kino et al. [266], pages 227–234.

[299] M. Laudet, D. Lacombe, and M. Schützenberger, editors. *Symposium on Automatic Demonstration, Held at Versailles/France, December 1968*, volume 125 of *Lecture Notes in Mathematics*. Springer-Verlag, 1970.

[300] F.W. Lawvere. Equality in hyperdoctrines and comprehension schema as an adjoint functor. In A. Heller, editor, *Applications of Categorical Algebra*, volume 17 of *Proceedings of Symposia in Pure Mathematics*, pages 1–14. AMS, 1970.

[301] D. Leivant. Assumption classes in natural deduction. *Zeitschrift für mathematische Logik und Grundlagen der Mathematik*, 25:1–4, 1979.

[302] D. Leivant. Reasoning about functional programs and complexity classes associated with type disciplines. In *Proc. IEEE Symposium on Foundations of Computer Science*, pages 460–469, 1983.

[303] D. Leivant. Syntactic translations and provably recursive functions. *Journal of Symbolic Logic*, 50(3):682–688, 1985.

[304] D. Leivant. Contracting proofs to programs. In Odifreddi [367], pages 279–327.

[305] D. Leivant. Discrete polymorphism. In *ACM Conference on Lisp and Functional Programming*, pages 288–297, New York, NY, USA, 1990. ACM Press.

[306] D. Leivant. Finitely stratified polymorphism. *Information and Computation*, 93(1):93–113, 1991.

[307] D. Leivant. Higher order logic. In D. Gabbay, C.J. Hogger, and J.A. Robinson, editors, *Handbook of Logic in Artificial Intelligence*, volume 2, pages 229–321. Oxford University Press, 1994.

[308] D. Leivant. Intrinsic reasoning about functional programs I: First-order theories. *Annals of Pure and Applied Logic*, 114(1-3):117–153, 2002.

[309] P. Lescanne. From $\lambda\sigma$ to $\lambda\upsilon$: A journey through calculi of explicit substitutions. In *Symposium on Principles of Programming Languages*, pages 60–109. ACM Press, 1994.

[310] P. Lescanne and J. Rouyer-Degli. Explicit substitutions with de Bruijn levels. In J. Hsiang, editor, *Rewriting Techniques and Applications*, volume 914 of *Lecture Notes in Computer Science*, pages 294–308. Springer-Verlag, 1995.

[311] J. Lipton. Realizability, set theory and term extraction. In de Groote [206], pages 257–364.

[312] R. Loader. Normalisation by calculation. Manuscript, 1995.

[313] M. Löb. Embedding first order predicate logic in fragments of intuitionistic logic. *Journal of Symbolic Logic*, 41(4):705–718, 1976.

[314] K. Lorenz. Dialogspiele als semantische Grundlage von Logikkalkülen. *Archiv für mathematische Logik und Grundlagenforschung*, 11:32–55,73–100, 1968. Reprinted in Lorenzen and Lorenz [316] pages 96–162.

[315] P. Lorenzen. Ein dialogisches Konstruktivitätskriterium. In *Infinitistic Methods*, pages 193–200. Pergamon Press, Oxford, 1961. Reprinted in Lorenzen and Lorenz [316], pages 9–16.

[316] P. Lorenzen and K. Lorenz. *Dialogische Logik*. Wissenschaftliche Buchgesellschaft, 1978.

[317] Z. Luo. A higher-order calculus and theory abstraction. *Information and Computation*, 90(1):107–137, 1991.

[318] H.G. Mairson. A simple proof of a theorem of Statman. *Theoretical Computer Science*, 103(2):387–394, 1992.

[319] H.G. Mairson. From Hilbert space to Dilbert space: Context semantics made simple. In M. Agrawal and A. Seth, editors, *FST TCS 2002: Foundations of Software Technology and Theoretical Computer Science*, volume 2556 of *Lecture Notes in Computer Science*, pages 2–17. Springer-Verlag, 2002.

[320] P. Mancosu, editor. *From Hilbert to Brouwer. The Debate on the Foundations of Mathematics in the 1920s*. Oxford University Press, 1998.

[321] Z. Manna and R. Waldinger. Toward automatic program synthesis. *Communications of the ACM*, 14(3), 1971.

[322] D. Marini and P.A. Miglioli. Characterization of programs and their synthesis from a formalized theory. In *Mathematical Foundations of Computer Science, Proceedings of Symposium and Summer School, High Tatras*, pages 259–266. Slovak Academy of Science, 1973.

[323] P. Martin-Löf. A construction of the provable wellorderings of the theory of species. Manuscript, 1970. Reprinted in C. A. A. Anderson and M. Zelëny, editors, *Logic, Meaning and Computation. Essays in Memory of Alonzo Church*, pages 343–351, Kluwer, 2001.

[324] P. Martin-Löf. Hauptsatz for the theory of species. In Fenstad [148], pages 217–233.

[325] P. Martin-Löf. An intuitionistic theory of types. Technical report, University of Stockholm, 1972. Reprinted in G. Sambin and J. Smith, editors, *Twenty-Five Years of Constructive Type Theory*, Oxford University Press, 1988, pages 127–172.

[326] P. Martin-Löf. An intuitionistic theory of types: predicative part. In H.E. Rose and J.C. Shepherdson, editors, *Logic Colloquium '73*, volume 80 of *Studies in Logic and the Foundations of Mathematics*, pages 73–118. North-Holland, 1975.

[327] P. Martin-Löf. *Intuitionistic Type Theory. Notes by Giovanni Sambin of a series of lectures given in Padua, June 1980*. Bibliopolis, 1984.

[328] P. Martin-Löf. Constructive mathematics and computer programming. In C.A.R. Hoare and J.C. Shepherdson, editors, *Mathematical Logic and Programming Languages*, pages 167–184. Prentice-Hall, 1985.

[329] R. Matthes. Monotone (co)inductive types and positive fixed-point types. *RAIRO—Theoretical Informatics and Applications*, 33(4/5):309–328, 1999.

[330] D. McAllester, J. Kučan, and D.F. Otth. A proof of strong normalization for F_2, F_ω, and beyond. *Information and Computation*, 121(2):193–200, 1995.

[331] C. McBride and J. McKinna. The view from the left. *Journal of Functional Programming*, 14(1):69–111, 2004.

[332] S. McCall, editor. *Polish Logic 1920–1939*. Clarendon Press, Oxford, 1967.

[333] D. McCarty. Completeness for intuitionistic logic. In Odifreddi [368], pages 301–334.

[334] J.C.C. McKinsey. Proof of the independence of the primitive symbols of Heyting's calculus of propositions. *Journal of Symbolic Logic*, 4(4):155–158, 1939.

[335] P.-A. Melliès. Typed λ-calculi with explicit substitution may not terminate. In Dezani-Ciancaglini and Plotkin [129], pages 328–334.

[336] P.-A. Melliès and B. Werner. A generic normalisation proof for Pure Type Systems. In E. Gimenez and Ch. Paulin-Mohring, editors, *Types for Proofs and Programs, International Workshop TYPES'96*, volume 1512 of *Lecture Notes in Computer Science*, pages 254–276. Springer-Verlag, 1998.

[337] E. Mendelson. *Introduction to Mathematical Logic.* Chapman & Hall, London, fourth edition, 1997.

[338] N.P. Mendler. Inductive types and type constraints in the second-order lambda calculus. *Annals of Pure and Applied Logic*, 51(1-2):159–172, 1991.

[339] C.A. Meredith and A.N. Prior. Notes on the axiomatics of the propositional calculus. *Notre Dame Journal of Formal Logic*, 4(3):171–187, 1963.

[340] A.R. Meyer and M.B. Reinhold. 'Type is not a type': Preliminary report. In *Symposium on Principles of Programming Languages*, pages 287–295. ACM Press, 1986.

[341] A.R. Meyer and M. Wand. Continuation semantics in typed lambda-calculi (summary). In R. Parikh, editor, *Logics of Programs*, volume 193 of *Lecture Notes in Computer Science*, pages 219–224. Springer-Verlag, 1985.

[342] R. Milner. A theory of type polymorphism in programming. *Journal of Computer and System Sciences*, 17:348–375, 1978.

[343] R. Mines, F. Richman, and W. Ruitenburg. *A Course in Constructive Algebra.* Universitext. Springer-Verlag, 1988.

[344] G. Mints. Proof theory in the USSR 1925–1969. *Bulletin of Symbolic Logic*, 56(2):385–424, 1991.

[345] G. Mints. Normal forms for sequent derivations. In Odifreddi [368], pages 479–492.

[346] G. Mints. *A Short Introduction to Intuitionistic Logic.* Kluwer Academic/Plenum Publishers, New York, 2000.

[347] G. Mints and T. Zhang. A proof of topological completeness of S4 in (0,1). *Annals of Pure and Applied Logic*, 133(1-3):231–245, 2005.

[348] J.C. Mitchell. Polymorphic type inference and containment. *Information and Computation*, 76(2-3):211–249, 1988. Reprinted in Huet [250], pages 153–272.

[349] J.C. Mitchell. *Foundations for Programming Languages.* The MIT Press, 1996.

[350] J.C. Mitchell. *Concepts in for Programming Languages.* Cambridge University Press, 2003.

[351] I. Moerdijk. Some topological spaces which are universal for intuitionistic predicate logic. *Nederl. Akad. Wetensch. Indag. Math.*, 44(2):227–235, 1982.

[352] J.D. Monk. *Mathematical Logic.* Springer-Verlag, 1976.

[353] J.D. Monk, editor. *Handbook of Boolean Algebras.* North-Holland, 1989.

[354] M. Moortgat. Categorial type logic. In J. van Benthem and A. ter Meulen, editors, *Handbook of Logic and Language*, pages 93–177. Elsevier, 1997.

[355] R. Murawski. *Recursive Functions and Metamathematics. Problems of Completeness and Decidability, Gödel's Theorems*, volume 286 of *Synthese Library*. Kluwer Academic/Plenum Publishers, New York, 1999.

[356] C.R. Murthy. An evaluation semantics for classical proofs. In *Logic in Computer Science*, pages 96–109. IEEE Computer Society Press, 1991.

[357] C.R. Murthy. A computational analysis of Girard's translation and LC. In *Logic in Computer Science*, pages 90–101. IEEE Computer Society Press, 1992.

[358] H. Nakano. A constructive formalization of the catch and throw mechanism. In *Logic in Computer Science*, pages 82–89. IEEE Computer Society Press, 1992.

[359] R. Nederpelt. *Strong normalization for a typed lambda calculus with lambda structured types*. PhD thesis, Eindhoven, 1973.

[360] R. Nederpelt, J.H. Geuvers, and R.C. de Vrijer, editors. *Selected Papers on Automath*, volume 133 of *Studies in Logic and the Foundations of Mathematics*. North-Holland, 1994.

[361] P.M. Neergard and M.H. Sørensen. Conservation and uniform normalization in lambda-calculi with erasing reductions. *Information and Computation*, 178(1):149–179, 2002.

[362] M.H.A. Newman. Stratified systems of logic. *Proceedings of the Cambridge Philosophical Society*, 39(2):69–83, 1943.

[363] K. Nielsen and M.H. Sørensen. Call-by-name CPS-translation as a binding-time improvement. In A. Mycroft, editor, *Static Analysis Symposium*, volume 983 of *Lecture Notes in Computer Science*, pages 296–313. Springer-Verlag, 1995.

[364] T. Nipkow, L.C. Paulson, and M. Wenzel. *Isabelle/HOL. A Proof Assistant for Higher-Order Logic*, volume 2283 of *Lecture Notes in Computer Science*. Springer-Verlag, 2002.

[365] B. Nordström, K. Petersson, and J.M. Smith. *Programming in Martin-Löf's Type Theory. An Introduction*. Oxford University Press, 1990.

[366] B. Nordström, K. Petersson, and J.M. Smith. Martin-Löf's type theory. In S. Abramsky, D.M. Gabbay, and T.S.E. Maibaum, editors, *Handbook of Logic in Computer Science*, volume V, pages 1–37. Oxford University Press, 2001.

[367] P. Odifreddi, editor. *Logic and Computer Science*. Academic Press, 1990.

[368] P. Odifreddi, editor. *Kreiseliana. About and Around Georg Kreisel*. A.K. Peters, 1996.

[369] I. Ogata. Cut elimination for classical proofs as continuation passing style computation. In J. Hsiang and A. Ohori, editors, *Advances in Computing Science—ASIAN'98*, volume 1538 of *Lecture Notes in Computer Science*, pages 61–78. Springer-Verlag, 1998.

[370] I. Ogata. A CPS-transform of constructive classical logic. In P.S. Thjagarajan and R. Yap, editors, *Advances in Computing Science—ASIAN'99*, volume 1742 of *Lecture Notes in Computer Science*, pages 266–280. Springer-Verlag, 1999.

[371] I. Ogata. Constructive classical logic as CPS-calculus. *International Journal of Foundations of Functional Programming*, 11(1):89–112, 2000.

[372] A. Ohori. A Curry-Howard isomorphism for compilation and program execution. In J.-Y. Girard, editor, *Typed Lambda Calculi and Applications*, volume

1581 of *Lecture Notes in Computer Science*, pages 280–294. Springer-Verlag, 1999.

[373] A. Ohori. The logical abstract machine: A Curry-Howard isomorphism for machine code. In A. Middledorp and T. Sato, editors, *Functional and Logic Programming*, volume 1722 of *Lecture Notes in Computer Science*, pages 300–318. Springer-Verlag, 1999.

[374] M. Okada. A uniform semantic proof for cut-elimination and completeness of various first and higher order logics. *Theoretical Computer Science*, 281(1-2):471–498, 2002.

[375] C.-H.L. Ong. A semantic view of classical proofs: Type-theoretic, categorical, and denotational characterizations. In *Logic in Computer Science*, pages 230–241. IEEE Computer Society Press, 1996.

[376] C.-H.L. Ong and C.A. Stewart. A Curry-Howard foundation for functional computation with control. In *Symposium on Principles of Programming Languages*, pages 215–227. ACM Press, 1997.

[377] J. van Oosten. Realizability: A historical essay. *Mathematical Structures in Computer Science*, 12(3):239–263, 2002.

[378] L. Paolini and S. Ronchi Della Rocca. *The Parametric λ-calculus: a Metamodel for Computation*. Texts in Theoretical Computer Science. An EATCS Series. Springer-Verlag, 2004.

[379] M. Parigot. On the representation of data in lambda-calculus. In E. Börger, H. Kleine Büning, and M.M. Richter, editors, *CSL'89. 3rd Workshop on Computer Science Logic*, volume 440 of *Lecture Notes in Computer Science*, pages 309–311. Springer-Verlag, 1990.

[380] M. Parigot. Free deduction: An analysis of "computations" in classical logic. In A. Voronkov, editor, *Logic Programming*, volume 592 of *Lecture Notes in Artificial Intelligence*, pages 361–380. Springer-Verlag, 1991.

[381] M. Parigot. $\lambda\mu$-calculus: An algorithmic interpretation of classical natural deduction. In A. Voronkov, editor, *Logic Programming and Automated Reasoning*, volume 624 of *Lecture Notes in Artificial Intelligence*, pages 190–201. Springer-Verlag, 1992.

[382] M. Parigot. Recursive programming with proofs. *Theoretical Computer Science*, 94(2):335–356, 1992.

[383] M. Parigot. Classical proofs as programs. In G. Gottlob, A. Leitsch, and D. Mundici, editors, *Computational Logic and Proof Theory*, volume 713 of *Lecture Notes in Computer Science*, pages 263–276. Springer-Verlag, 1993.

[384] M. Parigot. Strong normalization for second order classical natural deduction. In *Logic in Computer Science*, pages 39–46. IEEE Computer Society Press, 1993.

[385] J. Paris and L. Harrington. A mathematical incompleteness in Peano Arithmetic. In J. Barwise, editor, *Handbook of Mathematical Logic*, volume 90 of *Studies in Logic and the Foundations of Mathematics*, pages 1133–1142. North-Holland, 1977.

[386] Ch. Paulin-Mohring. Extracting F_ω's programs from proofs in the Calculus of Constructions. In *Symposium on Principles of Programming Languages*, pages 89–103. ACM Press, 1989.

[387] L.C. Paulson. *ML for the Working Programmer*. Cambridge University Press, second edition, 1996.

[388] G. Peano. *Arithmetices principia, nova methodo exposita*. Fratres Bocca, 1889. Reprinted in Peano [390, vol. II], pages 20–55. English translation: The principles of arithmetic, presented by a new method, in Heijenoort [218], pages 83–97.

[389] G. Peano. Sul concetto di numero. *Rivista di matematica*, 1:87–102, 256–267, 1891. Reprinted in Peano [390, vol. III], pages 80–109.

[390] G. Peano. *Opere scelte*. Edizioni Cremonese, 1957, 1958, 1959.

[391] S. Peyton Jones. *The Implementation of Functional Programming Languages*. Prentice Hall, 1987.

[392] F. Pfenning and C. Elliot. Higher-order abstract syntax. In *Programming Language Design and Implementation*, pages 199–208. ACM Press, 1988.

[393] B.C. Pierce. *Types and Programming Languages*. The MIT Press, 2002.

[394] A. Pitts. On an interpretation of second-order quantification in first-order intuitionistic propositional calculus. *Journal of Symbolic Logic*, 57(1):33–52, 1992.

[395] A. Pitts. Nominal logic: A first order theory of names and binding. *Information and Computation*, 186(2):165–193, 2003.

[396] G. Plotkin. Call-by-name, call-by-value and the λ-calculus. *Theoretical Computer Science*, 1(2):125–159, 1975.

[397] G. Plotkin. LCF considered as a programming language. *Theoretical Computer Science*, 5(3):223–255, 1977.

[398] J. van de Pol and H. Schwichtenberg. Strict functionals for termination proofs. In Dezani-Ciancaglini and Plotkin [129], pages 350–364.

[399] T. Połacik. Pitts' quantifiers are not topological quantification. *Notre Dame Journal of Formal Logic*, 39(4):531–544, 1998.

[400] T. Połacik. Propositional quantification in the monadic fragment of intuitionistic logic. *Journal of Symbolic Logic*, 63(1):269–300, 1998.

[401] R. Pollack. Closure under alpha-conversion. In Barendregt and Nipkow [30], pages 313–332.

[402] G. Pottinger. Normalization as a homomorphic image of cut-elimination. *Annals of Mathematical Logic*, 12:323–357, 1977.

[403] D. Prawitz. *Natural Deduction. A Proof Theoretical Study*. Almqvist & Wiksell, 1965.

[404] D. Prawitz. Some results for intuitionistic logic with second order quantification rules. In Kino et al. [266], pages 259–270.

[405] D. Prawitz. Ideas and results in proof theory. In Fenstad [148], pages 235–307.

[406] F. van Raamsdonk, P. Severi, M.H.B. Sørensen, and H. Xi. Perpetual reductions in λ-calculus. *Information and Computation*, 149(2):173–225, 1999.

[407] H. Rasiowa. Algebraic treatment of the functional calculi of Heyting and Lewis. *Fundamenta Mathematicae*, 38:99–126, 1951.

[408] H. Rasiowa and R. Sikorski. A proof of the completeness theorem of Gödel. *Fundamenta Mathematicae*, 37:193–200, 1950.

[409] H. Rasiowa and R. Sikorski. On the Gentzen theorem. *Fundamenta Mathematicae*, 48:57–69, 1959.

[410] H. Rasiowa and R. Sikorski. *The Mathematics of Metamathematics*. PWN, Warsaw, 1963.

[411] C. Rauszer and B. Sabalski. Notes on the Rasiowa-Sikorski lemma. *Studia Logica*, XXXIV(3):265–268, 1975.

[412] L. Regnier. Une équivalence sur les lambda-termes. *Theoretical Computer Science*, 126(3):281–292, 1994.

[413] N.J. Rehof and M.H. Sørensen. The λ_Δ calculus. In M. Hagiya and J. Mitchell, editors, *Theoretical Aspects of Computer Software*, volume 789 of *Lecture Notes in Computer Science*, pages 516–542. Springer-Verlag, 1994.

[414] J. Reynolds. Definitional interpreters for higher-order programming languages. In *Proceedings of the ACM Annual Conference*, volume 2, pages 717–740, 1972.

[415] J. Reynolds. Towards a theory of type structure. In B. Robinet, editor, *Proceedings of the Programming Symposium*, volume 19 of *Lecture Notes in Computer Science*, pages 408–425. Springer-Verlag, 1974.

[416] A. Rezus. Classical proofs: Lambda calculus methods in elementary proof theory, 1991. Manuscript.

[417] A. Rezus. Beyond BHK. In H. Barendregt, M. Bezem, and J.W. Klop, editors, *Dirk van Dalen Festschrift*, volume V of *Quaestiones Infinitae*. Utrecht University, 1993.

[418] J.A. Robinson. A machine-oriented logic based on the resolution principle. *Journal of the ACM*, 12(1):23–41, 1965.

[419] H. Rogers. *Theory of Recursive Functions and Effective Computability*. McGraw-Hill, 1967.

[420] J.B. Rosser. Highlights of the history of the lambda-calculus. *Annals of the History of Computing*, 6(4):337–349, 1984.

[421] L.E. Sanchis. Functionals defined by recursion. *Notre Dame Journal of Formal Logic*, VIII(3):161–174, 1967.

[422] J.C.S. do E. Santo. *Conservative extensions of the λ-calculus for the computational interpreation of sequent calculus*. PhD thesis, Laboratory for the Foundations of Computer Science, University of Edinburgh, 2002.

[423] M. Sato. Classical Brouwer-Heyting-Kolmogorov interpretation. In M. Li and A. Maruoka, editors, *Algorithmic Learning Theory*, volume 1316 of *Lecture Notes in Artificial Intelligence*, pages 176–196. Springer-Verlag, 1997.

[424] M. Sato. Intuitionistic and classical natural deduction systems with the catch and throw rules. *Theoretical Computer Science*, 175(1):75–92, 1997.

[425] A. Scedrov. On some non-classical extensions of second-order intuitionistic propositional calculus. *Annals of Pure and Applied Logic*, 27(2):155–164, 1984.

[426] E. Schechter. *Handbook of Analysis and Its Foundations*. Academic Press, 1997.

[427] M. Schönfinkel. Über the Bausteine der mathematischen Logik. *Mathematische Annalen*, 92:305–316, 1924. English translation:On the building blocks of mathematical logic, in Heijenoort [218], pages 355–366.

[428] A. Schubert. Second-order unification and type inference for Church-style polymorphism. In *Symposium on Principles of Programming Languages*, pages 233–244. ACM Press, 1998.

[429] A. Schubert. Type inference for first order logic. In J. Tiuryn, editor, *Foundations of Software Science and Computation Structures*, volume 1784 of *Lecture Notes in Computer Science*, pages 297–314. Springer-Verlag, 2000.

[430] H. Schwichtenberg. Minimal logic for computable functionals. In *Logic Colloquium 2005*. A.K. Peters. To appear.

[431] H. Schwichtenberg. Definierbare Funktionen im Lambda-Kalkül mit Typen. *Archiv für mathematische Logik und Grundlagenforschung*, 17:113–114, 1976.

[432] H. Schwichtenberg. Normalization. In F.L. Bauer, editor, *Logic, Algebra and Computation*, volume 79 of *NATO Advanced Study Institute, Series F: Computer and Systems Sciences*, pages 201–237. Springer-Verlag, 1991.

[433] H. Schwichtenberg. An upper bound for reduction sequences in the typed lambda-calculus. *Archive for Mathematical Logic*, 30:405–408, 1991.

[434] H. Schwichtenberg. Classifying recursive functions. In E.R. Griffor, editor, *Handbook of Computability Theory*, volume 140 of *Studies in Logic and the Foundations of Mathematics*, pages 533–586. North-Holland, 1999.

[435] H. Schwichtenberg. Refined program extraction from classical proofs: Some case studies. In F.L. Bauer and R. Steinbrüggen, editors, *Foundations of Secure Computation*, volume 175 of *NATO Science Series: Computer & Systems Sciences*, pages 147–180. IOS Press, 2000.

[436] D. Scott. Extending the topological interpretation to intuitionistic analysis. *Compositio Mathematica*, 20:194–210, 1968.

[437] D. Scott. Constructive validity. In Laudet et al. [299], pages 237–275.

[438] J.P. Seldin. Normalization and excluded middle. *Studia Logica*, XLVIII(2):193–217, 1989.

[439] J.P. Seldin. Curry's anticipation of the types used in programming languages. In *Proceedings of the Annual Meeting of the Canadian Society for History and Philosophy of Mathematics, Toronto, 24-26 May 2002*, pages 148–163, 2003.

[440] J.P. Seldin. Church and Curry: The lambda calculus and combinatory logic. In Gabbay and Woods [159]. To appear.

[441] J.P. Seldin and J.R. Hindley, editors. *To H.B. Curry: Essays on Combinatory Logic, Lambda Calculus and Formalism*. Academic Press, 1980.

[442] P. Selinger. Control categories and duality: on the categorical semantics of the lambda-mu calculus. *Mathematical Structures in Computer Science*, 11:207–260, 2001.

[443] J.R. Shoenfield. *Mathematical Logic*. Addison-Wesley, 1967. Reprinted by ASL in 2000.

[444] H. Simmons. *Derivation and Computation. Taking the Curry-Howard Correspondence Seriously*, volume 51 of *Cambridge Tracts in Theoretical Computer Science*. Cambridge University Press, 2000.

[445] D. Skvortsov. Non-axiomatizable second-order intuitionistic propositional logic. *Annals of Pure and Applied Logic*, 86(1):33–46, 1997.

[446] R.M. Smullyan. *First-Order Logic*. Springer-Verlag, 1968. Reprinted by Dover in 1995.

[447] S.K. Sobolev. (Соболев, С.К.) Об интуиционистском исчислении высказываний с кванторами. *Математические Заметки*, 22(1):69–76, 1977.

[448] M.H. Sørensen. Strong normalization from weak normalization in typed λ-Calculi. *Information and Computation*, 133(1):35–71, 1997.

[449] M.H. Sørensen. Properties of infinite reduction paths in untyped λ-calculus. In J. Ginzburg, Z. Khasidashvili, E. Vogel, J.J. Lévy, and E. Vallduví, editors, *The Tbilisi Symposium on Language, Logic, and Computation: Selected Papers*, Studies in Logic, Language, and Computation, pages 353–367. CSLI Publications, 1998.

[450] M.H.B. Sørensen and P. Urzyczyn. Lectures on the Curry-Howard isomorphism. DIKU rapport 98/14, Department of Computer Science, University of Copenhagen, 1998.

[451] C. Spector. Provably recursive functionals of analysis: a consistency proof of analysis by an extension of principles formulated in current intuitionistic mathematics. In J.C.E. Dekker, editor, *Recursive Function Theory*, volume 5 of *Proceedings of Symposia in Pure Mathematics*, pages 1–27. AMS, 1962.

[452] Z. Spławski and P. Urzyczyn. Type fixpoints: Iteration vs. recursion. In *International Conference on Functional Programming*, pages 102–113. ACM Press, 1999.

[453] J. Springintveld. Lower and upper bounds for reductions of types in $\lambda\underline{\omega}$ and λP. In Bezem and Groote [45], pages 391–405.

[454] G. Stålmarck. Normalization theorems for full first order classical natural deduction. *Journal of Symbolic Logic*, 56(1):129–149, 1991.

[455] R. Statman. Intuitionistic propositional logic is polynomial-space complete. *Theoretical Computer Science*, 9(1):67–72, 1979.

[456] R. Statman. The typed lambda-calculus is not elementary recursive. *Theoretical Computer Science*, 9(1):73–81, 1979.

[457] R. Statman. Number theoretic functions computable by polymorphic programs (extended abstract). In *Proc. IEEE Symposium on Foundations of Computer Science*, pages 279–282, 1981.

[458] M. Steinby and W. Thomas. Trees and term rewriting in 1910: On a paper by Axel Thue. *Bulletin of the European Association for Theoretical Computer Science*, 72:256–269, 2000.

[459] S. Stenlund. *Combinators, Terms, and Proof Theory*, volume 42 of *Synthese Library*. D. Reidel, 1972.

[460] L.J. Stockmeyer and A.R. Meyer. Word problems requiring exponential time: Preliminary report. In *STOC*, volume 5, pages 1–9. ACM Press, 1973.

[461] T. Streicher and B. Reus. Classical logic, continuation semantics and abstract machines. *Journal of Functional Programming*, 8(6):543–572, 1998.

[462] G. Sundholm. Constructions, proofs and the meaning of logical constants. *Journal of Philosophical Logic*, 12:151–172, 1983.

[463] V. Švejdar. On the polynomial-space completeness of intuitionistic propositional logic. *Archive for Mathematical Logic*, 42(7):711–716, 2003.

[464] J. Swift. *Gulliver's Travels into Several Remote Nations of the World*. Project Gutenberg, 1997.

[465] M.E. Szabo, editor. *The collected papers of Gerhard Gentzen.* Studies in Logic and the Foundations of Mathematics. North-Holland, 1969.

[466] W.W. Tait. Infinitely long terms of transfinite type. In Crossley and Dummett [93], pages 176–185.

[467] W.W. Tait. Intensional interpretations of functionals of finite type I. *Journal of Symbolic Logic*, 32(2):190–212, 1967.

[468] W.W. Tait. A realizability interpretation of the theory of species. In R. Parikh, editor, *Logic Colloquium: Symposium on Logic held at Boston, 1972-73*, volume 453 of *Lecture Notes in Mathematics*, pages 240–251. Springer-Verlag, 1975.

[469] W.W. Tait. The completeness of Heyting first-order logic. *Journal of Symbolic Logic*, 68(3):751–763, 2003.

[470] M. Takahashi. Parallel reductions in λ-calculus. *Information and Computation*, 118(1):120–127, 1995.

[471] G. Takeuti. *Proof Theory*, volume 81 of *Studies in Logic and the Foundations of Mathematics*. North-Holland, second edition, 1987.

[472] A. Tarski. *Pojęcie prawdy w językach nauk dedukcyjnych.* Nakładem Towarzystwa Naukowego Warszawskiego, 1933. English translation: The concept of truth in the languages of the deductive sciences, in *Logic, Semantics, Metamathematics. Papers from 1923 to 1938 by Alfred Tarski*, pages 152–278, Clarendon Press, 1956.

[473] A. Tarski. Der Aussagenkalkül und die Topologie. *Fundamenta Mathematicae*, 31:103–134, 1938.

[474] Terese. *Term Rewriting Systems.* Cambridge University Press, 2003.

[475] J. Terlouw. Strong normalization in type systems: a model theoretical approach. *Annals of Pure and Applied Logic*, 73:53–78, 1995.

[476] H. Thielecke. Continuation semantics and self-adjointness. *Electronic Notes in Theoretical Computer Science*, 6, 1997. Proc. Mathematical Foundations of Programming Semantics XIII.

[477] W. Thomas. Languages, automata and logic. In G. Rozenberg and A. Salomaa, editors, *Handbook of Formal Languages*, volume 3, pages 389–455. Springer-Verlag, 1997.

[478] S. Thompson. *Haskell. The Craft of Functional Programming.* Addison-Wesley, second edition, 1999.

[479] P. Trigg, J.R. Hindley, and M.W. Bunder. Combinatory abstraction using B, B' and friends. *Theoretical Computer Science*, 135(2):405–422, 1994.

[480] A.S. Troelstra. Introductory note to *1941.* In Gödel [195, vol. III], pages 186–189.

[481] A.S. Troelstra. Introductory note to *1958* and *1972.* In Gödel [195, vol. II], pages 217–241.

[482] A.S. Troelstra. *Metamathematical Investigation of Intuitionistic Arithmetic and Analysis*, volume 344 of *Lecture Notes in Mathematics*. Springer-Verlag, 1973. Second, corrected edition, ILLC Prepublication Series X-93-05, University of Amsterdam, 1993.

[483] A.S. Troelstra. On the early history of intuitionistic logic. In P.P. Petkov, editor, *Mathematical Logic*, pages 3–17. Plenum Press, 1990.

[484] A.S. Troelstra. History of constructivism in the 20th century. ITLI Prepublication Series, University of Amsterdam, 1991.
[485] A.S. Troelstra. *Lectures on Linear Logic*, volume 29 of *CLSI Lecture Notes*. Center for the Study of Language and Information, Stanford, 1992.
[486] A.S. Troelstra. Realizability. In Buss [63], chapter VI, pages 407–473.
[487] A.S. Troelstra. From constructivism to computer science. *Theoretical Computer Science*, 211(1-2):233–252, 1999.
[488] A.S. Troelstra and D. van Dalen. *Constructivism in Mathematics. An Introduction, Volume I and II*, volume 121 and 123 of *Studies in Logic and the Foundations of Mathematics*. North-Holland, 1988.
[489] A.S. Troelstra and H. Schwichtenberg. *Basic Proof Theory*, volume 43 of *Cambridge Tracts in Theoretical Computer Science*. Cambridge University Press, second edition, 2000.
[490] J.L. Underwood. *Aspects of the Computational Content of Proofs*. PhD thesis, Cornell University, 1994.
[491] A.M. Ungar. *Normalization, Cut-Elimination and the Theory of Proofs*, volume 28 of *CLSI Lecture Notes*. Center for the Study of Language and Information, Stanford, 1992.
[492] C. Urban and C. Bierman. Strong normalisation of cut-elimination in classical logic. *Fundamenta Informaticae*, 45(1-2):123–155, 2001.
[493] P. Urzyczyn. Predicates as types. In H. Schwichtenberg and K. Spies, editors, *Proof Technology and Computation, Proc. Marktoberdorf Summer School 2003*. IOS Press. To appear.
[494] P. Urzyczyn. Type inhabitation in typed lambda calculi (a syntactic approach). In de Groote and Hindley [208], pages 373–389.
[495] P. Urzyczyn. Positive recursive type assignment. *Fundamenta Informaticae*, 28(1-2):197–209, 1996.
[496] P. Urzyczyn. Type reconstruction in F_ω. *Mathematical Structures in Computer Science*, 7(4):329–358, 1997.
[497] B. Venneri. Intersection types as logical formulae. *Journal of Logic and Computation*, 4(2):109–124, 1994.
[498] R. Vestergaard. Revisiting Kreisel: A computational anomaly in the Troelstra-Schwichtenberg G3i system. Manuscript, 1999.
[499] R. Vestergaard and J. Wells. Cut rules and explicit substitutions. *Mathematical Structures in Computer Science*, 11(1):131–168, 2001.
[500] A. Voronkov. Proof search in intuitionistic logic with equality or back to simultaneous rigid E-unification. *Journal of Automated Reasoning*, 21(2):205–231, 1998.
[501] R.C. de Vrijer. Exactly estimating functionals and strong normalization. *Indagationes Mathematicae*, 49:479–493, 1987.
[502] P. Wadler. The Girard-Reynolds isomorphism. *Information and Computation*, 186(2):260–284, 2003.
[503] P. Wadler. Call-by-value is dual to call-by-name — reloaded. In J. Giesl, editor, *Rewriting Techniques and Applications*, volume 3467 of *Lecture Notes in Computer Science*, pages 185–203. Springer-Verlag, 2005.

[504] P.L. Wadler. Proofs are programs: 19th century logic and 21st century computing. Manuscript, 2000.

[505] M. Wajsberg. Untersuchungen über den Aussagenkalkül von A. Heyting. *Wiadomości Matematyczne*, 46:45–101, 1938. English translation: On A. Heyting's propositional calculus, in *Mordchaj Wajsberg, Logical Works* (S.J. Surma, editor), Ossolineum, Wrocław, 1977, pages 132–171.

[506] J. Wells. Typability and type checking in the second-order λ-calculus are equivalent and undecidable. *Annals of Pure and Applied Logic*, 98(1-3):111–156, 1999.

[507] H. Xi. Weak and strong beta normalisations in typed λ-calculi. In de Groote and Hindley [208], pages 390–404.

[508] H. Xi. On weak and strong normalisations. Research Report 96-187, Department of Mathematical Sciences, Carnegie Mellon University, 1996.

[509] H. Xi and F. Pfenning. Dependent types in practical programming. In *Symposium on Principles of Programming Languages*, pages 214–227. ACM Press, 1999.

[510] M. Zaionc. Mechanical procedure for proof construction via closed terms in typed λ-calculus. *Journal of Automated Reasoning*, 4(2):173–190, 1988.

[511] J.I. Zucker. The correspondence between cut-elimination and normalization. *Annals of Mathematical Logic*, 7(1):1–112, 1974.

Index

abstraction, 1, 3, 64, 133, 218, 276, 277, 327, 343, 346
   combinatory, 114, 118, 120–121
absurdity, 29, 89, 280, 310
accumulation point, 364
Ackermann's function, 255, 266, 367
Ackermann, W., 224, 323
Ada, 277
address, 132
admissible rule, 34, 330
affine
   logic, 120
   term, 119
Ajdukiewicz, K., 100
Akama, Y., 118
algebra, 365
algebraic term, 195, 365
α-conversion, 6, 9, 22, 23, 87, 196, 200, 270, 278, 297, 327, 344
antecedent, 162
anti-symmetric relation, 362
application, 1, 3, 64, 109, 133, 218, 276, 277, 326, 343
applicative $\lambda\mu$-context, 134
arithmetic, 229
   Heyting, 97, 229, 240
   higher-order, 313
   non-standard model, 230, 247
   Peano, 229, 232, 245, 246
   Presburger, 232
   second-order, 311, 314
   standard model, 229, 306
arithmetical relation (function), 231, 306
arity, 303
Artemov, S., 49
atomic formula, 29, 95, 140, 196
atomic $\lambda\mu$-term, 147
Automath, 98, 225, 339, 358

axioms
   in natural deduction, 32
   logical and non-logical, 103, 104
   of arithmetic, 232, 246, 255
   of classical logic, 129
   of first-order logic, 203
   of propositional logic, 104, 107, 119, 120
   of PTS, 346
   of second-order logic, 307

Barendregt's cube, 343, 346, 348, 357
Barendregt, H.P., 22, 340, 357, 358
Barendregt-Geuvers-Klop conjecture, 358
basis, 115, 125
BCI and **BCK** logic, 119
Ben-Yelles algorithm, 99
Ben-Yelles, C.-B., 99, 100
van Benthem, J.F.A.K., 191
Berardi, S., 351, 357
Berarducci, A., 300
beta function, 230
β-conversion (equality), 10
$\beta\eta$-reduction, 11
β-reduction, 2, 10, 64, 87, 219, 297, 327, 344
   first-order, 219, 319, 320
Beth model, 50, 299
Beth, E.W., 50, 190
Bezem, M., 339
BHK interpretation, 29, 49, 56, 77, 86, 96, 197, 247, 249, 271, 308
binary valuation, 37
Bloo, R., 176
Böhm, C., 300
Boolean algebra, 36
   complete, 207

bound (upper and lower), 363
Breazu-Tannen, V., 299
Brouwer, L.E.J., v, viii, 29, 48, 49, 96
de Bruijn, N.G., viii, 22, 97, 98, 176, 190, 225, 339, 340, 357
Bull, R.A., 299

Calculus of Constructions, 98, 343, 346–351, 357
call-by-name, 2
CAML, 357
candidates, method of, 288, 297, 299
Cantor's theorem, 352
capture-avoiding substitution, 373
captured variable, 2, 4
case, 87
catch, 136, 156
Catch-22, 145, 195, 200
chain, 363
channel, 134
Chinese remainder theorem, 230
Church numeral, 20, 282
Church, A., 22, 23, 25, 122, 123, 224, 299
Church-Rosser property, 13, 75, 88, 110, 124, 139, 260, 289, 290, 298, 320, 335
    weak, 24, 70, 75, 280
Church-style system, 63
Chwistek, L., 73
classical logic, 27, 128, 153
closed set, 364
closure, 364
c_n, 20
co-product, 216
coercion, 398
combinator (closed λ-term), 8
    B,C, 120
    I,K, 2
    S,Y,Ω, 11
combinator (in CL), 108
    B,C,I,W, 109–110
    K,S, 109
combinatory
    abstraction, 114
    completeness, 115
    logic, 108
        illative, 23, 100, 123
    term, 109
commuting conversions, 193

compactness theorem, 227
comparable elements, 363
compatible relation, 10, 12
complement, 35, 36
complete discharge, 101, 382
complete lattice, 363
complete problem, 370
complete theory, 234
completeness
    of CPC, 41, 130
    of first-order logic, 209, 214, 224, 226
    of implicational logic, 47
    of IPC, 40, 46
    of second-order logic, 272, 275, 307
complexity classes, 370
composition, 366
computability method, 74, 111, 123, 256, 288
computable
    instance, 258, 289
    term, 256, 287, 290
computable function, 367
condensed detachment, 124
confluence, *see* Church-Rosser property
conjunction, 29, 86, 89, 155, 281, 310
conservation theorem, 19
conservativity, 172, 238
    of λP over first-order logic, 339
    of CPC over CPC($\rightarrow$), 152
    of IPC over IPC($\rightarrow$), 47
    of predicate logic over propositional logic, 202
consistency, 14, 22, 50, 78, 110, 131, 247, 248
    of arithmetic, 236, 241, 244, 264
Constable, R.L., viii, 98, 99, 225, 340
constant, 365
construction, *see* BHK interpretation
constructive mathematics, 50
constructivism, 48
constructor, 89, 215, 296, 326–328, 348
constructor variable, 296
constructor vs destructor, 87
continuation passing style, 141–143
contracting map, 219, 220, 225, 251, 260, 309, 319, 323, 333, 358

INDEX 433

contraction, 164
contraction-free system, 192
control operator, 134
converse to the pairing function, 367
conversion rule, 329
convex set, 35, 50
co-product, 86, 280
Coq, 98, 340, 358
Coquand, T., viii, 340, 357
CPC, 128
CPS, *see* continuation passing style
CR, 71
creative subject, 28
critical pair, 139
Curien, P.-L., 190
Curry's paradox, 22
Curry, H.B., v, viii, 22, 23, 73, 97, 103, 122, 156, 339
Curry-Howard isomorphism, v, 77, 81, 83, 87, 89, 96–100, 113, 134, 278
Curry-style system, 63
cut elimination, 169, 202
cut rule, 162, 169
cut term, 179

van Daalen, D.T., 74
van Dalen, D., 48, 50
De Morgan's law, 31, 199
decidability
    of equality of typed terms, 72
    of inhabitation, 79
    of IPC, 43, 79
    of prenex formulas, 226
    of propositional sequent calculi, 173
    of typability, 62, 336
    of type checking, 62, 340
decidable problem, 368
decision problem, 368
Dedekind, R., 229, 245
deduction theorem, 103, 106, 108, 116, 122, 130, 204
Deelen, A., viii
definability of logical connectives, 53, 155, 227, 281
definition (explicit vs. inductive), 232, 306
degree
    of a cut, 170
    of a redex, 68
Dekkers, W., 101
dense set, 42
dependency, 346
dependent types, 98, 215, 325
depth, 75
derivation, 32
derived rule, *see* admissible rule
destructor, 87, 89
detachment rule, *see* modus ponens
detour, 84, 85, 88
diagram chase, 13
Dialectica interpretation, 97, 99, 265, 266, 299, 323
dialogue, 89, 92, 95, 144, 145, 149, 153, 182, 187
disjunction, 29, 86, 89, 155, 280, 309
disjunction property, 46, 172, 202, 245
distance, 364
domain (of a function), 362
domain-free system, 357
dot notation, 2, 3, 64, 197
double arrow, 10
double negation elimination, 127
Dowek, G., 336
downward-closed set, 363
Dragalin, A.G., 50, 190, 247, 265, 299
Dyckhoff, R., 189, 190

elimination rule, 32
eliminator, 256
empty domains, 197, 337
empty type, 89, 280
Entscheidungsproblem, 23, 224
environment, 56, 218, 326, 327, 344
    first-order, 338
    infinite, 56
    raw, *see* raw expression
    valid, 328
ε_φ, 132
ϵ_0, 395
equality, 10
    extensional, 11, 117
    Leibniz, *see* Leibniz equality
    weak, 109, 118
equivalence, 29
equivalence class, 362
equivalence relation, 362
erasing map, 67, 284, 335, 357
η-reduction, 11, 65, 85, 398

Euclidean space, 364
ex falso, 33, 129
exceptional return, 136
exchange rule, 164
excluded middle, 127
existence property, 245
existential quantifier, 216, 219, 225, 281, 282, 300
existential type, *see* existential quantifier
explicit substitution, 176, 177
extended
    polynomial, 73, 75, 76
    simply typed lambda-calculus, 86
Extended Calculus of Constructions, 352
extension of a structure, 212
extensionality, 338

Felleisen, M., 156, 157
Felscher, W., 99
Feys, R., viii, 23, 97
field of sets, 36
Figaro, 195, 225
Filinsky, A., 157
filter, 45, 224
    prime, 45, 51
    proper, 45
Fine, K., 299
finite model property, 43, 49–51
first-order
    arithmetic, 229
    formula, 196
    logic, 195
first-order β-reduction, 219
Fitch, F.B., 123
Fitting, M., 50
fixed point, 307, 310, 363
fixed point combinator, 24
forcing relation, 44
formal parameter, 1
formalism, 48
formalizing proofs, 233
formula
    atomic, 29, 95, 196, 304
    closed, 196
    first-order, 196
    negative, 52
    open, 196
    propositional, 28, 270

    second-order, 270, 304
    special, 221
free variable, 87, 196, 304, 327, 344
Frege, G., 96, 224
Friedman's translation, 241, 242, 247
Friedman, H., 247
full comprehension, 271, 308
function, 362
    arithmetical, 231
    beta, 230
    definable, 20
    in system **F**, 282, 322
    in system **T**, 254, 266
    in $\lambda_\to$, 72
    domain and range, 362
    initial, 366
    monotone, 363
    partial recursive, 367
    partial vs. total, 362
    primitive recursive, *see* primitive recursive function
    provably total (recursive), *see* provably total function
    recursive, 366, 367
    representable in PA, 237
    Turing computable, 367
function type, 56
function variables, 323

Gabbay, D., 276, 299, 300
Gajda, M., 359
Gandy, R.O., 74
generalization, 203, 277, 284, 307
generation lemma, 58, 133, 286, 332
Gentzen's Hauptsatz, 169
Gentzen, G., 49, 50, 84, 99, 156, 161, 189–191, 247
geometry of interaction, 123
Geuvers, H., 299, 323, 358
Giannini, P., 300
Girard's paradox, 98, 356, 357
Girard, J.-Y., vi, viii, 99, 157, 190, 265, 270, 288, 289, 299, 300, 322, 323, 357
Glivenko's theorem, 42, 54, 199
Glivenko, V.I., 49, 156
Gödel, K., viii, 49, 97, 156, 224, 230, 231, 234, 236, 246, 247, 251, 265
Gödel number, 234, 367

Goodman, N.D., 97, 225
Görnemann, S., 227
Goto, S., 99
graph, 23
Grassmann, H., 245
Griffin, T.G., 127, 156, 157
de Groote, P., 157
Grzegorczyk's scheme, 199, 225, 227, 276
Grzegorczyk, A., 227, 265

HA, 240
halting problem, 369
hard problem, 370
Harper, R., 340
Harrington, L., 246
Harrop, R., 245
HAS and HAs, 314
Haskell, 23, 66
Hauptsatz, 169
Hayashi, S., 99
head reduction, 15
Helman, G.H., 123
Henkin model, 307, 323
Henkin, L., 213, 224, 274, 323
Herbelin, H., 190
Herbrand, J., 103
Heyting algebra, 39
    complete, 207, 209, 224, 226, 272
    of open sets, 41
Heyting Arithmetic, 97, 240
    second-order, 314
Heyting, A., v, viii, 29, 49, 96, 224
higher-order
    arithmetic, 313
    polymorphism, 296
Hilbert style, 104, 129, 157, 158, 203
Hilbert's programme, 48, 246
Hilbert, D., 48, 73, 224, 323
Hindley, J.R., 50, 73, 97, 124, 125, 300
Hoare logic, 232
HOL, 98
homomorphism, 365
Honsell, F., 300, 340
Houyhnhnms, 28, 29
Howard, W., v, viii, 49, 97, 98, 225
Howe, D.J., 358
Huet, G., viii, 74, 340
Hurkens, A.J.C., 358

I, 2
illative combinatory logic, *see* combinatory logic
image, 362
implication, 29
    as universal quantification, 98, 326, 327, 339, 346
implicational logic, 47, 78, 85, 150
impredicativity, 271, 308
in, 311
$\text{in}_i^{\varphi \vee \psi}$, 87
incompleteness of PA, 235, 246
independence proofs, 158
induction
    at work, 5, 6, 364, 393
    axiom, 246, 311
    scheme, 233, 252
inductive
    kind, 354, 359
    type, 283, 310–311, 324, 400
infimum, 363
inhabitation, 60, 223, 291, 339
initial functions, 366
instantiation, 277, 284
int, 72, 253
Int(a), 306
interior, 364
internal reduction, 15
intersection, 35
intersection types, 86, 288
introduction rule, 32
intuitionism, 48
intuitionistic logic, 28, 32
intuitionistic sequent, 165
inversion principle, 33, 100
invertible rule, 387
IPC, 32
Isabelle, 98
isomorphism, 365
It, 311
iterator, 255, 266

Jaśkowski, S., 49
join, 35
judgement, 32, 57, 328
jump, 134

K, 2, 11
Kalmár, L., 130
Kant, I., 48

Kaplan, D., 299
kind, 296, 326–328, 346, 348
Kleene's
    T predicate, 249, 367
    normal form, 239, 367
Kleene, S.C., viii, 22, 23, 49, 99, 174,
    189–191, 225, 243, 247, 361
Klop, J.-W., 358
knowledge, state of, 44
Kolmogorov's translation, 140, 154, 201,
    241, 308, 314, 324
Kolmogorov, A.N., v, viii, 29, 49, 156
Kreisel, G., viii, 97, 225, 242, 247, 260,
    265
Kremer, P., 276
Kripke model, 43, 212, 272
    finite, 51, 227
    full, 274
    principal, 276, 323
    with constant domains, 225, 227,
    276, 300
Kripke, S., 50
Krivine, J.-L., 25, 300, 323
Kronecker, L., 48
Kuratowski, K., 363
Kuznetsov, A.V., 53

ℓ, 367
Ladner, R., 99
λI-term, 19
λx, 177
lambda-calculus, 1
    polymorphic, 270
    simply typed, 56
    simply typed, extended, 86
    untyped, 1
λ-calculus, see lambda-calculus
λ-term, 1, 3, 7
    vector notation for, 15
λ-variable, 3
$\lambda\mu$-calculus, 132
$\lambda\mu$-term, 132
    atomic, 147
Lambek, J., 100
lattice, 35
    complete, 363
    distributive, 36, 39, 50, 51
Läuchli, H., viii, 97, 225
law of excluded middle, 127
Lawvere, F.W., viii, 100

LCF, 98, 225, 340
Lebesgue, H.L., 48
leftmost reduction, 15
Lego, 98
Leibniz equality, 306, 316, 324, 354
Leivant, D., 225, 300, 323
Leśniewski, S., 299
let, 219, 282
Lévy, J.-J., 74
lexicographic order, 364
LF, 340
liar's paradox, 235
Lindenbaum algebra, 39, 210, 391
linear
    logic, vi, 82, 100, 120, 123
    term, 119
linear order, 363
Lisp, 23
list (data type), 283, 310
LJ, 165
LK, 162
LM, 165
Löb, M.H., 299
logic, 27
    BCI and **BCK**, 119
    affine, 120
    classical, see classical logic
    first-order, 195
    implicational, see implicational logic
    intuitionistic, see intuitionistic logic
    linear, see linear logic
    minimal, 33
    propositional, see propositional logic
    relevant, 119
    resource-conscious, 82
    second-order, see second-order logic
    substructural, 118
logic programming, 99
logical embedding, 140
long form, 146
long normal form, 79, 96, 290
looping combinator, 357
Lorenz, K., 99, 191
Lorenzen, P., vi, 99, 191
lower bound, 363
LP, 165
ludics, 99, 123
Łukasiewicz, J., 49, 299

Marini, D., 99

Markov's
　principle, 199, 227
　rule, 248, 394
Martin-Löf's type theory, 98, 225, 357
Martin-Löf, P., viii, 22, 97, 98, 225,
　　299, 339, 340, 357
Maurey, B., 25
McKinsey, J.C.C., 49, 50, 224
meet, 35
Mellies, ,P.-A, 190
Mendelson, E., 158
Mendler, N.P., 323
Meredith, C.A., viii, 97, 123
Meredith, D., 124
metric space, 211, 364
Meyer, A.R., 156, 340
Miglioli, P.A., 99
Milner, R., 73, 98, 300
minimal logic, 33
minimal sequent, 165
minimization, 366
Mints, G.E., 49, 50, 190
miracle, 87
Mitchell, J.C., 299, 300
mix rule, 170
ML, 23, 66, 73, 277
modified realizability, 265
modus ponens, 104, 203
Moerdijk, I., 225
monotone function, 363
Mostowski, A., 224
μ-redex, 135
$\hat{\mu}$-reduction, 147
multiset, 125, 174, 362, 385
Murthy, C.R., 127, 156, 157

$\mathcal{N}$, 229
$\mathbb{N}$, 361
$\underline{n}$ and $\overline{n}$, 229, 254
natural deduction, 32, 128, 271, 308
　in traditional notation, 82
natural numbers, 361
Nederpelt, R., 74
negation, 29, 30, 191
Newman's lemma, 71, 74
Newman, M.H.A, 73
NJ, 32, 33, 166
NK, 128, 166
normal form, 10, 109
　long, *see* long normal form

normal proof, 83, 85
normal return, 136
normalization, 14, 67, 68, 74, 78, 156
　proofs of, 99, 259, 287, 290
　strong, *see* strong normalization
　undecidability, 22, 25
numeral, 20, 229, 254, 282
Nuprl, 98, 99, 340, 358

object variable, 3
Ω, 11
ω (term), 11
ω (type), 282
Ong, C.-H.L., 157, 159
open set, 364
ordered set, 362, 363
Orevkov, V.P., 226
orthodox Church-style, 63, 66
$\mathcal{O}(\mathcal{T})$, 364

PA, 232
pair, 19, 155, 218
pairing function, 367
paradox
　Girard's, 356, 357
　liar's, 235
　Russell's, 352, 357
　Schwichtenberg's, 249
parallel internal reduction, 15
parentheses, 3
Parigot, M., 156, 157, 190, 255, 300
Paris, J., 246
partial function, 362
partial order, 362
PAS, 312
PAs, 313
Pascal, 66
Paulin, Ch., 357
PCF, 74
Peano Arithmetic, 232, 245, 246
Peano axioms, *see* axioms of arithmetic
Peano, G., 96, 229, 245–247
Peirce type sequent, 165
Peirce's law, 31, 42, 78, 127–129, 132,
　　144, 164, 376
Peirce, Ch.S., 224
permutative conversions, 193
Pinto, L., 190
PL, 340

Plotkin, G.D., 156, 300, 340
Poincare, J.H., 48
polymorphism, 59, 270, 300
    explicit, 277
    higher-order, 296
    implicit, 277, 284
position, 183
possible position, 187
possible world, 43, 212
Post Correspondence Problem, 369
Prawitz, D., 49, 50, 74, 99, 156, 190, 259, 299, 357
pre-term, 3
predicate calculus (logic), 195
prenex normal form, 226
Presburger Arithmetic, 232
prime
    filter, 51
    set of formulas, 213
primitive recursion, 366
primitive recursive
    function, 240, 266, 366
    set (relation), 367
principal
    pair, 61
    solution, 365, 366
    type, 61, 73, 287
principal model, 276
PRL, 98, 99, 225, 340
problem
    decidable vs undecidable, 368
    hard, complete, 370
product, 86, 89, 265, 281, 343, 344, 346, 361, 362
program extraction, 98, 264, 303
projection, 374
proof
    Hilbert-style, 104, 204
    in natural deduction, 32
    sequence, 104
proof irrelevance, 338
proof normalization, 83, 85
proof variable, 217
proponent strategy, 183
propositional logic, 28, 32
provable judgement, 205
provably total function, 239, 242, 264–267, 312, 315, 322
prover-skeptic dialogue, *see* dialogue

Prucnal, W., 101
pseudo-Boolean algebra, *see* Heyting algebra
pseudo-complement, 38
PSN, 177, 178
PTS, 343, 345
pure type system, 343, 345
PX, 99

$\mathbb{Q}$, 361
QBF, 299, 370
quantified Boolean formulas, *see* QBF
quantifier, 197, 224, 227
    existential, *see* existential quantifier
    second-order, 301, 304
    universal, *see* universal quantifier
quasi-head reduction, 17
quasi-leftmost reduction, 17
quotient set, 362

$\mathbb{R}$, 361
$\mathbf{R}_\sigma$, 254
Ramsey, F.P., 73
range (of a function), 362
Rasiowa, H., 50, 190, 224
Rasiowa-Sikorski lemma, 224–226
rational numbers, 361
raw
    expression, 327, 344, 346
    term, 63, 276, 327
real numbers, 54, 225, 227, 361, 364
realizability, 225, 243–245, 247, 249
    modified, 251, 260–264
recursor, 253, 283
redex, 10, *see* reduction
    leftmost, 14
    rightmost, 68
reducibility, 225
reductio ad absurdum, 127
reduction, 10, 135, 346
    β, *see* β-reduction
    η, *see* η-reduction
    for Hilbert style proofs, 124
    head, 15
    internal, 15
    leftmost, 15
    multi-step, 10
    parallel internal, 15
    quasi-leftmost, 17

INDEX

rightmost, 14
  weak, 109
reduction sequence, 10
reduction strategy, 17
  normalizing, 17
  perpetual, 18
reflexive relation, 362
Rehof, N.J., 157
Reinhold, M.B., 340
relation
  as a 0,1-valued function, 362
  compatible, 10, 12
  properties, 361
relative pseudo-complement, 39
relatively prime numbers, 230
relevant logic, 119
representability in PA, 237, 248
representative, 362
restricted $\lambda\mu$-terms, 143
Reynolds, J., viii, 156, 270, 300
Rezus, A., 123
Robinson, J.A., 366
Ronchi Della Rocca, S., 300
Rose, K.H., 176
Rosser, J.B., 22, 23
rule
  ξ, 12, 116
  derived (admissible), 34
  invertible, 387
  of a PTS, 346
Russell's paradox, 352, 357
Russell, B., 48, 73, 299

S, 11
Sato, M., 99
Scheme, 23
Schönfinkel, M., viii, 96, 103, 122
Schütte, K., 299
Schwichtenberg's paradox, 249
Schwichtenberg, H., 73, 248
Scott, D., viii, 97, 98
second-order
  arithmetic, 311
  formula, 304
  quantifier, 301, 304
second-order logic, 269
  classical, 305
  intuitionistic, 269
  predicate, 269, 308
  propositional, 269, 299

Seldin, J.P., 339
Selinger, P., 159
sentence, 196
sequent, 162
sequent calculus, 162, 201
  one-sided, 193
  terms, 179
  versus natural deduction, 165
set
  closed, 364
  convex, 35
  dense, 42
  open, 364
  ordered, 362, 363
  primitive recursive, 367
  recursive, 367
  recursively enumerable, 367
  upward-(downward)-closed, 363
  well-founded, 363
signature, 195, 365
Sikorski, R., 50, 190, 224
simple type, 56
simply typed λ-calculus
  à la Church, 63
  à la Curry, 56
  extended, 86
Skolem, T., 230
Skolem-Löwenheim theorem, viii
Skvortsov, D., 276
SN, 14
Sobolev, S.K., 299, 397
sort, 344, 345
soundness, 210
space
  metric, 364
  topological, 41, 208, 364
special formula, 221, 291
species, 305
Spector, C., 265, 315
Spławski, Z., 323
Springintveld, J., 339
standard number, 306, 312
standard semantics, 305
state (of a Kripke model), 43
Statman, R., 72, 99, 101
Stenlund, S., viii, 97, 266, 299
Stone's representation theorem, 37
Stone, M.H., 49
Strachey, Ch., 300

strategy for reduction, *see* reduction strategy
strategy in a dialogue, *see* winning strategy
strong cut elimination, 179
strong normalization, 14, 69
    for system **F**, 322
    for η-reduction, 24
    for $\lambda_\to$, 70, 124
    for λP, 335
    for $\lambda\mu$, 138, 157
    for $\hat\mu$-reductions, 147
    for PTS, 358
    for system **F**, 287–290, 302
    for system **T**, 256, 259, 266
    for system $\mathbf{F}_\omega$, 297, 298
    for typed CL, 113, 116
    in first-order logic, 220
    in second-order arithmetic, 320
    undecidability, 22, 25
    with sums and products, 88, 159, 266
strong reduction and equality, 118
strong sum, 225, 357
structural closure, 187
structural rule, 164
structure (model), 205
subalgebra, 365
subformula, 29, 43, 202
subformula property, 172, 202
subject conversion, 74, 125
subject reduction, 59, 65, 137, 147, 180, 286, 333, 346
substitution, 2, 4, 8, 87, 132, 134, 135, 197, 270, 278, 304, 327, 344, 346, 365, 373
    explicit, 176
    simultaneous, 9, 197, 373
substructural logic, 118
succedent, 162
successor, 254
sum type, 86, 89, 155
supremum, 363
Švejdar, V., 99
symmetric relation, 362
system
    Ecc and Ecc$_\Pi$, 352
    F, 270, 276, 300, 345, 346
    $\mathbf{F}_\omega$, 296, 347

$\mathbf{F}_\omega^+$, 352
G1 and G2, 174
G3, 174, 191, 192
λH, 318
λH, 325
λ2, 346
$\lambda *$, 356, 358
$\lambda\underline\omega$, 347
$\lambda_\to$, 56
λx, 177
$\lambda\mu$, 132
λC, 343–351
LJ, LM, and LP, 165
LK, 162
λP, 326
    compact, 344
λP$_1$, 215, 325
N$^\omega$JP, 351
PX, 99
T, 247, 253
U and **U**$^-$, 352, 356, 358
Sørensen, M.H., 157

$T(k, \vec n, m)$, 367
Tait, W.W., viii, 22, 74, 97, 225, 246, 264–266, 288, 299
Takahashi, M., 22
Takahashi, M., 299
Takeuti's conjecture, 299
target, 293
Tarski, A., 49, 224, 235, 299, 390
Tarskian semantics, 305
tautology, 40, 207
Terlouw, J., 357
term, 1, 7, 328, 348
    λ**I**, 19, 119
    (strongly) normalizing, 14
    affine, 119
    algebraic, 195, 365
    Church-style, 64
    closed, 8
    combinatory, 109
    computable, *see* computable term
    $\lambda\mu$, 132
    linear, 119
    object vs. proof, 325, 337, 338
    raw, *see* raw term
    typable, 60, 278, 287, 335, 358
term model, 205
term rewriting, 239

INDEX 441

terms
    for first-order logic, 218
    for second-order arithmetic, 318
    for sequent calculus, 179
tertium non datur, 27, 42, 46, 127, 145, 203
    at work, 239
$t_f(\vec{n}, m)$, 367
theorem, 32, 40, 104, 205
    Chinese remainder, 230
    Church-Rosser, 14, 22
    compactness, 227
    completeness, *see* completeness
    conservation, 19
    deduction, 204
    Gödel incompleteness, 235, 246
        second, 236, 246
    Gelfond-Schneider, 50
    Glivenko's, 42, 199
    Knaster-Tarski fixed point, 363
    normalization, 68
    Skolem-Löwenheim, viii
    Stone's representation, 37
    subject reduction, *see* subject reduction
theory, 204
    $Th(\mathcal{A})$, 207
    complete, 234
throw, 136, 156
Thue, A., 74
topological semantics, 211
topological space, 41, 211, 364
total (linear) order, 363
total function, 362
transitive relation and closure, 362
translation
    between CL and Λ, 114, 115, 121
    Friedman's, 242, 247
    from λC to $\mathbf{F}_\omega$, 349
    from λP to $\lambda_\to$, 333
    from classical to intuitionistic logic, 140, 201, 299, 308
    from PA to HA, 241
    from PAS to HAS, 314
    Gödel-Gentzen, 159
    Kolmogorov's, *see* Kolmogorov's translation
Troelstra, A.S., 323
truth, 27, 29, 206, 235

Turing computable function, 367
Turing Machine, 361, 371
Turing, A., 74, 259
2, 20
two-counter automaton, 369
typability, 60, 62, 335, 357
type, 328, 346, 348
    dependent, 215
    non-empty, 60
    polymorphic, 276
    principal, *see* principal type
    simple, 56
    written as superscript, 66
type checking, 62, 340
type constructor, *see* constructor
type inhabitation, *see* inhabitation
type-erasure, *see* erasing map
Tyszkiewicz, J., 73

undecidability
    of (strong) normalization, 22, 25
    of arithmetic, 232, 248
    of first-order logic, 201, 202
    of realizability, 244
    of second-order propositional intuitionistic logic, 299
    of type checking, 287, 336, 340
    of type inhabitation, 296, 339
undecidable problem, 368
Ungar, A.M., 190
unification, 60, 75, 365
union, 35
union types, 86, 399
universal closure, 196
universal quantifier, 271, 277, 326
    as product, 216, 343
upper bound, 363
upward-closed set, 363
Urban, C., 190

validity, 40
valuation, 272, 288
    binary, 37
    in a Boolean algebra, 37
    in a Heyting algebra, 40
    individual, 206
value of a formula
    first-order, 206, 207
    propositional, 37, 40
    second-order, 306

OHIO UNIVERSITY LIBRARY

Please return this book as soon as you have finished with it. In order to avoid a fine it must be returned by the latest date stamped below. All books are subject to recall after two weeks or immediately if needed for reserve.

CF